Salinisation of Land and Water Resources

Human causes, extent, management and case studies

F Ghassemi, A J Jakeman and H A Nix

Centre for Resource and Environmental Studies
The Australian National University
Canberra ACT 0200
AUSTRALIA

Published in Australia and New Zealand by
UNIVERSITY OF NEW SOUTH WALES PRESS LTD
Sydney 2052 Australia
Telephone (02) 398 8900
Fax (02) 398 3408

and in the rest of the world by
CAB INTERNATIONAL
Wallingford Oxon OX10 8DE UK
Telephone (01491) 832111 Fax (01491) 833508
Telex 847964 (COMAGG G) E-mail: cabi@cabi.org

in association with
Centre for Resource and Environmental Studies
The Australian National University Canberra ACT 0200 Australia
Telephone (06) 249 4277 Fax (06) 249 0757

© Centre for Resource and Environmental Studies 1995

First published in 1995

This book is copyright. Apart from any fair dealing for the purpose of private study, research, criticism or review, as permitted under the Copyright Act, no part may be reproduced by any process without permission from the publisher.

National Library of Australia
Cataloguing-in-Publication entry:

Ghassemi, F. (Fereidoun), 1940– .
  Salinisation of land and water resources.

  Bibliography.
  Includes index.
  ISBN 0 86840 198 6.

  1. Salinization. 2. Salinization – Control.
  3. Salinization – Control – Case studies. I. Jakeman, A.J.
  (Anthony John), 1951– . II. Nix, H.A. (Henry Allan). III. Title

631.416

A catalogue record for this book is available from the British Library.

ISBN 0 86840 198 6 (UNSW Press)
ISBN 0 85198 906 3 (CABI)

Text designed and formatted by Di Zign Pty Ltd
Printed in Singapore by Kyodo Printing, Singapore

Contents

Figures vii
Tables ix
Foreword xiii
Preface xiv
Acknowledgements xvii

Part One — General aspects of salinisation 1

Section A: Global Resource Overview

A brief history of human-induced salinisation 2
World population 3
World water resources 4
World climate 7
World dryland areas 11
World arable land resources 11
Agriculture and food production 14
Extent of human-induced land degradation 16

Section B: Crop and Irrigation Aspects

Crop salt tolerance 20
Crop water requirements 22
Irrigation 23
   Irrigation methods 23
      Surface irrigation 24
      Sprinkler or spray irrigation 24
      Localised irrigation 25
      Selection of irrigation methods 26
   Irrigation efficiency 26
   Irrigation water quality 28

Section C: Salinisation Processes and Damage

Processes of salinisation 31
   Source of salt 32
   Mobilisation of salt 32
   Dryland salinity 35
Characteristics of saline and sodic soils 36
Salinisation of water resources 38
Extent of land salinisation 40
   Previous estimates 40
   Current estimate 41
Extent of water resources salinisation 43
Environmental damage 45
Economic and social damage 46

Section D: Management of Salinity — Engineering Options

Management of salinity problems 49
   Engineering options 49
      Drainage 49
      Conjunctive use of surface and groundwater 53
      Interception scheme 53
      Dilution flow 54
      Improving irrigation water efficiency 54
      Disposal of saline drainage water 57

Section E: Management of Salinity — Other Options and Aspects

Biological options 72
   Agricultural practices 72
   Revegetation of salt-affected lands 75

Policy options	79
Water pricing	79
Transferable water entitlement	82
Integrated and total catchment management	84
Reclamation of sodic soils	86
Mathematical models for salinity management	88
Salinity and conflict resolution	93
Colorado River Basin	94
Murray-Darling Basin	97
Community involvement in salinity management	103
Geophysical methods for soil salinity mapping	106
Ground methods	107
Resistivity method	107
Electromagnetic method	107
Airborne electromagnetic method	112
Remote sensing	113
References	115

Part Two — Salinity problems in selected countries 127

Chapter 1: Argentina 128

Introduction	128
Rainfall and climate	130
Water resources	131
Surface water	131
Groundwater	132
Land cover	133
Irrigation	133
Salinity	134
Irrigated land salinity	135
Dryland salinity	136
Salinisation in the north-west of the Province of Buenos Aires	138
Management options	141
References	142

Chapter 2: Australia 143

Introduction	143
Rainfall and climate	144
Water resources	146
Surface water	146
Groundwater	148
Land cover	149
Irrigation	150
Salinity	152
Dryland salinity in the south-west of Western Australia	154
Extent of dryland salinity	156
Extent of stream salinity	157
Management options	164
Salinity and waterlogging in the Murray-Darling Basin	176
Geology	178.
Palaeoclimate and salinity	179
Origin of salt	180
Early records of salinity	181
Irrigated land salinity and waterlogging	182
Dryland salinity	186
River salinity	188
Management options	192
Economic damage and cost	206
References	207

Chapter 3: China 213

Introduction	213
Rainfall and climate	216
Water resources	217
Surface water	217
Groundwater	220
Land cover	221
Irrigation	221
Salinity	222
Salinity in the Huang-Huai-Hai Plain	224
Hydrogeologic conditions	226
Extent of salinisation	228
Salinisation in other regions of China	231
Management options	232
Management of saline soils	233
Management of alkaline soils	236
References	237

Chapter 4: Commonwealth of Independent States 240

Introduction	240
Rainfall and climate	242
Water resources	243

Surface water	243
Groundwater	246
Land cover	248
Irrigation	249
Salinity	250
Salinity and environmental problems in Central Asia	252
Irrigated land salinisation	256
Dryland salinisation	257
Surface water salinisation	258
Management options	261
Aral crisis	262
Salinity problems in other areas of the CIS	264
References	267

Chapter 5: Egypt — 270

Introduction	270
Rainfall and climate	272
Water resources	273
Surface water	273
Groundwater	275
Land cover	279
Irrigation	279
Salinity	281
Irrigated land salinity	281
River salinity	283
Management options	284
Surface and subsurface drainage	284
Reuse of drainage water	286
Revegetation	287
Economic damage	288
References	288

Chapter 6: India — 291

Introduction	291
Rainfall and climate	293
Water resources	295
Surface water	295
Groundwater	297
Land cover	299
Irrigation	299
Salinity	300
Irrigated land salinity	302
State of Haryana	304
State of Punjab	308
State of Rajasthan	313
State of Gujarat	317
Salinity problems in other states	328

General aspects of management options	329
Stream salinity	334
References	335

Chapter 7: Iran — 338

Introduction	338
Rainfall and climate	339
Water resources	341
Surface water	341
Groundwater	342
Land cover	344
Irrigation	345
Salinity	346
Irrigated land salinity	347
The Moghan Irrigation Project	348
The Zarrineh-Rud Irrigation Project	351
The Haft Tappeh Irrigation Project	353
The Khalafabad Irrigation Project	354
The Doroudzan-Korbal Irrigation Project	358
The Zayandeh-Rud Irrigation Project	361
The Zabol Irrigation Project	362
Stream salinity	365
References	366

Chapter 8: Pakistan — 369

Introduction	369
Rainfall and climate	371
Water resources	372
Surface water	372
Groundwater	374
Land cover	377
Irrigation	377
Salinity	380
Management options	384
Management of saline soils	384
Management of alkaline soils	392
Reclaimed saline and alkaline soils	393
Economic damage	393
References	394

Chapter 9: South Africa 396

Introduction 396
Rainfall and climate 397
Water resources 399
   Surface water 399
   Groundwater 402
Land cover 403
Irrigation 404
Salinity 405
   Irrigated land salinity 405
   Stream and reservoir salinity 406
      Salinity problem in the Breede River 411
      Salinity problem in the Berg River 416
      Salinity problem in the Great Fish River 419
      Salinity problem in the Sundays River 420
      Salinity problem in other river systems 421
   Salinity in urban and industrial areas 422
   Management options 425
      Mathematical models for salinity management 426
   Economic damage 427
References 428

Chapter 10: Thailand 431

Introduction 431
Rainfall and climate 432
Water resources 434
   Surface water 434
   Groundwater 437
Land cover 439
Irrigation 439
Salinity 441
   North-east Thailand 441
      Geology and hydrogeology 442
      Salinity in north-east Thailand 444
      Management options 453
References 456

Chapter 11: United States of America 459

Introduction 459
Rainfall and climate 461
Water resources 462
   Surface water 462
   Groundwater 464
Land cover 466
Irrigation 466
Salinity 469
   Colorado River Basin 470
      Irrigation 473
      Irrigated land salinity 474
      River salinity 474
      Management options 479
      Economic losses 485
   San Joaquin Valley 485
      Surface water 485
      Groundwater 487
      Irrigation, drainage and salinity 488
      Selenium in the San Joaquin Valley environment 490
      Management options 491
   Dryland salinity in the Northern Great Plains 494
   Salinity problems in other areas 499
References 501

Appendix I Summary data 506

Appendix II
Some institutions involved in salinity investigations 508

Argentina 508
Australia 508
China 509
Commonwealth of Independent States 509
Egypt 510
India 510
Iran 510
Pakistan 510
South Africa 511
Thailand 511
United States of America 511

Glossary 513

Subject Index 518

Geographical Index 522

Plant Index 526

Figures

Part One — General aspects of salinisation

1:	An agroclimatic classification of world climate into 34 classes	10
2:	Drylands of the world	12
3:	Effects of land clearing on the watertable for two Western Australian catchments	34
4:	Increase in the chloride concentration of the Rhine River	39
5:	Yuma Desalting Plant water flow schematic	70
6:	Schematic land use map of the Mildura area and the location of 17 interception well fields	89
7:	Average nodal salt load from aquifer to the river with inclusion of three additional well fields	90
8:	Map of the River Murray and its tributaries showing the study area between Euston and Red Cliffs	92
9:	Historical versus predicted salinity profiles at Red Cliffs	93
10:	Colorado River Basin Salinity Control Project, Title I Division	96
11:	Management structure for the Murray-Darling Basin Initiative	99
12:	The current and the likely future salinity trends for the River Murray at Morgan and the impacts of the salt interception and drainage schemes and land use management	101
13:	Geonics electromagnetic terrain conductivity meters	109
14:	QUESTEM system geometry	112

Part Two — Salinity problems in selected countries

Chapter 1: Argentina

1-1:	Main physiographic features of Argentina	129
1-2:	Administrative map of Argentina	130
1-3:	Major areas of saline and/or alkaline soils in the eastern humid region of Argentina	137
1-4:	Area affected by flooding in the north-west of the Province of Buenos Aires	139
1-5:	Change in average annual rainfall of the Province of Buenos Aires over the periods 1911–1970 and 1971–1984	140

Chapter 2: Australia

2-1:	Physiographic division of Australia	144
2-2:	Median annual rainfall in Australia	145
2-3:	Drainage divisions of Australia	146
2-4:	Schematic map of the areas susceptible to or containing dryland salinity	153
2-5:	A comparison of the area affected by salinity in Western Australia and the area cleared for agriculture	157
2-6:	The distribution of surface water resources, salinities and rates of salinity change in relation to rainfall and forest cover in the south-west of Western Australia	159
2-7:	(A) River basins of the South-West Drainage Division (B) Stream salinity trends of a number of catchments	160
2-8:	Location map for experimental catchments in the Collie River Basin	163
2-9:	Location of experimental reforestation sites	171
2-10:	Dependence of rate of change of groundwater level under reforestation relative to pasture	173
2-11:	Observed and predicted inflow salinities to Wellington Dam for various salinity control measures	175
2-12:	Murray-Darling Basin in south-east Australia	177
2-13:	River Murray system, regulation works and irrigation regions	178
2-14:	Piezometric map and depth to watertable in Murray Basin	183
2-15:	Temporal change of watertable levels in the irrigated areas of the Murray Basin	184
2-16:	Irrigated and dryland salinity in Victoria	187
2-17:	Average monthly salinities in the River Murray at Morgan in South Australia	188
2-18:	Contribution of each reach of the River Murray to the average salinity at Morgan in $\mu S\ cm^{-1}$ and in percentage	189
2-19:	Predicted increase in river salinity at Morgan and Tailem Bend	191
2-20:	Relative costs of specific saline water disposal projects in Australian dollars and in January 1988 prices	200

Chapter 3: China

3-1:	Principal mountain ranges of China	214
3-2:	Administrative divisions of China	215
3-3:	China's annual precipitation	216
3-4:	Major rivers in China	217

3-5: Major rivers of the
 Huang-Huai-Hai Plain   225
3-6: Sketch-map of the depth of shallow
 aquifers in the Huang-Huai-Hai Plain 227
3-7: Salinised soil in the
 Huang-Huai-Hai Plain   229

Chapter 4: Commonwealth of Independent States

4-1: Principal physiographic features of
 the Commonwealth of
 Independent States   241
4-2: Average annual precipitation
 of the Commonwealth of
 Independent States   242
4-3: Major river basins of the
 Commonwealth of Independent
 States   244
4-4: Irrigation development in
 Central Asia   254
4-5: Salinity of surface water bodies in
 Central Asia   259

Chapter 5: Egypt

5-1: Main physiographic features and
 rainfall map of Egypt   271
5-2: The Nile Basin   273
5-3: Potentiometric contours of the main
 aquifer of the Nile Delta   276
5-4: Salinity map of the Nile Delta
 aquifer for 1978   278
5-5: Schematic diagram of hydraulic
 works in the Nile Valley
 and its Delta   280
5-6: Soil salinity distribution in the
 Nile Delta   282
5-7: Open drainage network in the
 Nile Delta   285

Chapter 6: India

6-1: Major physiographic features of
 mainland India   291
6-2: Political division in India   292
6-3: Average annual rainfall of India 294
6-4: Major drainage divisions of India 295
6-5: Isohyetal and surface drainage
 pattern of Haryana   304
6-6: Shallow groundwater quality map
 of Haryana   305
6-7: Simplified canal network and Doabs
 of Punjab   309
6-8: Representative hydrographs of
 south-western parts of the
 Old Sirhand Canal Tract   311
6-9: Physiographic map of Rajasthan 314
6-10: Main physiographic features
 of Gujarat   318

6-11: Major irrigation projects in
 Gujarat and the coastal areas
 affected by seawater intrusion 320
6-12: Watertable depth in Kakrapar
 irrigation project for the period of
 pre-irrigation 1957–58   322
6-13: Watertable depth in Kakrapar
 irrigation project for the period
 of pre-monsoon 1982   323
6-14: Salinity of major rivers of India 334

Chapter 7: Iran

7-1: Main physiographic features of Iran 338
7-2: Average annual precipitation in Iran 340
7-3: Main river basins in Iran   342
7-4: Watertable depth of the Moghan
 aquifer for May 1983   348
7-5: Groundwater electrical conductivity of
 the Moghan aquifer for May 1983 349
7-6: Zarrineh-Rud Irrigation Project 352
7-7: Jarrahi catchment irrigation districts 355
7-8: Soil salinity and alkalinity map of
 Khalafabad Irrigation District 357
7-9: Main features of the Dorudzan-Korbal
 Irrigation Project and the watertable
 depth of the aquifer for 1986 358
7-10 Main features of the Zayandeh-Rud
 Irrigation Project   361
7-11: Main features of the Zabol
 Irrigation Project and the
 distribution of lands with very
 high salinity and alkalinity   363

Chapter 8: Pakistan

8-1: Indus River and its tributaries
 in Pakistan   370
8-2: Average groundwater salinity to
 the depth of 110 m   375
8-3: Canal irrigated areas and rainfall
 distribution in Pakistan   379
8-4: Watertable profiles along line A-A1 381
8-5: Left Bank Outfall Drain project 388

Chapter 9: South Africa

9-1: Main physiographic features of
 South Africa   396
9-2: Generalised mean annual
 precipitation of South Africa and
 seasonal rainfall zones   398
9-3: Major rivers of South Africa 400
9-4: Main river basins of South Africa 400
9-5: Surface water salinity distribution
 for South Africa for the period
 1979 to 1988   408
9-6: Major reservoirs in South Africa 410
9-7: Breede River Catchment   412

9-8: Stepwise salination of the irrigation supply water in the Breede River between the measuring weirs 17 and 5   413
9-9: Observed mean monthly TDS content of the Poesjesnels River for the period 1978 to 1980   416
9-10: Change in the salt content (TDS) of the Upper and Lower Great Fish River from May 1977 to April 1980   419
9-11: Five-year moving average of salinity concentrations in the Vaal Barrage   423
9-12: Map of the Buffalo River catchment   424

Chapter 10: Thailand

10-1: Main physiographic regions of Thailand   432
10-2: Simplified average annual rainfall distribution of Thailand for 1951–75   433
10-3: Simplified hydrographic network of Thailand and large-scale irrigation projects completed by December 1984   435
10-4: Administrative map of north-east Thailand   442
10-5: Soil salinity distribution in the north-east of Thailand   444
10-6: Location, capacity and fluctuations in the electrical conductivity of water in six reservoirs, Maha Sarakham Province, Thailand   449
10-7: Rock salt deposits and rock salt mining operations in the north-east of Thailand   452

Chapter 11: United States of America

11-1: The main physiographic regions of the contiguous United States   460
11-2: Average annual precipitation of the contiguous United States   461
11-3: Water Resources Regions of the contiguous United States   462
11-4: Colorado River Basin   472
11-5: Salinity levels at Imperial Dam and Colorado River Simulation System projection to the year 2010   475
11-6: Salinity projections with and without further controls at Imperial Dam   475
11-7: Agricultural irrigation salt source areas and the Colorado River Basin salinity control units   483
11-8: San Joaquin Valley drainage problem area   486
11-9: Generalised hydrogeologic cross-section of the San Joaquin Valley   488
11-10: Area of potential saline-seep development on the Northern Great Plains   495

Tables

Part One — General aspects of salinisation

1: Estimated and projected world population (in millions) from 1950 to 2025   4
2: World water reserves   5
3: Global river run-off   6
4: World water use, 1900–2000   7
5: Major bioclimatic groups of the world, their natural vegetations and representative locations   9
6: World drylands (in million hectares)   11
7: Total land area, potential arable land and cultivated land of the world by continent   12
8: Increase in world population and irrigated area   13
9: Ten leading countries in area irrigated in 1987   13
10: Human-induced soil degradation for the world   18
11: Global extent of human-induced salinisation   19
12: Salt tolerance of selected plants with respect to the electrical conductivity of saturated-soil extract   21
13: Labour requirements and investment cost for various field irrigation systems   23
14: Calculated efficiencies in a number of irrigated areas of the world   28
15: Average annual water balance for a native vegetated and an agricultural catchment north-east of Newdgate (Western Australia)   35
16: Characteristics of saline and sodic soils   38
17: Irrigated land damaged by salinisation in the top five irrigators and the world, estimated for the mid-1980s   41
18: Global estimate of secondary salinisation in the world's irrigated lands   42
19: Estimates of damage to the economy of a few countries with secondary salinity problems   46
20: Global estimate of the average annual income loss due to land degradation, in 1990 prices   48
21: Reduction of evaporation potential of water with different salinity   64
22: Major characteristics of a number of large evaporation basins in the Murray Basin, Australia   65

23: Estimated annual average recharge under different agronomic systems 73
24: Total evapotranspiration and rainfall over the growing season for four species at Kondut and Cunderdin in Western Australia 74
25: Potential mean annual recharge under different cropping rotations for Kondut and Cunderdin in Western Australia 74
26: Tree species and their relative tolerance to soil salinity 76
27: A selection of tree species suitable for sites of low, moderate and high salinity in southern Australia 77
28: The major features of the three irrigation districts in Sunraysia region of Victoria 81
29: The First Mildura Irrigation Trust charges for irrigation water in 1991/92 82
30: Changes to the River Murray Waters Agreement, 1914 to 1981 98
31: Identified problems and aims of the Murray-Darling Basin Natural Resources Management Strategy 102
32: Geonics ground conductivity meters 108

2-6: Stream salinity trends of major rivers 158
2-7: Stream salinity trends of fully forested catchments in south-western Western Australia 161
2-8: The regional effects of agricultural clearing on stream salinity in annual rainfall classes 162
2-9: Characteristics of five experimental catchments and the water and chloride output to input ratio over the period 1974–1983 163
2-10: Characteristics of experimental reforestation sites 170
2-11: Expansion of the shallow watertable areas in the Riverine Plain 185
2-12: Estimates of the annual salt pick-up rate of the drainage water in a few irrigated areas of the Murray Basin 186
2-13: Expansion of the irrigated and dryland salt-affected land in Victoria over the next 30 years 187
2-14: Annual average operational data of salinity mitigation schemes in the Murray Basin 194
2-15: Impact of the interception schemes on the salinity of the River Murray at Morgan 195
2-16: Total outfall schemes' costs for the Riverine Plain Zone over a period of 50 years (1990–2040) 199
2-17: Number of trees planted for salinity control by organisations and individuals in Victoria during 1989 202

Part Two — Salinity problems in selected countries

Chapter 1: Argentina

1-1: Flow characteristics of some rivers in Argentina 131
1-2: Flow characteristics of hydrographic basins of Argentina 132
1-3: Irrigated and salt-affected areas in 21 provinces of Argentina 134
1-4: Groundwater salinity of 109 samples from Tulum Valley, San Juan Province 136
1-5: Boron content of 109 groundwater samples in Tulum Valley, San Juan Province 136

Chapter 2: Australia

2-1: Surface water resources of Australia 147
2-2: Divertible groundwater resources 149
2-3: Irrigated land in Australia 150
2-4: Length of the irrigation and drainage canals and the area of lands protected by drainage works in Australia for 1977–78 151
2-5: The extent of human-induced land salinity in Australia 152

Chapter 3: China

3-1: Characteristics of China's major rivers 218
3-2: Number and capacity of reservoirs in China (1985) 222
3-3: Salt- and alkali-affected soils in cultivated areas of the Huang-Huai-Hai Plain 228

Chapter 4: Commonwealth of Independent States

4-1: Major river basins of the Commonweath of Independent States 244
4-2: Characteristics of the major rivers of the Commonwealth of Independent States 245
4-3: Characteristics of selected major lakes of the Commonwealthof Independent States 245
4-4: Capacity of some of the major reservoirs of the Commonwealth of Independent States 246

4-5: Groundwater resources of the Commonwealth of Independent States and their utilisation   247
4-6: Expansion of the irrigated area in the Commonwealth of Independent States since 1913   249
4-7: Expansion of the drained area in the Commonwealth of Independent States   250
4-8: Characteristics of the major reservoirs in Central Asia   253
4-9: Increase in irrigated areas in the republics of Central Asia   255
4-10: Estimates of the moderate and strongly salt-affected soils and production losses in the irrigated regions of the five Central Asian republics for 1985   256
4-11: Estimates for 1983 of the required measures for the improvement of the irrigation and drainage systems in the Aral region and in Uzbekistan   261

Chapter 5: Egypt

5-1: Present annual groundwater extraction   277
5-2: Areas provided with open drains in the Delta and Upper Egypt   284

Chapter 6: India

6-1: Distribution of arid lands in India   294
6-2: Surface water resources of India   296
6-3: The current and projected water demand and source of supply for India from 1900 to 2025   297
6-4: Irrigated and cultivated area in India from 1900 to 1987   300
6-5: The geographic distribution of salt-affected soils in India   301
6-6: Extent of waterlogging and soil salinity in some irrigation projects of India   303
6-7: Distribution of areas under different ranges of watertable depth in Haryana from 1955 to 1977   306
6-8: Average rise in watertables in canal irrigated areas of Rajasthan   315
6-9: Frequency distribution of groundwater electrical conductivity in five districts of Rajasthan   316
6-10: High watertable area in major irrigation schemes in Gujarat State for the pre-monsoon period (May)   321
6-11: Watertable depths in the Kakrapar Weir system in pre-monsoon periods   323
6-12: Watertable depth in the Mahi command area before irrigation (1958) and after irrigation in 1981 and 1986   325
6-13: Requirement of gypsum according to pH and texture of the soil   329
6-14: Relative tolerance of crops to exchangeable sodium   331

Chapter 7: Iran

7-1: Distribution of precipitation in Iran   340
7-2: Characteristics of the major river basins in Iran during the water year 1988–89   341
7-3: Characteristics of dams constructed since 1957 in Iran   343
7-4: Groundwater extraction in Iran in 1991   344
7-5: Changes in the extent of land salinity within the command areas of Canal A and the Main Canal before and after the development of irrigation in Moghan   350
7-6: Changes in the extent of land alkalinity within the command areas of Canal A and the Main Canal before and after the development of irrigation in Moghan   350
7-7: Watertable depth in Khalafabad Irrigation District in April 1985   356
7-8: Extent of groundwater salinity in Khalafabad Irrigation District   356
7-9: Extent of land salinity and alkalinity in Zabol Irrigation Project   364

Chapter 8: Pakistan

8-1: Mean annual flow of the Indus and its tributaries   373
8-2: Features of some important dams in Pakistan   373
8-3: Extent of groundwater salinity in the irrigated areas of Pakistan   376
8-4: Irrigated soil salinity in different provinces of Pakistan (1977–1979)   382
8-5: Extent of salinity and alkalinity in soil profiles in different provinces of Pakistan (1977–79)   383
8-6: Area under various watertable depths in April/June 1987 (pre-monsoon period)   383
8-7: Area under various watertable depths in October 1987 (post-monsoon period)   383
8-8: Land salinity status of the SCARP I Project with a gross area of 493 700 ha between 1959 and 1977–78 as per cent of the gross area   385

Chapter 9: South Africa

9-1: The ten largest rivers in South Africa in order of run-off volume — 399
9-2: Pattern of water use in South Africa from 1965 to 2000 — 401
9-3: Use of groundwater in South Africa in 1980 — 402
9-4: Land area suitable for dryland crop production in South Africa — 403
9-5: Irrigated area and irrigation water consumption in South Africa (1980–81) — 404
9-6: Occurrence of waterlogging problems in a number of South African irrigation schemes as determined by the Department of Agriculture in 1982 — 406
9-7: Salinity status of a number of reservoirs in South Africa — 409
9-8: Estimated components of total irrigation return flow in South Africa — 422
9-9: Annual direct costs for increase in water salinity from 300 to 500 and 800 mg L^{-1} TDS on the Pretoria, Witwatersrand, Vereeniging and Sasolburg Complex community — 427

Chapter 10: Thailand

10-1: Mean annual flow of some major rivers in Thailand — 436
10-2: Regional distribution of the reservoirs in Thailand — 437
10-3: Irrigation projects in Thailand completed and under construction in 1984 by river basin — 440
10-4: Salt-affected soils in north-east Thailand — 445
10-5: Capacity and average electrical conductivity of water of 132 reservoirs in north-east Thailand — 450

Chapter 11: United States of America

11-1: Natural annual run-off in the 18 Water Resources Regions of the contiguous United States — 463
11-2: The estimated water withdrawal, source of supply and the consumptive use in the United States for 1980 and 1990 — 464
11-3: Estimated area irrigated with overpumped groundwater in the United States, 1982 — 465
11-4: Irrigation water withdrawal and source of supply in 16 states in 1990, in order of the volume of withdrawal — 467
11-5: Cropland and pastureland soils affected by salinity and sodicity in 20 Water Resources Regions in 1982 — 470
11-6: Salinity and irrigation in 12 selected Water Resources Regions in 1982 — 471
11-7: Irrigated area, irrigation water use efficiencies and volumes in the Colorado River Basin in 1982 — 473
11-8: Estimated sources of salinity in the Colorado River Basin — 477
11-9: Salinity criteria and observed salinity at three locations in the Lower Colorado River Basin — 480
11-10: Federal programs for salinity control in the Colorado River Basin — 482
11-11: Colorado River salinity control program funding for the fiscal year 1993 — 484
11-12: Concentrations of trace elements in drainage water samples — 491

Foreword

Land degradation is a principal constraint in meeting the needs of world food production. A major factor contributing to land degradation is soil and water salinization in arid zones. Other major factors are water and wind erosion of top soil due to overgrazing and deforestation. The salinization, particularly secondary salinization, occurs in both irrigated and dryland agriculture.

Projections were made decades ago that opportunities existed to expand irrigated agriculture to meet the food and fiber requirements of the 1950 world population of 2.5 billion doubling by 1990. But irrigated agriculture did not expand to the extent projected. In fact, there has been a declining rate of growth of irrigated lands from the 1960s and expansion is minimal today. The expected rate of expansion in agriculture, particularly new irrigation projects, did not materialise due in part to economic constraints and environmental and ecological considerations.

Preserving and sustaining productivity of the current 230 Mha of irrigated lands and 1474 Mha of dryland agriculture are of paramount concern. Unfortunately, about 77 Mha of cultivated lands are salt-affected to varying degrees from human-induced processes. Salt-affected lands are reflected as saline seeps in dryland agriculture and secondarily-salinized irrigated lands.

This compendium is a timely and valuable contribution. It brings together from numerous sources the extent of salinity globally and regionally. The data base synthesized and the concise description of the salinization processes and site-specific management options are indispensable to our challenge to sustain world food and fiber production.

I urge colleagues to become familiar with the contents of this compendium in addressing the goal of achieving food security, presently and in the near future.

Kenneth K. Tanji
Professor of Hydrologic Science
University of California, USA

Preface

The world's limited arable land and freshwater resources have been in a state of continuous development to feed increasing population and to raise living standards. At the end of the eighteenth century, when world population was below one billion, the total irrigated area was about 8 Mha. By the end of the nineteenth century, when world population was around 1.5 billion, it had increased fivefold to 40 Mha. A 1978 estimate indicated that the irrigated area of the world would expand to more than 310 Mha by 1985 and to 420 Mha by the year 2000. This scale of expansion has not been realised because of negative side effects of water and land resources development. These have been due to inadequate technical design and improper evaluation of the socioeconomic cost and benefits; creation of favourable conditions for waterborne diseases; the rapid siltation of reservoirs; impacts on biodiversity; social problems due to relocation and resettlement of populations; environmental degradation due to increased application of fertilisers, herbicides and pesticides; and waterlogging and salinisation due to lack of adequate drainage facilities. Indeed, the world's irrigated land area had reached only 230 Mha by 1990, when the world population was approximately 5.3 billion, with the most rapid increase having been since 1950. Because water has been considered a freely available commodity, it has been used excessively and inefficiently. Currently some 65 per cent of water consumption in the world is used for agricultural production, and much of this is applied inefficiently.

Development of the world's land and water resources has required huge investment and massive engineering efforts. For example, the gigantic Indus irrigation system in Pakistan with 16 Mha of irrigated land is the largest single irrigation system in the world. It contains some of the world's largest infrastructural works and includes three major storage reservoirs, 20 diversion structures, 12 link canals, 48 main canals and some 89 000 watercourses. The length of the irrigation network of canals is about 63 000 km. This is more than 1.5 times the length of the equator. However, this irrigation system, as well as many others around the world, is faced with serious problems of human-induced (also known as secondary) salinisation.

Human-induced salinisation is as old as irrigation. Salinisation forced early settlers in Mesopotamia, the Indus River Basin and China to abandon their land and move to non-salinised lands. This option is no longer possible. The countries affected by human-induced salinisation are predominantly located in arid and semiarid regions and include: Argentina, Australia, Brazil, China, Chile, Commonwealth of Independent States (CIS), Egypt, India, Iran, Iraq, Pakistan, Peru, Spain, Syria, Thailand, Turkey and the United States of America.

Although rapid and inadequate development of large-scale irrigation systems, particularly since the Second World War, is to be blamed for the expansion of secondary salinisation, other activities, such as land clearing and replacement of native trees with shallow-rooted crops, contributed to the development of so-called dryland salinity. Well-documented examples of this type of salinisation exist in Australia, the United States, Canada and Thailand.

Like land resources, water resources are also affected by salinisation resulting from human activities such as the discharge of saline agricultural and mining drainage water to river systems and the increased rate of natural discharge of saline groundwater to surface water resources in irrigated and non-irrigated areas. Major rivers that were originally fresh have become saline and unusable for human and livestock consumption and other uses.

In some coastal areas the extraction of groundwater has proceeded to the point where intrusion of saline seawater into aquifers has degraded the quality of these resources. Continued irrigation with such low quality groundwater has contributed to the expansion of land salinisation. The coastal area of the Gujarat State in India is an example.

Although human-induced salinity problems can develop rapidly, their solution can be very time consuming and expensive. For example, construction of an outfall scheme to discharge saline waters from the Murray Basin in south-east Australia to the sea would have cost US$1.6–4.8 billion in 1990 prices. A similar scheme for the irrigated areas of the Indira Gandhi Canal in India would have cost up to US$9 billion as estimated in 1991. This would discharge the drainage waters to the Arabian Sea through Gujarat.

Experience shows that while the processes of salinisation in irrigated and non-irrigated lands are reasonably well understood and preventive and remedial measures are generally well known, their implementation throughout the world, including the developed countries, is constrained by short-term socioeconomic and political considerations. Examples exist in many countries to substantiate this point.

This publication is in two parts. The first provides a brief history of secondary salinisation, followed by a description of the world's population, water and land resources, climate and the extent of human-induced land degradation. This part also provides a brief description of crop water requirements, irrigation methods and processes of land and water resources salinisation and their management. However, no attempt has been made to be expansive in this part. Extensive literature concerning the processes of salinisation and the physical and chemical characteristics of salt-affected soils and their management is already available. References to these publications are provided.

The second part describes secondary salinity problems in selected countries: Argentina, Australia, China, Commonwealth of Independent States, Egypt, India, Iran, Pakistan, South Africa, Thailand and the United States. These countries contain approximately 70 per cent of the world's irrigated areas and provide examples of irrigated and non-irrigated (dryland) human-induced land salinity, as well as salinisation of water resources. In the case study of each country, descriptions of physiographic features, rainfall and climate, land cover and irrigation are briefly provided to give an understanding of why salinity occurs in a particular part of the country.

Two appendixes have been included. The first summarises statistics on the size, population, water resources, cultivated and irrigated land, major salt-affected areas, and the adopted management options of the countries discussed in Part Two. The second appendix provides names and addresses of institutions involved in salinity investigations in each country, to facilitate communication between countries facing similar problems. A glossary of terms is also included.

We do not pretend that this publication is a complete and comprehensive review of the human-induced salinity problem globally or even for those countries discussed. A more comprehensive description and analysis of the problem would require a series of books, each describing all aspects of the problem within a country or countries. Our main objective is to alert land and water resource developers, managers, irrigation and drainage engineers, hydrologists, hydrogeologists, soil scientists, agronomists and researchers as well as students (undergraduate and postgraduate) in related disciplines, to the severity of salinity problems that arise from inadequate design, construction, monitoring, maintenance and management of land and water resource projects.

Resource managers must confront the fact that development of remaining land and water resources will be much more difficult and much more expensive than in the past. The most suitable sites have already been dammed and the most fertile soils have been developed. Also, competition between various users of these water and land resources will become much more severe. Thus, stabilisation of world population together with efficient use of developed resources are better long-term options than the development of new resources. In the latter case, small to medium-size projects are likely to be more appropriate than large ones.

In brief, we hope that this publication will help others to avoid the mistakes of the past and to promote informed use of the world's vital land and water resources.

F. Ghassemi, A.J. Jakeman and H.A. Nix

Centre for Resource and Environmental Studies
The Australian National University
Canberra ACT 0200 Australia

Acknowledgements

This reference book is the result of collaboration with the many experts and institutions listed below. The authors are grateful to all of them for providing information and reviewing various sections and chapters.

a Individuals:

Abrol, I.P. (Dr): Indian Council of Agricultural Research, New Delhi, India.

Abu-Zeid, M. (Dr): Water Research Center, Cairo, Egypt.

Arunin, S. (Dr): Soil Salinity Research Section, Land Development Department, Bangkok, Thailand.

Asfaw, G. (Dr): Office of the National Committee for Central Planning, Addis Ababa, Ethiopia.

Barr, N. (Mr): Department of Food and Agriculture, Bendigo, Victoria, Australia.

Bhatti, M.A. (Dr): International Irrigation Engineering Management Institute, Lahore, Pakistan.

Boone, S.G. (Dr): United States Department of Agriculture, Soil Conservation Service, Denver, Colorado, USA.

Casas, R.R. (Mr): Institute of Soils, INTA Centre for Investigations of the Natural Resources, Buenos Aires, Argentina.

Crabb, P. (Dr): Department of Geography and Oceanography, University College, University of New South Wales, Australian Defence Force Academy, Canberra, Australia.

Dukhovny, V.A. (Professor): Central Asian Research Institute of Irrigation, Tashkent, Uzbekistan.

du Plessis, H.M. (Dr): Water Research Commission, Pretoria, South Africa.

Eder, J.C. (Mr): Regional Centre for Groundwater, San Juan, Argentina.

El-Ashry, M.T. (Dr): The World Bank, Washington DC, USA.

Evans, R. (Dr): Rural Water Corporation, Armadale, Australia.

Evans, W.R. (Mr): Australian Geological Survey Organisation, Canberra, Australia.

Flügel, W.-A. (Professor): Geographische Institute, Bonn, Germany.

Forster, S.F. (Dr): Department of Water Affairs and Forestry, Pretoria, South Africa.

Francis, R. (Mr): Murray-Darling Basin Commission, Canberra, Australia.

Görgens, A. (Dr): Ninham Shand Consulting Engineers, Cape Town, South Africa.

Hedlund, J.D. (Dr): Soil Conservation Service, West National Technical Center, Portland, Oregon, USA.

Hoffman, G.J. (Dr): United States Department of Agriculture, Agricultural Research Service, Water Management Research Laboratory, Fresno, California, USA.

Howard, K.W.F. (Professor): Physical Sciences Division, Scarborough Campus, University of Toronto, Ontario, Canada.

Hussain, N. (Dr): Water and Power Development Authority, Lahore, Pakistan.

Hussain, T. (Professor): Department of Soil Sciences, University of Agriculture, Faisalabad, Pakistan.

Imhoff, E.A. (Dr): San Joaquin Valley Drainage Program, Sacramento, California, USA.

Jia, Y. (Mr): Institute of Hydrogeology and Engineering Geology, Hebei, China.

Khublarian, M.G. (Professor): Water Problems Institute, Moscow, Russia.

Khonsary, J. (Mr): Fars Regional Water Authority, Shiraz, Iran.

Marcar, N.E. (Dr): CSIRO Division of Forestry, Canberra, Australia.

Michelena, R.R. (Mr): Centre for Investigation of the National Resources, Buenos Aires, Argentina.

Palaniappan, S.P. (Dr): Centre for Soil and Crop Management Studies, Tamil Nadu Agricultural University, Coimbatore, India.

Paydar, Z. (Dr): Faculty of Agriculture, Tehran University, Karaj, Iran.

Pitney, K.A. (Dr): USDA Colorado River Basin Salinity Control Program, Denver, Colorado, USA.

Rao, K.G.V.K. (Dr): Central Soil Salinity Research Institute, Karnal, Haryana, India.

Rhoades, J.D. (Dr): United States Salinity Laboratory, Riverside, California, USA.

Richardson, D.P. (Dr): CSIRO Division of Water Resources, Canberra, Australia.

Schofield, N.J. (Dr): Land and Water Resources Research and Development Corporation, Canberra, Australia.

Singh, K.P. (Dr): Punjab State Council for Science and Technology, Chandigarh, India.

Street, G.J. (Mr): World Geoscience, West Perth, Australia.

Szabolcs, I. (Professor): Research Institute for Soil Science and Agricultural Chemistry, Hungarian Academy of Sciences, Budapest, Hungary.

Tanji, K.K. (Professor): University of California, Davis, USA.

Voronin, A.D. (Professor): Soil Science Department, Moscow State University, Moscow, Russia.

Williams, B.G. (Dr): CSIRO Division of Water Resources, Canberra, Australia.

Williams, W.D. (Professor): University of Adelaide, Adelaide, Australia.

Williamson, D.R. (Dr): CSIRO Division of Water Resources, Perth, Australia.

You, M. (Professor): Shijiazhuang Institute of Agricultural Modernization, Academia Sinica, Hebi, China.

Yu, R. (Professor): Institute of Soil Science, Academia Sinica, Nanjing, China.

Bellati, J.I., Kugler, W.F., Prego, A.J. and

b Institutions:

Bureau of Reclamation, Yuma Project Office, Yuma, Arizona, USA.

Colorado River Basin Salinity Control Forum, Bountiful, Utah, USA.

Department of Agriculture and Water Supply, Pretoria, South Africa.

Geonics Limited, Mississauga, Ontario, Canada.

Institute of Soil Science, Academia Sinica, Nanjing, China.

International Waterlogging and Salinity Research Institute, Lahore, Pakistan.

Mahab-Ghodss Consulting Engineers, Tehran, Iran.

Murray-Darling Basin Commission, Canberra, Australia.

Salt Action Victoria, Department of Conservation and Natural Resources, East Melbourne, Australia.

Yekom Consulting Engineers, Tehran, Iran.

Water and Power Development Authority, Lahore, Pakistan.

Our special thanks go to Dr Peter Crabb for writing the case of conflict resolution in the Murray-Darling Basin. We also thank Susan Kelo, Shelley Santoso and Margaret Kennedy for word processing of the manuscript, Valerie Lyon and Paul Ballard for graphics and McComas Taylor for his valuable editorial advice.

PART ONE

General Aspects of Salinisation

Section A: Global Resource Overview

A brief history of human-induced salinisation

Salinisation is the increase in concentration of total dissolved solids in soil and water. Land and water resources can be salinised by naturally physical and chemical processes or by human activities (secondary salinisation). Secondary salinisation of land and water resources is as old as the history of human settlement and irrigation. A well-documented case occurred in Mesopotamia where early cities in the valleys of the Tigris and Euphrates rivers flourished in the later part of the 4th millennium BC. The development of these settlements was based on irrigation farming and the main crop cultivated during the early days of this civilisation was wheat. However, in response to an increasing concentration of sodium chloride in the soil, wheat was replaced by the more salt-tolerant barley. Boyden (1987) reports that by about 3500 BC the proportions of wheat and barley under cultivation were nearly equal. A little more than 1000 years later, the less salt-tolerant wheat accounted for only one-sixth of the total wheat and barley production. By about 2100 BC wheat production had decreased further and it accounted for less than 2 per cent of the grain (wheat and barley) production. By 1700 BC the cultivation of wheat had been abandoned completely.

Concurrent with the shift to barley cultivation in Mesopotamia was a serious decline in soil fertility which for the most part can be attributed to salinisation. The effect of this decline was particularly devastating in the cities, where the needs of a considerable superstructure of priests, administrators, merchants, soldiers and craftsmen had to be met with surpluses from the agricultural productions (Jacobsen and Adams, 1958). As salt continued to accumulate it eventually reached such concentrations in the kingdoms of Sumer and Akkad that even the substitute barley would not grow and the population was forced to abandon the area. Consequently, these kingdoms became depopulated and power in the region shifted further north to Babylon, where salinisation was less of a problem (Boyden, 1987; Jacobsen,

1982). Historic evidence from other parts of the world, such as China, the Indus River Basin, South America and Arizona (Casey, 1972), indicates that the problem of human-induced or secondary salinisation permeates our civilised history. The process forced people to shift to other locations which, in turn, became salinised in many cases (Szabolcs, 1989).

Human-induced salinisation has not been confined to antiquity. Large-scale irrigation developments were undertaken in the later part of the nineteenth century by the British engineers in India and Pakistan to increase food production and reduce the risk of famine. These led to the rise of watertables and to extensive salinisation. The development of new irrigation schemes and the diversion of surface waters for irrigation have not occurred without severe resource and environmental damages. A spectacular example is the case of the Amu Darya and Syr Darya rivers in the Commonwealth of Independent States (CIS). In the 1950s these rivers were diverted from the Aral Sea for the development of irrigated lands. These lands produced 90 per cent of the former USSR's cotton output and 40 per cent of its rice production (Trofimenko, 1985). Due to these major diversions, the Aral Sea is shrinking. Between 1960 and 1989 its level dropped 12.9 m and its area decreased by 40 per cent. This shrinking has had devastating effects on the flora and fauna, fully degraded the ecosystem and caused health hazards and climatic changes in the region (Micklin, 1988 and 1991). Another example is the San Joaquin Valley in California. In addition to salinisation of its rich irrigated lands, migratory birds have been poisoned by the concentration of selenium in evaporation ponds (Tanji et al., 1986). These two modern examples of secondary salinisation and many others are treated in the second part of this publication.

World population

Partly as a result of progress made in public health practices, sanitary conditions and living standards, the world population has been growing at a high rate for many decades, particularly in the developing countries. According to the World Bank (1992: 25–26), world population growth peaked at 2.1 per cent per year during the period 1965–70 and then started to decline. Population growth is presently about 1.7 per cent or 93 million a year and is expected to decline to about 1 per cent per year by the year 2030.

World population has doubled in the past 40 years (Table 1) and may double again in the next century, perhaps approaching stability at about 11 billion by the year 2100 (World Resources Institute, 1992: 75). Most of this increase will take place in the developing world. Asia and Africa will have populations of 4.9 billion and 1.6 billion respectively by the year 2025.

In 1990, 4.1 billion (77 per cent) of the world's 5.3 billion people lived in the developing world, while 1.2 billion inhabited the industrialised countries. By the year 2025, some 7.1 billion (84 per cent) of a total

Table 1: Estimated and projected world population (in millions) from 1950 to 2025.

Region	1950	1970	Year 1990	2000	2025
Africa	222	362	642	867	1597
North America	166	226	276	295	332
Latin America	166	286	448	538	757
Asia	1377	2102	3113	3713	4912
Europe	393	460	498	510	515
Oceania	13	19	26	30	38
Former USSR	180	243	289	308	352
Total	2517	3698	5292	6261	8503

Source: World Resources Institute (1992: Table 6.1).

population of 8.5 billion will live in developing countries and 1.4 billion in the industrialised countries. Population growth increases the demand for food, goods and services. Many of the earth's new citizens will degrade natural resources and will not be offered the health and educational resources necessary to reach their potential (World Resources Institute, 1992: Ch. 6).

World water resources

The life of mankind and of almost all terrestrial flora and fauna depends on the availability of freshwater resources. However, the global distribution of water is highly uneven. Water is also limited by its accessibility and suitability. Of the earth's total water volume of about 1386 million km^3, some 96.5 per cent is saline ocean water, which is unsuitable for human use. Of the remaining 3.5 per cent, 35 million km^3 is considered to be fresh, but 24 million km^3 is stored in ice sheets and glaciers and 10.53 million km^3 constitute the groundwater reserves to a depth of 600 m below the earth's surface. Freshwater in lakes totals 91 000 km^3 and in rivers, 2120 km^3 (Table 2).

The average annual precipitation on the land surface of the earth is about 800 mm (Chow et al., 1988: 71). However, the hydrological cycle distributes water unevenly around the globe and indeed the spatial distribution is non-uniform within relatively small countries. Accordingly, the world can be divided into water surplus and water deficit regions. Water is in surplus when precipitation is high enough to satisfy the potential demand of the vegetation cover. When precipitation is lower than this potential demand, there is a water deficit. In general, most of Africa, much of the Middle East, the western United States, north-western Mexico, parts of Chile and Argentina and major parts of Australia are considered water deficit regions (World Resources Institute, 1986).

River run-off is one of the main sources of freshwater from which various water demands are satisfied. Through its continuous renewal by the hydrological cycle, river run-off represents the dynamic component of the earth's total water resources, compared to the less mobile volumes of water contained in lakes, groundwater reservoirs and glaciers (Shiklomanov, 1990).

As shown in Table 3, annual global river run-off is about 44 500 km^3, excluding Antarctica which has an annual flow of 2230 km^3. Of this volume, 43 500 km^3 flows to the oceans and the rest into inland seas and lakes with no outlet to the oceans (Caspian Sea, Aral Sea, Lake Chad, etc.). Table 3 also shows the distribution of river run-off by continent.

Table 2: World water reserves.

Form of water	Area covered (km^2)	Volume (km^3)	Share of world reserves (%)	
			of total water reserves	of reserves of fresh-water
World ocean	361 300 000	1 338 000 000	96.5	–
Total groundwater	134 800 000	23 400 000	1.7	–
Fresh groundwater	134 800 000	10 530 000	0.76	30.1
Soil moisture	82 000 000	16 500	0.001	0.05
Glaciers and permanent snow cover:	16 227 500	24 064 100	1.74	68.7
Antarctica	13 980 000	21 600 000	1.56	61.7
Greenland	1 802 400	2 340 000	0.17	6.68
Arctic islands	226 100	83 500	0.006	0.24
Mountainous areas	224 000	40 600	0.003	0.12
Underground ice in zones of permafrost	21 000 000	300 000	0.022	0.86
Water reserves in lakes	2 058 700	176 400	0.013	–
Freshwater	1 236 400	91 000	0.007	0.26
Salt water	822 300	85 400	0.006	–
Marsh water	2 682 600	11 470	0.0008	0.03
Water in rivers	148 800 000	2 120	0.0002	0.006
Biological water	510 000 000	1 120	0.0001	0.003
Atmospheric water	510 000 000	12 900	0.0001	0.04
Total water reserves	510 000 000	1 385 984 610	100	–
Freshwater	148 800 000	35 029 210	2.53	100

Source: Korzun et al. (1978).

Asia and South America have the highest annual stream flows (14 410 km^3 and 11 760 km^3, respectively), while Australia has the lowest (348 km^3). These surface water resources have been in continuous development. During the period 1950–1986 the number of large dams worldwide increased from 5268 to 36 327. In China alone, over the same period, the number of large dams increased from 8 to 18 820 (Biswas, 1992). Collectively the 36 327 dams stored some 5500 km^3 of water, of which nearly 3700 km^3 was useable, the rest being dead storage.

Table 3: Global river run-off.

Territory	Annual stream flow (mm)	Annual stream flow (km^3)	Percentage of global run-off (%)	Area (km^2 × 10^3)	Specific discharge (L s^{-1} km^{-2})
Europe	306	3 210	7	10 500	9.7
Asia	332	14 410	31	43 475	10.5
Africa	151	4 570	10	30 120	4.8
North and Central America	339	8 200	17	24 200	10.7
South America	661	11 760	25	17 800	20.9
Australia	45	348	1	7 683	1.4
Oceania	1 610	2 040	4	1 267	51.1
Antarctica	160	2 230	5	13 977	5.1
Total land area	314	46 768	100	149 022	10.0

Source: Shiklomanov (1990).

On the basis of hydrological cycle data, La Rivière (1989) argues that about 9000 km^3 of water are available for human exploitation worldwide, which is enough to sustain 20 billion people. Yet, because both the world's population and useable water are unevenly distributed, the local availability of water varies widely. Much of the Middle East and North Africa, parts of Central America and the Western United States are already short of water. By the year 2000, many countries will experience extreme scarcity of water due to increasing demands for water for agriculture, industry and domestic use.

Global water use has been continuously increasing due to the increase of world population and the rise in standard of living. Table 4 shows changes in world water use during the period 1900–1990 and its projection to the year 2000. It indicates that global water use has increased about tenfold during the twentieth century. By the year 2000 more than half of the 9000 km^3 available water supply will be in use. Agriculture is the largest consumer of water resources. Its share of total water use was about 90.5 per cent in 1900 and this is projected to decrease to 62.6 per cent by the year 2000.

Table 4:   **World water use, 1900–2000 (in km^3).**

Water use	1900	1940	1950	1960	1970	1980	1990	2000 [a]
				Year				
Agriculture	525	893	1130	1550	1850	2290	2680	3250
Industry	37	124	178	330	540	710	973	1280
Domestic	16	36	52	82	130	200	300	441
Reservoirs	0.3	4	7	23	66	120	170	220
Total	578	1057	1367	1985	2586	3320	4123	5191

(a) Projected   *Source:* Shiklomanov (1990).

Water use has not been efficient in many countries. Overexploitation of groundwater has led to the depletion of resources in some areas and to increased encroachment of saline waters into aquifers along coastal zones in many countries. There are fears that the rapid expansion of agriculture in desert areas may lead to overexploitation of groundwater for irrigation. Excessive irrigation has also led to waterlogging and salinisation, thereby accelerating land degradation. The lack of maintenance of water delivery systems and overuse of water for domestic, commercial and industrial purposes, especially in developing countries, have caused a host of socioenvironmental and economic problems. Water losses, which in some cases amount to more than 70 per cent of water delivered, put increasing and costly pressures on water works, which have to meet the increasing demand for water (UNEP, 1992a: Ch. 5).

Adequate supply is not the only water problem facing many countries throughout the world. Water quality is also a very important issue. Concerns about water quality have grown since the 1960s. At first, attention centred on surface water pollution from point sources. But more recently, groundwater pollution and surface water pollution from non-point sources have been found to be at least equally serious problems. Further information concerning the world's water resources is available in Gleick (1993) and Engelman and LeRoy (1993).

World climate

The global climatic classifications of Köppen (1923) and Thornthwaite (1931) were based on long-term monthly mean rainfall and temperature data and defined class limits that were broadly coincident with major vegetation formation boundaries. These classifications and their derivatives remain in use in general geographic texts. The later work of Thornthwaite (1948), Thornthwaite and Mather (1954) and more particularly Papadakis (1975) reflected a shift away from the descriptive to a more functional approach that considered the effects of temperature and soil moisture on plant growth. Papadakis (1975) developed a global classification that had particular relevance to agricultural systems.

In a new classification of global bioclimate, Hutchinson et al. (1992) used a simple plant growth model (Nix, 1981) and monthly climatic data from 4159 stations around the globe to derive relevant agroclimatic attributes and numerical taxonomic methods to produce groups that were defined objectively. Ten major bioclimatic zones were defined, with these further subdivided into 34 groups. A simple alphanumeric code labels each group. The natural vegetation and representative locations are listed in Table 5 and mapped in Figure 1.

As might be expected, major irrigation development is concentrated in those climatic zones where temperatures permit long growing seasons (E, G, H, I) but where rainfall is partially (E, H, I) or totally (G) limiting. In addition, extensive irrigation development has occurred in cool, dry continental climates (C) in Asia and North America where wheat, oats and barley are commonly grown in the colder months and maize and cotton in the warmer months.

Salinisation of soils and water is characteristic of the truly arid (G) climates and all major irrigation areas have existing or incipient problems. Most are dependent on major river systems (e.g. Nile, Indus, Colorado), but groundwater sources can be important. Poleward of the warm to hot, very dry (G) zones are continental climates with warm to hot summers, but cold winters (C). Salinisation is a serious problem in Central Asia, China and North America in the (C) climates. Soils often have moderate to high salt loads because rainfall is not sufficient for leaching and removal. Irrigation can leach harmful salts below the root zone, but great care is necessary to prevent the rise of watertables that can carry these harmful salts back towards the root zone.

The warm temperate to subtropical seasonal wet/dry (E) climates in general have sufficient rainfall for dryland cropping in either the cool, wet months (Mediterranean climates) or the warm, wet months (subtropical climates), but supplementary irrigation is widely practised. In addition, irrigation is used to produce crops in the dry season. Once again, rainfall is insufficient for leaching of salts much beyond the root zone and great care is necessary to prevent watertables rising with irrigation. Salinisation of soil and water resources under irrigation is a serious problem in every one of eight climates represented in this (E) zone.

The hot, dry (H) and hot, wet/dry (I) climates of the monsoonal tropics are home to hundreds of millions of people in South Asia (India, Pakistan, Bangladesh, Thailand), Africa (Nigeria, Chad, Sudan, Kenya) and South America (Brazil, Venezuela) who are dependent on irrigation for food crop production. The drier (H) climates have greater salinisation problems, but some regions within the wetter (I) climates have serious problems because of inherently saline substrates, for example the Khorat Plateau of north-east Thailand.

Table 5: Major bioclimatic groups of the world, their natural vegetations and representative locations.

Climatic groups	Natural vegetation	Representative location
A. Very cold (polar, alpine)	–	Nome, Alaska
B. Cold (very cold winters, short summers)		
B.1	Coniferous forest	Tomsk, CIS
B.2	Deciduous forest	Bremen, Germany
C. Cool, dry (cold winters, warm to hot summers)		
C.1	Cool, very dry desert	Yumen, China
C.2	Cool, steppe and prairie	Swift Current, Canada
C.3	Modified Mediterranean	Pendleton, USA
C.4	Warm steppe, prairie	Lubbock, USA
D. Cool, wet (short summers, cold winters)		
D.1	Deciduous forest/prairie	Harbin, China
D.2	Deciduous forest	Oklahoma City, USA
D.3	Deciduous/coniferous forest	Frankfurt, Germany
D.4	Deciduous forest	Louisville, USA
D.5	Sclerophyll forest	Rutherglen, Australia
E. Warm, wet/dry (long hot summers, mild winters)		
E.1	Sclerophyll woodland, heath, forest	Algiers, Algeria
E.2	Sclerophyll woodland, heath, forest	Roseworthy, Australia
E.3	Sclerophyll woodland, grassland	Laboulaye, Argentina
E.4	Sclerophyll open-forest, woodland	Biloela, Australia
E.5	Sclerophyll woodland	Lusaka, Zambia
E.6	Sclerophyll woodland, shrubland	Kalgoorlie, Australia
E.7	Sclerophyll open-forest, woodland	Bundaberg, Australia
E.8	Sclerophyll woodland	Bulawayo, Zimbabwe
F. Warm, wet		
F.1	Deciduous forest	Montgomery, Alabama, USA
F.2	Evergreen forest, savannah	Guangzhou, China
F.3	Evergreen forest, grassland	Buenos Aires, Argentina
F.4	Evergreen forest	Posso Fundo, Brazil
F.5	Evergreen forest	Kampala, Uganda
G. Warm to hot, very dry	Desert	Oodnadatta, Australia
H. Hot, dry	Semiarid tropical	Julia Creek, Australia
I. Hot, wet/dry	Monsoonal	
I.1	Monsoon forest, savannah	Khon Kaen, Thailand
I.2	Savanna	Hyderabad, India
I.3	Evergreen and semideciduous forest	Rangpur, Bangladesh
I.4	Evergreen and semideciduous forest	Goias, Brazil
J. Hot, wet (humid tropics)		
J.1	Evergreen forest (rainforest)	Enugu, Nigeria
J.2	Evergreen forest (rainforest)	Habana, Cuba
J.3	Evergreen forest (rainforest)	Bambesa, Zaire

Source of data: Hutchinson et al. (1992).

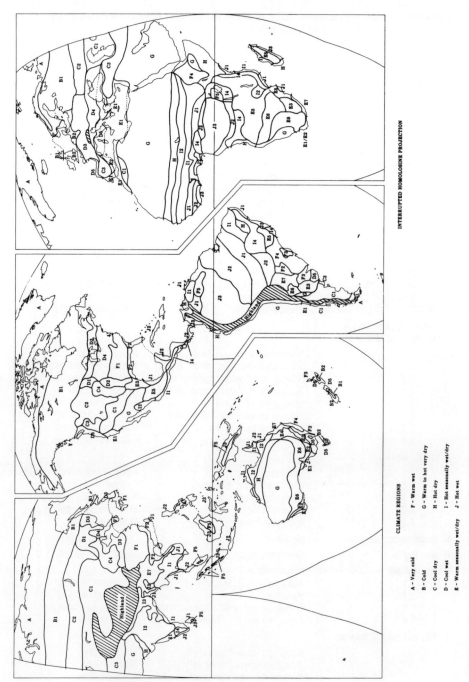

Figure 1: An agroclimatic classification of world climate into 34 classes (Hutchinson et al., 1992).

World dryland areas

The extent of the world's dryland areas has been recently estimated by using an aridity index (Dregne et al., 1991). The index is expressed as the ratio of precipitation over potential evapotranspiration. The various categories of dryland have the following aridity index ranges: hyperarid (<0.05); arid (0.05–0.20); semiarid (0.21–0.50); dry subhumid (0.51–0.65); moist subhumid and humid (>0.65). According to Dregne et al.'s method, the driest continent of the world is Australia where 75 per cent of its area is dry (Table 6). It is followed by Africa and Asia. Drylands comprise about one-third of the areas of Europe, North and South America. In absolute figures, however, the largest drylands occur in Africa (1959 Mha) and Asia (1949 Mha), totalling about 64 per cent of the world's drylands. The total area of drylands is about 6150 Mha, or 41 per cent of the total land area of the world. Of this nearly 978 Mha are hyperarid deserts and 5172 Mha are arid, semiarid and dry subhumid. Figure 2 shows the distribution of the world's dryland areas.

Table 6: World drylands (in million hectares).

	Africa	Asia	Australia	Europe	North America	South America	World total
Hyperarid	672	277	0	0	3	26	978
Arid	504	626	303	11	82	45	1571
Semiarid	514	693	309	105	419	265	2305
Dry subhumid	269	353	51	184	232	207	1296
Total	1959	1949	663	300	736	543	6150
% world total	32	32	11	5	12	8	100
% continent area	66	46	75	32	34	31	41

Source: Dregne et al. (1991).

World arable land resources

Like the freshwater resources, the potentially arable lands of the world are also limited and unevenly distributed. As shown in Table 7, some 3190 Mha of the world's land area are potentially arable, with 734 Mha in Africa and 681 Mha in South America. But in much of this area the soil is marginal or rainfall is unreliable (World Resources Institute, 1990: 87–88). Currently 1474 Mha or 46.2 per cent of these arable lands are cultivated (451 Mha in Asia, 274 Mha in North America and 233 Mha in the former USSR) and the

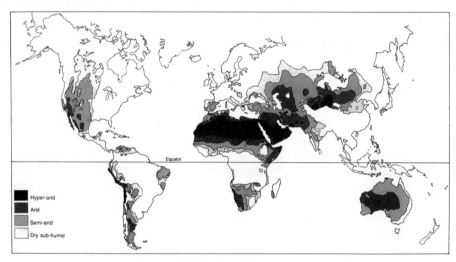

Figure 2: Drylands of the world (UNEP, 1992b).

remaining 1716 Mha are to be developed (549 Mha in Africa and 539 Mha in South America). In other words, slightly more land is available for reclamation in the future than is cultivated at present. On the other hand, most of the best land has already been developed.

Irrigated land constitutes 227 Mha or 15.4 per cent of cultivated lands and 7.1 per cent of potential arable lands. Asia has the largest area of irrigated land (142 Mha) while Oceania has the lowest (2 Mha).

Table 7: Total land area, potential arable land and cultivated land of the world by continent.

Continent	Land area[a] (Mha)	Potential arable land[a] (Mha)	Cultivated land[b] (Mha)	Irrigated land[b] (Mha)
Africa	2 964	734	185	11
Asia	2 679	627	451	142
Oceania	843	153	49	2
Europe	473	174	140	17
North America	2 138	465	274	26
South America	1 753	681	142	9
Former USSR	2 227	356	233	20
Total	13 077	3190	1474	227

Source: (a) US Report (1967) quoted in Buringh (1977); (b) FAO (1989).

The population and irrigated areas of the world have been continuously increasing throughout history. Table 8 shows these increases from 1800 to 1987. Around 1800 the area under irrigation in the world was some 8 Mha. It increased fivefold to 40 Mha by 1900. The area had extended to about 95 Mha by 1950 and in 1987 it reached 227 Mha.

Table 8: Increase in world population and irrigated area.

Year	Population (billion)	Irrigated area (Mha)
1800	1	8[a]
1900	1.5	40[a]
1950	2.5	95[b]
1987	5.1	227[c]

Source: (a) Framji et al. (1981); (b) Brown (1989); (c) FAO (1989).

Table 9 shows the ten leading countries in area irrigated in 1987. China has the largest tract of irrigated land (44.83 Mha). It is followed by India (42.10 Mha) and the former USSR (20.48 Mha).

Table 9: Ten leading countries in area irrigated in 1987.

Ranking	Country	Irrigated area (Mha)	Ranking	Country	Irrigated area (Mha)
1	China	44.83	6	Indonesia	7.40
2	India	42.10	7	Iran	5.70
3	Former USSR	20.48	8	Mexico	5.08
4	United States	18.10	9	Thailand	3.99
5	Pakistan	16.08	10	Romania	3.36

Source: FAO (1989).

The increase in world population has been accompanied, at least for a few consecutive decades since the Second World War, by increases in the harvested area of grain production and in area of irrigation. The spectacular rise in grain production has been due to a number of factors including the green revolution, the increased use of fertilisers and the expansion of irrigated lands (Brown, 1989). Statistics show that the growth rate of the world's harvested area of grain has declined since 1980. The growth rate of irrigated areas has also declined, to 8 per cent during 1980–85, compared with 40 per cent in 1950–60 (Brown, 1989). Although the world's irrigated area was expected to expand to 310 Mha in 1985 and to 420 Mha by the year 2000 (Framji et al., 1981, Table IX), it had reached only 227 Mha by 1987. The significant decline in the growth rate of the world's irrigated area is due to a number of factors. These include the limited supply of arable lands and pressure on their allocation for non-agricultural purposes such as town development, housing, industries, roads and national parks; the severe competition for water use in other sectors such as domestic and industrial water supply; the degradation of lands due to salinisation and to waterlogging caused by improper irrigation and drainage; the abandonment of eroded land; and the depletion of groundwater resources (Brown, 1989).

Agriculture and food production

The role of agriculture in food production is described in the United Nations Environment Programme (1992a: 85–97) and the World Resources Institute (1992: 93–110). The following account is mainly based on the first reference.

At the beginning of the 1990s, the worldwide average consumption of food products per capita was 2670 calories per day, a level considered nutritionally adequate. However, this global average has little significance so long as inadequate food consumption levels prevail in a large number of developing countries. There is a gap of 965 calories per capita between the developed and the developing countries (3399 and 2434 calories per capita respectively), and there are wide gaps between and within the developing countries themselves.

The worldwide disparity in per capita food availability has been created and aggravated by a combination of social, economic, environmental and political factors; these include a decrease in commodity prices, agricultural subsidies, agricultural trade barriers, inequitable access to resources and products and the often primitive conditions of production and processing of agricultural output in many areas. As a result, the number of chronically hungry people in the world increased from about 460 million in 1970 to about 550 million in 1990 and is expected to reach 600–650 million by the year 2000. Close to 60 per cent of the hungry people in the developing world live in Asia, about 25 per cent in Africa and some 10 per cent in Latin America and the Caribbean. Currently 1116 million people in developing countries are living in conditions of poverty and 630 million of them can be considered extremely poor. This last group is the most threatened by hunger and chronic malnutrition.

Agricultural output and food production increased in both developed and developing countries in the period 1970–1990. The annual rate of increase was higher in the developing countries (about 3 per cent) than in the developed (about 2 per cent). In developing countries there were major increases in Asia, a near stagnation in Latin America and a marked drop in Africa. The rate of increase in cereals production was higher in the developed countries (about 32 per cent in 1970 and 15 per cent in 1990). In the developed world the annual rate of growth in cereals production was higher than population growth (about twice as much), but in the developing countries it was much lower (about one-fifth as much). A wide gap, currently 529 kg per capita, continues to exist between annual cereal output of developed and developing countries as a whole (777 and 248 kg per capita respectively, in 1990).

About 12 per cent of the world's population is entirely dependent on livestock production for food. The production of meat in developing countries is much lower than in developed ones (68.7 million tonnes and 103.2 million tonnes respectively, in 1990). This is attributed mainly to the fact that most livestock in developing countries is in traditional, small-scale farming systems, where it is a source of subsistence and where animals supply the necessary power for agriculture. Also, additional income is generated by selling animal products.

As mentioned previously, the total area of potential arable land in the world is 3190 Mha, about 46 per cent of which (1474 Mha) is already under cultivation (Table 7). Worldwide, the area of arable land increased by only 4.8 per cent over the period 1970–1990: the increase was 0.3 per cent in developed countries and 9 per cent in developing countries. However, per capita arable land decreased from a worldwide average of 0.38 ha in 1970 to 0.28 ha in 1990, mainly due to population growth and loss of land for agriculture. The decrease was most noticeable in the developing countries: from 0.28 ha per capita to 0.20 ha per capita. It has been estimated that if the arable land area is maintained at the present level (1474 Mha worldwide) and no new land is brought under cultivation and no existing land goes out of production due to degradation or allocated to other use, the per capita arable land in the world will progressively decline to 0.23 ha in 2000, 0.15 ha in 2050 and 0.14 ha in the year 2100.

It has been said that large areas of new land could be brought under cultivation, but unused arable land is not always available to people who need it most, and opening up new areas remains an expensive means of increasing agricultural production. In fact, further expansion of agricultural land is constrained in many parts of the world. For example, in arid regions shortage of water for irrigation constitutes a major constraint on future expansion of cropland area.

With traditional agriculture the minimum dietary requirement per capita can be met from an estimated average of 0.6 ha of arable land. This means that the present area under cultivation in the world would only meet the minimum dietary requirements of less than half of the world population. Therefore, there has been no alternative but to increase the output of existing arable land through technological innovation. Efforts to do so have been successful; productivity gains have been achieved largely by using the 'green revolution' technology which requires the use of high-yield varieties of seeds and high inputs of water, fertilisers and pesticides.

The world's irrigated land increased from 168 Mha in 1970 to 227 Mha in 1987, an increase of 59 Mha in two decades. Although irrigated land at present accounts for 15.4 per cent of cultivated land, it produces one-third of the world's food (over twice the productivity of average rainfed land). However, the rate of expansion of irrigated land has declined in the 1980s because of the scarcity of additional irrigable land and of good quality water in many parts of the world. The scarcity of water resources is compounded by the inefficient use of irrigation water (see Irrigation efficiency).

Buringh (1977) argues that, taking into account the regional conditions of soils, climate and farm management, enough food could be produced for a population or five to ten times the present world population. However, it is clear that such development would require massive investment and substantial political will. Crosson and Rosenberg (1989) also argue that 100 years from now the earth may have 10 billion inhabitants, about twice as many as it has now. World food production could grow significantly (but more slowly than the current rate) and there would still be enough food for 10 billion people by the time they arrive. However, although the food supply must expand, it must expand in a way that does not destroy the natural environment. For that to happen, a steady stream of new technologies that minimise the erosion, desertification and salinisation of the soil and other environmental damage must be introduced. Crosson and Rosenberg (1989) are confident that these techniques will be developed if the strong system of agricultural research organisations already in place is provided with enough financial support and leadership.

It would be a mistake, however, to leave the impression that the world's food problem is simply a matter of supply and demand (Knutson et al., 1990: 28–29). It is in fact much more complex, requiring governmental systems, policies and programs of many nations of the world to be rationalised. Apart from the adequacy of global production capacity, the problem has at least two other major dimensions: distribution and trade. Reducing hunger and malnutrition is primarily a distribution problem. Properly distributed, there is enough food to feed the approximately 500 million malnourished people in the world. The causes of the distribution problem are low income, a lack of economic growth and underdevelopment. Keeping trade channels open has been a perpetual problem. While the benefits of free trade are evident, the benefits of protectionism are equally clear.

Extent of human-induced land degradation

The results of a Global Assessment of Soil Degradation (GLASOD) were published in 1991 by the International Soil Reference and Information Centre (ISRI). The assessment is based on the World Map of the Status of Human-

Induced Soil Degradation (Oldeman et al., 1991a). The map was prepared under financial agreement with the United Nations Environment Programme (UNEP) through a cooperative effort of about 250 soil scientists and soil degradation experts throughout the world, 21 regional coordinators and a number of international institutions (Winand Staring Centre, International Society of Soil Science, Food and Agriculture Organization of the United Nations and International Institute for Aerospace Survey and Earth Sciences).

A primary objective for creation of the soil degradation map was to generate awareness of the present status of soil degradation in the minds of policy makers, decision makers and the general public (Oldeman et al., 1991a). The GLASOD map covers 13 billion ha of the land surface between 72° N and 57° S. The GLASOD results are all the more alarming because, unlike other attempts to estimate land degradation, they do not include land degraded by ancient civilisations or even by colonial expansion; nor do they include land that is naturally barren. Soil scientists were asked to categorise only soils degraded over the past 45 years as a result of human intervention (World Resources Institute, 1992: 111–118).

The GLASOD considers two categories of human-induced soil degradation processes. The first deals with soil degradation by displacement of soil material and the second with physical and chemical soil deterioration. The two major types of soil degradation in the first category are water erosion and wind erosion. Water erosion includes loss of topsoil and terrain deformation. The most common phenomena of this degradation type are rill and gully formation. Wind erosion includes loss of topsoil, terrain deformation and overblowing. Chemical deterioration consists of loss of nutrient and/or organic matter, salinisation, acidification and pollution. Physical deterioration includes compaction, waterlogging and subsidence of organic soils (Oldeman et al., 1991a).

Water erosion is by far the most important type of soil degradation, occurring in 1094 Mha or 56 per cent of the total area affected by human-induced soil degradation (Table 10). On a global scale the area affected by wind erosion is 548 Mha (28 per cent); by chemical soil degradation, 239 Mha (12 per cent); and by physical soil deterioration, 83 Mha (4 per cent).

As shown in Table 10, four degrees of soil degradation are recognised. A light degree of soil degradation, implying somewhat reduced productivity of the terrain which is manageable by local farming systems, is identified for 38 per cent of all degraded soils. A larger percentage (46 per cent) has a moderate degree of soil degradation and greatly reduced productivity. Major improvements, often beyond the means of local farmers in developing countries, are required to restore productivity. Strongly degraded soils cover 296 Mha worldwide. These soils are no longer reclaimable at farm level and are virtually lost. Major engineering work or international assistance is required to restore these terrains. Extremely degraded soils are considered unreclaimable and beyond restoration. Their worldwide coverage is around 9 Mha.

Five types of human intervention were identified as resulting in soil degradation: deforestation and removal of natural vegetation (579 Mha); overgrazing of vegetation by livestock (679 Mha); improper management of agricultural land (552 Mha); overexploitation of vegetative cover for domestic use (133 Mha); and industrial activities leading to chemical pollution (23 Mha).

Table 10: **Human-induced soil degradation for the world.**

Type	Light (Mha)	Moderate (Mha)	Strong (Mha)	Extreme (Mha)	Total (Mha)	Total (per cent)
Loss of topsoil	301.2	454.5	161.2	3.8	920.3	
Terrain deformation	42.0	72.2	56.0	2.8	173.3	
WATER	343.2	526.7	217.2	6.6	1093.7	55.7
Loss of topsoil	230.5	213.5	9.4	0.9	454.2	
Terrain deformation	38.1	30.0	14.4	–	82.5	
Overblowing	–	10.1	0.5	1.0	11.6	
WIND	268.6	253.6	24.3	1.9	548.3	27.9
Loss of nutrients	52.4	63.1	19.8	–	135.3	
Salinisation	34.8	20.4	20.3	0.8	76.3	
Pollution	4.1	17.1	0.5	–	21.8	
Acidification	1.7	2.7	1.3	–	5.7	
CHEMICAL	93.0	103.3	41.9	0.8	239.1	12.2
Compaction	34.8	22.1	11.3	–	68.2	
Waterlogging	6.0	3.7	0.8	–	10.5	
Subsidence of organic soils	3.4	1.0	0.2	–	4.6	
PHYSICAL	44.2	26.8	12.3	–	83.3	4.2
Total (Mha)	749.0	910.5	295.7	9.3	1964.4	
(per cent)	38.1	46.1	15.1	0.5		100

Source: Oldeman et al. (1991b: Table 9).

Table 11 shows that globally more than 76 Mha of land is salt-affected, out of which 52.7 Mha (69 per cent) is in Asia, 14.8 Mha (19 per cent) in Africa and 3.8 Mha (5 per cent) in Europe. The four degrees of light, moderate, strong and extreme salt-affected land cover 34.6 Mha, 20.8 Mha, 20.4 Mha and 0.8 Mha respectively.

Table 11: Global extent of human-induced salinisation.

Continent	Light (Mha)	Moderate (Mha)	Strong (Mha)	Extreme (Mha)	Total (Mha)
Africa	4.7	7.7	2.4	–	14.8
Asia	26.8	8.5	17.0	0.4	52.7
South America	1.8	0.3	–	–	2.1
North and Central America	0.3	1.5	0.5	–	2.3
Europe	1.0	2.3	0.5	–	3.8
Australasia	–	0.5	–	0.4	0.9
Total	34.6	20.8	20.4	0.8	76.6

Source: Oldeman et al. (1991b: Tables 2 to 8).

Section B:
Crop and Irrigation Aspects

Crop salt tolerance

The salt tolerance of a plant can be defined as the plant's capacity to endure the effects of excess salt in the medium of root growth (Maas, 1990). The salt tolerance of a plant is not an exact value. It depends on many factors, conditions and limits including environmental factors (soil fertility, physical condition of the soil, salt distribution in the profile, irrigation method and climate) and biological factors (stage of growth, varieties and rootstocks).

The tolerance of a plant with respect to soil salinity can be described by the 'yield response function' which is a plot of the relative yield as a function of soil salinity. The yield response function can be represented by two linear lines, one a tolerance plateau (threshold) and the other a concentration-dependent line whose slope indicates the yield reduction per unit increase in salinity (Maas, 1990).

For soil salinities exceeding the threshold of any given crop, the crop yield can be given by the following linear equation (van Genuchten and Hoffman, 1984):

$$Y = Y_m - Y_m \, s(c - c_t)$$

where:

Y = crop yield
Y_m = the crop yield under non-saline condition
s = the slope of the line determining the yield decline per unit increase in salinity, beyond the threshold
c = the average root zone salinity
c_t = the salinity threshold

Maas (1990) provides salt tolerance data for a wide range of herbaceous crops (fibre, grain, fruit, vegetable, grass, forage crops and special crops) and woody crops (trees) and gives the limits of salt tolerance for 49 species of ornamental shrubs, trees and ground cover grown in Riverside, California. Table 12 shows the threshold and the yield decrease data for some 25 plants. Van Genuchten and Hoffman (1984) describe a computer program with 20 different options, based on a linear and two non-linear models that more accurately represent the yield response function. They also present several examples illustrating the type of results obtained with their program. Further information concerning plant salt tolerance is available in Staples and Toenniessen (1984), Abrol et al. (1988: 31–40) and Ayers and Westcot (1989: 29–41).

Table 12: Salt tolerance of selected plants with respect to the electrical conductivity of saturated-soil extract.

Plant	Threshold ($\mu S\ cm^{-1}$)	Yield decrease per 1000 $\mu S\ cm^{-1}$ increase in salinity (per cent)
Sensitive:		
Apricot	1600	24.0
Bean	1000	19.0
Carrot	1000	14.0
Grapefruit	1800	16.0
Orange	1700	16.0
Onion	1200	16.0
Peach	1700	21.0
Strawberry	1000	33.0
Moderately sensitive:		
Alfalfa	2000	7.3
Broad bean	1600	9.6
Corn (grain)	1700	12.0
Cucumber	2500	13.0
Lettuce	1300	13.0
Potato	1700	12.0
Rice	3000	12.0
Sugarcane	1700	5.9
Tomato	2500	9.9
Moderately tolerant:		
Barley (forage)	6000	7.1
Sorghum	6800	16.0
Soy bean	5000	20.0
Wheat	6000	7.1
Tolerant:		
Barley (grain)	8000	5.0
Cotton	7700	5.2
Sugar beet	7000	5.9
Wheat (semidwarf)	8600	3.0

Source: Maas (1990).

Crop water requirements

Plants consume water essentially for the two processes of photosynthesis and transpiration. They absorb water through the roots and primarily through the root hairs. Water is transported through the plant and then removed from the leaf surface via transpiration. Transpiration is controlled by the stomatal aperture and by the vapour pressure gradient from the leaf to the air (Blad, 1983). The crop transpires during its growth. At the same time evaporation takes place at the soil surface. The combined quantity of water used under conditions of optimum availability is known as consumptive use or evapotranspiration.

The amount of water required by plants for their growth depends on a number of factors including the type of plant, its stage of development, soil properties and meteorological conditions (temperature, radiation, humidity and wind). The demand for water is not evenly spread over the growing season. At the beginning of the season, consumptive use is low. It increases as the plant foliage develops and the days become warmer, peaks during flowering and fruit formation and rapidly decreases towards the end of the growing season.

The amount of water in the soil useable by plants lies between field capacity and wilting point. This portion is called the useable capacity. The field capacity is the maximum amount of water capable of being held by the soil in opposition to gravity. The wilting point corresponds to that amount of water that opposes the absorptive strength of the plant (Seemann, 1979). The amount of useable capacity depends on the soil type. It is low in sandy soil and high in loam and loess.

Optimum plant growth occurs when soil moisture is near field capacity. Wilting of plants occurs when the suction exerted by the roots on the moisture in the soil fails to maintain an adequate flow of water to the leaves. If the level of soil water is approaching wilting point, rainwater or irrigation is required to maintain plant growth. At the other limit, when saturation is reached, air is cut off from the roots and the plant growth stops. Excess water on farm lands may be caused by rain, excessive irrigation or by poor land grading. Excess water should be drained to maintain a healthy soil moisture for crop production.

A comprehensive description of the role of water in plants is available in Teare and Peet (1983). This publication provides a description of: the physical and physiological aspects of water transfer in the biosphere and atmosphere; factors that affect energy exchange in the biosphere, particularly evaporation; the soil factors that affect soil water status and water supply to crop roots; and the distribution, movement and function of crop-water in plant cells, tissues and organs. Teare and Peet (1983) also present a review of the crop-water relations of the 12 important food and fibre crops (alfalfa, bean, corn, cotton, peanuts, potato, rice, sorghum, soybean, sugar beet, sugarcane and wheat).

The evaluation of water requirements of crops to achieve full production at a particular location is based on the estimation of evapotranspiration. A simple method consists of converting the Class A pan evaporation to evapotranspiration by multiplying by a crop coefficient. The coefficient depends on the specific crop and the growth stage of the crop (Rural Water Commission of Victoria, 1988: 28–33; Seeman et al., 1979: 223–227).

It is important to note that because the amount of salt removed by crops is negligible, salt will accumulate in the root zone and will cause a loss in production. Consequently, salt must be leached by supplying more water than is required by the crops. The amount of leaching water needed depends mainly on the salt content of the irrigation water, soil and groundwater; and the salt tolerance of the crops. The ratio of the depth of drainage water to the depth of the applied water (irrigation plus rainfall) is called the leaching fraction. Hoffman (1990) describes the leaching fraction and the control of salinity at the root zone.

Irrigation

Irrigation methods

Irrigation methods are classified into three groups: surface irrigation, localised irrigation and sprinkler irrigation. Surface irrigation methods are labour intensive with relatively low capital cost, while localised and sprinkler irrigation are less labour intensive but require a high capital cost (Table 13).

Table 13: Labour requirements and investment cost for various field irrigation systems.

Irrigation method		Labour requirements (man-hour per irrigation per ha)	Capital costs of field system (US$ per ha)
Surface	Furrow	1–3	100–400
	Border	0.5–1.5	100–400
	Basin	0.1–1	100–400
Sprinkler	Hand-moved laterals	1–2.5	400–800
	Tractor-moved laterals	0.5–1	600–1000
	Self-moved systems	0.05–0.3	1000–1800
	Permanent systems	0.05–0.2	1800–2200
Drip	Orchard	0.1–0.3	1200–1800
	Row crops	0.1–0.3	1500–2500

Source: Jurriëns and Bos (1980).

The following is a brief description of irrigation methods. Further information concerning the irrigation methods is available from Kovda et al. (1973), Romita and Galbiati (1978), Dedrick et al. (1982), and Horton and Jobling (1984), while Petermann (1993) describes the environmental aspects of irrigation.

Surface irrigation

Surface irrigation is widely used and consists of the following types:

- *Border irrigation:* This is so called because of the type of preparation of the land surface required for distribution of water. In this method the land is divided into long, narrow parallel strips separated by earth banks. These are arranged lengthwise in the direction of the maximum gradient of the land. The water consigned to each irrigation unit from a watering conduit situated at the highest point flows down the gradient to the bottom, moistening the soil.

- *Basin irrigation*: The layout of basin irrigation is similar to that of border irrigation, the main differences being crossbanks constructed on the contour at regular intervals down the slope and a pipeline or channel to supply water to each basin. The spacing of the crossbanks is determined by the amount of the longitudinal slope. Once a basin is filled with water the flow is turned into another basin and the ponded water is allowed to soak into the soil.

- *Furrow irrigation*: Furrow irrigation has been practised in many parts of the world since ancient times and remains a very important method of irrigation today. The principles of the method are similar to those of border irrigation but land preparation differs because numerous furrows are used instead of the smooth surface of bays. As with other surface methods of irrigation, water is released from a supply point to head channel or pipeline located on the highest land. If an open head channel is used water is released into the furrows by siphon tubes, outlet pipes or shovel cuts. The furrows may be V-shaped or U-shaped.

Sprinkler or spray irrigation

A sprinkler system consists of four basic components: pumping unit; main lines; lateral lines; and sprinklers. The system sprays water over the land surface. The advantage of the system over surface irrigation is that it avoids uneven penetration of water and its subsequent waste. The system is suited to a wide range of slopes, soil and crops. However, because of high capital cost, the method is generally confined to crops that offer a high return, such as fruit, vines and tobacco.

A wide range of sprinkler systems is available; they can be classified into three types:

- Portable: refers to the technique where all equipment (pumping unit, main, submain and laterals) is portable and can be moved from one place to another.
- Semiportable or semipermanent: means that the pumping unit is fixed, the mains and submains are underground and only the laterals are portable.
- Permanent: means that all components (pumping unit, mains, submains, laterals) are permanently located.

Localised irrigation

Localised irrigation covers trickle (drip), microjet (microspray) and mini-sprinkler systems. Common to these three types of irrigation system is the frequent application of water at low rates, keeping the soil around the roots near field capacity. The advantages include: crops are watered with increased uniformity; soil structure is preserved; water is saved because of reduced evaporation; and the correct control of water quantities reaching plants. The disadvantages include: obstruction of small drippers because of water impurities, biological or chemical formations; creation of an area of permanently saturated or near-saturated soil favouring the development of plant or animal pests; and saline accumulation at the edges of moisture areas (Romita and Galbiati, 1978). Localised irrigation is almost exclusively used for orchards, vineyards, some vegetable crops, and occasionally sugarcane. Following is a brief description of the three localised irrigation systems:

- *Trickle (drip) irrigation*: Trickle or drip irrigation is the method of applying water directly to the soil around the plant root at low rates of flow but frequently enough to keep the soil around the roots at or near field capacity. The components of the system include: pump, filter, main, submains, laterals, and drippers. With drip irrigation, water should be free of physical impurities and organic matters and should be filtered to avoid the deposition of material in the lines and drippers. Fertiliser and pesticides can also be applied in the water.

    The interval between irrigations varies according to the soil texture and the evaporative demand. For very coarse soil with a small water-holding capacity, irrigation may be for a short period each day when plants are using the most water. For light sandy soil, the interval may vary from one to four days. For heavier loam and clay loam, the interval between watering may vary from two to eight days. The duration of application also varies with the type of soil. Heavy soil might be watered for 10 to 16 hours, light soils for 6 to 10 hours. Balogh and Gergely (1985) provide detailed information about this method and describe general aspects, design and operation of the system.

- *Microjet irrigation*: Microjet irrigation is essentially an under-tree method of irrigation. It differs from mini-sprinkler in as much as it has no moving spinner to distribute the water droplets. Instead a fine jet of water is directed onto a fixed surface to produce a spray.

- *Mini-sprinkler irrigation*: Mini-sprinkler irrigation is another method of under-tree irrigation. The conditions that govern the adoption of microjet irrigation also apply to this method. However, because mini-sprinklers deliver more water and over a greater area than microjets and because some sprinklers have a pressure compensation valve, they are often used on mature plantings in place of microjets. Microjets tend to be used on new plantings and young trees.

Selection of irrigation methods

Factors affecting the selection of irrigation methods include: nature of the soil; relief and land preparation requirements; size of the agriculture holding; salinity condition; groundwater and drainage; and water supply (Kovda et al., 1973). With respect to salinity, the basin irrigation method is well suited to all cases where leaching is necessary. Hence it will prove satisfactory in cases of surface salinity or where saline water is used. The furrow irrigation method must be practised with considerable care in saline and alkaline soils because water moves upwards from the furrow towards the surface of the beds. Sprinkler methods can prove suitable in most cases since they can provide controlled, uniform application. It should be noted that if saline water falls on leaves, considerable damage can be caused when the concentration exceeds a certain value. For citrus fruits, for example, burning of the foliage has been observed with water containing 800 to 900 mg L^{-1} total dissolved solids (TDS), 69 to 140 mg L^{-1} of Na^+ and 133 mg L^{-1} of Cl^{-1} (Kovda et al., 1973).

Irrigation efficiency

The operational aspects of farm irrigation and water supply systems are dominated not only by the physical characteristics of the irrigated area (climate, topography, soil type) but also by the socioeconomic conditions. The design of an efficient irrigation system requires access to the available information. According to Bos and Nugteren (1990), the lack of basic knowledge of water use efficiency has several serious drawbacks. Firstly, in the planning and design of irrigation systems a large safety margin is applied, as a consequence of which irrigation facilities like canals, structures and reservoirs are constructed with capacities that are too large; investments are thus considerably higher than would otherwise be necessary. Secondly, the limited water resources are not optimally distributed and used, as a result of which much water goes to waste and less land can be irrigated. Last but not least, the low overall irrigation efficiency creates harmful side effects such as rising watertables and soil salinisation. To control the watertable a costly subsurface drainage system may be necessary and this will seriously affect the economy of the project.

Bos and Nugteren (1990) define a number of efficiencies of irrigation water use, as follows:
- The conveyance efficiency (e_c) is the efficiency of canal and conduit networks from the reservoir, river diversion or pumping station to the offtakes of the distributary system. It can be expressed as:

$$e_c = \frac{V_d + V_2}{V_c + V_1}$$

where:
  V_c = volume diverted or pumped from the river (m³)
  V_d = volume delivered to the distribution system (m³)
  V_1 = inflow from other sources to the conveyance system (m³)
  V_2 = non-irrigation deliveries from the conveyance system (m³)

- The distribution efficiency (e_d) is the efficiency of the water distribution canals and conduits supplying water from the conveyance network to individual fields. It can be expressed as:

$$e_d = \frac{V_f + V_3}{V_d}$$

where:
  V_d = volume delivered to the distribution system (m³)
  V_f = volume of water furnished to the fields (m³)
  V_3 = non-irrigation deliveries from the distributary system (m³)

- The field application efficiency (e_a) is the relation between the quantity of water furnished at the field inlet and the quantity of water needed, and made available, for evapotranspiration by the crop to avoid undesirable water stress in the plants throughout the growing cycle. It can be expressed as:

$$e_a = \frac{V_m}{V_f}$$

where:
  V_f = volume of irrigation water furnished to the fields (m³)
  V_m = volume of irrigation water needed, and made available (m³).

- The tertiary unit efficiency (e_u) is the combined efficiency of the water distribution system and of the water application process. In other words, it is the efficiency with which water is distributed and consumptively used within the tertiary unit. The tertiary unit efficiency can be expressed as:

$$e_u = \frac{V_m + V_3}{V_d}$$

- The irrigation system efficiency (e_s) is the combined efficiency of the systems of water conveyance and distribution. It can be expressed as:

$$e_s = \frac{V_f + V_2 + V_3}{V_c + V_1}$$

- The overall or project efficiency (e_p) can be expressed as:

$$e_p = \frac{V_m + V_2 + V_3}{V_c + V_1}$$

Bos and Nugteren (1990) analysed data collected from 29 countries with a wide range of climatic and socioeconomic conditions. The survey covered 84 irrigated areas representative of about 5 Mha, or 2 per cent of the total irrigated area in the world. Irrigation methods used in the survey areas consisted of surface methods and sprinkler. The irrigated crops included cereals, rice, cotton, sugarcane, pasture, fruits and vegetables. Table 14 summarises the efficiencies calculated by Bos and Nugteren (1990) and indicates the low efficiency of the irrigation operations.

Table 14: Calculated efficiencies in a number of irrigated areas of the world.

Type of efficiency	Number of areas	Range of efficiency (per cent)	Average efficiency (per cent)
Project efficiency (e_p)	56	7–60	30
Tertiary unit efficiency (e_u)	46	12–94	44
Irrigation system efficiency (e_s)	49	22–93	58
Field application efficiency (e_a)	60	14–88	53
Distribution efficiency (e_d)	48	50–97	78
Conveyance efficiency (e_c)	48	26–98	73

Source: Bos and Nugteren (1990: 20–22).

Irrigation water quality

The quality of irrigation water is determined by the following chemical characteristics: salinity or total concentration of soluble salts; sodicity or concentration of sodium relative to other cations; anionic composition of the water, especially concentration of bicarbonate and carbonate anions; and concentration of boron and other elements that may be toxic to plant growth (Shainberg and Oster, 1978).

The salinity is the most important criterion for evaluating irrigation water quality. Total concentration is important because most crops respond to total concentration of ions in the growth medium rather than to any specific ion. Generally an increase in the salt content of irrigation water will result in an increase in the salinity of the soil water. The rate and extent of the increase will depend on a number of factors including: the leaching fraction, that is, the amount of water supplied by irrigation or rainfall in excess of that required to satisfy consumptive use of crops and leaching efficiency; the ionic composition of the irrigation water; and physical properties of the soil such as infiltration, moisture characteristics, drainage and water application.

Irrigation water classification schemes aimed at defining the salinity hazard for a given water are all based on broad generalisation regarding crop growth, climate, irrigation management and soil properties. According to Frenkel (1984), these types of classification are rigid and in many cases fail to recognise certain specific factors which are critical in determining the potential use of a given water. Therefore, any assessment of the suitability of water for irrigation must be made in relation to three main subjects: crop tolerance to salinity; how much leaching will be achieved; and what level of soil water salinity will be obtained when a certain type of irrigation water is used.

The sodicity or sodium hazard of irrigation and soil waters can negatively affect crop production. Unlike salinity hazard, excessive sodium does not impair the uptake of water by plants but does impair the infiltration of water into the soil. The growth of the plant is thus affected by an unavailability of soil water (see Pratt and Suarez (1990) and Frenkel (1984) for further information). Shainberg and Oster (1978) and Frenkel (1984) discuss the role of carbonates in irrigation water.

Toxicity normally results when certain ions are taken up with the soil water and accumulate in the leaves during water transpiration to an extent that results in damage to the plant. The usual toxic ions in irrigation water are chloride, sodium and boron. Ayers and Westcot (1989: 77–83) describe these toxicities and provide data on the tolerance of crops with regard to these toxic elements.

Some trace elements that occur in water and soil are essential for plant growth, but can become toxic at an elevated concentration. Pratt and Suarez (1990) provide data on the recommended maximum concentrates of 15 trace elements (arsenic, beryllium, cadmium, chromium, cobalt, copper, fluorine, lead, lithium, manganese, molybdenum, nickel, selenium, vanadium and zinc). Similar data are also available in Ayers and Westcot (1989: 95–96).

Rhoades (1990a) argues that numerous schemes for the classification of water for irrigation are essentially empirical and have some problems. For example, the substantial experience in using brackish water for irrigation shows that waters which would be classified by Ayers and Westcot (1989) as having a severe restriction for use have been successfully used in numerous places throughout the world under widely varying conditions of soil, climate, irrigation technique, cropping system, economics and cultural organisations. This fact shows that the actual suitability of a given water for irrigation depends very much on the specific conditions of use and on the relative economic benefits that can be derived from irrigating with that water compared to others. Rhoades (1990a) describes the application of a steady-state chemistry model named 'WATSUIT' for the assessment of water suitability for irrigation.

Section C: Salinisation Processes and Damage

Processes of salinisation

Salinisation is the process whereby the concentration of total dissolved solids in water and soil is increased due to natural or human-induced processes. Natural and human-induced salinisation are also called primary and secondary salinisation respectively. Salt-affected soils of primary origin are formed as a result of the long-term influence of natural processes leading to an accumulation of salts in a region. Accumulation of salts may be the result of a gradual accumulation of products of weathering or one-time submergence of soils under sea water. Primary salt-affected soils have a wide distribution on all continents. Szabolcs (1989) describes the extent and characteristics of these soils for many countries in Europe, Asia, Africa, North America, Central America, South America and Australia. According to his estimate, the global extent of primary salt-affected soils is about 955 Mha.

Secondary salinisation is the result of the salt stored in the soil profile and/or groundwater being mobilised by extra water provided by human activities such as irrigation or land clearing. The extra water raises watertables or increases the pressure of confined aquifers, creating an upward leakage to watertable aquifers. When the watertable becomes close to the soil surface, water is evaporated, leaving salts behind and causing land salinisation. The mobilised salt can also move laterally or vertically toward watercourses and increase their salinity.

Associated with soil salinisation is the secondary process of alkalisation, whereby the clay fraction of the soil becomes saturated with sodium. The effect of the sodium ions is to disperse the fine clay particles and cause the soil's desirable crumb structure to collapse. As a result, the soil tends to swell and its aggregates to slake down and clog the soil's pores, creating a less permeable condition that restricts water penetration and aeration (Hillel, 1990). Such a soil can become very difficult and expensive to reclaim. Soils with a high content of dispersible clay are naturally more prone to alkalisation than those that have low clay contents. The former soils are also more sensitive to waterlogging. For these reasons, irrigation of the soils with appreciable clay content must be treated with particular care. Further information about alkaline soils and alkalisation is available in Foth (1984), Abrol et al. (1988) and Szabolcs (1989).

Source of salt

As a source of salt, dissolved solids exist in rainwater, within the soil profile, in groundwater and in water used for irrigation. Rainwater generally contains 10–30 mg L^{-1} and sometimes more than 50 mg L^{-1} of salt. The geographic distribution of atmospheric salt decreases with increasing distance from the coast. Consider a salinity of 10 mg L^{-1} in rainwater. This would add 10 kg ha^{-1} of salt to the soil for each 100 mm of rainfall per year. Accumulation of this salt over the millennia would be considerable. Salt stored in the soil profile depends mainly on the nature of the soil and the average annual rainfall. Salt content is generally low for sandy soils and high for soils containing a high percentage of clay minerals. Salt content varies inversely with average annual rainfall. For example, in Western Australia, the salt content of a 40 m profile ranges from 170 to 950 t ha^{-1} for average annual rainfall of above 1000 mm and 600 mm respectively (see Chapter 2, Dryland salinity in the south-west of Western Australia).

Aquifers can store substantial amounts of salt, depending on their salinity, porosity, thickness and extent. For example, a 40 m thick aquifer with a porosity of 10 per cent and salinity of 1000 to 10 000 mg L^{-1} could store 40 to 400 tonnes of salt per hectare.

Irrigation water adds salt to the soil in proportion to the soluble salt content and the annual application of irrigation water. Even in the case of good quality irrigation water containing 200–500 mg L^{-1} of soluble salt, the amount of salt added per ha and per year can be substantial. For example, irrigation water with a salt content of 500 mg L^{-1} contains 0.5 tonnes of salt per 1000 m^3. Since crops require from 6000 to 10 000 m^3 of water per hectare each year, one hectare of land may receive 3 to 5 tonnes of salt.

Mobilisation of salt

Mobilising salt stored in the soil profile and aquifers due to human activities such as irrigation and land clearing is a major problem in arid and semiarid regions of the world. In non-arid areas where the rainfall and consequently the leaching of substances through the soil is substantial, salinisation can be avoided even without artificial drainage (Szabolcs, 1989). However, in arid and semiarid regions, lack of rainfall and high evaporation rates make the application of irrigation imperative.

Surface irrigation methods, which are used very widely, have a low efficiency. Moreover, the diversion of irrigation water to fields requires a network of canals which in the majority of cases are unlined. These have a high seepage rate and contribute significantly to the percolation of water to aquifers. Therefore, if in an irrigated area the natural or artificial drainage is not adequate, percolating water gradually accumulates and raises the watertable to within a few metres of the land surface. Consequently, after several years of using an irrigation system without adequate drainage facilities, the groundwater rises toward the surface. Once the watertable is close to the soil surface, appreciable upward movement of groundwater due to evaporation from the soil surface takes place and results in the accumulation of salt in the root zone (Abrol, 1986). The increase of evaporation rate with respect to the decrease in watertable depth is very significant and indicates that there is a critical depth of watertable above which there is a sharp increase in the evaporation rate and consequent soil salinisation.

The critical watertable depth was originally defined as the depth below which no upward movement to the surface can take place. Strictly, no such depth exists. Empirically it has been defined as the watertable depth from which a steady evaporation flux of 1 mm d^{-1} can be maintained (van Schilfgaarde, 1976 and 1984). Using a set of reasonable assumptions, Peck (1978) calculated that, for dryland conditions and a Mediterranean climate, a critical flux of 0.1 mm d^{-1} would be reasonable, compared to 1.0 mm d^{-1} for irrigated conditions. He then estimated the critical depth from the properties of ten soil types. For nine of the soil types he obtained a range of 0.9 m–2 m for irrigated and 1.6 m–6.3 m for dryland conditions, while for one soil type the estimated value was 6.6 m for irrigated and 31 m for dryland conditions. Peck (1978) also describes the limitations of the technique for estimating the depth of a saline watertable which will result in soil salinity in either dryland or irrigated conditions.

Unfortunately, the hazard of raising the watertable by irrigation is underestimated. Even when the natural watertable is 10–20 m below the surface, it can easily be elevated to within 1–2 m of the surface by irrigation systems, particularly in cases where drainage is poor. This simple rule has often been ignored during the planning and exploitation of many irrigation systems. In general, with the exception of places with good natural drainage (which are rather rare), the lack of artificial drainage in arid and semiarid regions leads, sooner or later, to salinisation (Szabolcs, 1989).

Land clearing for farming and pastoral activities can change the hydrologic balance and can be a major cause of rising watertables and subsequent salinity problems. In this case the mechanism is that there is less evapotranspiration from crops and pastures than from the native deep-rooted vegetations, so that clearing increases the amount of water percolating through the soil to recharge the aquifers (Peck, 1983). It is important to note that land clearing in the recharge areas of confined or semi-confined aquifers could increase their pressures. Subsequently the upward leakage from these aquifers to the shallow watertable aquifers contributes to land salinisation and the increase of salinity in watercourses. Figure 3 is an excellent illustration of the effect of land clearing. It shows that the watertable continues to drop in the non-cleared Salmon catchment, while it rises in the Wights catchment cleared in 1976–77.

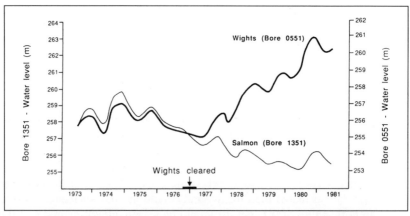

Figure 3: Effects of land clearing on the watertable for two Western Australian catchments: the Wights catchment was cleared in 1976–77, while the Salmon catchment remained forested (Peck, 1983).

More recently, Peck and Williamson (1987a) investigated the effects of forest clearing on groundwater systems in five catchments with individual surface areas ranging from 80 to 350 ha in the Collie River Basin of Western Australia (Figure 2-8) over the period of 1973 to 1984. Objectives of the study were to define characteristics of groundwater systems within these catchments and changes in these systems resulting from conventional practices of forest clearing and agricultural development. The study showed that, as a result of forest clearing, piezometer hydrographs have risen, indicating increased recharge. At one site in the wetter area of the Basin, the potentiometric surface has risen at a rate of 2.6 m yr^{-1} over three years, indicating an increased recharge rate of 65–110 mm yr^{-1} (6–12 per cent of the average local rainfall), depending on the specific yield used. Under different clearing strategies in the lower rainfall area, the potentiometric surface has risen 0.9 and 2.6 m yr^{-1}, corresponding to an increased recharge rate of 20–35 and 60–100 mm yr^{-1} respectively. The latter figure is 7–12 per cent of local average rainfall.

Results of a number of other investigations concerning the effects of land clearing on water and solute movement in the five catchments of the Collie Basin are documented in Peck and Williamson (1987b). Some of these results are reported in Chapter 2 (page 162).

In another study, McFarlane (1991) investigated the water balance of two catchments in the south-west of Western Australia. Both catchments were located in a low rainfall area where the annual average is 370 mm. One of the catchments had its native vegetation while the other was an agricultural catchment. Table 15 shows that in the agricultural catchment run-off and recharge were 18 mm and 26 mm per year respectively compared to nil for the catchment with the native vegetation, while the evapotranspiration decreased by 40 mm per year from 359 mm in the catchment with native vegetation to 319 mm in the agricultural catchment.

Table 15: Average annual water balance for a native vegetated and an agricultural catchment north-east of Newdegate (Western Australia).

	Rainfall (mm)	Run-off (mm)	Evapotranspiration (mm)	Interception (mm)	Recharge (mm)
Native vegetation	370	0	359	11	0
Agriculture	370	18	319	7	26

Source: McFarlane (1991).

Extensive removal of deep-rooted native vegetation in the south-west of Western Australia, South Australia and the south-east of Australia and its replacement with crops and pasture has raised watertables, mobilised salts stored in groundwater and the soil profile, caused soil salinisation, and facilitated seepage of saline groundwater to rivers, polluting the major sources of domestic and irrigation water supplies (see Chapter 2, particularly Table 2-6 and Figures 2-7 and 2-18).

Dryland salinity

Apart from irrigated areas, salinity poses a major management problem in many unirrigated areas where cropping relies on rainfed conditions. Dryland salinity has been a threat to land and water resources in several parts of the world although only in recent years has the seriousness of the problem become widely known. Problem soils range from a slightly saline soil condition which reduces crop growth to extensive areas where cultivation is almost impossible. Dryland salinity is an acute management problem in the southern half of the Australian continent and in the Great Plains region of North America. It occurs extensively in the prairie provinces of Manitoba, Saskatchewan and Alberta in Canada and in the states of Montana and North and South Dakota in the United States. Dryland salinity also occurs in South Africa, Turkey, Thailand, India and Argentina and it probably exists in other countries (Abrol et al., 1988). Saline spots or areas occurring in dryland fields have been known by several local names. In Australia, areas of dryland salinity are usually classified either as seepage salinity or scalds. According to

Shaw et al. (1987), scalded areas are caused when wind or water erosion removes topsoil from soils with saline or sodic subsoils. This usually occurs as a result of denudation of vegetation by overgrazing, drought or fire. When the vegetation is removed, wind and water can remove the topsoil and a crust forms at the topsoil-subsoil boundary. There are three types of scalds: saline, where erosion exposes saline subsoils; sodic, where erosion exposes sodic subsoils; and non-saline, where erosion exposes subsoils of limited fertility.

A saline seep is primarily the result of discharge of a saline groundwater system. The process is often accelerated by dryland farming which allows water to move through salt-laden substrata below the root zone. It refers to intermittent or continuous saline water discharge at or near the soil surface downslope from recharge areas under dryland conditions (Brown et al., 1983). It reduces or eliminates the growth of crops in the discharge area due to increased soluble concentration of salt in the root zone. The characteristics and causes of saline seeps are similar everywhere. Typically, native vegetation, which may include grasses, has been replaced with agricultural fields and cropping systems with lower potential evapotranspiration requirements.

Dryland salinity occurs in different geological, hydrogeological, geomorphological, agricultural and climatic settings. In some cases it could be considered a manifestation of salt accumulation in seepage spots at low points or side slopes in the landscape. In this case the groundwater flow is mainly lateral and downslope and occurs most often over a shallow, less permeable layer. In the discharge area, the groundwater rises to the soil surface, creating a seep. As the water evaporates from the seepage area, salt accumulates (Abrol et al., 1988). In some other cases textural changes in material constituting the aquifer and change in the slope of the topographic surface could cause the saline seepage. Topography of the basement rocks could play a major role in the development of dryland salinity. The movement of saline groundwater could be restricted by shallow basement rocks, dykes or other less permeable materials. The upward leakage from the deeper and more saline aquifer toward shallow aquifers is the major cause of dryland salinisation in Australia. In this case the recharge area may be located at a great distance from the discharge area. Further information concerning the origin and management of dryland salinity is available in Holmes and Talsma (1981), Brown et al. (1983) and Tanji (1990).

Characteristics of saline and sodic soils

Salt-affected soils are classified into two broad groups: saline and sodic (alkaline) soils. The principal criteria used to classify them are: salinity of the saturated soil extract (solution extracted from a soil sample after being mixed with sufficient water to produce a saturated past) as determined by the electrical conductivity (EC) at 25°C; sodium adsorption ratio (SAR); exchangeable sodium percentage (ESP); and the pH of the saturated soil extract.

The sodium adsorption ratio (SAR) is the relation between sodium and divalent cations (calcium plus magnesium) of irrigation water or saturated soil extract. It is used to express the relative activity of sodium in exchange reactions with soil. It is expressed as:

$$SAR = \frac{Na^+}{\sqrt{\frac{Ca^{++} + Mg^{++}}{2}}}$$

where the cations are expressed in milliequivalent per litre (Singer and Munns, 1987).

The exchangeable sodium percentage (ESP) of a soil is the percentage of exchangeable sodium ions to the total exchangeable cations of all types in the soil sample (Donahue et al., 1983: 371). It is expressed as:

$$ESP = \frac{\text{exchangeable sodium ions}}{\text{soil cation exchange capacity}} \times 100$$

where the ions are expressed in milliequivalent per 100 g of soil.

Saline soils are defined by the presence of an excess of soluble salts. The dominant soluble salts in saline soils mostly comprise chlorides, sulphates and bicarbonates of sodium, calcium and magnesium. Soil is usually considered saline when the electrical conductivity of an extract from the saturated soil (EC_e) exceeds 4000 µS cm^{-1}. This value is generally used worldwide, although the Soil Science Society of America has recommended that this limit be reduced to 2000 µS cm^{-1} because many crops can be damaged in the range of 2000–4000 µS cm^{-1} (Abrol et al., 1988: 13). The ESP value of these soils is less than 15, which is equivalent to an SAR value of 13. Saline soils have a pH value of 8.5 or less. The physical properties of saline soils are normally good. Plant growth in saline soils is impaired chiefly by the osmotic effects of excess soluble salts.

Alkaline soils have an electrical conductivity of their saturated soil extract of less than 4000 µS cm^{-1}, ESP of 15 or more, SAR of 13 or more and pH value of 8.5 to 10. In contrast to saline soils, alkaline soils are those which adversely affect plant growth due to an excessive amount of sodium on the exchange complex of the soil. The adverse effect of exchangeable sodium on plant growth is mainly associated with changes in the physical properties of the soil. High ESP causes dispersion of soil colloids which in turn results in blocking of soil pores. Consequently air and water movement is impeded.

Irrigation water or rainwater tends to stagnate, creating unfavourable conditions for plant roots to respire and absorb water and nutrients. Table 16 summarises the major characteristics of saline and sodic soils (see Foth (1984), Singer and Munns (1987), Abrol et al. (1988) and Szabolcs (1989) for further information about the characteristics of saline and sodic soils).

Table 16: Characteristics of saline and sodic soils.

Type of soil	Electrical conductivity (EC_e) ($\mu S\ cm^{-1}$)	Exchangeable sodium percentage (ESP) (per cent)	Sodium adsorption ratio (SAR)	pH
Saline	>4000	<15	<13	<8.5
Sodic	<4000	>15	>13	8.5–10

Source of data: Foth (1984).

Salinisation of water resources

The rising watertables of saline aquifers due to increased recharge or the upward leakage from deeper aquifers not only causes land salinisation, but will also increase the seepage of saline groundwater into rivers and watercourses and enhance their salinisation. The discharge of saline effluents from drainage schemes or from evaporation basins (ponds) and run-off from lands affected by salinisation can also contribute to increased salinity of the watercourses.

Construction of river structures such as dams, weirs and locks in regions where groundwater is saline can generate or aggravate salinisation problems. Such structures create a head of water of several metres which causes saline groundwater to flow into the rivers downstream of structures. The increased water level will raise watertable levels in surrounding regions leading to expansion of soil salinisation problems.

Discharge of industrial saline water can also cause increases in river salinity. In South Africa, for example, the salinity of effluents from heavy industry is about 1000 mg L^{-1} and mining effluents may have salinities of 10 000 mg L^{-1}, contributing substantially to the salinity of local rivers (Williams, 1987). Another example is the discharge of saline effluent from extraction and treatment of potash in south Alsace (eastern France) to the Rhine River. This mining and industrial activity increased the salinity of the river and the Rhine aquifer, which is the most important alluvial reservoir in Western Europe (Zilliox, 1989). Figure 4 shows the chloride concentration of the Rhine River estimated for an average discharge of 2000 m^3 s^{-1} (63 × 10^9 m^3 yr^{-1}).

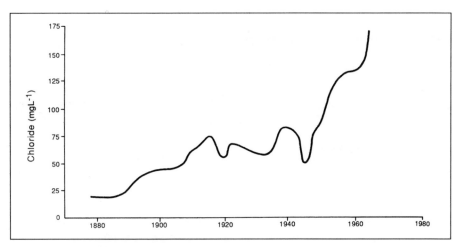

Figure 4: Increase in the chloride concentration of the Rhine River (World Resources Institute, 1990: 164).

Intrusion of saline seawater in coastal aquifers of the world and oceanic islands is another process leading to salinisation of groundwater resources. Countries with the problem of seawater intrusion inlcude Australia, Belgium, Egypt, France, Greece, Germany, Indonesia, Iran, Israel, Libya, Lebanon, Malaysia, Netherlands, Philippines, Thailand, Spain, United Kingdom, United States and Vietnam. In coastal aquifers freshwater usually overlies a transition zone which in turn overlies the saline seawater. Overpumping in the freshwater zone for domestic, agricultural and industrial water supplies changes the equilibrium between the fresh and saline water and causes the intrusion of seawater. Therefore management of groundwater resources is a delicate problem in these cases and requires special attention to minimise the extent of seawater intrusion into aquifers and upward movement of salt water near pumping stations (Ghassemi et al., 1990).

It is important to note that aquifer systems can be multi-layered, containing good quality groundwater in some layers and low quality in others. An aquifer can be single-layered with a gradual salinity increase along the flow lines. In these cases leakage of saline water from above or below, or its lateral migration due to excessive pumping, could have a devastating effect on the quality of groundwater resources. Therefore careful management is required to minimise the upward or downward leakages and the lateral movement of saline groundwater.

Finally, the annual application of millions of tonnes of de-icing agents, usually in the form of sodium chloride, in snow-belt regions of Europe, Canada and the United States contributes to the accumulation of salt in the soil profile and the salinisation of groundwater resources (Howard and Haynes, 1993).

Extent of land salinisation

Estimates of the area of land affected by salinity vary widely. In this section several previous estimates will be reviewed and an attempt will be made to estimate the extent of secondary salinisation based on analysis of data provided in the second part of this publication.

Previous estimates

The International Commission on Irrigation and Drainage (ICID) surveyed the irrigated and salt-affected lands in 24 countries where about 90 Mha were irrigated (Framji, 1976). Unfortunately, their data contains an error (23.5 Mha salt-affected soils for Iran), underestimation (0.23 Mha salt-affected soils for Egypt) and for a number of countries no data has been provided. However, by excluding the wrong data for Iran, the salt-affected area in surveyed countries was about 27 Mha, or about 30 per cent of the irrigated area.

The 1977 United Nations Conference on Desertification estimated that 22 Mha of the earth's area under irrigation was waterlogged (Holdgate et al., 1982: 267). Buringh (1977) calculated from various data sources that the world is losing at least ten hectares of arable land every minute. Of this amount, five hectares are lost from soil erosion, three from soil salinisation, one from other soil degradation processes and one from non-agricultural uses. According to this estimate, the world loses about 1.6 Mha of arable land to salinisation every year. White (1978) believed that 50 per cent of the irrigated soils in the Euphrates Valley in Syria, 30 per cent in Egypt and more than 15 per cent in Iran are affected by salt or waterlogging. Dukhovny (1978) estimated the total area of various saline lands under irrigation to be about 50 million ha, or nearly 20 per cent of all irrigated area.

Kovda (1983) estimated that soil salinisation, by both natural and human-induced causes, claims about 1 to 1.5 Mha per year. Much of this loss transforms fertile soils in irrigated croplands with high potential yields to barren lands. He also estimated that the total amount of land affected by salinisation from both natural and anthropogenic factors to be about 20–30 Mha.

Postel (1990) estimated the share of salt-affected soils for the five leading countries in area irrigated to be about 24 per cent. By assuming that the share of the remaining world irrigated area which is affected is the same as the share in the five leading irrigated countries, it was concluded that globally out of the 250 Mha irrigated 60 Mha are salt-affected (Table 17). There are two problems with this estimate: firstly, the salt-affected area for India is almost three times higher than the real value of about 7 Mha; secondly, their estimate of the world's irrigated land is higher than the real value of about 230 Mha.

Table 17: Irrigated land damaged by salinisation in the top five irrigators and the world, estimated for the mid-1980s.

Country	Area damaged (Mha)	Share of irrigated land damaged (per cent)
India	20.0	36
China	7.0	15
United States	5.2	27
Pakistan	3.2	20
Former Soviet Union	2.5	12
Subtotal	37.9	24
World	60.2	24

Source: Postel (1990).

Oldeman et al. (1991b) estimated that worldwide 76.6 Mha are affected by human-induced salinisation (Table 11), but they have not differentiated the extent of salt-affected lands in the irrigated and non-irrigated areas.

Dregne et al. (1991) estimated that about 43 Mha of irrigated land in the world's dry area are affected by various processes of degradation, mainly waterlogging, salinisation and alkalisation. They also estimated that the world is losing about 1.5 Mha of irrigated lands annually due to land degradation processes, mostly by salinisation.

Current estimate

Estimation of the extent of secondary salinisation is a very difficult task due to several problems. These include the dynamic nature of the problem, lack of data and inconsistencies between the data provided by various sources concerning the extent of irrigated lands, croplands and salt-affected lands. However, using the same methodology applied by Postel (1990) and the extent of the salt-affected soils reported in Part Two of this publication, an estimate can be made of the global salt-affected soils in irrigated areas. Table 18 provides the results of this estimation.

As shown in Table 18, globally about 20 per cent or 45.4 Mha out of a total 227 Mha of irrigated land are salt-affected. Although this estimate is less than that provided in Table 17, it still represents a large area. Obviously, loss of production due to such a large area of salt-affected soils is enormous and efforts should be made to reclaim these salt-affected soils and to prevent further salinisation in irrigated lands.

With respect to Table 18, a few points are worth mentioning:
- The 11 surveyed countries have a total irrigated area of 158.7 Mha or 70 per cent of the world's irrigated land. Therefore using the average value of

the share of salt-affected land in these countries for the estimation of the world's salt-affected lands in the irrigated area seems reasonable.
- In the case of Australia, although 0.16 Mha of the irrigated land are salt-affected, about 0.65 Mha have watertables shallower than 2 m. Moreover, saline seepage in non-irrigated lands affects 0.8 Mha and scalds affect an additional 3.8 Mha of croplands (see Chapter 2, Table 2-5).

Table 18: Global estimate of secondary salinisation in the world's irrigated lands.

Country	Cropped area[a] (Mha)	Irrigated area[a] (Mha)	Share of irrigated to cropped area (per cent)	Salt-affected land in irrigated area[b] (Mha)	Share of salt-affected to irrigated land (per cent)
China	96.97	44.83	46.2	6.70	15.0
India	168.99	42.10	24.9	7.00	16.6
Commonwealth of Independent States	232.57	20.48	8.8	3.70	18.1
United States	189.91	18.10	9.5	4.16	23.0
Pakistan	20.76	16.08	77.5	4.22	26.2
Iran	14.83	5.74	38.7	1.72	30.0
Thailand	20.05	4.00	19.9	0.40	10.0
Egypt	2.69	2.69	100.0	0.88	33.0
Australia	47.11	1.83	3.9	0.16	8.7
Argentina	35.75	1.72	4.8	0.58	33.7
South Africa	13.17	1.13	8.6	0.10	8.9
Subtotal	842.80	158.70	18.8	29.62	20.0
World	1473.70	227.11	15.4	45.4	20.0

Source: (a) Data for 1987 from FAO (1989); (b) Data for 1980s from different sources referred to in Part Two of this publication.

- In the case of South Africa, although the share of salt-affected soils in irrigated areas is relatively low, salinisation of reservoirs used for water supply is a major issue (see Chapter 9, South Africa).
- In the case of Thailand, about 2.85 Mha of lands are salt-affected due to human activities, but the share of salt-affected soils in irrigated lands is not well documented, so the figure of 0.4 Mha is a rough estimate.

A global total of 31.2 Mha can be attributed to secondary salinisation of non-irrigated lands if it is accepted that 76.6 Mha of land is affected by human-induced salinisation (see Table 11) and 45.4 Mha of land is affected in irrigated areas (see Table 18).

Extent of water resources salinisation

Like land resources, water resources are also affected by salinisation due to human activities. In almost all countries with major land salinisation, water salination is an accompanying problem. These countries include Australia, Argentina, China, Commonwealth of Independent States, India, Iran, Iraq, South Africa, Thailand and the United States. The main reason for water salinisation is the discharge of irrigation return flow or accelerated saline groundwater seepage to surface water systems. In South Africa, however, discharge of industrial and mining effluents contributes to surface water salinisation.

Estimates of the volume of water resources affected by salinisation and their salt load, as well as the contribution of natural and human-induced sources, are not available for most countries affected by this problem. However, the following few examples, mainly based on the information provided in the second part of this publication, will give an indication of the extent of the problem.

In south-east Australia, the River Murray has a salinity of less than 25 mg L^{-1} in its headwaters and about 480 mg L^{-1} downstream at Morgan. In an average year the river exports 5.5 million tonnes of salt to the sea and approximately 30 per cent of this load is human-induced, largely from groundwater inflow. The River Murray is a major source of irrigation water and makes an important contribution to the water supply of Adelaide, the capital of South Australia. On average about 35 per cent of the 200 × 10^6 m^3 yr^{-1} of water distributed through the Adelaide Metropolitan Water Supply system is pumped from the River Murray. This percentage rose to 90 per cent in the drought year of 1983. Periods of drought coincide with the periods of high salinity in the river. Therefore control of its salinity is a major management problem.

The Glenelg River Basin in the south-west of Victoria covers 12 660 km^2. Since 1837 two-thirds of the basin has been cleared for pasture to graze sheep and cattle. The mean annual flow of the basin is about 725 × 10^6 m^3. More than 95 per cent of this volume is classified as being brackish to saline. Because of the high river salinity, most of the water for domestic supply, stock and irrigation is derived from groundwater sources (Department of Water Resources Victoria, 1989: 293-300).

In the south-west of Western Australia, agricultural development has led to extreme stream salinisation to the extent that 36 per cent of the divertible surface water resources is no longer potable and a further 16 per cent is of marginal quality. Currently many rivers with high average annual run-off, such as the Blackwood River, have an average annual salinity of over 2000 mg L^{-1} and are therefore useless as water resources. In South Australia only 56 per cent of divertible surface water resources are fresh, 26 per cent are marginal and the remaining 15 per cent have a salinity exceeding 1500 mg L^{-1}.

In the Commonwealth of Independent States, the salinity of the Syr Darya River, discharging to the Aral Sea, increases from less than 300 mg L^{-1} to about 2000 mg L^{-1} downstream of Kzyl-Orda. In the same region the salinity of Amu Darya reaches about 1500 mg L^{-1} in the north of Nukus.

In Egypt, the salinity of the Nile River increases from 200 mg L^{-1} at Aswan High Dam to 300 mg L^{-1} at Delta Barrage near Cairo. At this location the salt load of the river is about 11 million tonnes per year. The salinity of the Nile River leaving the Nile Delta and discharging to the Mediterranean Sea is about 2300 mg L^{-1} and its salt load is 32 million tonnes per year. The 21 million tonnes annual increase in salt load between the Delta Barrage and the Mediterranean Sea is attributed mainly to upward seepage of saline seawater and partially to irrigation return flow, but the proportion between these two components is unknown.

In Iran, 6×10^9 m^3 of brackish water flow annually through its major rivers. In many cases salinisation is mainly due to the natural environment, such as the passage of the rivers through evaporites and salt domes which are abundant in the country, particularly in its south-western part. However, salinity increases along rivers such as the Karun, Dez, Zayandeh-Rud, Zarrineh-Rud and Kor are partly caused by the discharge of irrigation return flow.

In Iraq, the Tigris River has an average annual flow of 23.2×10^9 m^3 at Mosul and 37.7×10^9 m^3 at Baghdad (Ubell, 1971). Water quality measurements performed from September 1981 to August 1982 showed that the total dissolved solids (TDS) increased along the river from 292 mg L^{-1} in Mosul, to 469 mg L^{-1} in Baghdad and to 822 mg L^{-1} in Qurna, located about 60 km north-west of Basra (Al-layla, 1989). Salinity increase in this river is caused by natural sources as well as the development of irrigation projects.

In South Africa, the salinity of Lake Mentz, with a capacity of 10^6 m^3, exceeds 738 mg L^{-1} 90 per cent of the time. Also, the annual TDS concentration in the Vaal Barrage intake at Vereeniging increased from about 150 mg L^{-1} in 1940 to around 700 mg L^{-1} in 1980. In the Breede River, salinity increases from 230 mg L^{-1} TDS to 670 mg L^{-1} along a 40 km stretch of the river.

In north-east Thailand, salinity increase in rivers is due to seepage from non-irrigated saline lands. A survey of 132 reservoirs with capacities ranging from 50 000 m^3 to more than 1×10^6 m^3 showed that seven reservoirs had electrical conductivity (EC) values greater than 3000 µS cm^{-1}, 27 reservoirs had EC values of 700 to 3000 µS cm^{-1} and for 98 reservoirs EC values were less than 700 µS cm^{-1}.

In the United States, the Colorado River has a salinity of about 25 mg L^{-1} in its headwaters. The salinity concentration progressively increases downstream. In 1982 the salinity concentration averaged 825 mg L^{-1} at Imperial Dam. The salt load of the river is about 8.2 Mt (million tonnes) per year with a contribution of 47 per cent due to natural sources, 37 per cent due to irrigation, 12 per cent due to evaporation and the remaining 4 per cent due to other sources. In the San Joaquin River in California, salinity increased from 330 mg L^{-1} TDS in the 1930s to about 600 mg L^{-1} in the 1970s.

Environmental damage

Waterlogging, which is the forerunner of land salinisation in many cases, damages plant growth. Plant roots need water and oxygen and there must be a balance between the amount of air and water in the soil for healthy growth of the plant. If the soil is waterlogged, the plant's growth will be damaged and its production will be adversely affected. The intensity of waterlogging can be measured by the SEW_{30} index. The index is calculated by summing all daily values (in cm) of groundwater levels within 30 cm of the soil surface. For example, two days with water levels at 20 cm below the surface (i.e. 10 cm above the threshold) represents a waterlogging intensity of 20 cm days. McFarlane et al. (1992) used this index for the investigation of effects of waterlogging on crop and pasture production in a number of sites in southwest Western Australia.

Soil salinisation in its early stages of development reduces soil productivity, but in advanced stages kills all vegetation and consequently transforms fertile and productive land to barren land, leading to loss of habitat and reduction of biodiversity. River and stream salinisation has similar effects. Williams (1987) has provided a few examples in this respect: impoverishment of the fauna of the lower Avon River in Western Australia and replacement of the freshwater mussel by the brackish one; fish kills in North America from rivers in the Northern Great Plains; and in South Africa gradual elimination of salt-intolerant diatoms and their replacement by a less diverse but more salt-tolerant group of diatoms in the Fish and Sunday Rivers.

Environmental damage also occurs when, as part of a salinity mitigation program, saline water from drainage schemes or groundwater is stored in off-stream flood plain areas or wetlands, and either left to evaporate naturally or discharged when high flow rates in the main stream prevail (Williams, 1987). In the first case, damage occurs directly in the areas inundated with saline water and is often very severe. This is the situation with the Kesterson Reservoir and Natural Wildlife Refuge, where high levels of potentially toxic trace elements, particularly selenium, had been identified and were causing waterfowl death and deformity (Tanji et al., 1986). In the second case, damage may occur when pulses of highly saline water are allowed to flush downstream.

The Salinity Planning Working Group (1992) provides a preliminary list of species in Victoria (Australia) with conservation priority that have demonstrated low tolerance to increasing salinity. The list contains 97 flora, 38 terrestrial vertebrates and 12 fishes. The list is in its infancy and is generalised throughout Victoria. Some species may be at risk in some regions and not in others. A high proportion of listed rare and threatened flora species occur in areas at risk of rising watertables and increased salinity. Many species could be lost due to saline discharges. As salinity degrades vegetation communities (such as woodlands, wetlands and grasslands), it adds to the threats facing animals which rely on those communities for their survival. Also, all native freshwater fish are susceptible to changing salinity. However, some appear to be more susceptible than others.

Economic and social damage

No accurate global estimate of the damage caused by salinisation to the economy of salt-affected countries is available. However, values provided in Table 19 for a limited number of countries give an indication of the severity of the problem.

Table 19: Estimates of damage to the economy of a few countries with secondary salinity problems.

Country	Region	Estimated damage (million US$ per year)	Reference
Pakistan	Punjab and North-West Frontier Provinces	300[a]	Water and Power Development Authority (1988)
Australia	Murray-Darling Basin	208[a]	Murray-Darling Basin Ministerial Council (1989a)
	Murray-Darling Basin	52[b]	Simmons et al. (1991)
	South-west of Western Australia	50[a]	Western Australian Legislative Assembly (1991)
	South-west of Western Australia	72[c]	Western Australian Legislative Assembly (1991)
	South-west of Western Australia	32[d]	Water Authority of Western Australia (personal communication)
United States	Colorado River Basin	750[e]	Colorado River Basin Salinity Control Forum (1993)
	San Joaquin Valley, California	31[a]	El-Ashry et al. (1985)
South Africa	Pretoria, Witwatersrand, Vereeniging and Sasolburg complex	29[e]	Heynike (1981)

(a) Agricultural loss; (b) Water Supplies; (c) Waterlogging; (d) Stream salinity; (e) Total damage.

In Pakistan, a study was conducted by the Water Resources and Planning Division, Water and Power Development Authority (WAPDA) on the entire irrigated area of the Upper Indus Plain comprising the two provinces of Punjab and North-West Frontier. It concluded that the economy of the two provinces suffers a loss of approximately 4.3 billion rupees (US$300 million) annually from the decrease in farm production on soils slightly to moderately affected by salinity. The loss on a countrywide basis would be much higher (Water and Power Development Authority, 1988).

In Australia, it has been estimated that annual agricultural losses from salinisation in the Murray-Darling Basin amount to A$260 million (US$208 million). Costs of the effects of salinity on the quality of urban water supplies are estimated to be around A$65 million (US$52 million) a year. However, the total cost to the community from degraded lands, deteriorating water quality, rising groundwater and loss of natural habitat through the basin is likely to be many times greater (Murray-Darling Basin Ministerial Council, 1989a). In the south-west of Western Australia, the annual loss due to dryland salinity is about A$62 million (US$50 million) and the damage from waterlogging is A$90 million (US$72 million). In addition, costs associated with stream salinity are estimated to be A$40 million (US$32 million).

In the Colorado River Basin of the United States, the heavy annual salt load of about 8.2 Mt is costing users more than US$750 million per year. That estimate is expected to more than double by 2015 if controls are not instituted. In the San Joaquin Valley of California, crop yields have declined by 10 per cent or US$31.2 million annually since 1970. Because of high saline watertables, the losses are expected to increase to US$321.3 million by the year 2000 if no action is taken (El-Ashry et al., 1985). According to Reisner (1987: 483), it is estimated that each additional mg L^{-1} of salt in the water supply systems of the cities using Colorado River water causes US$300 000 worth of damage per year to the objects coming in contact with water (pipes, fixtures, machinery, etc).

In the Republic of South Africa, Heynike (1981) has estimated the annual economic damage for the communities of Pretoria, Witwatersrand, Vereeniging and Sasolburg complex (PWVS) due to an increase of salt content in the Vaal Barrage. An increase from 300 mg L^{-1} to 500 and 800 mg L^{-1} TDS will cost about US$29 million and $54 million respectively.

To overcome the problem of waterlogging and salinisation in the Nile Valley and its Delta, the Egyptian government is spending more than LE70 million (about US$30 million) annually on drainage. This sum represents a sizeable proportion of the country's investment in agriculture (Amer and de Ridder, 1989: VI).

In Ethiopia, the Amibara Melka Sadi area, which covers about 14 200 ha of net irrigable land in the Awash River Basin, encounters problems of salinisation and rising watertables to varying degrees. The estimated cost of the development program to introduce subsurface drainage, and thereby reduce salinity and rising watertable hazards, is about US$52.2 million (Office of the National Committee for Central Planning, 1988).

On a global scale, Dregne et al. (1991) estimated that the loss in production capacity, or what they call 'income forgone,' due to all processes of land degradation is about US$42.2 billion in 1990 prices (Table 20).

Table 20: Global estimate of the average annual income loss due to land degradation, in 1990 prices.

Degraded land	Extent (Mha)	Income loss per unit area (US$ ha^{-1})	Annual income loss (US$ billion)
Irrigated	43.15	250.0	10.8
Rainfed	215.56	38.0	8.2
Rangeland	3333.46	7.0	23.2
Total	3592.17	–	42.2

Source: Dregne et al. (1991: Tables 5 and 6).

The estimate of 45.4 Mha of salt-affected lands in irrigated areas and 31.2 Mha in non-irrigated lands, and the income loss values per unit area provided in Table 20, can be used to infer the global income loss due to salt-affected lands. These are about US$11.4 billion in irrigated areas and US$1.2 billion in non-irrigated areas. Taking into account damages caused to industrial users of water with a relatively high salt content and to water distribution systems, the total damage may exceed US$15 billion per year.

The social cost of salinisation is not easy to quantify. Salinisation causes occupational or geographic shifting of the farm population and reduction in aggregate regional income and expenditure. These events have social and economic repercussions on the region as a whole, including dependent country towns. The impacts are most apparent in rural hamlets and small towns because the opportunities for adjustment of the local economic base is more limited (Peck et al., 1983).

Section D: Management of Salinity — Engineering Options

Management of salinity problems

A wide range of management options is available for preventing salinisation and for management of irrigated and dryland (non-irrigated) salinity and salinisation of watercourses. However, implementation of any option depends on particular circumstances. Some of the options are more suitable for irrigated land salinisation, while others are suitable for dryland salinisation or reduction of salinity levels in watercourses. In general, while one option may be effective and feasible in one case, it may not be valid in another. Technical, economic, social and political considerations are the major influences on the implementation of management options. The question of who benefits and who pays is particularly critical. The situation becomes more complex when the limits of the problem cross jurisdictional (especially state or international) boundaries. Finally, it must be admitted that in some circumstances there is no suitable control option, leaving a do-nothing option to be best adopted. In the following sections a number of management options will be discussed, including engineering options, disposal of saline drainage water, biological options and policy options.

Engineering options

Drainage

Drainage is as necessary as irrigation for the maintenance of plant growth, prevention of waterlogging and soil salinisation. In an annotated bibliography Gupta and Gupta (1987) provide more than 1500 abstracts of papers published from 1960 to 1986 about this issue, while Pavelis (1987) describes farm drainage in the United States. The following sections describe the major drainage techniques.

Surface drainage

Surface drainage is the removal of excess water by shaping the land so as to make the water flow over the surface to furrows, ditches or waterways. Excess water on farmlands may be caused by excess irrigation, rain or poor land grading. Surface drainages are used to overcome these conditions. Surface drains may be deep enough to intercept groundwater, which will then enter the drain. The rate of groundwater flow into the drain depends on soil permeability and height of the watertable. The deeper a drain, the greater is the width of adjacent land affected by the drawdown of the groundwater. Although surface drains can serve to control the watertable, their use is being superseded by subsurface drains. Disadvantages of surface drains include loss of land, hindrance to farming operations and overland traffic and heavy maintenance requirements due to prolific weed growth or instability of banks (Zijlstra and van Someren, 1980). However, the benefits of surface drainage are seen as being provision of drainage outlets for farms, reduction of ponding of irrigation water and rainfall and hence of accessions to the watertable, and reduction in the watertable due to the flow of groundwater into the drains.

Surface drainage systems can be designed in flat and sloping areas. In flat areas drains can be parallel but need not necessarily be equidistant. Drain spacing depends on the hydraulic conductivity of soils, rainfall and topography. Random system drains follow depressions. In sloping areas drains are constructed parallel to contour lines (see Rural Water Commission of Victoria (1988) and Euroconsult (1989) for the design of surface drainage systems).

The run-off from farms contains a high level of manure and fertiliser. Levels are highest when irrigation run-off crosses irrigated land to enter drains. The steady flow of this water into drains during the summer months encourages weed growth which in turn reduces drain capacity.

Seepage interceptor drain

When a seep or spring is fed by a distinct water source, interceptor drains are used to cut off this excess groundwater before it reaches the problem area. Interceptor drains are usually a single tube or open drains placed between the water source and the problem area. Accurate location of the source of excess groundwater and proper placement of the drain are critical to the success of this form of drainage. However, investigation of subsoil conditions and groundwater flow is essential before drainage work begins (George, 1985). In cases where the interceptor drain is cut deeply into the soil, the upslope batter could erode if surface run-off is allowed to enter the drain. Such erosion causes silting of the drain and the need for frequent maintenance. To overcome this problem, reverse bank interceptor drains have been developed which allow surface water to be carried by a grassed strip above the drain (McFarlane et al., 1985).

The use of interceptor drains to reclaim saline areas has rarely been effective in Western Australia. One of the main reasons is that the groundwater flows to saline areas through deeply weathered soils. Interceptor drains cannot be installed deep enough to cut off this flow completely (George, 1985).

Horizontal subsurface drainage

Horizontal subsurface drainage is the technique of controlling the watertable and salinisation by installation of horizontal drains at a certain depth (about 1.5 to 2.5 m) below the surface. The pattern of drains allows the land to drain to collectors which remove the water from the land. The drains are installed at an appropriate depth and are spaced at intervals designed to ensure that the watertable in the intervening space does not rise above a given height. Drainage design parameters are related mainly to the hydraulic conductivity of the soil and the drainage criterion, which specifies the required discharge and the hydraulic head. Steady-state and transient methods can be used to calculate drain spacing. The steady-state approach assumes constant watertable, constant recharge and constant discharge, while under transient conditions, recharge and the watertable fluctuate with time (see van Schilfgaarde (1974), McWhorter (1977), FAO (1980), Euroconsult (1989) and Amer (1990) for the design of subsurface drainage systems). Flow into drains is induced by the lateral hydraulic gradient. Water flow in these artificial drains occurs by gravity. Subsurface drains have no effect and collect little water unless a watertable is developed higher than the level of the drains.

A mole drain is a simple subsurface drain. For its construction, a small mole (25 mm to 100 mm) is drawn through the ground at a predetermined level (0.4 to 0.7 m and spacing of 2 m to 5 m). In the right soil this mole then creates a cavity similar to the inside of a pipe that will allow the passage of water. Since no pipe or refill material is required, this is the lowest priced of all drainage techniques; likewise, it is expected to have the shortest length life and possibly the lowest efficiency (Hawkins, 1978). The practical application of a mole drain is limited to heavy clay soils. Mole drains have limited stability and the moling operation has to be repeated every few years. The frequency of remoling can only be determined by examination of the soil profile, local knowledge of the soil behaviour and a watch on drain performance. Castle et al. (1984: 129–138) provide detailed information concerning the historical background, design and construction of the mole drain.

According to Dierickx (1990) and Zijlstra and van Someren (1980), for many years tile and concrete pipes were predominantly used for subsurface drainage. In 1960, smooth plastic drainpipes with perforations in the form of longitudinal saw splits became available. They never found widespread use because corrugated plastic drainpipes with small perforations in the valleys of the corrugations became available in 1963. The corrugated form makes these pipes more resistant to deformation. They are manufactured in a series of diameters ranging from 50 to 200 mm and are delivered in coils of different lengths. They were so successful that they gradually began to replace the tile and concrete drainpipes. Corrugated plastic pipes are made of polyvinylchloride (PVC), polyethylene (PE) and polypropylene (PP). Preference for one of these materials is based on economic grounds.

Installation of tile and concrete drainpipes is very labour intensive. Drainage pipes can be installed manually or by specially developed machinery. The evolution in drainage materials has contributed to the development of high-speed machinery which increased installation capacities and reduced costs. The appearance of the laser system for automatic depth and grade control has improved the accuracy of drainpipe installation. (Dierickx, 1990; Zijlstra and van Someren, 1980). In the Netherlands, with a crew of three or four, installation rates of about 600 m of pipe per hour have become normal (Zijlstra and van Someren, 1980).

Subsurface drains are covered with a layer of envelope materials (Dierickx, 1990; Zijlstra and van Someren 1980). Envelope materials were originally intended to protect the drainpipes against soil particle invasion, but they must also facilitate water inflow by creating a more permeable environment surrounding the pipe. The material used can be classified as granular, organic or synthetic. Granular materials such as sand and fine gravel are still widely used. Organic envelopes consist of sieved fibrous peat, woodchips, sawdust, straw from cereal crops or heath bushes. Nowadays synthetic envelope materials are replacing organic envelope materials. They can easily be wrapped around drainpipes and do not decay once the drainpipes are installed. After the introduction of corrugated plastic drainpipes, techniques were developed to pre-wrap these pipes with an envelope material in the factory.

Vertical or tube-well drainage

Vertical drainage systems are basically water wells spaced on a grid which, like horizontal drainage, bring the watertable down to a predetermined level. Tube-well drainage may be an alternative to horizontal drainage in areas with productive aquifers at moderate depth, and in areas where surface layers of sufficient vertical hydraulic conductivity exist to allow percolation at the required infiltration rates without excessive build-up of perched watertable. Tube-well drainage offers the following advantages (Euroconsult, 1989): it can be applied on undulating land without extensive earthmoving and levelling for the installation of pipelines or main drain channels to interconnect the wells; it diminishes maintenance costs because of the smaller network of canals and/or drains that are necessary; the watertable can be drawn down to a much greater depth, reducing the risk of salinisation of the soil; it diminishes artesian pressure of aquifers underlying the top layer, which may reverse the direction of flow so that downward percolation of irrigation water becomes possible. In areas where groundwater is saline, the effluent of drainage wells is likely to be much more saline than that of horizontal drains, which makes vertical drainage less attractive or not acceptable for those areas.

In the case of aquifers with low transmissivity, well-point systems can be used for lowering of the watertable. Well-point systems are groups of closely spaced wells, usually connected to a header pipe and pump (Driscoll, 1989). During operation of a well-point system, a central pump lifts water from a number of wells by producing a partial vacuum in the header and riser pipes. The diameter of well-point systems is usually either $1\frac{1}{2}$ or 2 inches (38 mm or 51 mm). For a given aquifer the spacing of the well-points depends on the type of well-point used, transmissivity of the aquifer and the volume of water to be removed.

Hydrogeological investigations are necessary for the feasibility study and design of pumping wells or well-point systems. These must be aimed at the determination of the extent, depth and magnitude of the aquifer, its transmissivity and the vertical hydraulic conductivity of the covering layers and the effect of pumping from deeper aquifers on shallow watertables (see Driscoll (1989) for further information concerning the investigation, design and construction of tube-wells and well-point systems).

Conjunctive use of surface and groundwater

There is a growing awareness of the salinity benefits of groundwater pumping and the need for a fully integrated groundwater and surface water allocation policy to deal with salinity issues if the groundwater quality is suitable (Evans and Nolan, 1989). Irrigation wells are installed primarily to provide additional water for irrigation, either as insurance against drought or to provide a long-term supplementary source of water to the channel-supplied surface water. However, groundwater extraction wells can be installed to control conjunctive use of surface and groundwater to control salinity and waterlogging. Water pricing mechanisms are useful to encourage the implementation of this policy. Currently the conjunctive use of surface and groundwater is practised widely in Pakistan.

Interception scheme

In areas where a river drains a saline aquifer, interception schemes can be used to reduce the discharge of the saline groundwater to the river. The basic design principle assumes that the extraction of groundwater from an aquifer hydraulically connected to the river, using a line of pumping wells positioned close and roughly parallel to the river, will decrease the hydraulic gradient of the aquifer towards the river, thereby reducing the discharge of saline groundwater to the river. The design parameters of the scheme include the number of wells, their depths, spacings and pumping rates. Their values depend on the characteristics of the aquifer, particularly its hydrodynamic parameters (transmissivity and storage coefficient). The technique can be effective if facilities are available for disposal of the saline effluents.

In Australia a number of interception schemes are currently in operation or at various stages of construction along the River Murray in south-eastern Australia (Murray-Darling Basin Commission, 1990). Ghassemi et al. (1987 and 1988) describe the Mildura-Merbein interception scheme in the Sunraysia region of Victoria. The scheme consists of 17 pump sites, each comprising from 1 to 11 bores with depths of 14 m to 25.5 m and diameter of about 150 mm. The interception scheme prevented the discharge of 65 tonnes of salt per day (out of a natural discharge of 125 tonnes per day) to the river over the period of investigation from February 1980 to March 1983. Telfer (1989) describes the Woolpunda interception scheme in South Australia. The scheme has been designed to intercept saline groundwater which is discharging into the river from the regional watertable. It consists of about 50 wells along both sides of the River Murray. The Waikerie interception scheme in South Australia consists of a line of 17 interception bores drilled to depths varying between 90 and 125 m, approximately. The bores are located close to the river, thereby obviating the need to install bores on the northern side of the river (Engineering and Water Supply Department, 1990). Bish and Williams (1992) describe the Mallee Cliffs interception scheme in New South Wales. In this area a 3.75 km reach of the River Murray receives 105 tonnes of salt per day. The scheme consists of 7 interception wells. In all the above cases, saline effluents are disposed to evaporation basins.

Dilution flow

The release of freshwater from reservoirs in a river basin during periods of low river flow, which coincides with high river salinity, can reduce that salinity. Although this option is in practice in Australia, South Africa and some other countries, construction of new reservoirs for providing further dilution flow is not an economically feasible option.

Improving irrigation water efficiency

Irrigation channel lining or piping

In many irrigated areas, channel seepage is a significant contributor to the regional watertable. Eliminating channel seepage can therefore have some effect on the problems of shallow watertables and salinity. Hawkins (1978) describes different types of channel lining which include: concrete lining; masonry-type lining; synthetic membrane lining; bentonite membrane lining; and compacted earth lining. Channels can also be replaced by pipelines. Piping is usually considered an economic proposition in areas with permeable soils and high seepage losses.

Land-forming and grading

Land-forming is the process by which the existing land topography is changed by mechanical shifting of soil to create a defined slope. Grading or land-levelling is the process of smoothing the land to achieve a uniform surface. This can occur without significant land-forming. In irrigated areas both processes are now being aided by laser technology which improves accuracy and precision. In practice a laser control station is set up at a convenient location. A rotating horizontal laser beam creates a reference plan. The plan is correlated in height with a receiving device attached to the grader blade on the earth mover. The blade is automatically raised or lowered as necessary to form the land surface to the grade setting (Dedrick et al., 1982).

The overall result of land-forming and grading is that more efficient irrigation is possible. The application of water to the land surface is much more precise and the land can be drained more effectively. Accessions to the groundwater are thereby reduced by the elimination of waterlogged areas, resulting in a lowering of the watertable and greater control of salinity.

Changing method of irrigation

The method of irrigation affects both the efficiency of water use and the way salt accumulates (Ayers and Westcot, 1989). For surface flooding or sprinkler irrigation, which apply a uniform depth of water across the entire field, salt accumulation increases with depth. For furrow irrigation, which applies water to only part of the field surface, salt accumulates in the ridges between furrows. For localised irrigation methods, salt accumulates between emitters and the fringe of the wetted areas. Isolated pockets of accumulated salt frequently result where water does not infiltrate sufficiently to accomplish leaching. These can be raised areas, areas of more dense soil or areas not getting enough water during irrigation.

Each irrigation method has certain advantages and disadvantages and all known factors should be considered before attempting to improve salinity control by changing the method (Ayers and Westcot, 1989). With surface irrigation methods (border, basin and furrow), depth of applied water entering the soil varies with location in the field and depends on the infiltration rate and time available for infiltration. Differences in the rate of infiltration are caused by land slope, degree of compaction, textural changes and soil chemistry. The time during which infiltration can take place also varies; the upper end of the field nearest the water source usually has water on its surface for a much longer time than does the lower end. High spots in the field receive less water because they are covered by less water and for a shorter period.

Surface irrigation methods are usually not sufficiently flexible to apply less than an 80 to 100 mm depth per irrigation. As a result, irrigating more frequently to reduce possible water stress may also waste water and cause waterlogging and drainage problems. In order to relieve water stress, it may be easier to increase the frequency of irrigation with sprinklers or drip irrigation rather than with surface flooding. However, sprinkler and localised irrigation have their problems too and are not adapted to all conditions of water, soil, climate or type of crop (Ayers and Westcot, 1989).

Irrigation scheduling

Irrigation scheduling is the management practice of determining how much water to apply during an irrigation and the timing of the application. Irrigation scheduling requires a knowledge of the crop water requirement for particular growing conditions. The applied irrigation water depends on crop evapotranspiration, the cropping pattern, type of planting (density or spacing), leaching requirement, irrigation management and effective precipitation. Several methods are used for determining the time when a crop requires water. These include the calendar method, soil moisture status, plant stress indicators and the water budget method (Boyle Engineering Corporation, 1986; Heermann et al., 1990). However, the soil moisture status and the plant stress indicators are among the most frequently used.

According to Campbell and Campbell (1982), irrigation scheduling by soil moisture measurement is probably the oldest method in existence. Several devices and procedures have been used for obtaining soil moisture measurements. The most frequently used are the tensiometers and the neutron probe. Tensiometers consist of a metal or plastic shaft, the lower end of which is connected to a porous ceramic cell and the upper end to a pressure measuring device. Through the contact of the water-saturated porous cell with the soil, the soil water pressure is transmitted via the ceramic cell and the tensiometer fluid to the pressure indicator (Albert et al., 1987). The tensiometers have the advantage that they are relatively simple and inexpensive. Their disadvantage is that they measure the soil moisture only in the immediate vicinity of the unit, so that several tensiometers are needed to give a reliable spatial average.

The neutron probe operates by producing fast neutrons from a radioactive source. The emitted neutrons are scattered and moderated by the water in the soil. The probe is inserted into the soil through an access tube, and the water content is directly related to the number of neutrons scattered back to a detector. The measurement is therefore an average for the soil volume surrounding the access tube, so a single measurement gives the same information that several measurements from tensiometers would give. The calibration for the probe is relatively constant from soil to soil and, once it is known, it should not change with time. Readings are rapid and the instruments are reasonably portable (Campbell and Campbell, 1982).

Irrigation scheduling does not require a knowledge of the field average water content or potential for a field. Since the field is irrigated as a unit, a single representative monitoring site can be used to indicate the water status for the entire field. It is desirable to select a site for convenient access that is more than 10–20 m from any edge of the field and that is among healthy, vigorous plants, in soil of above-average water capacity for the field, and in a location exposed to normal climatic and irrigation variations. For irrigation scheduling it is necessary to determine the full and refill points of the field. The full point is the field capacity for the soil, while the refill point is the potential below which crop production is measurably reduced (Campbell and Campbell, 1982).

Since soil moisture measurements are generally not read on a continuous basis, moisture measurements should be plotted as a function of time such that soil moisture can be extrapolated to the point in time when irrigation is required. Irrigation should be scheduled so that the soil water content stays between the refill and full point values. If it goes above the full point, leaching will occur. If it goes below the refill point, production will be reduced (Campbell and Campbell, 1982; Boyle Engineering Corporation, 1986).

Plant stress indicators can be used for irrigation scheduling. Several observational methods can be used to determine plant stress, including changes in plant colour or wilting. Generally, when plant stress is observed, irrigation will be too late to prevent some suppression of plant growth and yield. Recent techniques such as the pressure chamber technique or the measurement of plant temperature with an infrared thermometer can be used to determine the plant stress (Boyle Engineering Corporation, 1986).

Leaf and canopy temperatures may be either warmer or cooler than the air, depending upon environmental factors. In humid climates canopy temperatures will be near to or higher than air temperature, with only a small range of temperatures. In arid areas, however, canopy temperatures may be more than 10°C below air temperature and have a range of perhaps 15°C (Jackson, 1982). It is in the arid areas where irrigation is practised that temperature techniques can work best and are most needed.

The tremendous advance in infrared (IR) technology allowed the production of lightweight hand-held IR thermometers that can be used to measure plant canopy temperature rapidly (Jackson, 1982). The plant temperature, the temperature of the air and the wet and dry bulb temperature are all used together to determine when a plant is stressed. The measurements take place inside the field where the temperature will not be affected by the field edge, which is drier and hence warmer. Measuring is best done between noon and 2 pm when the sun is higher overhead, shadows are at a minimum and transpiration is at its peak (Bureau of Reclamation, n.d.).

Disposal of saline drainage water
Reuse option
Drainage water can be used for the irrigation of less sensitive crops, depending on its level of salinity. According to Rhoades (1984), drainage water can be used again for irrigation to the extent that it still has value for

use by a crop of higher salt tolerance. This could be achieved by successive irrigation of a sequence of crops of increasing salt tolerance. When the drainage water quality is such that its potential for reuse is exhausted, then this drainage water should be disposed to evaporation ponds or by any other convenient means. Rhoades (1984) describes the technical feasibility of a 'cyclic' drainage water reuse strategy. The strategy is to irrigate moderately sensitive crops (lettuce, alfalfa, etc.) in rotation with river water and salt-tolerant crops (cotton, sugar beet, wheat, etc.) with drainage water. For the salt-tolerant crops, the switch to drainage water would usually occur after seedling establishment (preplant irrigations and initial irrigations being made with river water). The feasibility of this strategy is supported by the following: the maximum soil salinity in the root zone resulting from continuous use of drainage water will not occur when such water is only used for a fraction of the time; substantial alleviation of salt build-up resulting from irrigation of salt-tolerant crops with drainage water will occur during the time salt-sensitive crops are irrigated with river water; proper preplant irrigation and careful irrigation management during germination and seedling establishment leaches salts out of the seed area and from shallow soil depths; and data obtained in field experiments.

This strategy has been tested for two cropping patterns (Rhoades, 1990b) in a 20 ha field experiment, on a commercial farm in the Imperial Valley (California). One was a two-year successive crop rotation of wheat, sugar beet and cantaloupe (wheat, 1982; sugar beet and cantaloupe, 1983; wheat, 1984; sugar beet and cantaloupe, 1985). The Colorado River water with a salinity of 900 mg L^{-1} TDS was used for the preplant and early irrigations of wheat and sugar beet and for all irrigation of cantaloupe. The remaining irrigations were with drainage water of 3500 mg L^{-1} TDS. The other experiment was a four-year block rotation consisting of two years of cotton in 1982 and 1983, followed by wheat in 1984 and alfalfa in 1985.

The yields of wheat and sugar beet from substituting drainage water in either cycle of the rotation showed no significant losses. Also no significant yield loss was observed from growing cantaloupe using Colorado River water for irrigation in the land previously salinised from the irrigation of wheat and sugar beet using drainage water. The yields of each crop obtained in the block rotation also showed that there was no loss in lint yield in the first (1982) and second (1983) cotton crop from the use of drainage water for the irrigation following seedling establishment, which was accomplished using Colorado River water. But there was a significant and substantial loss of lint yield in the second season cotton crop where the drainage water was used solely for irrigation. This loss of yield was caused primarily by a loss of stand that occurred in 1983, because salinity was excessively high in the seedbed during the establishment period.

As described by Rhoades (1990b), the qualities of all these crops were never inferior, and were often superior, when they were grown using the drainage water for irrigation. Analysis of the amounts of water applied to each crop over the entire four-year period (1982–1985) shows that substantial amounts of drainage water were substituted for Colorado River water in the irrigation of these crops without yield loss. The results support the credibility of the cyclic crop and water strategy to facilitate the use of saline water for irrigation.

In another study reported by Rhoades (1990b), carried out in the San Joaquin Valley, California, drip irrigation was used to apply drainage water with a salinity of 8000 µS cm^{-1} to cotton (after the crops were established) for three consecutive years. This was followed by a wheat crop irrigated with high quality water and then by sugar beet irrigated with drainage water after stand establishment. Yields under these conditions were not less than those obtained from continuous irrigation with high quality water.

Rhoades (1990b) discusses the uncertainties associated with the long-term effects of the reuse strategy, which include reduction of the soil infiltration capacity, soil salinisation, and accumulation of certain elements (selenium, molybdenum, heavy metals...) in soils and plants that are toxic to the consumers of the crops (human and animal) and describes the management considerations of the reuse strategy, including the criteria for crop selection.

Drainage water can also be used for tree plantation and agroforestry. Schofield (1990) reports that, in the San Joaquin Valley of California, successful pilot wood lots have been implemented by the Californian Department of Food and Agriculture. From the beginning of the agroforestry program in 1985 to 1989, trees have been planted on about 70 ha on 21 farms. Eucalypts and casuarinas have been the predominant genera used, particularly *E. camaldulensis* for water use and wood products and *C. cunninghamiana* and *C. glauca* for windbreaks. Research shows that wood lots can reduce the volume of drainage water that needs to be evaporated or ultimately disposed of by 75 per cent. Several uncertainties with the agroforestry-evaporation pond approach that need further consideration have been identified. These include effects of salt accumulation in the soil and its impact on yields from wood lots; effects of salt and trace elements like selenium on the market suitability of the wood; and the entrance of selenium and other potentially harmful elements found in the drainage water into the wildlife systems.

The second reuse strategy is to blend supplies before or during irrigation. If the blending strategy is adopted there must be adequate control of this mixing of water supplies. Shalhevet (1984) describes two blending processes: network dilution and soil dilution. In network dilution different quality waters are blended in the irrigation conveyance system, making tailor-made water available for each crop and soil condition. Agronomically this option presents no particular problems. The salt tolerance function of the various crops must be known and the choice of water salinity level will depend on the tolerance threshold and other considerations, as with any other source of water. The choice of this option presupposes the availability of appropriate blending equipments to supply the desired mixture.

In the case of soil dilution, alternating good and poor quality water allows the mixing to take place in the soil. This option would be impractical for sprinkler irrigation systems if the poor quality water causes foliage damage. It may be advantageous, however, if the intermittent application of non-saline water provides better dilution by leaching the salt which was added during the application of saline water. Experience shows that leaching once or twice during the season with the same quality water as that used for irrigation is more efficient in removing salts than providing the leaching water in every irrigation. Thus using this option in preference to network dilution, when appropriate, will save the expense of construction and maintenance of the mixing facilities and may even result in better performance.

In most cases where dilution has been considered to reduce the salinity of drainage water, control of the quality of the diluted mix is desirable. Quality control could be imposed in principle for total salinity, control of certain toxic materials, control of specific ions or control of other water and salt content properties. Sinai et al. (1985a) discuss the application of automated flow and salinity control to dilution of saline irrigation water and describe two mixing methods. A three-way mixing valve, with two inlet valves for the waters to be mixed up and a third valve for the mixed water, is recommended where a precisely blended mix of the two source waters is needed. The second method uses dilution tanks which are open to the atmosphere and may have the following varieties: mixing tank with natural turbulence; mixing tank with induced turbulence; and large mixing tanks. When on-stream dilution control is required it is important to insure adequate and rapid dilution at the dilution point. Sinai et al. (1985b) describe an analytical approach to dilution network problems.

Heavy use of drainage water occurs in Egypt. According to Balba (1990), estimates for the drainage water from the Nile Delta range from 11 to 16×10^9 m^3 yr^{-1}. A number of drains in the east, middle and west Delta discharge 2.88×10^9 m^3 yr^{-1} of drainage water in either of the two branches of the Nile or in main canals to be reused downstream. The estimate of total drainage water to be used by the year 2000 is 5.38×10^9 m^3 yr^{-1}.

Disposal to rivers

This option is widely in practice. Disposal can be controlled or uncontrolled. In the case of controlled disposal, the volume, salinity and time of disposal are controlled. Effluents from the drainage facilities are collected in holding basins or evaporation basins and released to the river during periods of high river flow. The common practice of uncontrolled disposal of the drainage effluent merely serves to salinise the vital water supply for the less fortunate users who happen to be located downstream from the point of discharge. If they also drain their fields in a similar fashion, the river will undergo progressive salinisation and its lower reaches may become unfit as a water source for either human use or irrigation. The river then turns into a saline stream, with consequent effect upon its associated aquifer and estuary, or upon the lake or bay into which the river flows. If, in addition to agricultural drainage, domestic, municipal and industrial effluents are also discharged into the river, it can become in effect an open sewer which can endanger the entire population of the region (Hillel, 1990).

Disposal to seas and lakes

Disposal to open seas, inland seas and lakes is an attractive option, but mainly for economic reasons cannot be considered as a feasible option for areas far from the sea. With this option, saline effluents are disposed of after being concentrated in evaporation basins to reduce their volume or without concentration. Effects of the nutrients, pesticides and chemical components of the disposed waters on the aquatic life and algae growth should be carefully investigated during the period of feasibility study and environmental impact assessment. The following paragraphs describe or refer to some of the sea and lake disposal projects in operation, under construction or abandoned because of environmental effects or high cost.

In the San Joaquin Valley, California, the Interagency Drainage Program (IDP) was formed in 1975 to re-evaluate the alternatives for drainage management. Several alternatives were considered: no action; evaporation ponds within the valley; discharge of wastes to the San Joaquin River; discharge to the Pacific Ocean; and discharge to the Western Delta (Gardner, 1986). According to Boyle Engineering Corporation (1986), the IDP recommended a valley-wide master drain discharging to Suisun Bay near Chipps Island by 1981. Based on the available data this was considered the most economic and environmentally sound alternative. Included in the drainage system was a series of wetland marshes managed as both wildlife habitats and reservoirs (Kesterson Reservoir, consisting of a series of ponds) to regulate drainage water discharge to the Delta-Suisun Bay, by delaying the discharge peak from summer to winter and spring when the waste water would receive maximum dilution with waters flowing to the San Francisco Bay. The IDP recommendation addressed the water quality problems evident at the time: salinity and nitrate. The 1979 report of the IDP (San Joaquin Valley Interagency Drainage Program, 1979) concluded that the movement of the master drain outlet to a more westerly location because of lower project costs would not cause widespread salinity increase in the western Delta-Suisun Bay area and would not cause significant impact on algae growth. However, nitrate removal was considered as potentially necessary to ensure algae control. The primary concern identified in the IDP report was the potential impact of subsurface drainage water constituents on aquatic life. Concern was raised primarily for boron, chromium, iron, lead, mercury and certain pesticides as being at potentially toxic levels for aquatic life. More intensive drainage monitoring was recommended but selenium was not included (Boyle Engineering Corporation, 1986).

As a result of the 1979 IDP report, the US Bureau of Reclamation (USBR) initiated the required procedures needed to obtain a permit from the State Water Resources Control Board (State Board) to discharge drainage water flow to the Bay-Delta Estuary. The USBR also initiated the San Luis Unit special study in 1981 to supplement the previous environmental impact assessment, analyse alternative drainage management plans and develop information required by the State Board to establish drainage effluent criteria. The technical studies required by the State Board were largely complete when bird mortalities in the Kesterson Reservoir caused by selenium from the introduced drainage water were discovered in 1983 and raised concern for public health (see Chapter 11 for further information). In March 1985 the Department of Interior called for the cessation of all drainage flows into the Kesterson Reservoir by 30 June 1986 (Boyle Engineering Corporation, 1986). Subsequently the master drain outlet project was abandoned.

In Australia, disposal of saline effluents of the Murray Basin to the sea has been investigated and proved uneconomic considering the large expenditure involved and the poor economic returns (see Chapter 2). Egypt practices the disposal of its agricultural saline effluents to the Mediterranean Sea (see Chapter 5). In Pakistan, the Left Bank Outfall Drain Project is under construction for the disposal of saline agricultural waters of the Lower Indus River Basin to the Arabian Sea (see Chapter 8).

In the Commonwealth of Independent States, Lakes Sarykamysh and Aydarkul, with surface areas of 3000 km^2 and 2300 km^2 respectively, receive the drainage effluents of irrigated lands along the Amu Darya and Syr Darya Rivers (see Chapter 4).

In the United States, Salton Sea in southern California receives irrigation drainage water from the Imperial Valley and the Coachella Valley, located south-east and north-west of the sea respectively. Salton Sea was originally a salt-covered depression, known as the Salton Sink. It was formed during 1905–07 when the Colorado River broke through an embankment near Yuma (Arizona) and overflowed into California. The Salton Sea at first covered an area of approximately 1200 km^2 and had a maximum depth of 20 m. Its area has since decreased through evaporation. Currently, Salton Sea has an area of 880 km^2, elevation of 71.6 m below mean sea level, volume of 6.93×10^9 m^3, maximum depth of 12.5 m (Hammer, 1986) and salinity of about 40 000 mg L^{-1}.

Disposal to evaporation basins

Drainage water can be disposed to evaporation basins (ponds). These can be formed from natural depression, saline lakes and salina or can be constructed. Regulatory requirements may affect the salinity, construction and operation of the basins. They must be designed and operated with care since drainage water with high levels of trace elements can be considered hazardous or attain toxic levels in evaporation basins. Drainage water may be treated to remove toxic constituents before ponding, but this is costly. Moreover, the increasing concentration of toxic substances during evaporation stages may nullify such treatment (Lee, 1990).

Evans (1989) describes the disposal of saline effluents by different methods including disposal to evaporation basins. The following paragraphs are based on his account, unless other references are provided.

Evaporation basins vary in size from a few hectares serving individual farms to very large basins (thousands of hectares) serving large areas. Their advantages include: effectiveness, if they are properly designed, operated and maintained; salt harvesting potential; and the possibility of providing a habitat for a diversity of bird life. Their disadvantages consist of: local salinity and waterlogging; adverse effects due to lateral seepage; their finite life; their land requirements; and possible adverse environmental impacts.

One of the major problems associated with the operation of evaporation basins is leakage. There are two broad types: lateral and vertical leakage. Lateral leakage is often the principal undesired consequence because of its effects on surrounding land within a relatively short time, typically of the order of several years. Vertical leakage creates a groundwater mound beneath a basin which may raise groundwater levels close to the edge of the basin. Before the construction or selection of a natural evaporation basin, considerable geotechnical and hydrogeological investigation is required to evaluate leakage and its effects. Should excessive lateral seepage be expected during design, or observed after construction, the following techniques can intercept or reduce seepage: surface cut-off drain, typically 1 m to 2 m deep; interception bores located around the basin; cut-off walls in the surrounding embankment or down through relatively shallow aquifers to a low permeability layer; lining and treating the basin with impermeable lines like clay or synthetic materials. The traditional approach of locating evaporation basins where there is a substantial thickness of low-permeable clay may be good in theory but is difficult to find in practice.

Parameters involved in the choice of an evaporation basin include: horizontal and vertical hydraulic conductivity of any aquitards and aquifers; vertical hydraulic conductivity of the base of the basin; and salinity of the natural groundwater beneath the basin. Other factors which need to be considered in the design of evaporation basins are: size of the basin; shape and internal topography; landscape; location relative to drainage water resources; soil type; slope; visual amenity; and environmental and social effects.

One of the common criticisms of evaporation basins is that they occupy large areas. For example, the Wakool-Tullakol subsurface drainage scheme in New South Wales (Australia) protects 60 000 ha; the evaporation basin size is 2000 ha, which represents an area ratio of 1:30 (Evans, 1989). In the San Joaquin Valley, California, it has been estimated that evaporation basins, whether on-farm or district or regional facilities, would require a land area equal to about 20 per cent of the irrigated land to be drained (Gardner, 1986). Another problem with evaporation basins is the long-term concentration of salinity that occurs, progressively reducing the evaporative capacity of the basin (Table 21).

Evaporation basins provide only a relatively short-term solution to a long-term problem and when an evaporation basin is decommissioned, it represents a long-term environmental problem. The lifetime of an evaporation basin is primarily a function of the ability to dispose of salt. If no salt export occurs, gradually the salt crust will build up and the capacity of the basin will be reduced to zero. Methods of salt export include flushing to surface waters during periods of high river flow and salt harvesting.

Table 21: Reduction of evaporation potential of water with different salinity.

Salinity ($\mu S\ cm^{-1}$)	Percentage of freshwater evaporation (%)
Freshwater	100
50 000 (seawater)	99
250 000	90
500 000	66
535 000 (saturation)	59

Source: Gutteridge Haskins et al. (1983).

Much of the previous discussion concerned regional or subregional basins. The use of on-farm evaporation basins to dispose of saline water from tile drainage or groundwater pumping on individual farms also has several problems: effectiveness of basins located on the more permeable soils will be limited unless they are lined, and the potential to adversely affect neighbouring properties would be higher than that of large, public-scale basins which are usually sited on lower-permeability soils; basins might occupy valuable land; and investigation and construction costs associated with each basin could be high. However, the on-farm basins may be well used as short-term holding basins with release permitted during high flow periods.

Disposal to evaporation basins is practised in a number of countries including Australia and the United States. In Australia, where there is an abundance of natural depressions and salt lakes far from cultivated land, disposal to evaporation basins is an economical and convenient option. Table 22 shows the major characteristics of a number of large evaporation basins in the Murray Basin.

In the San Joaquin Valley, California, on-farm basins have been used for the disposal of drainage waters since the closure of the Kesterson Reservoir and drainage facilities. More than 6000 ha of on-farm basins (pounds) have been constructed and requests have been made to increase this amount significantly (Boyle Engineering Corporation, 1986).

Table 22: Major characteristics of a number of large evaporation basins in the Murray Basin, Australia.

Name	Area (ha)	Annual volume disposed (10^6 m^3 y^{-1})	Salinity of water disposed (1000 µS cm^{-1})	Range of salinity in the basin (1000 µS cm^{-1})	Type of basin
Wakool	2000	13.5	25	16–300	Constructed
Noora	1700	5	16	15–100	Natural
Tutchewop	1140	11	7	18–206	Natural
Woolpunda	750	7	37	90[a]	Natural with bank
Wargan	690	11.3	11	14–71	Natural
Chowilla	500	NA	30	100[b]	Natural with bank

(a) Mean value; (b) Approximate value. Source: Evans (1989).

Deep-well disposal

Deep-well disposal enables liquid wastes which cannot be conveniently or economically disposed of by other means to be injected into deep aquifers, where they do not represent a degradation of the existing sources of water supply or a threat to the environment. The technology for deep-well injection evolved during the 1930s in the petroleum industry of the United States and the potash mining industry of Germany. Brine produced with oil was injected back into the subsurface instead of discharging onto the land surface. Since then the petroleum industry has added injection wells for secondary and enhanced recovery of oil to an increasing number of brine disposal wells. According to Hickey and Vecchioli (1990), in 1983 there were more than 220 000 injection wells in the United States, with about 152 000 or 69 per cent used by the petroleum industry. By the 1970s deep-well injection had been adopted by other industries as a method of liquid waste management. LaMoreaux and Vrba (1990) describe different aspects of deep-well disposal of liquid waste which apply also to the injection of brine. The following paragraphs are based on their account.

The geology and hydrogeology of aquifers are the main consideration when assessing the technical and economic feasibility of a deep-well injection. To minimise environmental risks the thickness, impermeability and continuity of all confining units and the structural position of the aquifer system with respect to water-supply aquifers must be evaluated prior to the construction of an injection well. Moreover, injected waste should be physically and biochemically compatible with the host rock. Therefore comprehensive knowledge of the chemical and biological reactions between liquid waste, host rock and native water is required. Plugging of the pore space and gradual decrease in the permeability of the receiving unit is the most dangerous process that can occur during injection. Treatment of the waste prior to injection is most desirable.

Environmental impact of the deep-well injection should be assessed. Problems, failures and environmental consequences of poor deep-well injection are mostly related to errors in well construction, undetected pathways for fluid migration, operational errors and natural events such as earthquakes. Deep-well injection schemes should have adequate monitoring facilities to provide evidence that the injection well is operating correctly and that the injected fluids are being contained. Monitoring devices must be able to detect problems as soon as they develop.

With respect to the management of salinity problems, injection of brine into isolated deep aquifers can be a feasible disposal method in particular areas with adequate geologic and hydrogeologic characteristics. According to Gutteridge Haskins et al. (1983), specific problems arise in the injection of brine requiring compromise. Brines derived from evaporation basins are seldom free of organic matter, being home to a large variety of halophytic shrimps, diatoms and algae. In addition, because of their relatively high viscosity, sand and fine dust do not settle out readily, and at higher concentrations they can include a significant amount of microcrystalline salts. These characteristics make them difficult to inject, because they tend to clog the aquifer close to the well and the well interface with the aquifer. These problems are further aggravated by the highly reactive nature of the brines which can cause near instantaneous precipitation of calcite (calcium carbonate) and gypsum (calcium sulphate) from native groundwaters, along with siderite (iron carbonate) and jarosite (iron sulphate). Those precipitates tend to further clog the pore space and reduce the formation permeability. Also, clays in the formation may suffer rapid cation exchange and may expand to take up more space, thus reducing the permeability of those sands with clay contents. Many of these problems can be offset by the use of surface treatment to remove organic and inorganic solids.

In the United States, deep-well injection of brine is a key part of the Paradox Valley Unit of the Colorado River Basin Salinity Control Program. The Unit is located in south-western Colorado (see Figure 11.7) and has been designed to reduce the Valley's contribution to the salinity level in the Colorado River. According to Jensen and Leach (1984), the Dolores River, a tributary of the Colorado River, picks up an estimated 186 000 tonnes of salt annually as it crosses the 6.5 km wide Paradox Valley. Approximately 97 per cent of the salt enters the river through seeps and springs located in the bottom of its channel. The remaining three per cent comes from irrigation return flow and natural surface run-off in the valley. The Unit has the potential to remove 163 000 tonnes of salt annually and reduce the salinity of the Colorado River at Imperial Dam by 18.2 mg L^{-1}.

In 1979 a management plan and environmental impact statement were completed and approved. The plan recommended pumping brine from wells paralleling the Dolores River and disposing of it in a large evaporation basin. In March 1980, pump testing and monitoring was initiated. Brine production wells varying in depth from 15 m to 92 m were located on both sides of the river. The scheme included a large number of monitoring wells and a temporary brine holding basin. The testing program showed this plan to be a highly effective method of controlling brine inflow to the river.

As a result of brine disposal re-evaluation, deep-well injection was chosen as a more feasible alternative. The feasibility study confirmed that the Mississipian and Devonian formations present under the Paradox Valley have the desired characteristics for satisfactory long-term disposal of the brine. The proposed brine disposal system could be operated indefinitely with no significant impact on the environment. Successful long-term deep-well disposal of the brine will require filtering the brine, which contains approximately 1.0 mg L^{-1} suspended solids. A diatomaceous earth filtering system was chosen as the most desirable type. Deep-well injection as a method of disposal will reduce the original plan's project costs by one-half to one-third, a saving of about US$60 million to the American taxpayers (Jensen and Leach, 1984).

Another example is the deep-well injection of agricultural drainage waters of the San Joaquin Valley in California (see Chapter 11 for a description of the problem). URS Corporation was contracted by the US Bureau of Reclamation to undertake an appraisal of the problem. Based upon existing information, URS Corporation (1986) proposed that the most attractive alternative for disposal of 38 000 m^3 d^{-1} of drainage water would be deep-well injection. The proposal consists of injection through triple-cased wells at a rate of 60 L s^{-1} per well into a Tertiary sandstone aquifer with an assumed 300 m thick injection zone. The system should utilise filtration to avoid mechanical formation clogging and chlorination to reduce the chance of biological fouling. The Tertiary aquifer is highly saline, lies below all useable groundwater resources and is covered by a very extensive, thick and continuous impermeable layer, the bottom of which lies roughly 1500 m below the San Joaquin Valley land surface.

According to URS Corporation (1986), the major uncertainty in the feasibility of deep-well injection is in the chemical and biological stability of agricultural drain waters interacting with formation fluids. The formation waters are estimated to be a sodium chloride water low in calcium and sulphate. The agricultural drain waters are of sodium sulphate type, high in calcium chloride and nitrogen. The total dissolved solids are estimated to be 40 000–50 000 mg L^{-1} for the formation waters and 10 000 mg L^{-1} for drain water. The blend of these waters can cause chemical precipitation and formation clogging by gypsum (calcium sulphate) or calcite (calcium carbonate). Possible formation clogging due to biological activities include biomass plugging and sulphide precipitation. Clogging might extend sufficiently far into the injection horizon to be beyond the range of remedial measures. The URS Corporation (1986) report does not conclude that deep-well injection is feasible. Instead it suggests that feasibility can only be evaluated by pilot testing.

In Australia, the feasibility of brine disposal into deep aquifers of agricultural areas has been investigated but the technique is not currently undertaken. Woolley and Kalf (1979) have investigated the feasibility of brine and bittern (generated at the Wakool Evaporation Basin in New South Wales) injection into a sand aquifer at a depth of 310–330 m with a transmissivity of 387 m^2 d^{-1} and low salinity of 1700 mg L^{-1}. They showed that technically it would be possible to inject bitterns with a high concentration of 430 000 mg L^{-1} and specific gravity of 1.3 at a rate of 10 L s^{-1} into the aquifer, using only the head caused by the column of fluid and not requiring any additional pumping pressure. However, they concluded that the project would cause contamination of a large part of an aquifer containing water which is only marginally unsuitable by present day standards for domestic and industrial purposes and which may otherwise become of economic importance in the future. They estimated that, after 200 years of operations, the volume of water stored in the aquifer within the area likely to be affected would be of the order of 0.5×10^9 m^3.

Deep aquifer injection of brine or bittern was also considered for the Woolpunda and Mallee Cliffs schemes (Evans, 1989), but was rejected for a variety of reasons associated with technical and operational difficulties, which include aquifer clogging, bore corrosion and the enhancement of upward leakage.

Desalination

Desalination is the process of separating the dissolved salt content of saline or brackish water, to render it suitable for domestic, industrial or other purposes. Various desalination processes are available (Porteous, 1983). They include: reverse osmosis (RO), electrodialysis (ED), ion exchange (IE), multistage flash distillation (MSF), vapour compression (VC) and solar distillation. A brief description of these processes now follows. In reverse osmosis, water is pumped through a membrane, leaving the dissolved salts behind. Electrodialysis is a process in which ions are transported through a membrane from one solution to another under the influence of a direct current electrical potential. Ion exchange involves passing the saline water through two beds of resins, one of which removes the anions and the other the cations. Pure water emerges from the second bed. After a time the beds become exhausted and are regenerated. In MSF, the heated saline water is passed through a series of chambers under a gradually increasing vacuum. The water thereby 'flashes' into a steam-water mixture and the steam is separated and condensed to pure water. In VC, steam produced from boiling water is compressed mechanically to produce freshwater. Solar distillation in its simplest form consists of a basin with a heat-absorbing surface (usually coloured black) containing the saline water. The basin is covered by a sloping transparent glass or clear plastic which transmits sunlight and condenses vapour evolved from the saline water. The condensate trickles down the slope of the cover to the edge and is collected in a trough.

All desalination processes produce, as well as freshwater, a by-product with high salt content which needs to be disposed of. Plants on the seaboard can conveniently discharge their by-product to the sea, but for inland plants disposal is a problem. A possible solution could be disposal to evaporation ponds or salt lakes.

Desalination has been widely used for the production of freshwater for human consumption on a large scale since the 1950s. In 1969 the total world capacity of desalination plants was about 1×10^6 m^3 d^{-1}. By December 1989, a worldwide total of 7536 desalting units with a total capacity of 13.3×10^6 m^3 d^{-1} had been installed or contracted (Wangnick, 1991). The major production of desalting capacity is installed in Saudi Arabia (26.8 per cent), followed by the United States (12 per cent), Kuwait (10.5 per cent) and the United Arab Emirates (10 per cent). In relation to the number of units, the RO process ranks first (55.2 per cent), followed by MSF (14.1 per cent) and ED (13.7 per cent). Proportion of feed-water treated consists of sea water (65 per cent), brackish (28 per cent) and others (7 per cent). Desalination plants are increasingly being used for the treatment of waters other than seawater or brackish water. The current areas of application lie in the treatment of effluent waters, river water and groundwater that has been polluted by nitrates and pesticides (Wangnick, 1991).

In the United States, the Yuma Desalting Plant has been in operation since 7 May 1992. It was designed for the treatment of irrigation drainage water from the Wellton-Mohawk Irrigation and Drainage District (WMIDD) in Arizona to reduce the salinity of water delivered to Mexico (see the section on salinity and conflict resolution, pp. 93-97). The following is a brief description of this plant, summarised mainly from the Bureau of Reclamation (1992). The Yuma plant is the world's largest reverse osmosis facility. It has an installed capacity to produce 72.4 million gallons (274 000 m^3) of desalted water per day from about 100 million gallons (378 500 m^3) of drainage water per day with a salinity of about 3000 mg L^{-1}. If the need later exists, the plant design permits expansion to 96 million gallons (363 000 m^3) per day. The plant product water has a salinity of 295 mg L^{-1}.

The drainage water flows to the plant in a concrete-lined canal and before being desalted passes through several pre-treatment processes to remove all solids and micro-organisms which otherwise would quickly clog the expensive desalting membranes (Trompeter and Suemoto, 1984). The plant has a total of 9000 membrane elements of cellulose acetate.

The product water is blended with the untreated irrigation drainage water before being discharged to the Colorado River. The desalting plant produces about 100 000 m^3 per day of concentrate water with a salinity level of 10 000 mg L^{-1}. This concentrate water flows into the bypass drain that carries the concentrate to the Gulf of California at the Santa Clara Slough (Figure 5).

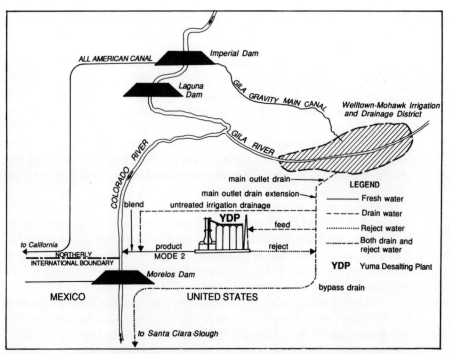

Figure 5: Yuma Desalting Plant water flow schematic (Bureau of Reclamation, 1992).

Inside the Yuma Desalting Plant there is another small desalting plant capable of desalting one million gallons of water (3800 m^3) per day. It is a test plant serving several functions. It is used to train operators; test new types of reverse osmosis membranes; try out different chemical treatments of the water; develop and test new hardware and software; and experiment with changes to the desalting process before using those changes on a large scale in the main plant.

The Yuma Desalting Plant power requirement is 22 megawatts in full-load daily operation. The investment cost was US$264.5 million (in 1992 prices) and the total cost of the recovered water is about US$495–567 per 1000 m^3 (Bureau of Reclamation, 1992). Obviously such a high investment and production cost cannot be justified solely on an economic basis. In this case fulfilment of the United States' obligations towards Mexico has been the driving force behind the project.

In Australia, the technical and economic aspects of disposal of three classes of saline water in Victoria have been considered. These are: Shepparton region groundwater, with a salinity of 1000–4000 µS cm^{-1}; Barr Creek, with a salinity of 4000–7500 µS cm^{-1}; and Kerang region groundwater, with salinity of 20 000–70 000 µS cm^{-1} (Evans, 1989). According to Gutteridge Haskins et al. (1983), the production cost per 1000 m^3 per day ranged approximately from A$500 (for Shepparton groundwater) to A$3800–7800 (for Kerang region groundwater) in 1983 prices. The studies clearly showed that, even ignoring the cost of disposing of the highly saline waste stream, the costs of desalination are far above the value of water produced (Evans, 1989).

Section E: Management of Salinity — Other Options and Aspects

Biological options

Agricultural practices

The water balance of a catchment can be manipulated to some extent by altering agronomic practices. Varying agronomic practices should result in different amounts of water being removed from the landscape via evapotranspiration. There is a range of agricultural practices which can reduce the rate of recharge to the groundwater system. Specific techniques include the replacement of shallow-rooted agricultural species with deep-rooted ones such as alfalfa, lupins, lucerne and perennials.

The recharge rate under different crops depends, among other things, on soil type and the average annual rainfall and its seasonal distribution. Table 23 provides some estimates of the annual recharge under different agronomic systems for a number of sites in Western Australia and Victoria. Much of the difference between estimated recharges beneath the same species is due to variation in seasonal rainfall quantity and distribution. However, common to all these results is the relatively low water use of clover-based annual pasture and consequently high rates of groundwater recharge beneath it (Schofield et al., 1989).

The successful demonstration of salinity control by revegetation is rare and circumstantial (Greenwood, 1988). The best documented example is the replacement of shallow-rooted grasses with the deep-rooted alfalfa on recharge areas of the Northern Great Plains of the United States. Saline seeps affected a large area of productive dryland agriculture in this region. Control of saline seepage has been achieved with an intensive cropping system and deep-rooted crops. Alfalfa, when grown on about 80 per cent of the recharge area, reduced the deep percolation of soil water and provided hydrologic control within one year after its establishment (Halvorson and Reule, 1980). As the perched watertable reduced, the soil surface in the seepage area dried, soil salinity decreased and crops grew better in the seepage area. Halvorson and Reule (1980) and Halvorson (1990) argue that alfalfa can be used to gain hydrologic control of saline seeps because of its greater water requirements and because it can extract soil water from deeper depths than the grain crops (they observed water extraction to a depth of 360 cm by alfalfa).

Table 23: Estimated annual average recharge under different agronomic systems.

Treatment	Annual recharge (mm)	Site	Approximate average annual rainfall (mm)
Clover	162	Cunderdin (WA)[a]	350
Lupins	61	Cunderdin (WA)	350
Wheat	139	Cunderdin (WA)	350
Barley	83	Cunderdin (WA)	350
Clover	125	Kondut (WA)	350
Lupins	80	Kondut (WA)	350
Wheat	47	Kondut (WA)	350
Barley	9	Kondut (WA)	350
Clover	17	Wongan Hills (WA)	390
Bare-fallow	20	Wongan Hills (WA)	390
Wheat	9	Wongan Hills (WA)	390
Annual pasture	50	Axe Creek (Vic)[b]	650
Lucerne	22	Axe Creek (Vic)	650
Perennial pasture	30	Axe Creek (Vic)	650
Clover	341	Perth (WA)	800
Perennial pasture	176	Perth (WA)	800

(a) Western Australia; (b) Victoria.   *Source:* Schofield et al. (1989).

The potential to manipulate the water balance of catchments by altering agronomic practices in the south-west of Western Australia has been discussed by a number of researchers. According to Nulsen and Baxter (1982), it is possible to use a large part of the excess water by changing the cropping strategy. The changes required are relatively minor and thus could be acceptable to the farmers. Changes include identification of areas of preferential recharge and the growing of lupins in the rotation instead of the traditional clover. The additional water used by the alternative rotation will at least reduce the rate of salinisation.

Nulsen (1984) describes the results of evapotranspiration measurements at two sites during 1980 and 1981. The 1980 site at Kondut was characterised by a loamy sand soil-type. Species grown were wheat, barley, lupins and clover. During 1981 another site was selected at Cunderdin. Total seasonal evapotranspiration and rainfall for both sites are presented in Table 24.

The differences in evapotranspirations measured at both sites can partly be explained by differences in root depth and the leaf area index. Although root depths were not measured in this experiment, studies have shown that for the same soil type, clover roots extend to 0.6 m–0.8 m, wheat

Table 24: Total evapotranspiration and rainfall over the growing season for four species at Kondut and Cunderdin in Western Australia.

Site	Rainfall (mm)	Evapotranspiration from:			
		Wheat (mm)	Barley (mm)	Lupins (mm)	Clover (mm)
Kondut	162	115	153	82	37
Cunderdin	258	119	175	193	96

Source: Nulsen (1984).

roots to 1.0 m–1.2 m and lupin roots have been found to extend to 2.5 m–3.0 m (Nulsen, 1984). Table 25 shows the potential mean annual recharge under a number of likely rotations.

It is evident from Table 25 that while none of the rotations eliminates recharge, rotations containing lupins instead of clover contribute much less recharge. Nulsen (1984) concludes that, despite being limited in his experiments to non-perennial agricultural species, it appears that agronomic manipulation can reduce recharge to the groundwater system and restrict the rate of salinisation.

Table 25: Potential mean annual recharge under different cropping rotations for Kondut and Cunderdin in Western Australia.

Rotation	Mean annual recharge at:	
	Kondut (mm)	Cunderdin (mm)
Wheat-clover	86	150
Wheat-clover-clover	99	154
Wheat-wheat	47	139
Wheat-lupins	64	100
Barley-lupins	44	77

Source: Nulsen (1984).

Perennial pasture can be very effective in reducing recharge as indicated by the results of experiments over a period of 12 months from December 1984. Under a rainfall of 384 mm, a rotationally grazed stand of lucerne used 433 mm of water compared with 231 mm for an adjacent wheat crop (Western Australian Department of Agriculture, 1988).

It is well known that the level of nutrition influences plant growth, transpiration rate and water use. Halse et al. (1969) sowed wheat on a loamy sand. Nitrogen was applied as urea at three rates: 0 kg N ha^{-1}; 56 kg N ha^{-1} at sowing; and 336 kg N ha^{-1} (112 kg N ha^{-1} at sowing plus 112 kg N ha^{-1} five weeks after sowing and another 112 kg N ha^{-1} ten weeks after sowing). The corresponding grain yields were 887, 1770 and 2980 kg ha^{-1}. Using a simplistic approach, Nulsen (1983) estimated that 100 mm of rainfall is required for grain set and each additional mm produces 10 kg ha^{-1} of grain.

The respective water requirements for the three treatments were 188 mm, 277 mm and 398 mm.

Revegetation of salt-affected lands

Research has now been in progress for many years in several countries investigating the revegetation of recharge areas and saline lands and development of agricultural production systems using salt-tolerant crops, shrubs and trees. Ismail et al. (1990), in their publications funded by the Australian Centre for International Agricultural Research, provide some 500 abstracts and bibliographical references in this area, published from 1970 to 1990 by researchers from about 30 countries. More than 80 per cent of the entries originate from Australia, United States, India, Pakistan, United Kingdom and Israel. This section provides a brief account of examples in a few countries for revegetation options.

The National Research Council (1990) examines some of the plants that may be suitable for economic production in saline environments in developing countries. The four sections of this publication highlight the salt-tolerant plants that may serve as food, fuel, fodder and other products such as essential oils, pharmaceuticals and fibre. In each of these sections, plants are described that have potential for productive use. Each section also contains an extensive list of recent papers and other publications that contain additional information on these plants.

The Nuclear Institute for Agriculture and Biology (NIAB) has successfully demonstrated the potential of kallar grass (*Leptochloa fusca*) for the saline, sodic and waterlogged soils of Pakistan (Malik et al., 1986). Kallar grass is a deep-rooted, highly salt-tolerant, perennial grass which grows well even under waterlogged and sodic conditions. It has been growing in saline areas of Pakistan for quite a long time. It has also been employed in certain areas of the central Punjab by progressive farmers who find it to be economically attractive. Myers (1990) describes the results of a field trial of kallar grass in Tatura (Victoria, Australia) and provides a list of grass species which may be grown on saline lands.

In India, soil salinity is widespread in many states including Uttar Pradesh, Gujarat, Punjab and Haryana (see Chapter 6). The Central Soil Salinity Research Institute (CSSRI) in Karnal, the premier salinity research group in India, has been involved in developing saline land farming technologies, selection of salt-tolerant trees and development of improved plantation techniques. Gill and Abrol (1991) have recently summarised data relating to tree performance on salt-affected land and the effect of site preparation methods on the performance of *Eucalyptus tereticornis* and *Acacia nilotica* in highly alkaline soils. They also consider the effects of afforestation on improvement of the soil environment (soil pH, salinity and organic carbon content) and provide a list of tree species and their relative tolerance to soil salinity (Table 26).

Table 26: Tree species and their relative tolerance to soil salinity.

Highly tolerant	Moderately tolerant
Tamarix articulata	*Acacia cyanophylla*
Tamarix gallica	*Acacia decurrens-dealbata*
Casuarina equisetifolia	*Parkinsonia aculeata*
Casuarina glauca	*Eucalyptus camaldulensis*
Casuarina cunninghamiana	*Eucalyptus citriodora*
Prosopis juliflora	*Eucalyptus tereticornis*
Acacia nilotica	*Azadirachta indica*
Acacia decurrens-mollis	*Albizia lebbek*
Acacia catechu	*Dalbergia sissoo*
Zizyphus jujuba	*Leucaena leucocephala*
Zizyphus spina-vulgaris	*Populus euphratica*
Butea monosperma	*Pinus halepensis*
Ailanthus excelsa	
Terminalia arjuna	
Capparis aphylla	

Source: Gill and Abrol (1991).

In Australia, which has a native flora rich in halophytes, some very valuable research and demonstration programs have shown the potential of agricultural production systems using salt-tolerant plants. According to Barrett-Lennard et al. (1990), the growth of forage on saltland is now an established technology. More than 20 years of research and experience in Western Australia has shown that farming (mainly grazing) on saline soils is not only possible, but is also profitable and readily incorporated into whole farm systems. Malcolm (1990) describes the establishment of 1200 ha of saltbush (*Atriplex*) in Western Australia. It has produced a net return of A$125 (US$100) per ha per annum on land which was previously regarded as wasteland by the farmers. Saltbush technology is being actively promoted also on salt-affected land in Victoria. Malcolm (1992) gives criteria, field test methods and guidelines for the selection of shrub species suitable for forage production on saline land.

As described by Marcar (1990 and 1992), tree-based land management strategies can contribute significantly both to minimising the spread of salinity and to the productive use of salt-affected land. Considerable scope exists in the choice of tree species for agroforestry or woodlot schemes on both dryland and irrigated saline land as well as land irrigated with saline effluent. However, except for soils of low salinity (electrical conductivity of saturated soil paste extract, 'EC_e', less than 8000 µS cm^{-1}) with little chance of waterlogging, there seems to be little potential for commercial wood (timber or pulpwood) production on saline land. Highly salt-tolerant tree species (e.g. *Eucalyptus occidentalis*, *Melaleuca halmaturorum* and *Casuarina glauca*) have great potential in saline drainage water reuse schemes.

Considerable knowledge has accumulated, particularly over the past 5–10 years, on the extent and nature of salt tolerance within Australian tree and shrub genera. Certain native tree species, such as *E. camaldulensis*, have been extensively examined, and salt-tolerant clones developed. Significant scope exists for intraspecific selection within species of proven commercial value, for example, *E. globulus*. Techniques such as mounding and mulching, to improve survival and growth on saline sites, have been investigated.

Table 27 attempts to categorise the performance of a range of species in terms of tolerance to low, moderate and high saline soils, on the basis of the results of a range of formal and informal species evaluation trials and demonstrations. Salinity classes are expressed in terms of conductivity of a saturated soil paste extract (EC_e) and are defined as low, 4000–8000 µS cm^{-1}; moderate, 8000–15 000 µS cm^{-1}; and high, 15 000–30 000 µS cm^{-1}.

According to Schofield (1992), tree planting, in combination with other vegetation treatments, is regarded as a leading solution to dryland salinity and has a potential role in controlling irrigation salinity. Research has now shown that planting trees can significantly lower watertables, and thereby reverse the causal process of salinisation. Substantial progress has been made toward answering such basic questions as which species to plant, how to plant, where to plant and what density to plant. However, serious constraints are apparent, relating primarily to cost, uncertainty and attitude.

Table 27: A selection of tree species suitable for sites of low, moderate and high salinity in southern Australia.

Low	Moderate	High
Acacia acuminata [A,C]	*A. saligna* [I?]	*A. ampliceps* [A,I?]
A. mearnsii [A,C,G]	*A. salicina* [A,I?]	*A. stenophylla* [A,C,D,I?]
A. melanoxylon [A,E,F,G]	*Casuarina cunninghamiana* [A,C,H]	*C. glauca* [A,C,H*]
Eucalyptus astringens [A]	*E. camaldulensis* [A,C,E,F,G*]	*C. obesa* [A*]
E. camphora [A,C]	*E. largiflorens* [A,C,F,G]	*E. kondininensis*
E. cladocalyx [A,B,E,I]	*E. leucoxylon* [A,C,E?]	*Melaleuca halmaturorum* [A]
E. globulus [A,B,D,E,I]	*E. microtheca* [A,C,F,G]	*Tamarix* spp. [A,H]
E. grandis [A,B,D,E,F,G]	*E. occidentalis* [A,C?,F*]	
E. ovata [A,C,E,F,G]	*E. sargentii* [*]	
E. sideroxylon [A,C,E,F]	*E. spathulata*	
Populus spp. [C]	*E. tereticornis* [A,C,E,F,G]	
Salix spp.		

*Significant provenance differences in salt and combined salt and waterlogging tolerance have been identified in these species and clones are being or have been developed.

[A-G]: species suitable for the following products:

A	Firewood (all regions)	F	Honey
B	Preserved posts	G	Pollen
C	Durable posts, rails etc. (most regions)	H	Attractive to bees
D	Pulpwood (> 600 mm annual rainfall)	I	Fodder
E	Sawlogs (> 500 mm annual rainfall)	?	Indicates uncertainty about product output

Source: Marcar (1992).

Perhaps overriding all the technical uncertainties has been the question of cost and benefit. Sustainable rehabilitation can be expensive, and the benefits are not always substantial or immediate in financial terms. A commonly expressed difficulty is the concept of replanting prime agricultural land to large blocks of trees to restore downslope salinised areas, or at least to control the spread of salinisation. Unless the planted trees have high value and can be harvested, or it can be demonstrated that the land to be reforested will ultimately go saline, landholders often see little benefit in giving up precious, productive land. However, the multiple benefits deriving from innovative configurations of trees combined with agriculture, such as wide timberbelts, are now being realised. Such systems can improve the productivity of crops, pasture and livestock, often by providing shelter, can provide products from the trees and can also control salinity. In Western Australia, for instance, about 20 000 ha of farmland has been planted to timberbelts of *Eucalyptus globulus* by farmers wishing to control salinity and produce pulpwood.

In deciding which tree species to plant, three important criteria (adaptation, water use and multiple use) should be addressed (Schofield, 1992):

- Adaptation is the primary criterion to be satisfied and includes adaptation to such factors as climate, soil, pests, waterlogging and salinity.
- Knowledge of the potential water use of species could allow trees to be selected to minimise the area required for tree planting. This is particularly important where there are strong reasons for maintaining land under agriculture.
- Multiple use refers to the range of beneficial uses of trees other than salinity control which may influence selection. These include commercial tree planting for timber, pulp, firewood, fodder and other products, shelter and shade, wind and water erosion control, waterlogging control and aesthetics. Although salinity control may be the prime objective, the attractiveness and feasibility of tree planting may depend strongly on some of these other beneficial uses.

The above criteria guide the tree selection process. To help determine the species from which to choose, a tree performance database is being developed to store and retrieve data from trials on salt land and recharge areas (Walker et al., 1991). In Western Australia, Victoria, South Australia and New South Wales extensive trials to identify tree species tolerant of saline conditions have been conducted in universities and by government agencies under both glasshouse and field conditions. The importance of provenance variation became apparent during these trials, for example with *Eucalyptus camaldulensis* (Thomson, 1988). In Western Australia more than 100 species of Australian woody plants have been screened for salt/waterlogging tolerance under glasshouse conditions (van der Moezel and Bell, 1990). At present 22 of the most tolerant species have been targeted to cover a range of climatic conditions throughout Australia (and overseas) and to serve a range of economic uses (Schofield, 1992).

According to Marcar (1992), successful establishment of suitable salt-tolerant trees and shrubs on saline sites requires the use of appropriate pre- and post-planting strategies for reducing site environmental stress. Any strategy should aim to minimise the shock to transplanted seedlings or direct seeded plants by ensuring as low a soil salinity and waterlogging regime as possible. This will mean that the time of planting will need to coincide with the period of maximum leaching (usually late winter or early spring). Several strategies have been shown to be effective or are being evaluated, which include: construction of mounds, particularly double-ridge mounds; application of mulch, particularly straw, newspaper or plastics; and pre-conditioning of seedlings to salt and waterlogging.

A full analysis of the costs and benefits to landowners and the community is necessary to assess who should pay for reforestation and agroforestry strategies to prevent salinisation or reclaim salt-affected lands. Costs may include loss of some agriculturally productive land, preparation of land for tree planting, seedlings, tree planting itself, fencing, management of stands, and logging and transport (in the case of commercial wood plantations). Landowner benefits consist mainly of the value of the forest, increased agricultural productivity and maintenance of the natural resource base (Schofield, 1992).

About 16 per cent of farmers in Australia establishing trees are doing so for salinity control (Prinsley, 1991). In all, 7 per cent of farmers are planting to prevent salinity, while 9 per cent are reclaiming saline land. In many cases the farmers' successes and failures are known locally but the information is not published. It is difficult therefore to know how successful their efforts are, or what specific strategies they are using.

Policy options

Water pricing

Both national governments and international agencies are deeply concerned about policies for pricing irrigation water, particularly in terms of repayment, yet many problems exist in implementing a system of water charges. Water charges are usually decided within the context of overall government agricultural development policy, which may involve food subsidies (Easter and Welsch, 1986).

In most countries irrigation water has an artificially low price and is highly subsidised by governments. This has led to an inefficient use of limited water resources and development of side effects including waterlogging and salinisation. For example, in Australia, both the state and the federal governments have subsidised irrigation in the Murray-Darling Basin to achieve regional development goals. The basin has an irrigated area of

1.27 Mha and uses 6.36×10^9 m^3 yr^{-1} of surface water for irrigation (Australian Water Resources Council, 1987). Currently the level of subsidisation is of the order of A$300 million or about US$240 million per year (Simmons et al., 1991).

With low water prices there is little incentive to improve irrigation efficiency. Higher prices may have the following effects (OECD, 1987): less water could be applied to a given crop; farmers might utilise a more efficient irrigation technology and water application practice; farmers might choose a different cropping pattern and consequently different water use.

There are at least six general methods for assessing water charges (Easter and Welsch, 1986): direct charges based on measured volume of water; direct charges per share of stream or canal flow, or per irrigation; direct charges per hectare irrigated or potentially irrigated; indirect charges on crop outputs marketed or on inputs purchased such as fertiliser; development rebates or promotional water charges; and a general land or property tax.

Volumetric charges are only possible if water delivered to farmers can be measured. Charges based on shares received are best suited for rotating irrigations where water is delivered to the users along a canal in turn according to some prearranged schedule. Charges per hectare are best fitted for continuous flow irrigation, where water flows continually in the main canal and farmers are free to take whatever quantity they need. This charge can either be the same for all farmers or varied by type of crop grown or by season. Indirect charges are used when ease of collection is an important objective. Development or promotional fees are used to encourage greater water utilisation with lower fees at the start of a project. Finally, taxes or fees levied on all lands and property in the irrigated area are used when the objective is to distribute the cost of the project among all direct beneficiaries. The idea behind this tax is that irrigation increases economic activity throughout the areas and therefore everyone should pay for the benefits. A more restricted land tax would be one just on the irrigated land. This is sometimes called a betterment levy and is based on the increase in land value due to irrigation.

Water pricing can be considered as a means of increasing irrigation efficiency, reducing irrigation water percolation and preventing watertable rise and salinisation. The rate structures can be adjusted to make water quantities exceeding efficient irrigation needs more costly (El-Ashry, 1980).

The following paragraphs provide some information concerning the price of irrigated water in the United States and a number of irrigated areas in the state of Victoria (Australia) affected by salinity.

In the western United States, the availability of inexpensive irrigation water was a key aspect of economic development. Water prices charged to irrigators in most western states have historically been one to two orders of magnitude less than the price paid by municipal and industrial users. Irrigators who pay exceptionally low water prices have little economic incentive to reduce their use of water. Also, the fees paid by irrigators to discharge drainage and run-off are too low to create an incentive to reduce pollution load by improving management of on-farm irrigation (Willey, 1990).

In the Upper Colorado River Basin, irrigation water costs about US$3.50 per acre-foot (1233 m^3), which is subsidised by the Federal Government (Reisner, 1987). In California the price for delivered irrigation water ranges from less than US$2 to more than US$200 per acre-foot (1233 m^3). Such differences reflect the date of completion of the system, topography, type of ownership and extent of subsidy (OECD, 1987).

Water prices and use in the Central Valley of California irrigation districts depend largely on the water rights or entitlement held by a district. By 1985 districts formed prior to 1940 had paid off the loans and bonds used to finance the construction of main canals and distribution systems. Irrigation districts are non-profit public corporations and some, such as Modesto Irrigation District in the San Joaquin Basin, charge nominal water costs because they also have hydroelectric power revenues. This district (Modesto) charged US$2.50 per ha for an entitlement of 12 000 m^3 ha^{-1} (Moore and Howitt, 1988).

In the Riverine Plain of Northern Victoria, gravity surface water currently costs A$12.10 per 1000 m^3 per year and is rising at the rate of inflation plus 2 per cent per year towards the full cost of A$22.50 at 1989 prices (Evans and Nolan, 1989). In the three irrigation districts of Robinvale, Red Cliffs and Merbein, supply rates for pumped water per 1000 m^3 per year are A$56.11, A$51.32 and A$44.16 respectively (Table 28). These irrigation districts operate a two-tier tariff structure. The base tier requires irrigators to pay for a set water entitlement whether they use it or not. The second tier involves a volumetric charge for extra water used.

Table 28: The major features of the three irrigation districts in Sunraysia region of Victoria.

Irrigation district	Area irrigated (ha)	Average annual irrigation diversion 1985/86–1990/91 (10^6 m^3)	Basic water entitlement (m^3 ha^{-1})	Supply rate (A$ per 1000 m^3)	Drainage rate (A$ per 1000 m^3)
Robinvale	2400	20.42	7600	56.11	4.80
Red Cliffs	2510	41.73	9100	51.32	5.38
Merbein	3520	32.75	9100	44.16	3.30

Source: Sunraysia Community Salinity Working Group (1991).

In the Mildura irrigation district, which has an irrigated area of 3520 ha, the First Mildura Irrigation Trust (FMIT) has a three-tiered rating structure (Table 29). There is a fixed maintenance rate for each hectare of land with water rights. There is also a fixed irrigation rate. The irrigation rate entitles each irrigator to apply up to four furrow irrigations. The third tier is an excess rate for each extra furrow irrigation. In the area serviced by FMIT, water is not metered and an application of 150 mm is assumed per irrigation for each hectare watered. The 1991/92 FMIT irrigation charges are shown in Table 29.

Table 29: The First Mildura Irrigation Trust charges for irrigation water in 1991/92.

	Rate (A$ ha^{-1})	Approximate rate (A$ per 1000 m^3)
Maintenance rate	71.00	11.61
Irrigation rate	281.00	45.97
Drainage rate	24.00	3.93
Total	352.00	61.51
Excess rate	24.00	15.7

Source: Sunraysia Community Salinity Working Group (1991).

The existing tariff structures for both metered districts (Robinvale, Red Cliffs and Merbein) and the unmetered district (Mildura) do not reward efficient irrigation, nor do they discourage overirrigation. For this reason the tariff structure is under review; ideally irrigators should pay only for the water they use (Sunraysia Community Salinity Working Group, 1991). This pricing policy is expected to increase irrigation efficiency and to alleviate salinity problems.

Transferable water entitlement

In many countries the days of cheap water supply development have passed. It is now considered vital that existing water supplies be allocated more efficiently. There is a need for both short-term and long-term flexibility in water supply. Water should be easily shifted from use to use (e.g. among agricultural users and between the agricultural sector, municipalities and industries) and from place to place as economic, demographic, climatic and other conditions change over time. Exchange through markets is one way of allocating water. In many settings this method possesses a number of desirable attributes (e.g. flexibility, security, predictability, fairness) to a greater degree than alternative allocation mechanisms. Markets have shortcomings, too, especially related to quantity and quality return flow effects, but these can be mitigated through changes in the administrative framework of the water rights system (Howe et al., 1986; Howe, 1990).

The Transferable Water Entitlement (TWE) is a policy adopted in areas of the western United States and in some states of Australia to allow irrigation water to be allocated in a more economically efficient way. It is a mechanism by which a market for water can operate by allowing entitlements to be bought and sold without the necessity of buying and selling accompanying land. With the introduction of TWE, water use will tend to change in the following ways (Western Australian Water Resources Council, 1989): transfer of water to higher value uses; decreased use on land which is poorly suited to irrigation (e.g. waterlogged and salinised soils); and increased adoption of water-saving technologies because water saved can be sold.

Transferable water entitlement is a means of more flexible water allocation which involves a reduced government role and increased decision making by individual users. Individuals are free not to participate in the market. They will participate only if they judge themselves to be better off buying or selling water at the market price.

According to Anderson (1990), the United States has a long history of water transfer. Some of the reasons and facts are as follows:

- Ten per cent of irrigated land receives only government supplied water. Another 11 per cent gets some supplemented water supply from government projects. The remaining 79 per cent receives privately developed water.
- Water right is obtained by appropriating a quantity of water from a stream and having the right validated by the court. Prior appropriation means first in time, first in right. This is a use right that can be traded.
- Non-government irrigation systems are operated by farmer irrigators as mutual irrigation companies or irrigation districts.
- In most western states of the United States, water is not attached to land, with the exceptions of Wyoming and Arizona.
- Seasonal water rental, water exchanges and water rights sales are of long standing, particularly in Colorado, California, and other states.
- Agriculture to agriculture transfers have occurred for a long time. Now transfers are from agriculture to municipal-industrial use. This can dry up irrigated areas and can cause local community disruption.

Saliba (1987) describes five markets for transferring water entitlements (rights) in Arizona, Colorado, Nevada, New Mexico and Utah and evaluates market processes using transactions data from each study area. This study shows that water markets are functioning throughout the western United States where economic incentives for water transfers outweigh costs associated with market transactions. Examination of market prices and transfer patterns in five areas suggests that markets are working relatively well in allocating water between agricultural, municipal and industrial uses.

In Australia, traditionally water entitlements have been tied to a specific parcel of land. Until recently the only way in which water could be readily transferred from one area or one purpose to another was by purchase of the land to which a water entitlement was attached. Such an approach, although appropriate during earlier stages of resource development in Australia, imposes constraints on the efficient use of water and on economic growth. Since the early 1980s, several water management agencies in Australia have considered the adoption of more flexible arrangements of transfer of water (e.g. South Australia since 1982/83, New South Wales since 1983, Queensland since 1987/88, Victoria since 1987/88 and Tasmania since 1989/90). In most states transferability of water entitlements was tried on a one-year basis (Delforce et al., 1990).

With respect to the management of salinity problems, TWE allows water to be transferred from a saline environment of a basin to a non-saline part of the basin. This mechanism has the advantage of alleviating the salinity problem in one part of a basin and permitting the use of water for agricultural production or other uses in another part of the basin. According to Delforce et al. (1990), for landholders wishing to move out of irrigation, but to remain in dryland agriculture, TWE allows the water to be sold separately from the land. Prior to TWE, the options for such irrigators were either to cancel their water rights licence (or not renew) and receive nothing for the right, or sell the entire irrigation holding and buy a dryland property elsewhere. Further information concerning different aspects of TWE in the United States and Australia is available in Pigram and Hooper (1990).

Integrated and total catchment management

Catchments are naturally occurring units of the landscape which contain a complex array of interlinked and interdependent resources and activities, irrespective of political boundaries. A catchment is a dynamic and integrated biophysical, social and economic system containing natural resources (land, water, flora and fauna), agriculture, industry, communities, services and recreational facilities. Since a catchment is a logical planning unit with readily identifiable boundaries and characteristics, its resources are ideally suited to coordinated planning and management (Irwin and Williams, 1986).

According to Laut and Taplin (1989: 4–5) it is commonly accepted that catchment management should be a holistic activity that should involve consideration of all aspects of biophysical and socioeconomic environments which affect the catchment and its use. This view has produced two management concepts:

- The integrated catchment management (ICM) is the integration of water and land management activities and of government agencies involved in these activities within a catchment.
- The concept of total catchment management (TCM) implies consideration of all resources and their uses within a catchment. Total catchment management involves the coordinated use and management of land, water, vegetation and other physical resources and activities within a catchment to ensure minimal degradation and erosion of soils and minimal impact on water yield and quality and on other features of the environment (Cunningham, 1986).

It could be said that ICM involves only those parts of a catchment management system which are directly concerned with achieving a given water supply or water use end, whilst TCM involves all aspects of resource use and production to achieve minimum adverse impact on the catchments. In Australia, since the early 1980s an emerging enthusiasm for the concept of integrated management of water and land resources on a catchment-wide basis has become evident. This can be attributed to a number of causes (Burton, 1992).

Firstly, there has been an increasing concern for the management of water quality on a river basin scale, triggered largely by the river salinity problems which have developed in the south-west of Western Australia and the Murray-Darling Basin. In addition there has been rapidly increasing evidence of dryland salinity as a major catchment management problem in Western Australia, Victoria and more recently along the western slopes of the Great Dividing Range in New South Wales and Queensland.

Secondly, there has been an increasing concern from state and federal governments for a variety of catchment-based problems, and the need for an integrated approach to such problems. The latter has been evident through the introduction of a range of policy documents which have included the National Conservation Strategy, the various state conservation policies, National Soil Conservation Program, the National Tree Program and the Decade of Landcare Program. Of the greatest significance is the cooperative federal-state project which has come to be known as the Murray-Darling Basin Initiative, which is specifically concerned with integrated natural resources and catchment management on a scale unprecedented in Australia and perhaps elsewhere in the world (see also Landsberg, 1992). Some of the Australian states' commitments to catchment management are as follows:

In New South Wales (NSW), the Total Catchment Management Policy was announced in 1984 (Burton, 1992). It is implemented through Total Catchment Management Committees (TCMCs), whose principal task is to ensure an integrated approach by state government agencies to problems associated with catchment management. At present, TCMCs appear to operate at two levels: firstly, they react to public issues of catchment management by finding the agencies which can respond to the issue; and secondly, they raise state agencies' awareness of the implications of their activities for catchment management. The adoption of TCM as a policy has been politically attractive in NSW. As yet its implementation is considerably short of the concept of total catchment management (Laut and Taplin, 1989).

During 1987 a milestone in integrated catchment management development was put in place by the Victorian Government. This was the adoption of the state-wide 'Salt Action: Joint Action' program aimed at managing the salinity problems which have affected water and land resources throughout the state. The Victorian 'Salt Action: Joint Action' strategy is based solidly on an integrated catchment management approach (Burton, 1992; Hunter and David, 1992).

In Western Australia, the government recognised a need for a coordinated approach to catchment management and in 1987 adopted the policy of implementing an integrated catchment management approach. The objective of the state strategy for integrated catchment management is to achieve land and water allocation, use and management which are compatible in relation to long-term community interests, resource suitabilities, reducing unwanted impacts of land and water use, and matters of equity and community values. The first application of the ICM approach to salt-affected catchments has been developed for the Denmark River catchment. The Denmark River catchment is a major potential water resource for the southern coastal region of south-west Western Australia. Average stream salinities were less than 600 mg L^{-1} TDS during the 1970s, but they have increased to average 890 mg L^{-1} TDS over the period 1982–86 (see Table 2-6 and Figure 2-7). Clearing of the native forest for agriculture has greatly increased stream salinity in this region. In addition, significant areas of salinised and waterlogged soils exist in the catchment. The project aim is an integrated catchment strategy to enhance farm productivity and profitability and to reduce the salinity of the Denmark River (Schofield et al., 1989).

In Queensland, the state government is now planning to adopt a state-wide integrated catchment strategy. A document detailing this strategy has recently been released for general public comment. The implementation of the new catchment management policy is likely to be facilitated by the fact that the state government agencies most likely to play a major role in catchment management are now all part of the Department of Primary Industries, which will have the overall responsibility for the coordination and implementation of the strategy (Burton, 1992).

Reclamation of sodic soils

Reclamation of sodic soils requires the replacement of exchangeable sodium by calcium ions. This can be accomplished in many ways, depending on local conditions, available resources and the kinds of crops to be grown on the reclaimed soils (Abrol et al., 1988). Sodic soils can be reclaimed at a slow rate by the application of ample irrigation water coupled with good farming practices. However, the reclamation process may be accelerated through the application of various chemicals. Foth (1984: 118–119) and Abrol et al. (1988: 47–65) describe the chemical amendments suitable for the reclamation of sodic soils. These chemicals consist of gypsum ($CaSO_4$), calcium chloride ($CaCl_2$), iron sulphate ($FeSO_4$), aluminium sulphate ($Al_2(SO_4)_3$), sulphuric acid (H_2SO_4), sulphur(S) and pyrite (FeS_2). The suitability of each amendment depends on the nature of the soil and cost considerations. The following paragraphs are based on Abrol et al. (1988) unless other references are cited. However, further information relating to this issue is available in Keren and Miyamoto (1990).

The chemical reaction between the amendment and soils leads to formation of the leachable sodium components such as sodium sulphate (Na_2SO_4) and sodium chloride (NaCl). The choice of an amendment depends on its effectiveness, relative cost and the time required for the amendment to react with the soil. Calcium chloride, iron sulphate and aluminium sulphate are usually expensive and have not been used for large-scale improvement of sodic soils. Amendments like sulphur and pyrite must first be oxidised to sulphuric acid by soil micro-organisms before being available for reaction. Therefore they act slowly. Sulphuric acid has been used extensively in some parts of the world, particularly in the western United States and parts of the Commonwealth of Independent States.

Gypsum is the most widely used amendment because it is the cheapest and the most abundantly available. It comes both from mines and as a by-product of the phosphate fertiliser industry. Gypsum is soluble in water to the extent of about 0.25 per cent at 25°C and is a source of soluble calcium. However, according to Keren and Miyamoto (1990), under comparable conditions the rate of dissolution of industrial gypsum is higher than that of mined gypsum. Gypsum reacts with both sodium carbonate (Na_2CO_3) and absorbed sodium and produces sodium sulphate (Na_2SO_4), which is leachable.

The gypsum requirement is the amount of gypsum required to reduce the sodium saturation to some acceptable level for a given quantity of soil. According to Foth (1984: 219), for each milliquivalent of exchangeable sodium per 100 g of soil, 0.086 g of gypsum is required. Keren and Miyamoto (1990) describe a mathematical formulation for the calculation of gypsum requirements which takes into account the initial and the desired final exchangeable sodium percentage, soil bulk density, soil exchange capacity and the depth of soil to be reclaimed.

Mined gypsum requires grinding before application in sodic soil reclamation. The fineness to which gypsum must be ground is an economic consideration. Very fine grinding entails higher cost. Although it is often maintained that finer gypsum particles are more effective, Abrol et al. (1988) demonstrate that gypsum passed through a 2 mm sieve and with a wide particle size distribution is likely to be more efficient for the reclamation of sodic soils having appreciable quantities of sodium carbonate.

Amendments like gypsum are normally applied on the soil surface and then incorporated with the soil by disking or ploughing. Gypsum mixed with the surface 15 cm of soil is more effective in the removal of exchangeable sodium than gypsum applied on the soil surface.

Mathematical models for salinity management

Mathematical models which describe and quantify basic hydrological processes and phenomena under a range of conditions are very useful in salinity investigation, particularly in the assessment of the efficiency of remedial and preventive measures. Jakeman et al. (1987) delineate the types of models required and discuss some problems in their construction and calibration.

Groundwater models are useful if a management strategy is required to consider the effects of rainfall, irrigation, cropping activity, groundwater pumping and land use behaviour on groundwater levels and on land and stream salinity.

Hydrologic routing and surface water quality models can be used to predict downstream salinity concentrations from upstream data, for management strategies such as salinity dilution by release of additional discharge from upstream storages; to provide advance warning of salinity levels which are too high for irrigation usage; and for quantifying saline accessions within reach of a river.

Other models include steady-state soil salinity models to predict salinity in the root zone from irrigation application and associated water salinity (Oster and Rhoades, 1990); models which simulate the dynamics of salt movement (accumulation and leaching) in the soil profile and plant response to salinity (Hutson et al., 1990); models which simulate irrigation return flow and its salinity (Aragües et al., 1990); and models which simulate solute transport in the saturated zone (Konikow and Bredehoeft, 1978; Konikow, 1981; Huyakorn et al., 1986).

The success of model development and subsequent simulation depends upon a number of interrelated factors, including the objectives of the modelling exercise; the complexity of variables dominantly controlling the behaviour of the system and the level of understanding and knowledge of system structure; the mathematical properties of the derived model; and the quantity and quality of data available and the modelling approach taken.

The complexity of model development for particular cases and the lack of required field data are major reasons for some models being used more frequently than others. For example, groundwater, stream routing and surface water quality models are used more than complex groundwater solute transport models and unsaturated flow and transport models.

In this section two examples relating to the application of groundwater and stream salinity models will be described. They illustrate how relevant simplifying assumptions must be made to deal with the problems associated with salinity modelling. Further examples are available in DeCoursey (1990) and Ghassemi et al. (1991).

Ghassemi et al. (1987 and 1989) have simulated the Mildura aquifer located on the southern bank of the River Murray in Australia. The objectives of their study were to quantify the salt load from the aquifer to the river, evaluate the effectiveness of the groundwater interception scheme and provide options to improve its effectiveness. As shown in Figure 6, the interception scheme consists of 17 pump sites or well fields. The groundwater generally flows from the south toward the river and groundwater salinity ranges from 1200 to 84 000 mg L^{-1} TDS. Saline effluents from the scheme are disposed first to Lake Ranfurly, then to the evaporation basins, 13 km from Mildura.

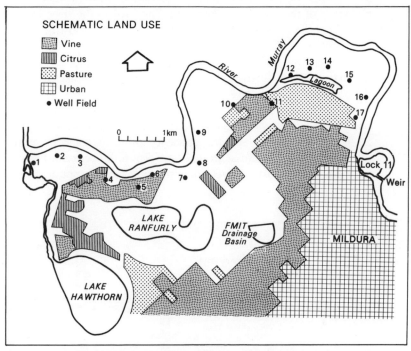

Figure 6: Schematic land use map of the Mildura area and the location of 17 interception well fields (Ghassemi et al., 1989).

The flow of water in the aquifer was simulated with steady-state and transient models using a finite-difference approximation of the two-dimensional partial differential equation for horizontal groundwater flow in a confined aquifer subject to recharge and pumping:

$$\frac{\partial}{\partial x}\left[T\frac{\partial h}{\partial x}\right] + \frac{\partial}{\partial y}\left[T\frac{\partial h}{\partial y}\right] = S(x,y)\frac{\partial h}{\partial t} + Q(x,y,t) \qquad (1)$$

Here the transmissivity T and storativity (storage coefficient) S represent physical aquifer properties in space, h is the hydraulic head, Q is the source/sink term, and (x,y) represent a cartesian coordinate system. Note that the temporal derivative $\frac{\partial h}{\partial t}$ is zero in the steady-state case.

For both transient and steady-state models, saline water flow from the aquifer to the river was estimated by assuming constant salinity in the aquifer. Salt fluxes to the river were obtained by multiplying flow rates computed at nodes along the river by corresponding groundwater salinity values at those nodes. This circumvents the tremendous difficulties and uncertainties associated with the development of a subsurface solute transport model. This transient model simulation computes the nodal discharge and salt flow at each time step, and their average values for the whole period of simulation. The results of the mathematical simulation showed that without the operation of the pumping scheme, the salt load to the river would be 125 tonnes per day, while the average salt load during the period of simulation from 1980 to 1983 was about 65 tonnes per day.

Using the transient model simulations, the salt load from the aquifer to the river under different operation strategies was estimated. By increasing the pumping rate from 64 to 107 L s^{-1} in existing pump sites, the salt load would decrease to 36 tonnes per day. Finally, by inclusion of three additional pump sites in the scheme (sites 18, 19 and 20), salt load would decrease to 24.4 tonnes per day and the efficiency of the scheme in terms of salt interception would increase from 49 per cent to 80 per cent. Figure 7 shows the nodal salt load to the river under this management option.

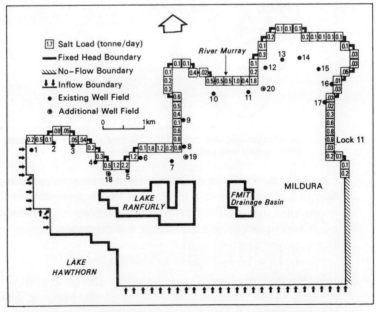

Figure 7: Average nodal salt load from aquifer to the river with inclusion of three additional well fields 18, 19 and 20 (Ghassemi et al., 1989).

Dietrich et al. (1989) and Jakeman et al. (1989) have developed a dynamic model for the transport of a conservative solute along a stream connected to an aquifer contaminated by the solute. It relates downstream concentration to upstream concentration, stream discharge and piezometric levels in the aquifer. Also, submodels for solute travel time in the stream and aquifer inflow to the stream have been developed. They then applied the model to a reach of the River Murray.

Two objectives that the model was to achieve in the presence of scarce data were firstly to provide a daily estimate of stream salinity at a fixed downstream location during periods of low to medium flows in response to upstream flows and upstream salt concentration as well as to lateral inflows from salty aquifers, and secondly to quantify the salt load discharged from the aquifer into the stream in response to stream flow levels.

The first objective stems from the requirement to estimate downstream levels of salt concentration which if too high would be damaging to irrigation activities. Such events usually occur during periods of low to medium flows since in such instances dilution is low and groundwater gradients towards the stream can be high. The second objective results from the need to assess management strategies aimed at reducing the influx of aquifer salinity into the stream.

One approach to achieve these objectives is to model the salt transport using the advection-diffusion equation:

$$\frac{\partial c}{\partial t} + u \frac{\partial c}{\partial x} = \frac{\partial}{\partial x}\left(D \frac{\partial c}{\partial x}\right) + r + r_{gw} \qquad (2)$$

where c is the stream salinity, u is the averaged cross-sectional advective velocity, D is a diffusion coefficient, r_{gw} denotes the source of salt coming from the groundwater, and r represents all sources or sinks of salt other than r_{gw}.

Solution of the latter equation requires estimates of the quantities u, D, r and r_{gw}. It may be relatively easy to make measurements of D and r, and the advective velocity u could be obtained independently from physical representations of stream flow such as the Saint Venant equations. However, the salt inflow r_{gw} is likely to be very difficult to estimate and solution of the Saint Venant equations is not straightforward as they are non-linear and of hyperbolic type. In addition to these problems, their observations contain measurement and sampling errors and possible missing values, where the time series may be long but the spatial distribution is sparse.

In view of these limitations, Jakeman et al. (1989) made some simplifying assumptions, transforming the governing equation from its differential form to a more global representation that attempts to retain enough physically-based structure to be sensitive to relevant variables required by the modelling objectives. The major assumptions were that the solute concentration in the aquifer is high in comparison with that of the stream, that advection dominates diffusion in the stream, that aquifer properties do not vary along the stream, that there is a single-valued relationship between wetted cross-sectional areas of the stream and stream discharge, and that sources or sinks of water along the river are small enough not to affect significantly the downstream propagation of river flow.

The developed models have been applied to a 207 km reach of the River Murray between Euston and Red Cliffs. This stretch is characterised by a significant amount of salt intrusion from adjacent aquifers. The location of the study area and other geographical features associated with the River Murray and its tributaries are shown in Figure 8.

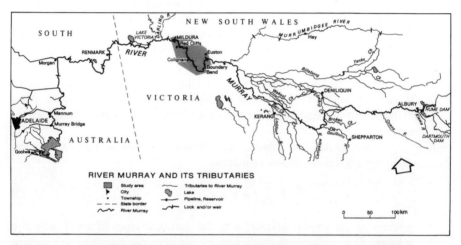

Figure 8: Map of the River Murray and its tributaries showing the study area (shaded) between Euston and Red Cliffs (Jakeman et al., 1989).

For the study reach, the following hydrological data were available: stream discharge at Euston and Colignan, upper pool level at Mildura weir, and river salinity at Euston and Red Cliffs. Stream stage height at Colignan and upper pool level at Mildura weir are instantaneous measurements collected daily. Salinity measurements are instantaneous (no continuous recordings were available) but not always available on a daily basis. Geologic evidence indicates that throughout the study area the shallow aquifer is hydraulically connected to the River Murray.

Figure 9 compares the historical and predicted salinity profile at Red Cliffs. The predictions are regarded as quite accurate given the level of error in the data. They also estimated the steady-state load per unit stage height to be 292 tonnes per day.

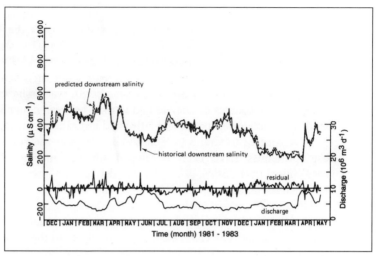

Figure 9: Historical (continuous) versus predicted (dash) salinity profiles at Red Cliffs (Jakeman et al., 1989).

Salinity and conflict resolution

Water sharing and water quality including salinity is often the source of conflicts and disputes at various levels, from disputes between neighbouring farmers and landowners, to disputes between states within a federation, to disputes between independent nations. According to UNEP (1992a), on a global scale 214 river basins are shared by more than one country: 155 of these are shared between two countries; 36 among three countries; and the remaining 23 among four to 12 countries. About 50 countries have more than 75 per cent of their total land area falling within international river basins, and an estimated 35–40 per cent of the world population lives in international basins.

Generally water users in the upper parts of a basin have the privilege of enjoying higher quality water. Along the river, water quantity diminishes and its quality deteriorates due to water diversion and discharge of reused and drainage waters to the river systems. Water quality degradation may reach such a level that it could cause major problems for the water users in the lower part of the basin. This type of problem exists in river basins of many countries and some of the river basins shared by more than one jurisdiction. In Australia, the Murray-Darling Basin is shared between the four states of Queensland, New South Wales, Victoria and South Australia (see Chapter 2). The Colorado River Basin is another example where seven states (Wyoming, Colorado, Utah, New Mexico, Arizona, Nevada and California) share the basin, and outflow of the river from the United States enters Mexico (see Chapter 11). In the Commonwealth of Independent States, the Aral Sea Basin with its two major rivers (Amu Darya and Syr Darya) is shared between Kazakhstan, Uzbekistan, Turkmenistan, Tadjikistan and Kyrgyzstan (see Chapter 4). In these examples and many others not cited here (see Clarke, 1991, Chapter 7; Bulloch and Darwish, 1993), water sharing and water quality are the sources of dispute and tension between countries, states and jurisdictions requiring special legislative and institutional arrangements.

According to Crabb (1988a and 1991), over the years a number of basic principles have been developed for the management of inter-jurisdictional river basins, principles that derive from international situations but which are no less relevant to intra-national situations. These principles are as follows: 'The river basin is the basic hydrological management unit: acceptance of this principle has implications for virtually all of the other principles; no state can claim exclusive sovereignty to the waters of an inter-state river basin; reasonable and equitable participation in the management and control of the resource; reasonable and equitable apportionment of the water resources of the basin; protection and non-abuse of the resources of the basin; acceptance of the interrelationships of natural resources, especially water and land and surface and groundwater' (Crabb, 1991). 'Nevertheless whilst these principles have gained wide acceptance, finding the right institutional arrangements by which to put them into practice has not been easy' (Crabb, 1988a).

The following sections describe the adopted arrangements to resolve the conflicts for the management of the Colorado River Basin in the United States and the Murray-Darling Basin in Australia.

Colorado River Basin

Although the Colorado River is neither the biggest nor the longest river in the world, it is the most legislated, most debated and the most litigated in the entire world. The long history of negotiations, disputes, court actions and even military expedition for the settlement of water-sharing disputes between the seven states is documented by Reisner (1987) and Hundley (1975). Here, the conflict resolution between the United States and Mexico regarding water sharing and its salinity will be described, mainly based on Bureau of Reclamation (1992).

In 1922, negotiation of the Colorado River Compact took place in New Mexico. The delegations from the seven states arbitrarily divided the river basin into the Upper and Lower Basin (see Figure 11-7). They also divided the 21.58×10^9 m³ average annual flow of the river as follows: 9.25×10^9 m³ yr^{-1} for each of the Upper and the Lower Basins; 1.85×10^9 m³ yr^{-1} were reserved for Mexico; and 1.23×10^9 m³ yr^{-1} were apportioned as a bonus to the Lower Basin. The compact was signed by the delegates in November 1922, but was not ratified by the legislators of their respective states (Reisner, 1987).

According to the Bureau of Reclamation (1992), in 1944, the United States and Mexico signed a treaty requiring the United States to deliver 1.85×10^9 m^3 of Colorado River water to Mexico annually. The treaty did not address the salinity of the delivered water. In 1961, the salinity of the water delivered to Mexico increased sharply, from a range of about 700–900 mg L^{-1} to about 1340 mg L^{-1} TDS. In some months, the salinity exceeded 2500 mg L^{-1}. The increased salinity was caused by: the discharge to the Colorado River of saline irrigation drainage pumped from newly constructed wells in the Wellton-Mohawk Irrigation and Drainage District (WMIDD) in Arizona; a reduction in Colorado River flows to Mexico, resulting from the construction and closure of Glen Canyon Dam; and construction of Painted Rock Dam, which significantly reduced the less saline Gila River flood-flows to the Colorado River below Imperial Dam. In November 1961, Mexico filed a formal protest with the United States, claiming that the increased salinity was damaging crops in the Mexicali Valley, a violation of international law.

The United States and Mexico pursued a series of temporary solutions to the salinity problem throughout the 1960s and early 1970s. Then, in 1972, President Nixon appointed Herbert Brownell (a former US Attorney General) to study and recommend a permanent, definitive and just solution to the problem. Among other things, Brownell recommended constructing a desalting plant to treat WMIDD irrigation drainage and allowing for a specified differential between the salinity of the waters arriving at Imperial Dam, near Yuma, Arizona, and the salinity of the waters delivered to Mexico.

The Presidents of the United States and Mexico approved Brownell's recommendations in the form of Minute No. 242. The most important provision requires that the average annual salinity of the Colorado River water (approximately 1.68×10^9 m^3) delivered upstream from Morelos Dam would not exceed the average annual salinity of the water arriving at Imperial Dam by more than 115 mg L^{-1}, plus or minus 30 mg L^{-1}. This value is known as the salinity differential.

The United States made several commitments to secure the Basin States' support for Minute No. 242. Among these were that the Basin States would not bear the cost of fulfilling any agreement with Mexico and that the United States would recognise replacement of the water lost to the desalting plant reject stream and any bypassed WMIDD irrigation drainage as a national obligation.

Public Law 93-320 was passed on 24 June 1974, to enable the United States to comply with its obligations under Minute No. 242 without depriving the Basin States of any of their apportioned water (Title I) and to authorise projects to control the salinity of the water delivered to users in the United States (Title II). Major Title I works are the Yuma Desalting Plant, the

WMIDD irrigation drainage reduction program, concrete lining of the first 79 km of the Coachella Canal in California, and a well field in Arizona known as the Protective and Regulatory Pumping Unit (Figure 10).

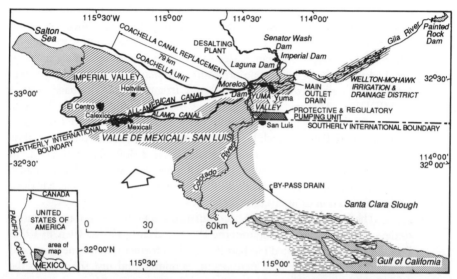

Figure 10: Colorado River Basin Salinity Control Project, Title I Division (Bureau of Reclamation, 1992).

The Yuma Desalting Plant (YDP) uses a reverse osmosis technology and will recover the majority of WMIDD irrigation return flow for deliveries to Mexico (see the section on desalination, p. 68). The WWIDD irrigation drainage reduction program has reduced WMIDD irrigation drainage pumping and, thus, the required size of the YDP by improving irrigation efficiencies and reducing irrigated acreage. Before the program was initiated, irrigation efficiency was about 56 per cent. While the program was active, irrigation efficiency met and exceeded the target efficiency of 72 per cent. Irrigation drainage dropped from 271×10^6 m^3 to a low of 146×10^6 m^3 per year, and irrigated lands were reduced by 4000 ha.

Concrete lining of the first 79 km of the Coachella Canal was completed in 1982. Since then, water conserved as a result of lining the canal (an estimated 163×10^6 m^3 per year) has been credited toward the replacement of WMIDD irrigation drainage bypassed to comply with the salinity differential.

The Protective and Regulatory Pumping Unit (PRPU) was constructed to manage and conserve groundwater for the United States and for delivery to Mexico. By June 1992, 21 of the 35 planned wells and associated structures had been completed. With 35 wells, the PRPU would be capable of producing about 197×10^6 m^3 of water per year. Ultimately, approximately 154×10^6 m^3 of water from the PRPU, combined with 18.5×10^6 m^3 of water from wastewaters and drains in Yuma Valley, would furnish 172.5×10^6 m^3 of Mexico's total 1.85×10^9 m^3 annual entitlement. The water will be delivered at the Southerly International Boundary near San Luis, Arizona.

Murray-Darling Basin

Prior to Australian Federation in 1901, the River Murray was the source of considerable conflict between the colonies of New South Wales, Victoria and South Australia. The fact that much of the boundary between New South Wales and Victoria was located on the top of the bank on the Victorian side of the river only added to the difficulties. Each of the colonies pursued its own settlement and development, without any regard to what was happening in the others. In particular, each tried to gain maximum access to the very important river trade, the Murray-Darling system being the major means of transport in the pre-railway period. As a result, the initial concerns about the river focused particularly on navigation, concerns over water for irrigation coming somewhat later.

The disputes continued well after Federation, as did the conferences, inquiries and Royal Commissions concerned with the Murray. After many years of negotiations, however, the River Murray Waters Agreement (RMWA) was signed in 1914 by the Federal Government of Australia, state governments of New South Wales, South Australia and Victoria. In many respects, it was a significant achievement (Crabb, 1988a). However, it was only signed 'because the Commonwealth could not be induced to use its superior legislative power and because other initiatives to effect a limited peace between the combatants had failed. Indeed, the Agreement always has been a fragile and limited peace treaty which achieved only very narrow goals' (Clark, 1982). What was perhaps the most significant conclusion of the 1902 Inter-state Royal Commission on the River Murray, namely that 'the river and its tributaries must be looked on as one' (quoted in Clark, 1983), was ignored and the Agreement limited to the main stream of the Murray. In terms of functions, it was simply 'a formula for sharing the available water in the Murray between the three states, guaranteeing certain minimum flows to South Australia, and confirming the rights of New South Wales and Victoria to use water in their tributaries' (Clark, 1982). 'To implement the Agreement and carry out the required works program, the River Murray Commission (RMC) was established, though it did not come into being until 1917. Over the years, the very limited powers of the Commission were gradually extended, both by amendment and informal practice (Table 30). However, even these limited changes often involved bitter disputes, both between and within states, with that over the Dartmouth and Chowilla dams in the late 1960s–early 1970s one of the worst' (Crabb, 1988b).

Table 30: Changes to the River Murray Waters Agreement, 1914 to 1981.

Matters beyond the powers of the River Murray Commission in 1914:
　　Problems arising on tributary rivers
　　Problems caused by adjacent land-use
　　Problems of flood mitigation and protection
　　Problems of erosion and catchment protection
　　Problems of water quality and pollution from agricultural and other sources
　　Problems of influent and defluent waters
　　The needs of flora and fauna
　　Possible recreation, urban or industrial use
　　The environmental or aesthetic consequences of particular proposals

Matters permitted by previous amendments and informal practice before 1976:
　　Limited powers of catchment protection
　　Power to initiate future proposals
　　Provision of certain dilution flows to maintain water quality
　　Lock maintenance work, improving navigability
　　Provision of recreational facilities
　　Expenditure on salinity investigations
　　Expenditure on redesigned works to protect fish life
　　Construction and operation of storages on tributaries

Principal innovations in agreement reached in 1981:
　　Power to consider any or all relevant water management objectives, including water quality, in the investigations, planning and operation of works
　　Power to monitor water quality
　　Power to coordinate studies concerning water quality in the River Murray
　　Power to recommend water quality standards for adoption by the states
　　Power to make recommendations to any government agency or tribunal on any matter which may affect the quantity or quality of River Murray waters
　　Power to make representations to any government agency concerning any proposal which may significantly affect the flow, use, control or quality of River Murray waters
　　Power to recommend future changes to the Agreement

Source: Clark (1982).

In spite of the growing seriousness of water quality problems in the Murray through the 1960s and 1970s and the numerous studies that called for urgent action, conflict continued between the signatories to the RMWA. However, in 1976, interim agreement was reached that permitted the RMC to take water quality matters into account in its activities, though the formal amendements to the RMWA were not agreed to until 1981 (Crabb, 1988b).

> Through the early 1980s, the continuing deterioration of the environment of the Murray-Darling Basin (particularly problems of widespread land degradation and salinity), the growing community awareness of the problems, their severity and consequences, and the inability of the River Murray Waters Agreement and the River Murray Commission to tackle the problems (though through no fault of their

own), brought renewed calls for action. The widespread concerns and pressures reached a point where they could be ignored no longer by the government concerned. In late 1985, there was a meeting of ministers responsible for water, land and environmental resources from the governments of New South Wales, Victoria, South Australia and the Commonwealth (Australia). For various reasons, Queensland limited itself to observer status at some of the early discussions. The 1985 meeting was followed by two years of countless meetings, discussions and negotiations involving the ministers and bureaucrats from the four governments. The outcome was the Murray-Darling Basin Agreegment signed in October 1987 (Crabb, 1991).

The Agreement, in the form of an amendement to the RMWA, passed through the Commonwealth and States' parliaments and came into effect on 1 January 1988.

The original Agreement and the RMWA have since been replaced by a totally revised Murray-Darling Basin Agreement, which came into operation on 1 July 1992. The prime objective of the Agreement remains the same, 'to promote and co-ordinate the equitable, efficient and sustainable use of the water, land and other environmental resources of the Murray-Darling Basin' (Murray-Darling Basin Ministerial Council, 1992: C1.1). To achieve this objective, the Agreement established political, bureaucratic and community participation institutions, namely the Murray-Darling Basin Ministerial Council (MDBMC), the Murray-Darling Basin Commission (MDBC), and the MDBMC Community Advisory Committee (CAC) (Figure 11).

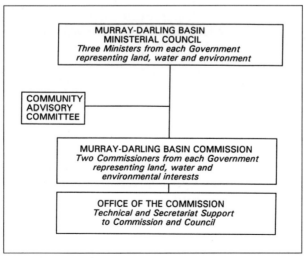

Figure 11: Management structure for the Murray-Darling Basin Initiative (Blackmore, 1991).

- The Ministerial Council is made up of three ministers from each of the signatory governments, representing the portfolios of water, land and the environment.
- The Commission is the executive arm of the Council and has two senior bureaucrats from each of the signatory governments (representing water, land and environmental resources) and an independent President, and a technical executive known as the Office of the Commission. The Office of the Commission carries out a number of statutory functions with regard to the operation of the Commission's water storages, the sharing of water between the three southern states, and the development of long-term resources management plans for the Council. In particular, it has 'to advise the Ministerial Council in relation to the planning, development and management of the water, land and other environmental resources of the Murray-Darling Basin' (Murray-Darling Basin Ministerial Council, 1992: C1. 17.(1)(a)). This is not a discretionary power, but a mandatory one.
- The Community Advisory Committee is made up of both regional and special interest group representatives drawn from across the Basin and beyond. Its role is to advise the Council on the effectiveness of policies and programs being developed and to act as a two-way means of communication between the Council and the residents of the Basin. It reports directly to the Council, but its role is limited to an advisory one.

As well as continuing the essential thrust of the first Agreement, the revised document marked a number of very important changes. Queensland is a signatory to the 1992 Agreement, though on somewhat different terms to the other states as it is not a party to the sharing of River Murray water. Thus the Agreement now covers the whole hydrologic basin.

The most critical of the problems that gave rise to the Murray-Darling Basin Initiative was salinity and it is not surprising that this was given the highest priority by the new management structure. A period of intensive negotiations resulted in the Salinity and Drainage Strategy, which was approved in 1989 (Murray-Darling Basin Ministerial Council, 1989b). A coordinated plan to tackle the problems of salinity, waterlogging and land salinisation, the Strategy has three objectives:

- to improve water quality in the River Murray (and the South Australian portion in particular) for all users;
- to control land degradation, prevent further land degradation and, where possible, rehabilitate land resources, especially in the Murray and Murrumbidgee valley irrigation areas; and
- to conserve the natural environments of these valleys.

A critical goal of the Strategy is to maintain River Murray salinity, as measured at Morgan in South Australia (Figure 2-13), below 800 µS cm^{-1} for 95 per cent of the time for a 20-year period from 1990 (Figure 12). To achieve this, a number of salt interception schemes are being constructed to reduce the flow of highly saline groundwater to the Murray. Further, in recognition of their contributions to the cost of the salt interception works, New South Wales and Victoria can each increase river salinity by 15 µS cm^{-1}. These 'salinity credits' will enable the states to drain at least some of their high value production irrigation areas to the Murray. However, given that the credits are very small in relation to the total demand for salt disposal, land management and drainage schemes for irrigation areas are also essential elements of the Strategy. Such schemes are being put into place in a number of the irrigation areas in New South Wales and Victoria. The overall outcome of the Strategy is that there will be a net 80 µS cm^{-1} reduction in salinity at Morgan.

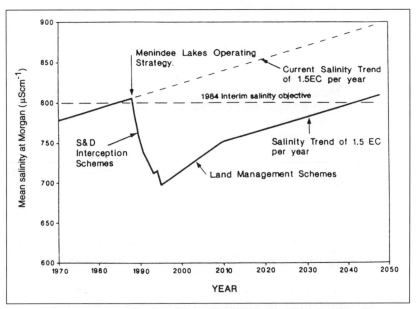

Figure 12: The current and the likely future salinity trends for the River Murray at Morgan and the impacts of the salt interception and drainage schemes and land use management (Engineering and Water Supply Department 1990).

The Salinity and Drainage Strategy was the first major action by the Murray-Darling Basin Ministerial Council because it dealt with the most difficult and contentious issues. However, it has to be seen as part of the Natural Resources Management Strategy (NRMS). Described as the 'cornerstone' of the Basin's management, the NRMS is not a document with a solution to every resource problem, but rather 'the framework for cooperative and coordinate Community and Government action to address the Basin's natural resource management problems on a long-term, integrated basis' (Murray-Darling Basin Ministerial Council 1990: 1). The problems identified by and the aims of the NRMS are set out in Table 31.

Table 31: Identified problems and aims of the Murray-Darling Basin Natural Resources Management Strategy.

Problems:
- Rising salinity level in soils and streams
- Deteriorating quality of water supplies
- Land degradation (e.g. soil erosion and acidification)
- Decline and loss of native vegetation
- Loss of native habitats
- Overcommitment of, and competing demands for, water supplies
- Cultural losses (e.g. Aboriginal heritage sites)

Aims:
- Prevent further degradation of natural resources
- Restore degraded resources
- Promote sustainable user practices
- Ensure appropriate resource use planning and management
- Ensure a long-term viable economic future for those dependent on the Basin
- Minimise adverse effects of resource use
- Ensure community and government cooperation
- Ensure self-maintaining populations of native species
- Preserve cultural heritage
- Conserve recreational values

Source: Murray-Darling Basin Ministerial Council (1990).

Another important achievement of the new management structure is a new method of sharing River Murray water. Agreement was reached on this in 1990 after many years of negotiating. 'Continuous accounting' provides a more equitable system of water sharing, a greater level of security for individual states and, at the same time, the flexibility to permit water trading between the states. This new method of water sharing forms part of the new Agreement, while the Salinity and Drainage Strategy is included as a Schedule to the MDBA.

The revised Agreement and the particular issues outlined above are the initial substantive achievements of the new management structure. They also indicate what is perhaps an even more important achievement, namely that the MDBA has established a process for decision making and conflict resolution on inter-jurisdictional management issues. This could be even more important in the future with the increasing pressures being placed on the Basin's limited natural resources.

The Murray-Darling Basin Agreement, in spite of its flaws, is, by world standards, a unique and innovative agreement, especially with respect to the institutional arrangements that have been put in place (Crabb, 1991). It is an agreement from which many other countries have much to learn.

Community involvement in salinity management

Controlling rising regional watertables and their subsequent salinisation of land and water resources cannot be achieved by individual landholders working in isolation taking local measures. Also, government action in the form of isolated management programs run by different agencies is not the most effective way of tackling the problem. Instead, governments should develop a coordinated plan of action with the participation of their agencies and the communities. Government support for community education and their participation in development, implementation, monitoring and cost-sharing of action plans is crucial for effectiveness and success of the management plans.

In Australia the concept of community participation in resource management is well established. Examples include the National Soil Conservation Program, the One Billion Trees Program and the Save the Bush Program (Hawke, 1989). In management of salinity problems and other natural resources of the Murray-Darling Basin (MDB), the Community Advisory Committee is one of three major management structures of the Basin (Figure 11). The Committee's role is, firstly, 'to provide advice to Council on community perceptions, identification of issues, expectations and proposed solutions to the management of the land, water and environmental resources of the Murray-Darling Basin' and secondly, 'to provide a two-way channel of communication between the council and the Murray-Darling Basin Community' (Burton, 1989).

The problems identified and given priority in the Natural Resources Management Strategy of the MDB include: rising salinity levels in soils and streams; deteriorating quality of water supplies; and land degradation. The strategy provides for the implementation of on-ground works and measures to be largely the responsibility of individuals and communities. It provides for the recognition or establishment of community groups. According to Blackmore (1991), the fundamentals for community involvement in the strategy are for the community to: identify local natural resources issues which need management/intervention; enlist government support funds to complement its own resources for its activities, through the funding program which supports the Natural Resources Management Strategy and any other sources: develop and implement action-based management plans for its locality; promote the adoption of improved management practices; and communicate to government its aspirations and concerns for the management of natural resources at the local, regional and basin-wide level.

Based on the experience of the State of Victoria where salinity management has a well coordinated structure, with community participation, Stone (1991) describes that genuine and effective community consultation and involvement requires as much careful planning and expert attention as any technical process. Stone (1991) provides the following step-by-step process leading to community planning with the Victorian salinity program:

- *Subregion identification:* the interdepartmental State Salinity Planning Working Group responds to a community or departmental request for a 'subregional' salinity plan to be developed in a particular area.
- *Departmental input:* the Interdepartmental Committee then nominates the lead agency. This is always the department with the greatest involvement in land and water use management in the area. The term 'lead' reflects a department's responsibility for resourcing, convening and acting as the secretariat for the Community Working Group, preparing and managing budget, employing personnel, and managing of the technical input. It does not mean it has the ownership of the plan. That belongs to the Working Group.
- *Funding identified:* at the outset, the lead agency applies for the necessary subregional planning resources through the annual state salinity budget process.
- *Election:* while the request for subregional planning status may come from the local community, it is usually the case that the lead government agency organises the local election of the Community Working Group.
- *Redefining of boundaries:* after its formation and election of a chairperson, the Community Working Group typically considers, then redefines, the original subregional boundaries roughly identified by the lead agency.
- *Membership:* the membership of current subregional working groups varies between those which have community and local government representatives only, to those which include state government officers who live and work in the area. Committee sizes range from 12 to 19 and committees include representatives from different parts of the subregion, plus representatives of social, environmental, commercial and local government interests.

Each subregional Community Working Group has the assistance of a team of government technical specialists which may include economists, agricultural scientists, engineers, hydrogeologists, environmental scientists and sociologists from government departments. The Community Working Group has the following responsibilities: identification of initial issues and determining the scope of the study; using research or study findings from the specialist support group or consultants to help formulate the plan; keeping abreast of progress being made by other adjoining subregional working groups; communicating progress to the Cabinet Committee; keeping the planning as close as possible to its agreed budget deadline for submission to the government; and presenting a final comprehensive, community-endorsed plan, along with implementation recommendations, to the Cabinet Committee.

In a comprehensive document, Wilkinson and Barr (1993) have evaluated different aspects of community involvement in salinity management in Victoria, where five years of experiences were available. These aspects include formation of the working groups, planning processes, the role of departmental officers, and community education. They have evaluated eight major projects and in their conclusions measured the performance of the Victorian Salinity Program against each of the points quoted by Syme and Eaton (1989). The points were derived from the work of the Institute for Participatory Planning (1981) as a list of characteristics of effective community planning. These points are: the process of public participation should be agreed upon between the agency and participants; public participation should start early in the decision-making process; the objectives of public participation need to be clearly stated and people need to be aware of the level of power being offered; efforts should be made by the agency to identify all interested parties; information should be available to all participants; participants should know how their submission will be processed; and adequate resources should be made available for the required tasks and meetings.

The community planning process explored by Wilkinson and Barr (1993) has produced a range of outcomes. Some successful plans have emerged with widespread community support and strong government support. Some plans were developed in response to government initiatives, rather than community pressure. In some cases the most technically feasible options for salinity control were shown to be uneconomic. A deliberate decision was therefore made to take no action for the time being. One plan was rejected by the local community because of the unwanted compulsory costs to the community.

Wilkinson and Barr's (1993) conclusions include the following points:
- Different community planning tools are needed for different planning scenarios.
- Planning could be improved if the likely planning scenario were known as early as possible. This would lead to more efficient use of government funds and would reduce the possibility of a community backlash.
- The challenge for the planners is to find out enough about each local community as early as possible to determine the likely nature of the plan.
- Nobody can provide recipes which will ensure successful and effective community participation. Different approaches are needed for different areas.
- The final decision on which methods are appropriate to maximise the chance of successful community-based planning rests with the people who will be organising the participation and conducting the consultation.

Finally, they concluded their report with a plea for realism in expectations and care with rhetoric. In their own words:

Early Salinity Program rhetoric implied community involvement would lead to community ownership of both the problem and the solutions, and increased adoption of salinity control measures. After five years, experience has shown these early hopes and expectations to be unrealistic. With hindsight, much of the early optimism was naive. This is not to imply that this grand experiment in widespread community planning was a failure. It should not be judged by the unrealistic expectations of earlier years. It should be judged by what is achieved on the ground. To date there is much to be proud of. Yet the final judgement will not be made in this report, but by the historians of following decades.

Geophysical methods for soil salinity mapping

Soil salinisation is a dynamic process due to the influence of a wide range of parameters including soil permeability, climatic and hydrogeological conditions and management practices. Soil salinity information is quickly out of date as management, watertable depth and climatic conditions change. Monitoring of soil salinity and preparation of soil salinity maps are essential tools for the management of salt-affected lands. However, the expenditure of time and effort to characterise the salinity conditions of a large area with conventional sampling and laboratory analysis procedures becomes prohibitive. It is thus obvious that a practical means of measuring soil salinity directly in the field is advantageous, if not essential, to obtain timely information required for making appropriate management decisions (Rhoades, 1990c).

Instruments for assessing bulk soil conductivity and soil salinity and its distribution in the soil profile by means of electric current and electromagnetic induction (EM) have advanced since 1971 (Rhoades, 1990c). These methods have been developed on the basis of the geophysical techniques used by geophysicists and geologists for mineral exploration and hydrogeological investigations (see Parasnis, 1986 and Milson, 1989). They estimate the soil salinity via the measurement of bulk electrical conductivity of the soil, which depends on the salinity of soil solution, porosity and the type and amount of clay in the soil.

According to Rhoades (1990c), reliable estimates of soil salinity are obtainable from measurements of bulk soil electrical conductivity, estimates of soil water content and clay percentage. This statement is based on the results of an intensive study (more than 700 sites tested) in the San Joaquin Valley of California. Rhoades (1990c) also provides an example of the measured and predicted salinity map, where a high correspondence between predicted and measured salinity magnitude and distribution within the study area can be seen. As described by Kandiah (1990: 5), global positioning systems have become available for accurate and rapid mapping. These techniques, linked with salinity monitoring in a geographic information system, can be used for salinity mapping purposes. The following sections briefly describe ground and airborne geophysical methods that allow rapid field-wide measuring of soil salinity.

Ground methods

Resistivity method

A number of devices are available for soil salinity measurements using the resistivity method (Rhoades, 1990c), including the four electrode sensor, which has an array of four electrodes a few centimetres apart. The four electrodes may be equally spaced (the Wenner array) or have a variable spacing. The two external electrodes are connected to the source of electric current and the voltage difference is measured between the two internal electrodes. The source of electrical current may be either a hand-cranked generator type or a battery-powered type. Current electrodes are made of stainless steel, copper, brass or any other corrosion-resistant metal. Potential electrodes are made of stainless steel. Electrodes have a diameter of 1 to 1.25 cm and a length of 45 cm. The probe provides the measurement of soil conductivity at the measuring point at a depth which depends on the distance between the electrodes. In the case of the Wenner array the depth sampled is roughly equal to the electrode spacing. For survey or traverse work the array electrodes may be mounted in a board with a handle so that the soil conductivity measurements can be made quickly for a given inter-electrode spacing. These fixed-array units save the time involved in positioning the electrodes. For most purposes an inter-electrode spacing of 30 or 60 cm is adequate and convenient.

Electromagnetic method

The theory of operation of ground electromagnetic (EM) conductivity meters is fully described by McNeill (1980) and summarised in McNeill (1986). They consist of a small transmitter coil, energised with an alternating current. This current generates a primary time-varying magnetic field in the ground which, in turn, induces small currents which generate their own secondary magnetic field. A receiver coil in the vicinity of the transmitter coil responds to both primary and secondary magnetic field components. In general, the secondary magnetic field, which is a small fraction of the primary magnetic field, is proportional to conductivity. This feature forms the basis of the ground conductivity meters. The EM technique offers the outstanding advantage of permitting measurement of ground conductivity without ground contact, that is, without the use of any probes. All problems of attempting to obtain good electrical connection with the ground are avoided and, since there are no or fewer cables to manipulate, the speed at which a conductivity survey can be carried out is now substantially increased. Furthermore, since these instruments automatically average the measurement over a lateral area which is approximately equal to the depth of exploration, they give an accurate value for the bulk conductivity of the soil and are able to detect very small variations in this quantity. For these reasons, electromagnetic ground conductivity meters are replacing conventional resistivity techniques for many survey applications. Although the initial expense is higher, the savings resulting from faster, more accurate surveys usually quickly offset this factor (McNeill, 1986).

Although both the transmitter and receiver coils could be used in a horizontal position (the so-called vertical dipole mode of operation), it is also possible to employ this method in the horizontal dipole mode, in which case both coils are vertical and coplanar. It is a useful fact that these two different modes of operation produce quite different responses from material at different depths. In the vertical dipole mode the relative response to near-surface material is essentially zero, and the response increases with depth, becoming a maximum at a depth of about 0.4 times the intercoil spacing. It then decreases slowly thereafter. However, for the horizontal dipole mode the relative response with depth is quite different, being a maximum for material very near the surface and decreasing slowly thereafter with depth (McNeill, 1986).

A number of electromagnetic conductivity meters are manufactured by Geonics, Ltd. Major specifications of these instruments are summarised in Table 32.

Table 32: **Geonics ground conductivity meters.**

Instrument	Intercoil spacing (m)	Depth of exploration	
		HD[a] mode (m)	VD[b] mode (m)
EM38	1.0	0.75	1.5
EM31	3.7	3.0	6.0
EM34-3	10.0	7.5	15.0
	20.0	15.0	30.0
	40.0	30.0	60.0
EM39	0.5	Radial Distance = 0.9m	

(a) Horizontal Dipole; (b) Vertical Dipole. *Source:* McNeill (1986).

The EM38 (Figure 13A) is very lightweight (2.5 kg) and only 1 m long. It was designed to be particularly useful for agricultural surveys for soil salinity. The EM38 can cover large areas quickly. It provides a depth of exploration of 1.5 m in the vertical dipole mode and 0.75 m in the horizontal dipole mode. Measurement is normally made by placing this instrument on the ground and noting or recording the meter reading. However, it is also possible to mount the EM38 on a wooden sled which is towed at walking speed across the ground while data are recorded essentially continuously (about every 0.3 m) using a digital data recorder. In this mode of operation 3000 data points can be measured easily in one hour to provide extremely dense data acquisition.

(A)

(B)

Figure 13:   Geonics electromagnetic terrain conductivity meters: (A) near surface EM38; (B) two-man variable depth EM34-3 (Courtesy of Geonics Ltd, Mississauga, Ontario, Canada).

The EM31 has an intercoil spacing of 3.7 m. This one-man portable instrument measures the conductivity to a depth of several metres. It can be used in a reconnaissance mode, where it is carried on a sling over the shoulder and operated continuously, or it can also be laid on the ground for horizontal and vertical dipole measurements.

The EM34-3 (Figure 13B) is the largest of the ground conductivity meters. This two-man instrument can be operated at intercoil spacings of 10 m, 20 m and 40 m so as to vary the depth of exploration, as shown in Table 32. To make a measurement, the transmitter operator should be positioned at the survey station. The receiver operator has two meters, one of which measures the intercoil spacing, the other the ground conductivity. Coil location is adjusted for the correct spacing and the conductivity reading recorded.

The EM39 conductivity logger provides measurement of the electrical conductivity of the soil and rock surrounding a borehole or monitoring well, using an inductive electromagnetic technique. The unit employs an intercoil spacing of 50 cm. Measurement is unaffected by conductive borehole fluid in the monitoring well or by the presence of plastic casing. The instrument operates to a depth of 200 metres.

Electromagnetic case studies

Electromagnetic ground conductivity methods have been used for mapping soil salinity in both detailed and reconnaissance investigations. Cameron et al. (1981) surveyed soil salinity in a 16 ha severely saline dryland site at Gull Lake, in south-western Saskatchewan. They compared the soil salinity map with soil salinity maps prepared with a Wenner array, an EM38 and an EM31. The three soil salinity maps prepared by the three instruments are highly correlated.

van der Lelij (1983) describes the application of an EM38 for the measurement of soil conductivity along a profile and preparation of a few soil salinity maps in an irrigated area near Griffith in Australia. The EM readings were calibrated against the measurements on soil samples. A total of 101 sites were examined. The EM readings were taken in both horizontal and vertical modes with the instrument at the soil surface. Soil samples were taken at depths of 0 m, 0.05 m, 0.3 m, 0.6 m, 1.0 m and 1.5 m. Samples were analysed for total salinity and the soil moisture content. A good correlation was found between soil salinity at 0.3 m depth and readings of the EM38. The salinities at other depths were also well correlated with the horizontal and vertical readings of EM38.

Soil conductivity measurements along a 5 km profile at 20 m intervals correlated quite well with the principal terrain features. The effects of micro-topography account for some but not all of the conductivity variations. Topographic highs generally had been associated with low conductivities. van der Lelij (1983) also provides three examples of farm soil salinity surveyed with EM38 based on a 100 m grid. They reported that the process of mapping a 200–300 ha farm takes about 2 days.

Engel et al. (1989) describe the application of geophysical techniques including electromagnetic terrain conductivity for saline seep investigations in Hardie's catchment in Western Australia. The catchment has an area of 160 ha and about 18 ha of land in the catchment is salt-affected. For the electromagnetic induction measurement an EM34-3 has been used. The survey was carried out on north-south lines 100 m apart and at measuring points of 20 m intervals.

The EM34-3 terrain conductivity contour map of the catchment shows a clear association between the conductivity readings and geological and landscape features of the catchment. The main features are the area of high conductivity located in the centre of the catchment, which is associated with the position of two dolerite dykes. Low terrain conductivities occur in most upland areas where there are coarse textured soils and the depth to bedrock is greater than 25 m. Engel et al. (1989) concluded that the application of geophysical techniques to this catchment demonstrated a rapid means for providing a significant contribution to land use planning.

Williams and Arunin (1990) surveyed an area of 3500 ha in Bung Phan Dung, located in the north-east region of Thailand. They describe the use of a multi-frequency EM system for delineating recharge/discharge zones at a regional scale, and compared their results with those obtained from a piezometric study of the same area. Williams and Arunin (1990) used a Geonics EM34-3 conductivity meter on a 500 m × 500 m grid, with frequencies of 6400 Hz, 1600 Hz and 400 Hz providing nominal depths of penetration of 7.5 m, 15 m and 30 m. Comparing the three isoconductivity maps at those sounding depths showed the extensive accumulation of salt in the near surface part of the soil profile, particularly in the valley floor. They concluded that, in some landscape systems at least, multifrequency EM surveys can provide a rapid means for delineating the major recharge and discharge areas.

Richardson and Williams (1994) interpreted measurements made by an EM34-3 in terms of actual or potential recharge/discharge areas within a landscape. Their method depends on characterising the salt distribution in actual discharge zones, and then using this information to identify areas of potential discharge. Use is made of the fact that groundwater discharge sites have salt concentrations that increase towards the upper part of the soil profile because of the upward movement of the salts, while the recharge areas are characterised by increasing salt concentrations with depth down the profile. The shape of the conductivity profile can then be expressed as an index. Richardson and Williams (1994) have developed an empirical discharge index (DI) defined as:

$$DI = (V10 / V40) \times V20$$

where V10, V20 and V40 are the EM values measured in vertical coil mode (i.e. horizontal dipole) for coil separations of 10 m, 20 m and 40 m, corresponding to penetration depths of 7.5 m, 15 m and 30 m. High DI values (80–100) generally correspond to a potential saline discharge area. Low DI values indicate low conductivity close to the surface and increasing with depth, or a recharge area.

Richardson and Williams (1994) applied their methodology to mapping potential discharge zones over a dryland agricultural area of 40 000 ha in the state of New South Wales in Australia. In their field trial DI values greater than 90 were taken as representing potential groundwater discharge areas. They showed that there is a high degree of correspondence between areas with DI values greater than 90 and areas affected by waterlogging/salinisation. They concluded that the multi-frequency survey with the EM34-3 instrument appears to be of considerable value in delineating areas prone to groundwater discharge.

Airborne electromagnetic method

The airborne electromagnetic method can be used for soil salinity investigations. Street (1992) and Street and Duncan (1992) described the QUESTEM (a registered name) airborne electromagnetic mapping system. The transmitter antenna is slung between the extended nose, wingtips and tail of an aircraft and consists of six horizontal turns, each of 186 m^2. The receiver is a horizontal axis coil towed behind the aircraft. The aircraft normally operates at an altitude of about 120 m (Figure 14). QUESTEM samples the ground response very shortly after each transmitter pulse. This ensures that information reflecting the distribution of conductivity in the near surface is measured.

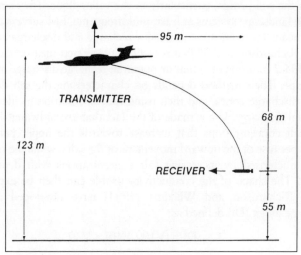

Figure 14: QUESTEM system geometry (Street, 1992).

Measurements made of the electromagnetic fields at rapid sampling rates are subject to a large signature from the aircraft, which is itself a good conductor. A stable technique for the separation of aircraft and ground response has been designed and implemented for work in highly saline areas. Electromagnetic interference caused by atmospheric discharge has also been reduced. Powerlines interference, a significant problem for EM surveys in farming areas, has also been eliminated.

Street (1992) describes the results of an airborne EM survey of the East Yornaning catchment, located 170 km south-east of Perth in Western Australia. The catchment has an area of 13 700 ha. Within the catchment a total of 1305 ha or 9.5 per cent was assessed as being salt-affected. The assessment is based on an interpretation of aerial photographs in 1990. The electrical conductivity distribution maps exhibited a strong correlation with topography. Areas high in the catchment have a lower overall conductivity than the lower valleys around the drainage lines. Comparison of the airborne electromagnetic results within the same area covered with the ground-based system shows a similar pattern and similar values of conductivity. The difference in resolution of the ground-based system can mostly be explained by the closer line spacing of the ground survey (100 m) versus the airborne survey (200 m). The other major difference in the two surveys is the cost. Commercial acquisition costs for the area covered by the ground survey (1900 ha) would be similar to the cost of the airborne survey over the entire catchment (13 700 ha).

According to Street and Duncan (1992), chloride concentrations obtained from drill hole data confirm that the electromagnetic apparent conductivity is responding to variations in the salt storage in the regolith. Statistically correlating the apparent conductivity against salt storage gives a correlation coefficient of 0.73, while correlating regolith thickness with apparent conductivity gave a correlation coefficient of 0.51. Using these data, Street and Duncan (1992) derived an empirical relationship for apparent salt storage and estimated the salt storage across the catchment to be around 870 tonnes per ha.

Remote sensing

In its broadest sense, remote sensing is the measurement or acquisition of information of some property of an object or phenomenon by a recording device that is not at physical or intimate contact with the object or phenomenon under study (Johannsen and Sanders, 1982). Remote sensing has a wide range of applications in investigation and management of natural resources including water, land and forest resources (Johannsen and Sanders, 1982; Colwell, 1983; Lo, 1986). While Mulders (1987) describes the application of remote sensing in soil science, there is no comprehensive publication concerning the application of remote sensing in soil salinity monitoring and investigations. However, Robbins and Wiegand (1990) describe the following remote sensing methods for salt-affected soils.

- *Aerial photography*: Of the different types of film used for remote sensing, colour infrared film has produced the most useful data on how the salinity-induced plant stress varies. With this film, dark-green vegetation produces a bright red image; light-green foliage produces a pink image; barren saline soil produces a white image; non-saline soil a grey, bluish-grey or greenish-grey image; salt-stressed crops a reddish-brown to reddish-black image; clear water a very dark-blue to black image, and sediment-laden water a blue image. Thus, plant responses to salinity, soil and water can be readily identified. Modern image analysis systems can digitise infrared photographs, cluster the scene into classes of salinity severity, and produce fractional area estimates that fall into each class. The response of crops to salinity severity usually varies from season to season and from crop to crop and depends mainly on patterns of precipitation and crop tolerance to salinity. Infrared photography is also very useful for recording year-to-year and crop-to-crop responses to salinity and changes in management practices.
- *Videography*: Video cameras, because of their recording ease, immediate playback, real time display and declining prices, are replacing traditional photographic cameras. Videography for agricultural application started about 1982 and since then interest has rapidly increased. The application of videography for soil salinity investigation was reported for the first time in 1987. The video signals have been digitised and an image-processing system has been used to distinguish between saline and non-saline areas and to estimate the area of land affected by salinity. Although resolution is lower than with photographic cameras, videography adequately surveys and documents salinity patterns.
- *Infrared thermometry*: Temperatures of the crop canopy can be measured by infrared thermometers and correlated with the crop water stress index to monitor the water stress caused by soil salinity. Infrared thermometry can also be used to monitor saline seeps, since the temperatures of wet and dry soil differ. Infrared thermometers can be hand-held or flown across fields in transects.
- *Multispectral scanners*: These systems sense in multiple bands or wavelength intervals. The visible band (0.4 µm to 0.7 µm) responds to vegetative ground cover as affected by salinity. Reflective infrared (0.75 µm to 1.35 µm) responds to the development of plants expressed as leaf area index. Water-affected bands (1.4 µm to 2.5 µm) contrast living vegetation with dry soil and wet versus dry soil. The thermal bands (5 µm to 20µm) measure the temperature of the canopy. The information from the various

bands is complementary and extensively documents the effects of salinity on plant growth. The costs of operating and processing data are high for multispectral scanners (MSS), so they are best suited to gathering data for large areas. They are more cost-effective when the data is used for multiple purposes and several users share the costs. The thematic mapper (TM) of the LANDSAT satellite series is the best instrument for continental or global surveys. It has the appropriate wavelengths, and makes six observations in each band on each area of the earth's surface every 16 days.

- *Microwave sensors*: Measurements of microwaves can be related to soil salinity because soil water and solutes affect the natural passive microwave emissions of the soil. Data on microwaves is currently being applied to salt-affected soils, but results are only qualitative.

References

Abrol, I.P. 1986. Salt-affected soils: Problems and prospects in developing countries. In. Swaminathan, M.S. and Sinha, S.K. eds. *Global Aspects of Food Production.* Oxford: Tycooly International. 283–305.

Abrol, I.P., Yadav, J.S.P. and Massoud, F.I. 1988. *Salt-affected Soils and Their Management.* Rome: FAO (Soils Bulletin 39). 131 pp.

Albert, W., Bramm, A. and Gonsowski, P. 1987. Employment of tensiometers for irrigation control. In. *Thirteenth International Congress on Irrigation and Drainage, Rabat, Morocco, September 1987.* New Delhi: International Commission on Irrigation and Drainage. Transactions, v.1B: 1061–1080.

Al-layla, M.A. 1989. Impact of lakes on water quality. In. *Regional Characterization of Water Quality.* (IAHS Publiction no. 182). 159–171.

Amer, M.H. 1990. Design of drainage system with special reference to Egypt. In. *Symposium on Land Drainage for Salinity Control in Arid and Semi-Arid Regions. February 25th to March 2nd 1990, Cairo, Egypt.* Delta Barrage, Cairo: Drainage Research Institute. v.1 (Keynotes): 39–60.

Amer, M.H. and de Ridder, N.A. eds. 1989. *Land Drainage in Egypt.* Delta Barrage, Cairo: Drainage Research Institute. 377 pp.

Anderson, R.L. 1990. A comparison of water transfer in Australia and the United States. In. Pigram, J.J. and Hooper, B.P. eds. *Transferability of Water Entitlements.* Armidale, Australia: Centre for Water Policy Research, University of New England. 155–158.

Aragües, R., Tanji, K.K., Quilez, D. and Faci, J. 1990. Conceptual irrigation return flow hydrosalinity model. In. Tanji, K.K. ed. *Agricultural Salinity Assessment and Management.* New York: American Society of Civil Engineers. 504–529.

Australian Water Resources Council. 1987. *1985 Review of Australia's Water Resources and Water Use.* Canberra: Australian Government Publishing Service. v.1: Water Resources Data Set, 158 pp; v.2: Water Use Data Set, 114 pp.

Ayers, R.S. and Westcot, D.W. 1989. *Water Quality for Agriculture.* Rome: Food and Agriculture Organization of the United Nations. 174 pp.

Balba, A.M. 1990. Agricultural reuse of drainage water. In. *Symposium on Land Drainage for Salinity Control in Arid and Semi-Arid Regions, February 25th to March 2nd 1990, Cairo, Egypt.* Delta Barrage, Cairo: Drainage Research Institute. 113–124.

Balogh, J. and Gergely, I. 1985. *Basic Aspects of Trickling Irrigation.* Budapest: Vizügyi Dokumentációs Szolgáltató Leányvállalat nyomdája. 280 pp.

Barrett-Lennard, E.G., Warren, B.E. and Malcolm, C.V. 1990. Agriculture on saline soils — directions for the future. In. Myers, B.A. and West, D.W. eds. *Revegetation of Saline Land.* Proceedings of a workshop held at the Institute for Irrigation and Salinity Research, 29–31 May 1990. Tatura, Victoria: Institute for Irrigation and Salinity Research. 37–45.

Bish, S. and Williams, R.M. 1992. *Mallee Cliffs Salt Interception Scheme: Installation of Production Bores and Review of Hydrogeological Conditions*. Parramatta, New South Wales: Department of Water Resources, Technical Services Division. 40 pp. plus Appendices.

Biswas, A.K. 1992. Sustainable water development: A global perspective. *Water International*. 17(2): 68–80.

Blackmore, D.J. 1991. Murray-Darling Basin Initiative: A case study in integrated natural resources management. *Water and the Environment*. (Newsletter of the Water Research Foundation of Australia). 305: 2–8.

Blad, B.L. 1983. Atmospheric demand for water. In. Teare, I.D. and Peet, M.M. eds. *Crop-Water Relations*. New York: John Wiley & Sons. 1–44.

Bos, M.G. and Nugteren, J. 1990. *On Irrigation Efficiencies*. Fourth edition. Wageningen: International Institute for Land Reclamation and Improvement (ILRI). 117 pp.

Boyden, S. 1987. *Western Civilization in Biological Perspective: Patterns in Biohistory*. Oxford: Oxford University Press. 370 pp.

Boyle Engineering Corporation. 1986. *Evaluation of On-Farm Agricultural Management Alternatives*. (Prepared for the San Joaquin Valley Drainage Program). Fresno, California: Boyle Engineering Corporation. (Various pagings, about 300 pp).

Brown, L.R. 1989. Reexamining the world food prospect. In. *State of the World 1989*. A World Watch Institute Report on Progress Toward a Sustainable Society. Washington DC: World Watch Institute. 41–58.

Brown, P.L., Halvorson, A.D., Siddoway, F.H., Maylan, H.F. and Miller, M.R. 1983. *Saline-seep Diagnosis, Control and Reclamation*. Washington DC: US Department of Agriculture, Agricultural Research Service. (Conservation Research Report no. 30). 22 pp.

Bulloch, J. and Darwish, A. 1993. *Water Wars: Coming Conflicts in the Middle East*. London: Victor Gollancz. 224 pp.

Bureau of Reclamation. (n.d.). *Irrigating the Lower Colorado Region*. (Waters of the Colorado Series, Fact Sheet #3). Yuma, Arizona: Yuma Project Office. 4 pp.

Bureau of Reclamation. 1992. *Title I Program Colorado River Basin Salinity Control Act*. (Report to the Secretary of the Interior and the Congress). Boulder City, Nevada: Bureau of Reclamation, Lower Colorado Region. 38 pp. plus Appendices.

Buringh, P. 1977. Food production potential of the world. In. Radhe Sinha. ed. *The World Food Problem: Consensus and Conflict*. Oxford: Pergamon Press. 477–485.

Burton, J. 1989. The community advisory committee in action — experiences and achievements. In. *Proceedings of the First Community Conference of the Murray-Darling Basin Ministerial Council's Community Advisory Committee*. Albury, 16–17 November 1988. Canberra: Department of Primary Industries and Energy. 1–6.

Burton, J. 1992. Catchment management in Australia — an historical review. In. *Catchments of Green: A National Conference on Vegetation and Water Management*, Adelaide, 23–26 March 1992. Canberra: Greening Australia Ltd. Conference Proceedings. v.A: 1–8.

Cameron, D.R., De Jong, E., Read, D.W.L. and Oosterveld, M. 1981. Mapping salinity using resistivity and electromagnetic inductive techniques. *Canadian Journal of Soil Sciences*. 61: 67–78.

Campbell, G.S. and Campbell, M.D. 1982. Irrigation scheduling using soil moisture measurements: Theory and practice. In. Hillel, D. ed. *Advances in Irrigation*. New York: Academic Press. v.1: 25–42.

Casey, H.E. 1972. *Salinity Problems in Arid Lands Irrigation: A Literature Review and Selected Bibliography*. Tucson, Arizona: University of Arizona, Office of Arid Land Studies. 300 pp.

Castle, D.A., McCunnall, J. and Tring, I.M. 1984. *Field Drainage: Principles and Practices*. London: Batsford Academic and Educational. 250 pp.

Chow, V.T., Maidment, D.R. and Mays, L.W. 1988. *Applied Hydrology*. New York: McGraw-Hill. 572 pp.

Clark, S.D. 1982. The River Murray Waters Agreement: Down the drain or up the creek? *Civil Engineering Transactions*. 24: 201-208.

Clark, S.D. 1983. Inter-governmental quangos: The River Murray Commission. In. Curnow, G.R. and Saunders, C.A. eds. *Quangos: The Australian Experience*. Sydney: Hale & Iremonger Ltd. 154–172.

Clarke, R. 1991. *Water: The International Crisis*. London: Earthscan Publications Ltd. 193 pp.

Colorado River Basin Salinity Control Forum. 1993. *1993 Review: Water Quality Standards for Salinity, Colorado River System.* Bountiful, Utah: Colorado River Basin Salinity Control Forum. (Various pagings, about 115 pp).

Colwell, R.N. ed. 1983. *Manual of Remote Sensing.* Second edition. Falls Church, Virginia: American Society of Photogrammetry. v.1 and v.2. 2440 pp.

Crabb, P. 1988a. *The Murray-Darling Basin Agreement.* Canberra: Centre for Resource and Environmental Studies (CRES), The Australian National University. (CRES working paper 1988/6). 30 pp.

Crabb, P. 1988b. Managing the Murray-Darling Basin. *Australian Geographer.* 19: 64–88.

Crabb, P. 1991. Resolving conflicts in the Murray-Darling Basin. In. Handmer, J.W., Dorcey, A.H.J. and Smith, D.I. eds. *Negotiating Water: Conflict Resolution in Australian Water Management.* Canberra: Centre for Resource and Environmental Studies (CRES), The Australian National University. 147–159.

Crosson, P.R. and Rosenberg, N.J. 1989. Strategies for agriculture. *Scientific American.* September: 78–85.

Cunningham, G.M. 1986. Total catchment management: Resource management for the future. *Journal of Soil Conservation, New South Wales.* 42(1): 4–5.

DeCoursey, D.G. ed. 1990. *Proceedings of the International Symposium on Water Quality Modeling of Agricultural Non-Point Sources.* Logan, Utah State University, June 19–23, 1988. US Department of Agriculture, Agricultural Research Service, ARS–81. US Government Printing Office. Part 1: 1–421, Part 2: 423–875.

Dedrick, A.R., Erie, L.J. and Clemmens, A.J. 1982. Level-basin irrigation. In. Hillel, D. ed. *Advances in Irrigation.* New York: Academic Press. v.1: 105–145.

Delforce, R.J., Pigram, J.J. and Musgrave, W.F. 1990. Impediments to free market water transfer in Australia. In. Pigram, J.J. and Hooper, B.P. eds. *Transferability of Water Entitlements.* Armidale, Australia: Centre for Water Policy Research, University of New England. 51–64.

Department of Water Resources Victoria. 1989. *Water Victoria: A Resource Handbook.* Melbourne: VGPO. 311 pp.

Dierickx, W. 1990. Developments and shortcomings in drainage technology. In. *Symposium on Land Drainage for Salinity Control in Arid and Semi-Arid Regions.* February 25th to March 2nd 1990, Cairo, Egypt. Delta Barrage, Cairo: Drainage Research Institute. v.1 (Keynotes): 93–109.

Dietrich, C.R., Jakeman, A.J. and Thomas, G.A. 1989. Solute transport in a stream-aquifer system. 1. Derivation of a dynamic model. *Water Resources Research.* 25: 2171–2176.

Donahue, R.L., Miller, R.W. and Shickluna, J.C. 1983. *An Introduction to Soil and Plant Growth.* Fifth edition. Englewood Cliffs, New Jersey: Prentice Hall. 667 pp.

Dregne, H., Kassas, M. and Razanov, B. 1991. A new assessment of the world status of desertification. *Desertification Control Bulletin.* (United Nations Environment Programme). 20: 6–18.

Driscoll, F.G. 1989. *Groundwater and Wells.* Second edition. St. Paul, Minnesota: Johnson Filtration Systems Inc. 1089 pp.

Dukhovny, V.A. 1978. Saline and alkaline soils; their use, improvement and related problems. In. *State-of-the-Art Irrigation Drainage and Flood Control.* New Delhi: International Commission on Irrigation and Drainage. no. 1: 391–458.

Easter, K.W. and Welsch, D.E. 1986. Implementing irrigation projects: Operational and institutional problems. In Easter, K.W. ed. *Irrigation Investment Technology and Management Strategies for Development.* Boulder, Colorado: Westview Press. 33–56.

El-Ashry, M.T. 1980. Groundwater salinity problems related to irrigation in the Colorado River Basin. *Ground Water.* 18(1): 37–45.

El-Ashry, M.T., Schilfgaarde, J.V. and Schiffman, S. 1985. Salinity pollution from irrigated agriculture. *Journal of Soil and Water Conservation.* January-February: 48–52.

Engel, R., McFarlane, D.J. and Street, G.J. 1989. Using geophysics to define recharge and discharge areas associated with saline seeps in south-western Australia. In. Sharma, M.L. ed. *Groundwater Recharge.* Rotterdam: A.A. Balkema. 25–39.

Engelman, R. and LeRoy, P. 1993. *Sustaining Water: Population and the Future of Renewable Water Supplies.* Washington DC: Population Action International. 56 pp.

Engineering and Water Supply Department. 1990. *Waikerie Salt Interception Scheme: Public Environment Report.* Adelaide: Engineering and Water Supply Department. 95 pp. plus Appendices.

Euroconsult, 1989. *Agricultural Compendium for Rural Development in the Tropics and Subtropics.* Third edition. Amsterdam: Elsevier. 740 pp.

Evans, R.S. 1989. Saline water disposal options in the Murray Basin. *BMR Journal of Australian Geology & Geophysics.* 11(2/3): 167–185.

Evans, R.S. and Nolan, J. 1989. A groundwater management strategy for salinity mitigation in Victorian riverine plain, Australia. In. *Groundwater Management: Quanity and Quality.* Proceedings of the Benidorm Symposium, October 1989. (IAHS Publication no. 188). 487–499.

Food and Agriculture Organization of the United Nations (FAO). 1980. *Drainage Design Factors: 28 Questions and Answers.* Rome: FAO. 52 pp.

Food and Agriculture Organization of the United Nations (FAO). 1989. *Production Yearbook.* Rome: FAO. v.42. 350 pp.

Foth, H.D. 1984. *Fundamentals of Soil Science.* Seventh edition. New York: John Wiley & Sons. 435 pp.

Framji, K.K. ed. 1976. *Irrigation and Salinity: A World-wide Survey.* New Delhi: International Commission on Irrigation and Drainage. 106 pp.

Framji, K.K., Garg, B.C. and Luthra, S.D.L. 1981. *Irrigation and Drainage in the World: A Global Review.* Third Edition. New Delhi: International Commission on Irrigation and Drainage. v.I. 491 pp.

Frenkel, H. 1984. Reassessment of water quality criteria for irrigation. In. Shainberg, I. and Shalhevet, J. eds. *Soil Salinity Under Irrigation: Processes and Management.* Berlin: Springer-Verlag. 143–172.

Gardner, B.D. 1986. Assessing salinity and toxic-element disposal problems in the San Joaquin Valley. In. Horner G.L. ed. *Decision Criteria for Residual Management in Agriculture.* Davis: University of California, The Agricultural Issues Centre. 3–35.

George, P.R. 1985. Sub-surface drainage methods for salinity control. *Journal of Agriculture—Western Australia.* 26(4): 112–114.

Ghassemi, F., Jakeman, A.J. and Jacobson, G. 1990. Mathematical modelling of sea water intrusion, Nauru Island. *Hydrological Processes.* 4(3): 269–281.

Ghassemi, F., Jakeman, A.J. and Nix, H.A. 1991. Human induced salinisation and the use of quantative methods. *Environment International.* 17(6): 581–594.

Ghassemi, F., Jakeman, A.J. and Thomas, G.A. 1987. *Groundwater Modelling and Simulation of Salinity Management Options in the Sunraysia Region of Victoria.* Canberra: Centre for Resource and Environment Studies (CRES), The Australian National University. 257 pp.

Ghassemi, F., Jakeman, A.J. and Thomas, G.A. 1989. Ground-water modelling for salinity management: An Australian Case Study. *Ground Water.* 27(3): 384–392.

Ghassemi, F., Thomas, G.A. and Jakeman, A.J. 1988. Effect of groundwater interception and irrigation on salinity and piezometric levels of an aquifer. *Hydrological Processes.* 2(4): 369–382.

Gill, H.S. and Abrol, I.P. 1991. Salt affected soils, their afforestation and its ameliorating influence. *The International Tree Crops Journal.* 6(4): 239–260.

Gleick, P.H. ed. 1993. *Water in Crisis: A Guide to the World's Freshwater Resources.* New York: Oxford University Press. 473 pp.

Greenwood, E.A.N. 1988. The hydrologic role of vegetation in the development and reclamation of dryland salinity. In. Allen, E.B. ed. *The Reconstruction of Disturbed Arid Lands.* Boulder, Colorado: Westview Press. 205–233.

Gupta, S.K. and Gupta, I.C. 1987. *Global Research on Drainage in Agriculture: An Annotated Bibliography, 1960-1986.* New Delhi: Concept Publishing Company. 661 pp.

Gutteridge Haskins & Davey Pty Ltd, ACIL Australia Pty Ltd, Australian Groundwater Consultants Pty Ltd and Melbourne University School of Agriculture and Forestry. 1983. *The Application of Salinity Control Techniques in Victoria.* Melbourne: Gutteridge Haskins & Davey. 131 pp.

Halse, N.J., Greenwood, E.A.N., Lapins, P. and Boundy, C.A.P. 1969. An analysis of the effects of nitrogen deficiency on the growth and yield of a Western Australian wheat crop. *Australian Journal of Agricultural Research.* 20(6): 987–998.

Halvorson, A.D. 1990. Management of dryland saline seeps. In. Tanji, K.K. ed. *Agricultural Salinity Assessment and Management.* New York: American Society of Civil Engineers. 372–392.

Halvorson, A.D. and Reule, C.A. 1980. Alfalfa for hydrologic control of saline seeps. *Soil*

Science Society of America Journal. 44: 370–374.

Hammer, U.T. 1986. *Saline Lakes Ecosystems of the World*. Dordrecht: Dr W. Junk Publishers. 616 pp.

Hawke, R.J.L. 1989. *Our Country Our Future*. (Statement on the Environment by the Prime Ministry of Australia, Canberra). Australian Government Publishing Service. 66 pp.

Hawkins, G.P. 1978. Modern trends in mechanisation of construction of irrigation and drainage projects. In. Framji, K.K. ed. *State-of-the-Art Irrigation Drainage and Flood Control*. New Delhi: International Commission on Irrigation and Drainage. no. 1. 107–254.

Heermann, D.F., Martin, D.L., Jackson, R.D. and Stegman, E.C. 1990. Irrigation scheduling controls and techniques. In. Stewart, B.A. and Nielsen, D.R. eds. *Irrigation of Agricultural Crops*. Madison, Wisconsin: American Society of Agronomy. 509–535.

Heynike, J.J.C. 1981. *The Economic Effects of the Mineral Content Present in the Vaal River Barrage on the Community of the PWVS Complex (A desk study)*. Pretoria: Water Research Commission. 131 pp.

Hickey, J.J. and Vecchioli, J. 1990. Subsurface injection of liquid waste with emphasis on injection practices in Florida. In. La Moreaux, P.E. and Vrba, J. eds. *Hydrogeology and Management of Hazardous Waste by Deep-Well Disposal*. (International Contribution to Hydrogeology, v.12). Hannover: Heise. 99–116.

Hillel, D. 1990. Ecological aspects of land drainage for salinity control in arid and semi-arid regions. In. *Symposium on Land Drainage for Salinity Control in Arid and Semi-Arid Regions*. February 25th to March 2nd 1990, Cairo, Egypt. Delta Barrage, Cairo: Drainage Research Institute. v.1 (Keynote): 125–135.

Hoffman, G.J. 1990. Leaching fraction and root zone salinity control. In. Tanji, K.K. ed. *Agricultural Salinity Assessment and Management*. New York: American Society of Civil Engineering. 237–261.

Holdgate, M.W., Kassas, M. and White, G.F. eds. 1982. *The World Environment 1972-1982*. Dublin: Tycooly International. (Published for the United Nations Environment Programme). 637 pp.

Holmes, J.W. and Talsma, T. eds. 1981. *Land and Stream Salinity*. Papers presented at the Land and Stream Salinity Seminar and Workshop, Perth, Western Australia, November 1980. Amsterdam: Elsevier. 392 pp.

Horton, A.J. and Jobling, G.A. eds. 1984. *Farm Water Supplies Design Manual (v.II: Irrigation Systems)*. Second edition. Brisbane: Queensland Water Resources Commission. (Various pagings).

Howard, K.W.F. and Haynes, J. 1993. Groundwater contamination due to road de-icing chemicals: Salt balance implications. *Geoscience Canada*. 20(1):1–8.

Howe, C. 1990. An analytical framework for water transfers. In. Pigram, J.J. and Hooper, B.P. eds. *Transferability of Water Entitlements*. Armidale, Australia: Centre for Water Policy Research, University of New England. 43–48.

Howe, C.W., Schurmeier, D.R. and Shaw, W.D. 1986. Innovative approaches to water allocation: The potential for water markets. *Water Resources Research*. 22(4): 439–445.

Hundley, N. 1975. *Water and the West: The Colorado River Compact and the Politics of Water in the American West*. Berkeley: University of California Press. 395 pp.

Hunter, G. and David, G. 1992. Victoria's catchment-based salinity program, achievements and challenges. In. *Catchments of Green: A National Conference on Vegetation and Water Management, Adelaide, 23-26 March 1992*. Canberra: Greening Australia Ltd. Conference Preceedings. v.A. 15–21.

Hutchinson, M.F., Nix, H.A. and McMahon, J.P. 1992. Climate constraints on cropping systems. In. Pearson, C.J. ed. *Ecosystems of the World: Field Crop Ecosystems*. Amsterdam: Elsevier. 37–58.

Hutson, J.L., Dudley, L.M. and Wagenet, R.J. 1990. Modeling transient root zone salinity. In. Tanji, K.K. ed. *Agricultural Salinity Assessment and Management*. New York: American Society of Civil Engineers. 482–503.

Huyakorn, P.S., Ward, D.S., Rumbaugh, J.O., Lester, B.H., Broome, R.W. and Siler, A.K. 1986. *SEFTRAN: A Simple and Efficient Two-Dimensional Groundwater Flow and Transport Model*. Herndon, Virginia: GeoTrans. Inc. 144 pp.

Institute for Participatory Planning. 1981. *Citizen Participation Handbook for Public Officials and Others Serving the Public*. Fourth edition. Laramie: Wyoming.

Irwin, F. and Williams, I.R. 1986. Catchments as planning units. *Journal of Soil Conservation, New South Wales*. 42(1): 6–10.

Ismail, S., Malcolm, C.V. and Ahmad, R. eds. 1990. *A Bibliography of Forage Halophytes and Trees for Salt-Affected Land: Their Use, Culture and Physiology*. Karachi: Department of Botany, University of Karachi. 258 pp.

Jackson, R.D. 1982. Canopy temperature and crop water stress. In. Hillel, D. ed. *Advances in Irrigation*. New York: Academic Press. v.1. 43–85.

Jacobsen, T. 1982. Salinity and irrigation agriculture in antiquity, Diyala Basin archaeological projects: report on essential results, 1957–58. *Bibliotheca Mesopotamica*. Malibu, California: Undena Publications. v.14. 107 pp.

Jacobsen, T. and Adams, R.M. 1958. Salt and silt in ancient Mesopotamian agriculture. *Science*. 128(3334): 1251–1258.

Jakeman, A.J., Dietrich, C.R. and Thomas, G.A. 1989. Solute transport in a stream-aquifer system. 2. Application of model identification to the River Murray. *Water Resources Research*. 25(10): 2177–2185.

Jakeman, A.J., Thomas, G.A., Ghassemi, F. and Dietrich, C.R. 1987. Salinity in the River Murray Basin: Management and modelling approaches. *Search*. 18(4): 183-188.

Jensen, E.G. and Leach, R.W. 1984. Salinity control by pumping and deep well injection. The Paradox Valley Unit. In. French, R.H. ed. *Salinity in Watercourses and Reservoirs: Proceedings of the 1983 International Symposium on State-of-the-Art Control of Salinity*. 13-15 July 1983, Salt Lake City, Utah. Boston: Butterworth Publishers. 349–358.

Johannsen, C.J. and Sanders, J.L. eds. 1982. *Remote Sensing for Resource Management*. Ankeny, Iowa: Soil Conservation Society of America. 665 pp.

Jurriens, M. and Bos, M.G. 1980. Developments in planning of irrigation projects. In. Wiersma-Roch, M.F.L. ed. *Land Reclamation and Water Management*. Wageningen: International Institute for Land Reclamation and Improvement. 99-112.

Kandiah, A. ed. 1990. *Water, Soil and Crop Management Relating to the Use of Saline Water*. Rome: Food and Agriculture Organization of the United Nations. 193 pp.

Keren, R. and Miyamoto, S. 1990. Reclamation of saline, sodic, and boron-affected soils. In. Tanji, K.K. ed. *Agricultural Salinity Assessment and Management*. New York: American Society of Civil Engineers. 410–431.

Knutson, R.D., Penn, J.B. and Boehm, W.T. 1990. *Agricultural and Food Policy*. Second edition. Englewood Cliffs, New Jersey: Prentice Hall. 437 pp.

Konikow, L.F. 1981. Role of solute transport models in the analysis of groundwater salinity problems in agricultural areas. In. Holmes, J.W. and Talsma, T. eds. *Land and Stream Salinity*. Amsterdam: Elsevier. 187–205.

Konikow, L.F. and Bredehoeft, J.D. 1978. *Computer Model of Two-Dimensional Solute Transport and Dispersion in Ground Water*. (Techniques of Water Resources Investigations of the United States Geological Survey, Book 7, Chapter C2). Washington DC: US Government Printing Office. 90 pp.

Köppen, W. 1923. *Die Klimate der Erde; Grundriss der Klimakunde*. Berlin: Walter de Gruyter.

Korzun, V.I., Sokolov, A.A., Budyko, M.I., Voskresensky, K.P., Kalinin, G.P., Konoplyantsev, A.A., Korotkevich, E.S., Kuzin, P.S. and Lvovich, M.I. eds. 1978. *World Water Balance and Water Resources of the Earth*. Paris: UNESCO. 633 pp.

Kovda, V.A. 1983. Loss of productive land due to salinisation. *Ambio*. 12(2): 91–93.

Kovda, V.A., Berg, C. and Hagan, R.M. eds. 1973. *Irrigation, Drainage and Salinity: An International Source Book*. London: Hutchinson & Co. (For FAO/UNESCO). 510 pp.

Landsberg, J. 1992. Role of the Murray-Darling Basin Commission in integrated catchment management. In. *Catchment of Green: A National Conference on Vegetation and Water Management, Adelaide, 23-26 March, 1992*. Canberra: Greening Australia Ltd. Conference Proceedings. v. A. 9–14.

LaMoreaux, P.E. and Vrba, J. eds. 1990. *Hydrogeology and Management of Hazardous Waste by Deep-well Disposal*. (International Contribution to Hydrogeology, v.12). Hannover: Heise. 136 pp.

La Riviere, J.W.M. 1989. Threats to the world's water. *Scientific American*. September: 48–55.

Laut, P. and Taplin, B.J. 1989. *Catchment Management in Australia in the 1980s*. Canberra: CSIRO Division of Water Resources (Divisional Report 89/3). 252 pp.

Lee, E.W. 1990. Drainage water treatment and disposal options. In. Tanji, K.K. ed. *Agricultural Salinity Assessment and*

Management. New York: American Society of Civil Engineers. 450–468.

Lo, C.P. 1986. *Applied Remote Sensing.* Harlow, Essex: Longman Scientific and Technical. 393 pp.

Maas, E.V. 1990. Crop salt tolerance. In. Tanji, K.K. ed. *Agricultural Salinity Assessment and Management.* New York: American Society of Civil Engineers. 262–304.

Malcolm, C.V. 1990. Saltland agronomy in Western Australia – an overview. In. Myers, B.A. and West, D.W. eds. *Revegetation of Saline Land.* Proceedings of a Workshop held at The Institute for Irrigation and Salinity Research, 29-31 May 1990. Tatura, Victoria: Institute for Irrigation and Salinity Research. 21–25.

Malcolm, C.V. 1992. Selecting forage shrubs for productive use of saline land. In. *National Workshop on Productive Use of Saline Land.* Waite Agricultural Research Institute, 22–24 September 1992. Adelaide: Department of Agriculture. 52–58.

Malik, K.A., Aslam, Z. and Naqvi, M. 1986. *Kallar Grass: A Plant for Saline Land.* Faizalabad, Pakistan: Nuclear Institute for Agriculture and Biology. 93 pp.

Marcar, N.E. 1990. Tree option for utilisation of salt-affected land. In. Myers B.A. and West, D.W. eds. *Revegetation of Saline Land.* Proceedings of a workshop held at the Institute for Irrigation and Salinity Research, 29-31 May 1990. Tatura, Victoria: Institute for Irrigation and Salinity Research. 53–59.

Marcar, N.E. 1992. Trees for salt-affected land. In. *National Workshop on Productive Use of Saline Land.* Waite Agricultural Research Institute, 22-24 September 1992. Adelaide: Department of Agriculture. 12–23.

McFarlane, D.J. 1991. A review of secondary salinity in agricultural areas of Western Australia. *Land and Water Research News.* (Western Australian Steering Committee for Research on Land Use and Water Supply). 11: 7–16.

McFarlane, D.J., Negus, T.R. and Cox, J.W. 1985. Drainage to control waterlogging. *Journal of Agriculture-Western Australia.* 26(4): 122–125.

McFarlane, D.J., Wheaton, G.A., Negus, T.R. and Wallace, J.F. 1992. *Effects of Waterlogging on Crop and Pasture Production in the Upper Great Southern, Western Australia.* South Perth: Department of Agriculture, Western Australia. (Technical Bulletin no. 86). 44 pp.

McNeill, J.D. 1980. *Electromagnetic Terrain Conductivity Measurement and Low Induction Numbers.* Mississauga, Ontario: Geonics Limited. (Technical Note TN-6). 15 pp.

McNeill, J.D. 1986. *Rapid, Accurate Mapping of Soil Salinity Using Electromagnetic Ground Conductivity Meters.* Mississauga, Ontario: Geonics Limited. (Technical Note TN-18). 15 pp.

McWhorter, D.B. 1977. Drain spacing based on dynamic equilibrium. *Journal of the Irrigation and Drainage Division.* v.103, IR2: 259–271.

Micklin, P.P. 1988. Desiccation of the Aral Sea: A water management disaster in the Soviet Union. *Science.* 241: 1170–1175.

Micklin, P.P. 1991. *The Water Management Crisis in Soviet Central Asia.* Pittsburgh: University of Pittsburgh, Center for Russian and East European Studies. (The Carl Beck Papers in Russian and East European Studies, no. 905). 120 pp.

Milson, J. 1989. *Field Geophysics.* Milton Keynes, England: Open University Press. 182 pp.

Moore, C.V. and Howitt, R.E. 1988. The Central Valley of California. In. El-Ashry, M.T. and Gibbons, D.C. *Water and Arid Lands of the Western United States.* New York: Cambridge University Press. 85–126.

Mulders, M.A. 1987. *Remote Sensing in Soil Science.* Amsterdam: Elsevier. 379 pp.

Murray-Darling Basin Commission. 1990. *1990 Annual Report.* Canberra: Murray-Darling Basin Commission. 130 pp.

Murray-Darling Basin Ministerial Council. 1989a. *Draft: Murray-Darling Basin Natural Resources Management Strategy.* Canberra. Murray-Darling Basin Ministerial Council. 19 pp.

Murray-Darling Basin Ministerial Council. 1989b. *Salinity and Drainage Strategy.* Canberra: Murray-Darling Basin Ministerial Council. 6 pp.

Murray-Darling Basin Ministerial Council. 1990. *Murray-Darling Basin Natural Resources Management Strategy: Towards a Sustainable Future.* Canberra: Murray-Darling Basin Ministerial Council. 24 pp.

Murray-Darling Basin Ministerial Council. 1992. *Murray-Darling Basin Agreement.* Canberra: Murray-Darling Basin Ministerial Council. 81 pp.

Myers, B. 1990. Grasses for saline land. In. Myers, B.A. and West, D.W. eds.

Revegetation of Saline Land. Proceedings of a workshop held at the Institute for Irrigation and Salinity Research, 29–31 May 1990. Tatura, Victoria: Institute for Irrigation and Salinity Research. 27–35.

National Research Council. 1990. *Saline Agriculture, Salt-Tolerant Plants for Developing Countries.* Washington DC: National Academy Press. 143 pp.

Nix, H.A. 1981. Simplified simulation models based on specified minimum data sets: The CROPEVAL concept. In. Berg, A. ed. *Application of Remote Sensing to Agricultural Production Forecasting.* Rotterdam: A.A. Balkema. 151–169.

Nulsen, R.A. 1983. Manipulation of recharge by agronomic techniques. In. *Papers of the International Conference on Groundwater and Man.* Sydney, 5–9 December 1983. Australian Water Resources Council (Conference Series no. 8). Canberra: Australian Government Publishing Service. v.2. 317–325.

Nulsen, R.A. 1984. Evaporation of four major agricultural plant communities in the south-west of Western Australia measured with large ventilated chambers. *Agricultural Water Management.* 8: 191–202.

Nulsen, R.A. and Baxter, I.N. 1982. The potential of agronomic manipulation for controlling salinity in Western Australia. *The Journal of the Australian Institute of Agricultural Science.* 48(4): 222–226.

Office of the National Committee for Central Planning. 1988. Personal communication. Addis Ababa. Ethiopia.

Oldeman, L.R., Hakkeling, R.T.A. and Sombroek, W.G. 1991a. Second revised edition. *World Map of the Status of Human-Induced Soil Degradation: An Explanatory Note.* Wageningen: International Soil Reference and Information Centre (ISRIC). 34 pp.

Oldeman, L.R., van Engelen, V.W.P. and Pulles, J.H.M. 1991b. The extent of human-induced soil degradation. In. Oldeman, L.R., Hakkeling, R.T.A. and Sombroek, W.G. *World Map of the Status of Human-Induced Soil Degradation: An Explanatory Note.* Wageningen: International Soil Reference and Information Centre (ISRIC). 27–33.

Organization for Economic Cooperation and Development (OECD). 1987. *Pricing of Water Services.* Paris: OECD. 145 pp.

Oster, J.D. and Rhoades, J.D. 1990. Steady-state root zone salt balance. In. Tanji, K.K. ed. *Agricultural Salinity Assessment and Management.* New York: American Society of Civil Engineers. 469–481.

Papadakis, J. 1975. *Climates of the World and their Potentialities.* Buenos Aires: J. Papadakis. 200 pp.

Parasnis, D.S. 1986. *Principles of Applied Geophysics.* Fourth edition. New York: Chapman and Hall. 402 pp.

Pavelis, G.A. ed. 1987. *Farm Drainage in the United States: History, Status and Prospects.* Washington DC: US Department of Agriculture, Economic Research Service. 170 pp.

Peck, A.J. 1978. Note on the role of a shallow aquifer in dryland salinity. *Australian Journal of Soil Research.* 16(2): 237–240.

Peck, A.J. 1983. Response of groundwaters to clearing in Western Australia. In. *Papers of the International Conference on Ground Water and Man.* Sydney, 5–9 December 1983. Australian Water Resources Council (Conference Series, no. 8). Canberra: Australian Government Publishing Service. v.2. 327–335.

Peck, A.J., Thomas, J.F. and Williamson, D.R. 1983. *Salinity Issues, Effect of Man on Salinity in Australia.* (Water 2000 Consultants Report, no.8). Canberra: Australian Government Publishing Service. 78 pp.

Peck, A.J. and Williamson, D.R. 1987a. Effects of forest clearing on groundwater. *Journal of Hydrology.* 94(1/2): 47–65.

Peck, A.J. and Williamson, D.R. eds. 1987b. Hydrology and salinity in the Collie River Basin, Western Australia. *Journal of Hydrology.* 94(1/2). 198 pp.

Petermann, T. 1993. *Irrigation and the Environment.* Eschborn: Deutsche Gesellschaft für Technische Zusammenarbeit (GTZ) GmbH. Part I (Influence of irrigation on the environment and vice-versa, about 205 pp). Part II (Environmental considerations in planning and operation, about 285 pp).

Pigram, J.J. and Hooper, B.P. eds. 1990. *Transferability of Water Entitlements.* Armidale, Australia: Centre for Water Policy Research, University of New England. 289 pp.

Porteous, A. ed. 1983. *Desalination Technology: Developments and Practice.* London: Applied Science Publishers. 271 pp.

Postel, S. 1990. Saving water for agriculture. In. *State of the World 1989.* A World Watch Institute Report on Progress Toward a Sustainable Society. Washington DC: World Watch Institute. 39–58.

Pratt, P.F. and Suarez, D.L. 1990. Irrigation Water Quality Assessments. In. Tanji, K.K. ed. *Agricultural Salinity Assessment and Management.* New York: American Society of Civil Engineers. 220–236.

Prinsley, R.T. 1991. *Australian Agroforestry: Setting the Scene for Future Research.* Canberra: Rural Industries Research and Development Corporation. 90 pp.

Reisner, M. 1987. *Cadillac Desert: The American West and its Disappearing Water.* New York: Penguin Books. 582 pp.

Rhoades, J.D. 1984. Reusing saline drainage waters for irrigation: A strategy to reduce salt loading of rivers. In. French, R.H. ed. *Salinity in Watercourses and Reservoirs: Proceedings of the 1983 International Symposium on State-of-the-Art Control of Salinity.* Salt Lake City, Utah, 13-15 July 1983. Boston: Butterworth Publishers. 455–464.

Rhoades, J.D. 1990a. Assessing suitability of water quality for irrigation. In. Kandiah, A. ed. *Water, Soil and Crop Management Relating to the Use of Saline Water.* Rome: Food and Agriculture Organization of the United Nations. 52–70.

Rhoades, J.D. 1990b. Strategies to facilitate the use of saline water for irrigation. In. Kandiah, A. ed. *Water, Soil and Crop Management Relating to the Use of Saline Water.* Rome: Food and Agriculture Organization of the United Nations. 125–136.

Rhoades, J.D. 1990c. Measuring and monitoring soil salinity. In. Kandiah, A. ed. *Water, Soil and Crop Management Relating to the Use of Saline Water.* Rome: Food and Agriculture Organization of the United Nations. 71–88.

Richardson, D.P. and Williams, B.G. 1994. *Assessing Discharge Characteristics of Upland Landscapes Using Electromagnetic Induction Techniques.* Canberra: CSIRO Division of Water Resources (Technical Memorandum 94/3). 10 pp. plus Figures and Tables.

Robbins, C.W. and Wiegand, C.L. 1990. Field and laboratory measurements. In. Tanji, K.K. ed. *Agricultural Salinity Assessment and Management.* New York: American Society of Civil Engineers. 201–219.

Romita, P.L. and Galbiati, G.L. 1978. Improvement of irrigation methods. In. Framji, K.K. ed. *State-of-the-Art Irrigation Drainage and Flood Control.* New Delhi: International Commission on Irrigation and Drainage. no. 1. 255–329.

Rural Water Commission of Victoria. 1988. *Irrigation and Drainage Practice.* Armadale, Victoria: Rural Water Commission of Victoria. 262 pp.

Saliba, B.C. 1987. Do water markets "work"?: Market transfers and trade-offs in the southwestern states. *Water Resources Research.* 23(7): 1113–1122.

Salinity Planning Working Group. 1992. *Regional Salinity Impact (Draft).* Melbourne: Salinity Bureau, Department of the Premier and Cabinet. 42 pp.

San Joaquin Valley Interagency Drainage Program. 1979. *Agricultural drainage and salt management in the San Joaquin Valley.* (Final report including recommended plan and first-stage environmental impact report). US Bureau of Reclamation, California Department of Water Resources and California State Water Resources Control Board.

Schofield, N.J. 1990. Salinity problems and remedies (with particular reference to agroforestry) in the San Joaquin Valley, California. *Land and Water Research News.* (Western Australian Steering Committee for Research on Land Use and Water Supply). 6: 14–17.

Schofield, N.J. 1992. Tree planting for dryland salinity control in Australia. *Agroforestry Systems.* 20(1/2): 1–23.

Schofield, N.J., Loh, I.C., Scott, P.R., Bartle, J.R., Ritson, P., Bell, R.W., Borg, H., Anson, B. and Moore, R. 1989. *Vegetation Strategies to Reduce Stream Salinities of Water Resources Catchments in South-West Western Australia.* Leederville: Water Authority of Western Australia, Water Resources Directorate. (Report no. WS33). 81 pp.

Seemann, J. 1979. Water requirements of plants. In. Seemann, J., Chirkov, Y.I., Lomas, J. and Primault, B. *Agrometeorology.* Berlin: Springer-Verlag. 294–297.

Seemann, J., Chirkov, Y.I., Lomas, J. and Primault, B. 1979. *Agrometeorology.* Berlin: Springer-Verlag. 324 pp.

Shainberg, I. and Oster, J.D. 1978. *Quality of Irrigation Water.* Bet Dagan, Israel: International Irrigation Information Center. 65 pp.

Shalhevet, J. 1984. Management of irrigation with brackish water. In. Shainberg, I. and Shalhevet, J. eds. *Soil Salinity Under Irrigation: Processes and Management.* Berlin: Springer-Verlag. 298–318.

Shaw, R.J., Hughes, K.K., Thorburn, P.J. and Dowling, A.J. 1987. Principles of landscape, soil and water salinity — processes and management options, Part A. In. *Landscape, Soil and Water Salinity*. Proceedings of the Brisbane Regional Salinity Workshop, Brisbane, May 1987. Brisbane: Queensland Department of Primary Industries (Conference and Workshop Series QC87001).

Shiklomanov, I.A. 1990. Global water resources. *Nature & Resources*. 26(3): 34–43.

Simmons, P., Poulter, D. and Hall, N.H. 1991. *Management of Irrigation Water in the Murray-Darling Basin*. Canberra: Australian Bureau of Agricultural and Resource Economics. (Discussion Paper 91.6). 42 pp.

Sinai, G., Jury, W.A. and Stolzy, L.H. 1985a. Application of automated flow and salinity control to dilution of saline irrigation water. *Irrigation Science*. 6: 179–190.

Sinai, G., Koch, E. and Farbman, M. 1985b. Dilution of brackish waters in irrigation networks: An analytic approach. *Irrigation Science*. 6: 191–200.

Singer, M.J. and Munns, D.N. 1987. *Soils: An Introduction*. New York: Macmillan. 492 pp.

Staples, R.C. and Toenniessen, G.H. eds. 1984. *Salinity Tolerance in Plants: Strategies for Crop Improvement*. New York: John Wiley & Sons. 443 pp.

Stone, S. 1991. Community-led land and water use planning: Some Victorian experiences. In. Handmer, J.W., Dorcey, A.H.J. and Smith, D.I. eds. *Negotiating Water: Conflict Resolution in Australian Water Management*. Canberra: Centre for Resource and Environmental Studies, The Australian National University. 219–236.

Street, G. 1992. Airborne geophysics — a tool to identify strategic areas for revegetation. In. *Catchments of Green: A National Conference on Vegetation and Water Management*. Adelaide, 23–26 March 1992. Canberra: Greening Australia Ltd. Conference Proceedings. v.B. 43–53.

Street, G.J. and Duncan, A.C. 1992. The application of airborne geophysical surveys for land management. In. *Proceedings, 7th International Soil Conservation Organisation*. Sydney, 27–30 September 1992. v.2. 762–770.

Sunraysia Community Salinity Working Group. 1991. *Sunraysia Draft Salinity Management Plan*. Mildura, Victoria: T & V James Printing. 94 pp.

Syme, G.J. and Eaton, E. 1989. Public involvement as a negotiation process. *Journal of Social Issues*. 45(1): 87–107.

Szabolcs, I. 1989. *Salt-Affected Soils*. Boca Raton, Florida: CRC Press. 274 pp.

Tanji, K.K. ed. 1990. *Agricultural Salinity Assessment and Management*. New York: American Society of Civil Engineers. 619 pp.

Tanji, K.K., Lauchli, A. and Meyer, J. 1986. Selenium in the San Joaquin Valley. *Environment*. 28(6): 6–11 and 34–39.

Teare, I.D. and Peet, M.M. eds. 1983. *Crop-Water Relations*. New York: John Wiley & Sons. 547 pp.

Telfer, A. 1989. Groundwater-riverwater density contrast: Its effect on the pattern of groundwater discharge to the River Murray. *BMR Journal of Australian Geology and Geophysics*. 11(2/3): 227–232.

Thomson, L. 1988. *Salt Tolerance in Eucalyptus camaldulensis and Related Species*. Ph.D. Thesis, Melbourne: University of Melbourne. 307 pp.

Thornthwaite, C.W. 1931. The Climates of North America according to a new classification. *Geographical Review*. 21(4): 633–655.

Thornthwaite, C.W. 1948. An approach to a rational classification of climate. *Geographical Review*. 38(1): 55–94.

Thornthwaite, C.W. and Mather, J.R. 1954. Climate in relation to crops. *Meteorol. Monographs*. 2(8): 1–10.

Trofimenko, S. 1985. The state and density of the Aral Sea. *Ambio*. 14(3): 181–182.

Trompeter, K.M. and Suemoto, S.H. 1984. Desalting by reverse osmosis at Yuma Desalting Plant. In. French, R.H. ed. *Salinity in Watercourses and Reservoirs: Proceedings of the 1983 International Symposium on State-of-the-Art Control of Salinity*. Salt Lake City, Utah, 13–15 July 1983. Boston: Butterworth Publishers. 427–437.

Ubell, K. 1971. Iraq's water resources. *Nature & Resources (UNESCO)*. 7(2): 3–9.

United Nations Environment Programme (UNEP). 1992a. *Saving Our Planet: Challenges and Hopes*. Nairobi: UNEP. 200 pp.

United Nations Environment Programme (UNEP). 1992b. *World Atlas of Desertification*. London: Edward Arnold. 69 pp.

URS Corporation. 1986. *Deep-well Injection of Agricultural Drain Waters: An Appraisal Level Study with Application to Kesterson Reservoir Problem (Summary Report)*. Sacramento, California: URS Corporation. 140 pp.

US Report. 1967. The world food problem, a report of the President's Science Advisory Committee. Vol II, Report of the panel on the world food supply. Washington, The White House.

van der Lelij, A. 1983. *Use of an Electromagnetic Induction Instrument (Type EM-38) for Mapping of Soil Salinity*. Sydney: Water Resources Commission, New South Wales. (Internal Report, Research Branch). 21 pp.

van der Moezel, P.G. and Bell, D.T. 1990. Saltland reclamation: Selection of superior Australian genotypes for discharge sites. *Proceedings of the Ecological Society of Australia*. 16: 545–549.

van Genuchten, M.Th. and Hoffman, G.J. 1984. Analysis of crop salt tolerance data. In. Shainberg, I. and Shalhevet, J. eds. *Soil Salinity Under Irrigation: Processes and Management*. Berlin: Springer-Verlag. 258–271.

van Schilfgaarde, J. ed. 1974. *Drainage for Agriculture*. Madison: American Society of Agronomy. 700 pp.

van Schilfgaarde, J. 1976. Reprinted in 1984. Water management and salinity. In. *Prognosis of Salinity and Alkalinity*. Rome: FAO (FAO Soil Bulletin 31). 53–67.

van Schilfgaarde, J. 1984. Drainage design for salinity control. In. Shainberg, I. and Shalhevet, J. eds. *Soil Salinity Under Irrigation: Processes and Management*. Berlin: Springer Verlag. 190–197.

Walker, J., Williams, B.G., Hatton, T.J., O'Loughlin, E.M., Jupp, D.L.B., Booth, T.H., Marcar, N.E., Jovanovic, T. and Vercoe, T. 1991. Use of trees for dryland salinity control. In. *The Role of Trees in Sustainable Agriculture, A National Conference*. Sept. 30 – Oct. 3, 1991. Albury: New South Wales. (Sponsored by the Rural Industries Research and Development Corporation, Canberra). Salinity: 81–94.

Wangnick, K. 1991. 1990 worldwide desalting plants inventory: The development of the desalination market. *Desalination*. 81: 19–37.

Water and Power Development Authority (WAPDA). 1988. Private communication. WAPDA, 94 Ferozepur Road, Lahore, Pakistan.

Western Australian Department of Agriculture. 1988. *Salinity in Western Australia: A Situation Statement*. Perth: Western Australian Department of Agriculture, Division of Resource Management (Technical Report no. 81). 116 pp.

Western Australia Legislative Assembly. 1991. *Select Committee into Land Conservation, Final Report*. Perth: Legislative Assembly. 171 pp.

Western Australia Water Resources Council. 1989. *Transferable Water Entitlements in Western Australia*. Leederville, Western Australia: WAWRC (Publication no. 8/89). 23 pp.

White, G.F. 1978. *Environmental Effects of Arid Land Irrigation in Developing Countries*. (MAB Technical Note 8). Paris: UNESCO. 67 pp.

Wilkinson, R. and Barr, N. 1993. *Community Involvement in Catchment Management: An Evaluation of Community Planning and Consultation in the Victorian Salinity Program*. Melbourne: Department of Agriculture. 197 pp.

Willey, Z. 1990. Environmental quality and agricultural irrigation reconciliation through economic incentives. In. Tanji, K.K. ed. *Agricultural Salinity Assessment and Management*. New York: American Society of Civil Engineers. 561–583.

Williams, W.D. 1987. Salinisation of rivers and streams: An important environmental hazard. *Ambio*. 16(4): 180–185.

Williams, B.G. and Arunin, S. 1990. *Inferring Recharge/Discharge Areas from Multifrequency Electromagnetic Induction Measurements*. Canberra: CSIRO Division of Water Resources (Technical Memorandum 90/11). 17 pp.

Woolley, D.R. and Kalf, F.R. 1979. Feasibility of disposal of bitterns or brine by injection into an aquifer, Tullakool irrigation area, New South Wales. In. Lawrence, C.R. and Hughes, R.J. eds. *Proceedings of the Groundwater Pollution Conference*. Perth, 1979. (Australian Water Resources Council, Conference Series; no. 1). Canberra: Commonwealth Government Printer. 83–100.

World Bank. 1992. *World Development Report 1992: Development and the Environment*. New York: Oxford University Press. 308 pp.

World Resources Institute. 1986. *World Resources 1986*. A Report by the World Resources Institute and the International Institute for Environment and Development. New York: Basic Books. 353 pp.

World Resources Institute. 1990. *World Resources 1990-91*. A report by the World Resources Institute in collaboration with the United Nations Environment Programme and the United Nations Development Programme. New York: Oxford University Press. 383 pp.

World Resources Institute. 1992. *World Resources 1992–93*. A report by the World Resources Institute in collaboration with the United Nations Environment Programme and the United Nations Development Programme. New York: Oxford University Press. 385 pp.

Zijlstra, G. and van Someren, C.L. 1980. Development in subsurface drainage techniques. In. Wiersma-Roche, M.F.L. ed. *Land Reclamation and Water Management: Development, Problems and Challenges*. Wageningen: International Institute for Land Reclamation and Improvement (ILRI). 171–180.

Zilliox, L. 1989. Industrial impact on the quality of groundwater in a large basin: Case of the Rhine aquifer. *Water International*. 14(2): 62–68

PART TWO

Salinity Problems in Selected Countries

Chapter 1: Argentina

Introduction

Argentina occupies much of the southern part of South America, covering a triangular area of 2 776 888 km² (Figure 1-1). It is the second largest South American country after Brazil and stretches about 3700 km north to south, with a maximum width of about 1450 km. The country's coastline measures 2665 km.

Argentina comprises a diverse landscape of mountains, upland areas and plains. It can be divided into four regions (Figure 1-1):

- **The Andes** are the great mountain system of the South American continent which form the western boundary of Argentina. Argentina has within its borders, or shares with Chile, 16 Andean peaks exceeding 6100 m. The nation's highest peak is Mount Aconcagua with an elevation of 6959 m, which is the highest mountain in the Western Hemisphere.

- **The Northern Plains** consist of the southern portion of the South American region known as the Chaco, covering the northern part of Argentina, eastern Bolivia, Paraguay and Western Brazil. The Chaco, at an elevation of 180 to 240 m, is a vast alluvial lowland formed from erosion of the Andes. The Argentinian Chaco is crossed by the Pilcomayo, Bermejo, Salado del Norte and Dulce Rivers. Mesopotamia is a humid lowland area between the Paraná and Uruguay Rivers, and is a continuation of the Chaco.

- **The Pampas** are the great level plains lying south of the Chaco and southeast of the Sierra de Córdoba. A distinction is commonly made between the humid Pampa of the east and dry Pampa of the west. The plains are composed of deep accumulations of loess, wind-deposited material and alluvium, covering an old hilly landscape of granite and other ancient rocks now buried as much as 300 m below the surface. The Pampas are the agricultural heart of the nation, with grain growing areas in the west and cropland and pasture in the east.

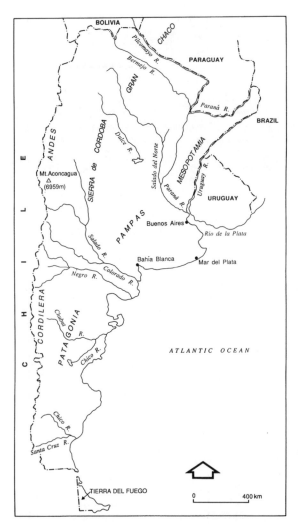

Figure 1-1: Main physiographic features of Argentina (Framji et al., 1981).

- **Patagonia** is a cool, dry, windswept plateau region south of the Rio Colorado. It extends from the Atlantic coast in the east to the foothills of the southern Andes. It contains many glacial lakes near the Andes. The plateau is composed mostly of horizontal sedimentary strata and, in places, of dark coloured lavas.

Argentina had an estimated population of 32 million in 1990, with a growth rate of 1.2 per cent over the period 1980–90 (World Bank, 1992: 269). Argentina is divided politically into 22 provinces, a territory and a federal district that consists only of the capital, Buenos Aires (Figure 1-2).

Figure 1-2: Administrative map of Argentina (Bram et al., 1983).

Rainfall and climate

The northern plains have a humid subtropical type of climate characterised by a heavy summer rainfall diminishing in amount toward the west. To the west and south the climate of the Andean highlands and northern Patagonia is much drier. This gives way to desert and semidesert climate in the interior southern lowlands. The Andes have a moderate subpolar climate.

Precipitation in Argentina is marked by wide regional variations. More than 1500 mm falls annually in the extreme north and 2000 mm in Mesopotamia. In the vicinity of Buenos Aires annual rainfall is about 1000 mm. However, conditions change gradually towards the south and west as these regions become semiarid to arid. Although the average annual rainfall of the country is about 515 mm, only 10 per cent of the country receives an annual precipitation of more than 1000 mm, 50 per cent less than 500 mm and 30 per cent less than 200 mm (Framji et al., 1981). About 75 per cent of the country is arid to semiarid with an average annual rainfall of 100–700 mm per year (Maddaloni, 1986).

Water resources

Surface water

Argentina's hydrology is dominated by the rivers that drain into the Rio de la Plata. This is the second largest river system in South America after the Amazon River. About one-third of the Rio de la Plata Basin lies within the boundaries of Argentina, and the rest lies in Brazil, Paraguay, Bolivia and Uruguay. Rivers that contribute to the discharge of the Rio de la Plata include the Paraná, Uruguay and Pilcomayo. The Paraná River traverses the north central portion of the country. The Uruguay River forms part of the boundary with Uruguay and the Pilcomayo River forms the boundary with Paraguay. The other important rivers of Argentina are the Colorado, Salado del Norte and Negro rivers. Table 1-1 shows the flow characteristics of the major rivers in Argentina, while Table 1-2 summarises the drainage area and the average annual run-off of the hydrographic basins in Argentina. The average annual run-off of the country is 914×10^9 m^3 which is equivalent to 330 mm or 64 per cent of the annual rainfall.

Table 1-1: Flow characteristics of some rivers in Argentina.

River	Basin Area (km^2)	Flow Mean (m^3 s^{-1})	Flow Max (m^3 s^{-1})	Flow Min (m^3 s^{-1})	Run-off (L s^{-1} km^{-2})
Paraná	2 302 000	14 900	22 380	6500	6.5
Pilcomayo	110 000	166	1 150	1	1.5
Salado del Norte	40 000	20	400	0	0.5
Uruguay	239 000	4 660	37 000	92	19.7
Salado del Atuel	812	12	58	2	14.2
Colorado	22 300	134	830	32	6.0
Negro	95 000	930	3	420	759.3
Chubut	16 400	49	540	4	3.0
Santa Cruz	15 550	748	2 090	194	48.1

Source: van der Leeden (1975).

No recent estimate of the water use in the country is available. Therefore, by considering an annual per capita consumption of 800 m³ of water (Framji et al., 1981) including irrigation, drinking water supplies, industrial usage, etc., and a population of 32 million, the volume of water use will be about 25.6×10^9 m³.

Table 1-2: Flow characteristics of hydrographic basins of Argentina.

Hydrographic basin	Drainage area (km²)	Average annual run-off (10^9 m³)	Per cent of total (%)
Rio de la Plata	950 000	800	87.5
Atlantic Ocean basins	1 000 000	74	8.1
Pacific Ocean basins	30 000	34	3.7
Interior basins	800 000	6	0.7
Total	2 780 000	914	100.0

Source: Framji et al. (1981).

Groundwater

The extensive arid and semiarid regions of Argentina are dependent on groundwater resources, owing to a lack of surface water supplies. The annual extraction of groundwater and number of wells are estimated to be about 4.7×10^9 m³ and 300 000 respectively (Economic Commission for Latin America and the Caribbean, 1985). Large-scale groundwater use for irrigation is practised in the Provinces of San Juan and Mendoza. In both provinces, about 1×10^9 m³ per year are consumed and in the rest of the country it is estimated that 0.8×10^9 m³ or more is used for irrigation (United Nations, Department of Economic and Social Affairs, 1976). The following brief description of the hydrogeology of Argentina is summarised from the United Nations, Department of Economic and Social Affairs (1976).

The Miocene basin of Bahia Blanca has artesian aquifers in sandstones and conglomerates. The aquifers are at a depth of 700 m and are overlain by upper Miocene and Pliocene clays. In the upper Miocene and Pliocene layers, saline groundwaters are found throughout the Argentine Plain, from Formosa in the north to La Pampa and the Province of Buenos Aires. In the north of the Province of Buenos Aires and in part of Entre Rios and Santa Fe Provinces, the Pliocene sands aquifer is recharged by good quality water from the Paraná River. This aquifer is probably the largest and most developed in the country. Substantial parts of the urban centres of Greater Buenos Aires and southern Santa Fe are supplied by this aquifer. The unconsolidated Pliocene and Quaternary sediments are also major aquifers covering vast areas of the country.

Saline groundwater occurs in many locations. In the zones bordering on the Cordilleras and in the intermountain basins, aquifer discharge is mainly through evapotranspiration, with the result that the upper aquifers are increasingly saline. In these zones shallow groundwater containing 5000 to 10 000 mg L^{-1} TDS is common, but deeper levels are less saline and contain water suited for agriculture and even for human supply.

Although many areas of Argentina are dependent on groundwater resources, excess water causes problems in agricultural areas which lack an adequate drainage system, such as the agricultural areas of San Juan and Mendoza Provinces. In some areas along the Atlantic coast, salinity problems have developed as a result of the intrusion of marine waters in heavily pumped aquifers, as in the case in Mar del Plata and Bahia Blanca.

Conjunctive use of surface water and groundwater is practised in many areas. In a number of places, joint utilisation permits the improvement of water quality. For example, in Santa Fe Province, artificial recharge experiments using waters of the Paraná River resulted in the freshening of groundwater previously salty and not fit for consumption.

Land cover

According to FAO (1989), in 1987, out of a total land area of 273.7 Mha (total area of the country excluding areas under inland water bodies), about 35.8 Mha were under cultivation. Of this cultivated land, 1.72 Mha were irrigated and the remaining 34.08 Mha non-irrigated. Permanent pasture covered 142.5 Mha, forest and woodlands 59.5 Mha and other lands (such as potentially productive lands, wastelands and roads) 35.9 Mha.

Irrigation

Argentina is a crop and livestock producing country. The main crops consist of wheat, corn, sorghum, soybeans, sunflower, sugarcane and vegetables. Increasing pressure to step up agricultural production for both food and industrial raw materials for internal consumption as well as for export is reflected in expansion of the cultivated area on one hand and increasing intensity of use on the other hand. The cultivated area of the country has increased from 27.2 Mha in 1937 to 30 Mha in 1975 (Framji et al., 1981) and to 35.8 Mha in 1987 (FAO, 1989). The irrigated area has increased from 0.9 Mha in 1937 (Framji et al., 1981) to 1.54 Mha in 1986 (Michelena, 1988). About 1.29 Mha of irrigated land is in the arid and semiarid regions of the country and 0.254 Mha in the humid regions. As shown in Table 1-3, large-scale irrigation systems operate in the Provinces of Mendoza (443 500 ha), Buenos Aires (176 500 ha), Tucuman (140 700 ha), Rio Negro (117 100 ha) San Juan (96 100 ha) and Santiago del Estero (54 300 ha).

Table 1-3: Irrigated and salt-affected areas in 21 provinces of Argentina.

No.	Province	Irrigated area (ha)	Salt-affected area (ha)	Salt-affected area % of IA[a]	Area with drainage problems (ha)	Area with drainage problems % of IA[a]
1.	Buenos Aires	176 500	12 500	7.1	43 750	24.8
2.	Catamarca	26 884	1 517	5.6	?	?
3.	Chaco	4 700	500	10.6	?	?
4.	Chubut	26 404	12 646	47.9	20 969	79.4
5.	Cordoba	55 863	3 747	6.7	?	?
6.	Corrientes	52 310	?	?	?	?
7.	Entre Rios	56 800	?	?	?	?
8.	Formosa	5 200	?	?	?	?
9.	Jujuy	90 514	11 500	12.7	10 000	11.1
10.	La Pampa	3 964	1 982	50.0	2 500	63.1
11.	La Rioja	13 456	1 200	8.9	700	5.2
12.	Mendoza	443 523	255 940	57.7	255 310	57.6
13.	Neuquen	14 527	3 938	27.1	4 367	30.1
14.	Rio Negro	117 106	46 423	39.6	52 975	45.2
15.	Salta	129 000	57 791	44.8	17 584	13.6
16.	San Juan	96 133	76 566	79.6	55 000	57.2
17.	San Luis	8 797	2 436	27.7	2 250	25.6
18.	Santa Cruz	2 000	?	?	?	?
19.	Santa Fe	20 500	1 600	10.6	?	?
20.	Santiago del Estero	54 273	33 370	61.5	33 370	61.5
21.	Tucuman	140 734	60 393	42.9	51 941	36.9
	Total	1 539 188	584 049	37.9	554 716	36.0

(a) Irrigated Area. *Source:* Instituto Nacional de Tecnologia Agropecuaria (1986).

No estimate of the irrigation water use in the country and its efficiency is available. However, considering a total irrigated area of 1.54 Mha and an estimated water demand of 12 000 m^3 per hectare and per year in the irrigated lands of the country (van der Leeden, 1975) gives a total irrigated water use of about 18.5×10^9 m^3 per year.

Salinity

In Argentina, there are 32 Mha of soils with saline and/or alkaline problems. These soils occupy 11 per cent of the total land area and occur in arid, semiarid and humid regions (Maddaloni, 1986). The following sections describe salinity problems within the irrigated and non-irrigated areas of the country.

Irrigated land salinity

In irrigated areas where the soil is deep and well drained, there are no salinisation problems and the groundwater stays deep. In other cases where the soil is poorly drained, problems exist due to the high watertable and salinisation.

As shown in Table 1-3, about 584 000 ha of a total irrigated area of 1 540 000 ha are affected by salinity problems. In other words, some 37.9 per cent of the country's irrigated area is salt-affected, which represents a high degree of salinisation. The degree of salinisation varies in different provinces — for example, the irrigated areas of the Provinces of San Juan, Santiago del Estero and Mendoza are 79.6, 61.5 and 57.7 per cent salt-affected, respectively. The province of Mendoza, with its 255 940 ha of salt-affected land, is the most affected province.

Lack of adequate drainage facilities is the main cause of salinisation. As shown in Table 1-3, about 554 700 ha or 36 per cent of irrigated land have a drainage problem. For example, in the Province of Mendoza, 255 310 ha or about 57.6 per cent of the irrigated land has a drainage problem.

Tulum Valley and Jachal Valley, both from San Juan Province, are examples of salt-affected areas. Tulum Valley, in San Juan River Valley, is the main agricultural area in the San Juan Province, with some 70 000 ha under irrigation. Irrigation water comes both from the San Juan River (90 per cent) and the Tulum Aquifer (10 per cent), which is a Quaternary alluvial aquifer up to 600 m thick. The average salinity of San Juan River water, which is the main source of the aquifer's recharge, is about 500–600 $\mu S\ cm^{-1}$ (Eder, 1990).

Groundwater levels are measured by a network of piezometers. The number of piezometers was 190 in 1987 and another 60 were planned for 1988. Analysis of the piezometric data shows that the groundwater depth is less than 3 m in a major part of the area and even reaches to less than 1 m. Groundwater levels are deep in February and March and shallow about August and September (Poblete and Guimaraes, 1988).

Salinisation of the aquifer is due to a high watertable and to irrigation practices. Groundwater salinity is in the range of less than 1000 $\mu S\ cm^{-1}$ to 79 000 $\mu S\ cm^{-1}$. Table 1-4 summarises the results of salinity measurements on 109 groundwater samples. It shows that about 85 per cent of the samples have a salinity higher than 2000 $\mu S\ cm^{-1}$ and 38 per cent of samples have a salinity of higher than 10 000 $\mu S\ cm^{-1}$. Obviously such a high level of salinisation poses a serious threat to the agricultural production and economy of the region.

Apart from salinity, the boron content of groundwater is also high. Table 1-5 summarises the results of boron measurements on 109 groundwater samples. The maximum measured boron content was 36 mg L^{-1}.

Table 1-4: Groundwater salinity of 109 samples from Tulum Valley, San Juan Province.

Salinity range ($\mu S\ cm^{-1}$)	Number of samples	Percentage
< 1000	5	4.6
1000–2000	11	10.1
2000–5000	41	37.6
5000–10 000	11	10.1
10 000–20 000	25	22.9
20 000–40 000	10	9.2
40 000–60 000	4	3.7
60 000–80 000	2	1.8
Total	109	100.0

Source: Poblete and Guimaraes (1988).

Table 1-5: Boron content of 109 groundwater samples in Tulum Valley, San Juan Province.

Boron content range ($mg\ L^{-1}$)	Number of samples	Percentage
< 1	31	28.4
1–2	21	19.3
2–5	17	15.6
5–10	13	11.9
10–20	19	17.4
20–30	6	5.5
30–40	2	1.9
Total	109	100.0

Source: Poblete and Guimaraes (1988).

In the Jachal Valley, the average river salinity expressed in electrical conductivity is about 1900–2000 $\mu S\ cm^{-1}$ for an average flow rate of 7 $m^3\ s^{-1}$. Its boron content is also quite high. The cultivated area in Jachal Valley is about 5000 ha. The aquifer here is also a Quaternary alluvial formation and the Jachal River is the main source of recharge. In this valley, both boron and salinity pose problems (Pedro and Ernesto, 1985; Eder, 1990).

Dryland salinity

In the eastern humid region of Argentina, the salt-affected soils are concentrated in several lowland areas. The humid region lies to the east of the 700 mm rainfall isohyet. The region is predominantly flat or undulating with two large lowland areas, Pampa Deprimida (Province of Buenos Aires) and

Bajos Submeridionales (Province of Santa Fe). Both areas have serious flooding and salinity problems, and overall about 20 per cent of the soils are affected by these problems. According to Maddaloni (1986), the major salt-affected areas are as follows (Figure 1-3):

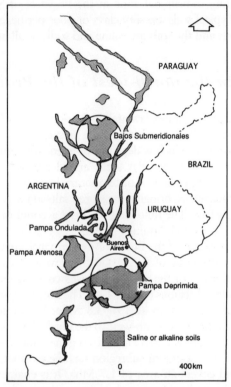

Figure 1-3: Major areas of saline and/or alkaline soils in the eastern humid region of Argentina (Maddaloni, 1986).

- *Pampa Deprimida (Lowland Pampa)*: The area is located in the central part of the province of Buenos Aires and covers 9.5 Mha. About 50 per cent of this area is affected by flooding and salinity. The salt, which is of oceanic origin, falls in the rain and has accumulated over thousands of years because of the poor regional drainage.
- *Pampa Ondulada (Waved Pampa)*: The area lies to the north of the Pampa Deprimida. It is one of the best drained areas in the eastern humid region but includes large areas of alkaline soils.
- *Pampa Arenosa (Sandy Pampa)*: The area is located to the north-west of the Pampa Deprimida. The soils in this area are of aeolian origin and, where the subsoil contains 23–31 per cent of clay, sodic soils have developed. The watertable fluctuates around 1.5 m and extensive saline soils have

developed in a large part of the area. The expansion of the salinity is caused by climatic changes expressed as significant increases in average annual rainfall. This issue will be described in the following section.
- *Bajos Submeridionales (Submeridional Lowland)*: The area is located in the subtropical temperate transitional zone in the Province of Santa Fe and covers 1.5 Mha. Typical soils are silty clays of poor permeability. Salinity increases with depth and the soils are saline and sodic in all horizons.

Salinisation in the north-west of the Province of Buenos Aires

The extraordinary increase in annual rainfall in the north-west of the Province of Buenos Aires since 1972 has increased the frequency of flooding and waterlogging in a vast area which is deprived of adequate natural drainage. Accumulation of rainwater in the flat areas and depressions over a long period of time has created permanent or temporary lagoons. This has led to the regional rise of watertables, bringing the stored salts in the soil profile and groundwater to the surface. The following is a brief account of the problem in this area based on Casas and Pittaluga (1990).

The north-west region of the Buenos Aires Province covers an area of 5.5 Mha. It has a gentle slope of 0.25/1000 from west to east. However, based on the characteristics of the aeolian deposits covering the region it is possible to identify two subregions (Figure 1-4). The northern subregion with longitudinal sand-dunes covers an area of 3.8 Mha. Sand-dunes form the concentric arcs directed from the south-west to the north. Their length is more than 100 km and their width is 2 to 5 km. Between the dunes there are plains of 0.5 to 3 km wide. The southern subregion has the parabolic or crescent shape sand-dunes and covers an area of 1.7 Mha. It is characterised by its greater thickness of the aeolian deposits than the northern subregion.

The topography of the region has prevented the development of an effective natural surface drainage system. The resultant excess rainwater creates permanent or temporary lagoons. The region has dominantly a temperate-subhumid climate with alternate dry and humid periods. The average annual rainfall of the region ranges between 700 and 850 mm and decreases from east to west. Rainfall is unevenly distributed over the year. Spring and summer are the rainy periods while winter is dry. The average annual temperature of the region is about 16°C.

Soils have been developed from the sandy materials deposited over the fine textured, low permeable clay materials. Percolated waters into the soil accumulate on the top of the low permeable materials and form a watertable aquifer. The watertable depth depends on the thickness of the sandy layer. It is deep in areas where the sandy layer is thick and is shallow where the layer of low permeability comes close to the land surface.

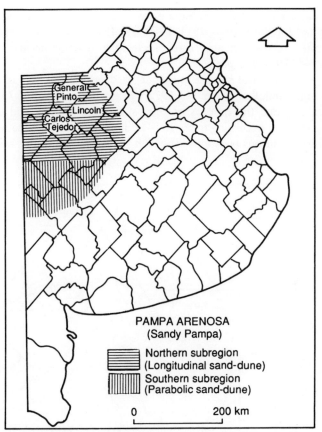

Figure 1-4: Area affected by flooding in the north-west of the Province of Buenos Aires (Casas and Pittaluga, 1990).

Considering the position of the soils in the topography of the landscape it is possible to define a range of soil types. Where the topography is undulating and the sandy cover is thick, soils with sandy texture have developed. These soils have acid or neutral reaction and are well drained. In areas where soils are mainly used for agricultural production, watertables are deep and soils are not affected by their fluctuations. Major reserves of fresh groundwater have developed and soils are not affected by salinity and/or alkalinity. In areas where the landscape is flat or forms a depression, soils are not adequately drained and are susceptible to salinisation after being affected by prolonged flooding, waterlogging or by capillary rise.

From the beginning of the 1970s, the region has entered a period of humid climate which is characterised by annual rainfalls exceeding the historic records and concentration of high rainfall events in a small number of days. Figure 1-5 shows the isohyets corresponding to the mean annual rainfall for the periods 1911–1970 and 1971–1984. The north-west of the province shows an appreciable increase in average annual rainfall which exceeds 1000 mm per year.

Analysis of satellite images showed early indications of areas affected by high rainfall at the end of 1972 and the beginning of 1973. The affected area included the districts of Carlos Tejedor, Lincoln and General Pinto (Figure 1-4). According to the 1978 estimates, about 1.5 Mha was covered by water on various occasions. The high rainfall period of the 1970s continued

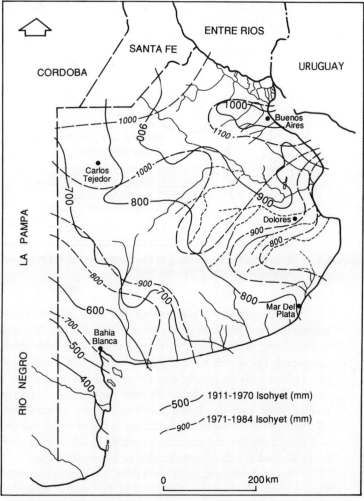

Figure 1-5: Change in average annual rainfall of the Province of Buenos Aires over the periods 1911–1970 and 1971–1984 (Casas and Pittaluga, 1990).

in the 1980s. In 1985 and 1986 high rainfall affected the parabolic sand-dune areas for the first time, impacting on 1 Mha. This phenomenon continued during 1987 and 1988, adding to the gravity of the situation.

The major change in the rainfall regime of the region and the inadequate natural drainage system caused the regional rise of watertables, which generally have a high salt content. Hence watertables reached a critical depth which affected the normal development of the vegetation, by waterlogging and salinising large areas used for livestock grazing.

The salt content of the watertable aquifers is highly variable. The electrical conductivity of some samples in depressions between the sand-dunes reaches 10 000–12 000 µS cm^{-1}, representing a salt content of 7000 to 8500 mg L^{-1} TDS. The predominant salts are chlorides, sulphates and bicarbonates of sodium and magnesium.

During the normal period of rainfall in the region, watertables are high from April until August with a depth of around 0.7 m. From September, watertables begin to decline and reach a depth of 1.3 m. During the 1986 to 1988 high rainfall period, watertables mostly maintained a high level, oscillating between zero and 70 cm depth. When rainfall decreases, the accumulated waters in the flooded and waterlogged areas will be slowly eliminated by infiltration and evaporation, causing salinisation. The salt content of the watertable aquifers and their depth are detrimental in the salinisation process. In the sandy soils of the region the capillary rise intensifies when the watertable is below 1 m depth. The maximum capillary flux ranges from 1 to 10 mm d^{-1} depending of the texture of the soils and the depth of watertable. The high salt content part of the soil is in the top 15 cm of the soil profile. The salt concentration in the soil surface reaches its maximum value during the spring-summer periods (September to the next February) and the soil salinity even exceeds 4 to 6 times that of the watertable aquifer. In the field it is possible to detect areas affected by salinisation. When the electrical conductivity of the saturated soil extract (ECe) is about 8000 µS cm^{-1} vegetation is less developed and its normal composition is modified. When the EC_e reaches about 12 000 µS cm^{-1}, it prevents plant development, and patches without vegetation appear.

Management options

In the irrigated areas, major management options consist of improving drainage facilities and irrigation systems. However, no published information about the salinity mitigation schemes and management of the boron problem is available.

In the non-irrigated areas of the north-west of the Province of Buenos Aires, salinity is caused by climatic changes. Although the evolution of these soils and their improvement depend on the climatic conditions, a number of measures would accelerate the improvement processes. Casas and Pittaluga (1990) describe these measures based on the results of experiments in the

district of Carlos Tejedor. Some of these measures are as follows:
- To undertake recovery works when watertables are deeper than one metre.
- Fencing of affected areas will permit the establishment of natural vegetation species adapted to high levels of soil salinity.
- During the period of recovery it is useful to mulch the soil surface with straw. This has the effect of reducing the capillary rise of the saline aquifers.
- Sowing of affected areas by salt-tolerant forage species like *Melilotus albus*, *M. officinalis*, tall wheat grass (*Agropyron elongatum*) and *A. scabrifolium*. Fencing the sown field during the first year would prevent grazing and would help establishment of grasses.
- Adequate management of natural and cultivated pasture, crop management, crop rotation and afforestation will contribute to increased water use and improvement of the salt-affected soils.

References

Bram, L.L., Phillips, R.S. and Dickey, N.H. eds. 1983. *Funk & Wagnalls New Encyclopedia*. USA. v.2. 304–5.

Casas, R.R. and Pittaluga, A. 1990. Anegamiento y salinizacion de suelos en el noroeste de la Provincia de Buenos Aires. In: Bellati, J.I., Kugler, W.F., Prego, A.J. and Sabella L.J. eds. *Manejo de Tierras Anegadizas*. Buenos Aires: Fundacion Para La Education, La Ciencia y La Cultura. 259–278.

Economic Commission for Latin America and the Caribbean (ECLAC). 1985. *The Water Resources of Latin America and the Caribbean and their Utilization*. Santiago, Chile: United Nations. 142 pp.

Eder, J.C. 1990. Centre Regional de Agua Subterranea, San Juan, Argentina. (Personal communication).

Food and Agriculture Organization of the United Nations (FAO). 1989. *Production Yearbook*. Rome: FAO. v.42. 350 pp.

Framji, K.K., Garg, B.C. and Luthra, S.D.L. 1981. *Irrigation and Drainage in the World: A Global Review*. Third edition. New Delhi: International Commission on Irrigation and Drainage. v.I. 491 pp.

Instituto Nacional de Technologia Agropecuaria (INTA). 1986. *Documento Básico para el Programa de Riego y Drenaje*. Buenos Aires: INTA.

Maddaloni, J. 1986. Forage production on saline and alkaline soils in the humid region of Argentina. *Reclamation and Revegetation Research*. 5: 11–16.

Michelena, R.R. 1988. Centro de Investigation de Resources Naturales. Buenos Aires, Argentina (Personal communication).

Pedro, L. and Ernesto, G. 1985. *Analisis Hidroquimico de las Possibilidades de Mejorar la Calidad de las Aguas del Rio Jachal Estudio de la Hipotesis de Eliminar el Aporte del Rio Salado*. San Juan, Argentina: Centro Regional de Agua Subterránea. 30 pp.

Poblete, M. and Guimaraes, R. 1988. *Comportamiento de la Freatica en el Valle de Tulum, Provincia de San Juan — Ano 1987*. San Juan, Argentina: Centro Regional de Agua Subterránea. 12 pp. Plus Tables, Figures and Maps.

United Nations, Department of Economic and Social Affairs. 1976. *Ground Water in the Western Hemisphere*. (Natural Resources/Water Series no. 4). New York: United Nations. 337 pp.

van der Leeden, F. ed. 1975. *Water Resources of the World: Selected Statistics*. Port Washington, New York: Water Information Center. 568 pp.

World Bank. 1992. *World Development Report 1992: Development and the Environment*. New York: Oxford University Press. 308 pp.

Chapter 2: Australia

Introduction

The Australian land mass covers 7 682 300 km², with maximum dimensions of 3680 km from north to south and 4000 km from east to west. Excluding minor indentations, the coastline of Australia extends approximately 36 800 km. The average altitude of the surface of this land mass is only 300 m, with approximately 87 per cent of the land mass having an altitude of less than 500 m and 99.5 per cent less than 1000 m.

Australia is divided into three physiographic units (Australian Geographic Society, 1988) as shown in Figure 2-1. These are:

- **The Eastern Highland Belt** extends along the east coast of Australia. It is a series of ranges of varying height and includes the main divide of the Great Dividing Range. The divide passes through Australia's highest point, Mt Kosciusko, with an elevation of 2228 m, while in parts it is less than 300 m above mean sea-level. The western slopes of the Great Dividing Range are generally gradual, the eastern descent to the coastal lowlands being much steeper.

- **The Central Eastern Lowlands** stretches across the continent from the Gulf of Carpentaria to the south-east coast of South Australia. Most of the lowlands are below 150 m and at Lake Eyre descend to about 14 m below sea-level.

- **The Great Western Plateau** covers most of Western Australia and the Northern Territory. The average elevation is 300 m. Over much of Western Australia the surface is very flat and uniform, diversified by wide shallow valleys which become deeper and more defined towards the coast.

Australia is divided into the six states of New South Wales, Victoria, Queensland, South Australia, Western Australia and Tasmania, two mainland territories, the Northern Territory and the Australian Capital Territory, and several small island territories.

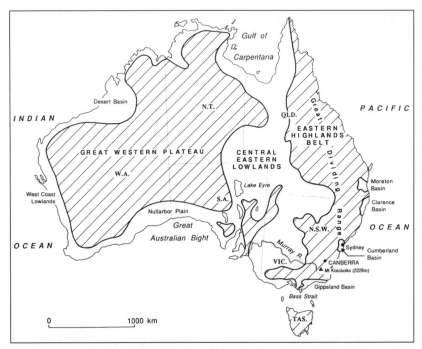

Figure 2-1: Physiographic division of Australia (Australian Geographic Society, 1988).

In 1992 Australia supported a growing population of more than 17.4 million. Most of the Australian population is concentrated in two widely separated coastal regions. By far the largest of these, in terms of area and population, lies in the south-east, stretching in an unbroken crescent from South Australia through Victoria, Tasmania and New South Wales to Queensland. The smaller of the two regions is the south-west of Western Australia. Neither region ever extends inland by more than two or three hundred kilometres. In both coastal regions the population is further concentrated into capital cities, other major cities and towns. In June 1988, 70.6 per cent of the Australian population lived in the combined state and territory capitals (Australian Bureau of Statistics, 1990: 111). The population density of Australia is about 2 persons per square kilometre.

Rainfall and climate

The island continent of Australia features a wide range of climatic zones, from the tropical regions of the north, to the arid expanses of the interior, to the temperate regions of the south.

The land mass is relatively arid with 80 per cent of its area receiving an annual average rainfall lower than 600 mm, 50 per cent lower than 300 mm and 30 per cent lower than 200 mm. Less than 4 per cent of the continent has a rainfall higher than 1200 mm per year. The area of lowest rainfall is in the vicinity of Lake Eyre in South Australia, where the average annual rainfall is about 100 mm (Australian Bureau of Statistics, 1990). The region with the highest rainfall is the east coast of Queensland between Cairns and Cardwell, where the median annual rainfall recorded at Tully over 63 years to 1987 was 4048 mm (Figure 2-2). The mountainous region of western Tasmania also has a high annual rainfall with Lake Margaret having a median of 3565 mm (63 years to 1987). In the mountainous area of northeast Victoria and some parts of the east coastal slopes there are small pockets with median annual rainfall greater than 2500 mm. The Snowy Mountains area in New South Wales also has a particularly high rainfall. The highest median annual rainfall for this region is 3200 mm. The rainfall pattern of Australia is strongly seasonal in character with a winter rainfall region in the south and a summer rainfall region in the north.

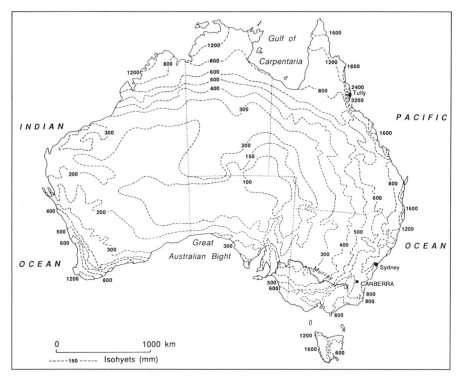

Figure 2-2:   Median annual rainfall in Australia (Parkinson, 1986).

Water resources

Surface water

Australia is considered to be the driest continent on earth in terms of the ratio of rainfall to run-off. The average rainfall over the continent is 465 mm and this results in 52 mm of run-off. In other words, only 11 per cent of the rainfall in Australia emerges as run-off and almost all the remaining 89 per cent (with the exception of a minor fraction which recharges the aquifers) returns to the atmosphere by evaporation and plant transpiration. In South America, North America and Asia, for example, run-off values are 57, 52 and 48 per cent of the rainfall, respectively (Brown, 1983).

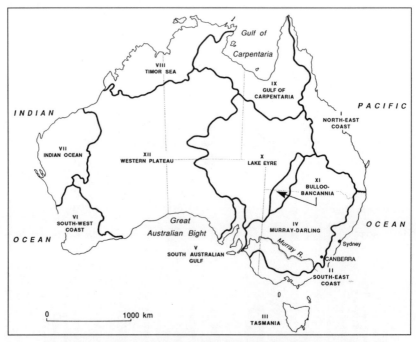

Figure 2-3: Drainage divisions of Australia (Australian Water Resources Council, 1987).

Australia is divided into 12 drainage divisions (Figure 2-3 and Table 2-1). The longest river system is the Murray-Darling, which drains part of Queensland, the major part of New South Wales, a large part of Victoria and part of South Australia. The length of the Murray is about 2520 km and the Darling and Upper Darling together are also just over 2500 km long (Australian Bureau of Statistics, 1990).

Table 2-1: Surface water resources of Australia (in 10^9 m^3).

Drainage division		Area (km^2)	Mean annual run-off	Mean annual outflow	Major divertible resources					Developed resources
					Fresh[a]	Marginal[b]	Brackish[c]	Saline[d]	Total	
North-east Coast	I	451 000	83.90	83.90	22.90	0	0	0	22.90	3.54
South-east Coast	II	274 000	41.90	41.90	14.70	0.24	0.11	0.02	15.07	4.28
Tasmania	III	68 200	52.90	52.90	10.90	0	0	0	10.90	1.02
Murray-Darling	IV	1 060 000	24.30	12.20	12.30	0.04	0.03	0	12.37	10.00
South Australian Gulf	V	82 300	0.88	0.77	0.16	0.07	0.03	0.01	0.27	0.12
South-west Coast	VI	315 000	6.67	6.60	1.39	0.47	0.85	0.16	2.87	0.39
Indian Ocean	VII	519 000	3.96	3.84	0.23	0.05	0.01	0.01	0.30	0.03
Timor Sea	VIII	547 000	80.70	80.70	22.00	0	0	0	22.00	1.98
Gulf of Carpentaria	IX	641 000	92.50	92.50	13.20	0	0	0	13.20	0.08
Lake Eyre	X	1 170 000	6.31	0	0.20	0	0	0	0.20	0.03
Bulloo-Bancannia	XI	101 000	1.09	0	0.04	0	0	0	0.04	0
Western Plateau	XII	2 450 000	1.58	0	0.10	0	0	0	0.10	0
Total		7 680 000	397.00	375.00	98.12	0.87	1.03	0.20	100.22	21.50

(a) TDS < 500; (b) 500 < TDS <1500; (c) 1500 < TDS < 5000; (d) TDS > 5000; (TDS values in mg L^{-1}).

Source: Australian Water Resources Council (1987).

The total continental run-off of 397×10^9 m^3 is unevenly distributed across Australia. About 95 per cent is contained in the six drainage divisions along the north and east coasts (Bergman, 1989). Moreover, 98.1×10^9 m^3 of surface run-off can be diverted on a sustained basis into conventional water supply systems, utilising existing storages and potential dam sites. Currently, only 21.5×10^9 m^3 of these resources have been developed. About 98 per cent of the divertible surface water resources of the country are fresh. But out of 2.12×10^9 m^3 of marginal, brackish and saline divertible surface water resources, nearly 1.48×10^9 m^3 (70 per cent) belong to the south-west coast division. Its salinity is mainly human-induced.

Statistics published by the Australian Water Resources Council (1987) show that annual water use in Australia is 14.64×10^9 m^3, of which 10.24×10^9 m^3 (70 per cent) is used for irrigation. Domestic, industrial and commercial sectors use 3.06×10^9 m^3 (21 per cent), while the rural sector uses just 1.34×10^9 m^3 (9 per cent) for non-irrigation purposes.

As a source of water supply, surface water with an annual volume of 12.04×10^9 m^3 contributes 82 per cent, while groundwater provides 2.60×10^9 m^3 or 18 per cent of water use in Australia. Although the contribution of groundwater to total water use seems relatively small, it should be noted that many areas are totally or at least heavily dependent on groundwater as a source of water supply. These areas occur mainly in arid or semiarid zones but also in temperate and tropical zones.

Groundwater

The groundwater resources of Australia occur in unconsolidated surficial materials (surficial aquifers), porous and permeable rocks (sedimentary aquifers) and fractured geological formations (Jacobson et al., 1983). Based on the nature of its aquifers, Australia has been divided into 61 provinces. The annual groundwater abstraction of Australia is 2.6×10^9 m^3 and is distributed among the three types of aquifers as follows:

- sedimentary aquifers: 1.2×10^9 m^3 or 46 per cent
- surficial aquifers: 1.1×10^9 m^3 or 42 per cent
- fractured rock aquifers: 0.3×10^9 m^3 or 12 per cent

The estimated number of bores in Australia is about 360 000. The maximum concentration is in the Perth province (93 800), followed by the Great Artesian Basin (39 300). Bore yields depend on the type, thickness and extent of the aquifer, hydrodynamic parameters and design specifications. Generally, for the sedimentary and surficial aquifers, bore yields range from 1–100 L s^{-1}, while for the fractured rock aquifers bore yield is less than 3 L s^{-1} (Jacobson et al., 1983).

Analysis of recent data (Australian Water Resources Council, 1987) shows that the divertible resources, or the average annual volume of water which could be abstracted on a sustainable basis from the nation's major and minor groundwater resources, is of the order of 31.3×10^9 m^3. The water quality of these resources has been defined in four salinity classes (Table 2-2). These estimates indicate the overall scope for additional groundwater development, although certain regions are stressed (Jacobson and Lau, 1988).

Table 2-2: Divertible groundwater resources.

Quality class	Total dissolved solids (TDS) (mg L^{-1})	Volume (10^9 m^3)		
		Major sources	Minor sources	Total
Fresh	TDS < 500	4.8	4.1	8.9
Marginal	500 < TDS < 1500	7.0	6.3	13.3
Brackish	1500 < TDS < 5000	1.9	4.0	5.9
Saline	TDS > 5000	1.9	1.3	3.2
Total		15.6	15.7	31.3

Source: Australian Water Resources Council (1987).

In summary, it seems on the one hand that the country enjoys a considerable volume of undeveloped water resources per head of population. However, after acknowledging the temporal variability of surface water supply, the uneven distribution of surface water and groundwater resources and their considerable distance from population centres and irrigation areas, it becomes clear that water resource managers have a difficult task in harnessing these sources.

Land cover

According to FAO (1989), the total Australian land area of 761.8 Mha was covered in the following way in 1987: about 436 Mha (57.2 per cent) was under permanent pasture and used for grazing; 106 Mha (13.9 per cent) was covered by forest and woodland; 47.1 Mha (6.2 per cent) was under extensive cropping and sown pastures with respective coverages of 19.76 Mha and 27.34 Mha (Australian Bureau of Statistics, 1988); and 172.7 Mha (22.7 per cent) was devoted to unused but potentially productive land, barren land, cities, etc. Nix (1988) argues that the area of agricultural productivity is limited to 77 Mha or 10 per cent of the land surface. Of this, 55 Mha are already under cultivation or are occupied by developments such as cities and roads. Only 22 Mha are potentially available for development.

The extent of tree clearing in Australia in the name of economic development is one of the most striking features of the country's development and its subsequent impacts on environmental degradation and salinity problems. Before European settlement, forest and woodland covered 33 per cent of the country or about 250 Mha. It has been estimated that 50 per cent of the original tall and medium forests and 35 per cent of the woodlands have now been cleared or severely modified (Prinsley, 1991).

Irrigation

Irrigation is one of the main factors of agricultural development in semiarid regions of Australia. The first significant attempts at irrigation began in the Murray Basin in the 1870s, but variability in the flow of the River Murray and its tributaries proved to be a major limitation. Therefore, storages have been required for the regulation of river flow to enable a stable irrigated agriculture to develop (Smith et al., 1983). With gradual development of water resources, the irrigated area has increased. As shown in Table 2-3, about 1.7 Mha was irrigated in 1984. Pasture, which covers 846 000 ha, is the dominant consumer of irrigation water, followed by crops (685 000 ha) and horticulture (167 000 ha). About 1.27 Mha of irrigated land is located in the Murray-Darling Basin. The uneven distribution of irrigated land in the country is illustrated by the fact that the states of New South Wales, Victoria and Queensland have 712 000, 558 000 and 276 000 ha of irrigated land, respectively, constituting about 91 per cent of the total irrigated land in Australia.

Table 2-3: Irrigated land in Australia.

State	Pasture (ha)	Crop (ha)	Horticulture (ha)	Total (ha)
New South Wales	282 000	392 000	38 200	712 000
Victoria	456 000	50 400	51 900	558 000
Queensland	28 700	215 000	32 300	276 000
Western Australia	17 600	5 420	8 890	31 900
South Australia	44 100	2 520	33 700	80 300
Tasmania	18 400	19 800	1 890	40 000
Northern Territory	0	82	12	96
Total	846 000	685 000	167 000	1 700 000

Source: Australian Water Resources Council (1987: v.2, Table 2).

Recent statistics published by the Australian Bureau of Statistics (1988) show the irrigated area had increased to 1 836 500 ha in 1987. According to these statistics, New South Wales has 855 000 ha of irrigated land, followed by Victoria (550 400 ha), Queensland (275 400 ha), South Australia (91 800 ha), Western Australia (24 000 ha), Tasmania (38 200 ha) and Northern Territory (1700 ha). The source of irrigation water was also determined by the Australian Bureau of Statistics census as follows: 1 538 100 ha (83.8 per cent) was irrigated by surface water, 266 500 ha (14.5 per cent) by groundwater, 6950 ha (0.4 per cent) with town or country reticulated water supply and for 24 950 ha (1.3 per cent) the source of water supply was unspecified. In terms of irrigation method, furrow and flood irrigation were dominant, accounting for 71 per cent (1 307 300 ha), followed by sprays (excluding microsprays) at 22 per cent (408 850 ha) and the remaining 7 per cent (120 350 ha) by trickle, microspray and other methods.

Smith et al. (1983) have reviewed the development of and prospects for irrigation in Australia. They provided some estimates of the efficiency of water distribution and application in a number of irrigation areas of Australia. According to their data, distribution efficiency varies from 70 per cent in northern Victoria to 90 per cent in Bundaberg in Queensland. Distribution efficiency has been increasing over the past few decades through the concrete lining of open canals, the use of pipelines and better control structures and management. Application efficiency varies from 40 per cent in the Namoi Valley of northern New South Wales to about 69 per cent in the Murrumbidgee Irrigation Area of southern New South Wales. Application efficiency is also increasing because of the use of more efficient irrigation methods.

According to Grieve (1990), surface drains were installed early in the life of irrigation areas in Kerang (Victoria), since 1914, and in the Murrumbidgee Irrigation Area (MIA) (New South Wales), since 1912. Subsurface drainage involving tile drains and tube-wells have been used wherever conditions were suitable. Tile drains have been used from the 1930s onward in Sunraysia and later in the MIA to protect horticultural crops. Tube-wells have been installed in the MIA, Goulburn Valley, Curlwaa, Wakool and Berriquin (New South Wales). Table 2-4 shows the length of the irrigation and drainage canals for 1977–78.

Table 2-4: Length of the irrigation and drainage canals and the area of lands protected by drainage works in Australia for 1977–78.

State	Irrigation canal (km)	Drainage canal (km)	Protected area (ha)
New South Wales	5 944	2 710	220 000
Victoria	8 238	4 376	300 302
Queensland	994	450	25 921
South Australia	371	339	21 247
Western Australia	1 005	2 432	366 183
Tasmania	103	800	55 000
Total	16 655	11 107	988 653

Source: Framji et al. (1981).

Salinity

Primary or naturally occurring salinity is extensive in Australia. According to Williamson (1990), 29 Mha of non-irrigated lands in Australia are naturally salt-affected, of which 14 Mha are covered by salt marshes, salt flats and salt lakes. All are associated with highly saline groundwater and often with internal drainage. A further 15 Mha of land in arid and semiarid regions have naturally saline subsoils, but no groundwater in the profile.

Secondary salinity has developed since settlement by Europeans and the introduction of agriculture and domestic grazing animals little more than 200 years ago. Extensive clearing of indigenous eucalypt forest, woodland and savanna scrublands for establishment of dryland agriculture has resulted in a significant increase in surface soil salinity and stream salt loads, particularly in the southern half of Australia (Williamson et al., 1987). Table 2-5 summarises the extent of current human-induced salinisation of soils in Australia.

Table 2-5: The extent of human-induced land salinity in Australia.

State	Area of human-induced salinised soils			Area of shallow groundwater (<2 m) in irrigated lands (ha)
	Scalds (ha)	Saline seeps (non-irrigated) (ha)	Irrigated saline soil (ha)	
New South Wales	920 000	14 000	10 000	260 000
Victoria	60 000	100 000	144 000	385 000
Queensland	580 000	8 000[a]	1 000	500
South Australia	1 200 000	225 000	500	4 500
Western Australia	340 000	443 000	500	0[b]
Tasmania	0	8 000	0	0
Northern Territory	680 000	0	0	0
Total	3 780 000	798 000	156 000	650 000

(a) and (b) Updated figures 10 000 ha and 11 400 ha, respectively (see the text). *Source:* Williamson (1990).

Dryland salinity is emerging as a major form of land and water degradation in Western Australia, South Australia and Victoria and to a lesser extent in New South Wales, Queensland and Tasmania (Figure 2-4). In Western Australia the average rate of increase of salt-affected soil has been 11 000 ha per year over the period 1955–1989, with a higher rate of 18 000 ha per year between 1979 and 1989 (George, 1990). In the irrigated areas of south-west Western Australia, 11 400 ha are affected by shallow watertable (updated figure from an initial estimate by Consultative Committee for the Irrigation Strategy Study, 1990). The estimated area of saline seeps in South

Australia has increased by 170 000 ha since 1982, although part of this may be a consequence of better assessment methods being applied (Williamson, 1990). In Victoria, non-irrigated salinity is expanding at about 2 per cent per year. In Queensland, a survey conducted during 1990 showed that 10 000 ha were affected by human-induced salinity with a concentration in the central and south-east regions. The survey also indicated that a further 73 000 ha are susceptible to induced salinisation (Gordon, 1991). Based on limited data it was predicted in 1982 that the area of non-irrigated salt-affected land in Australia would be about 900 000 ha by the year 2000. However, given current trends, it seems this will be exceeded (Williamson, 1990).

Salinisation of the limited surface water resources of the South-west Coast and South Australian Gulf drainage divisions is particularly serious. As shown in Table 2-1, in the South-West Coast drainage division, out of a mean annual run-off of 6.67×10^9 m^3, only 2.87×10^9 m^3 (43 per cent) are classified as divertible for water supply purposes. Of these, 1.39×10^9 m^3 (48 per cent) are fresh, 0.47×10^9 m^3 (16 per cent) are marginal, 0.85×10^9 m^3 (30 per cent) are brackish and 0.16×10^9 m^3 (6 per cent) are saline. In other words, agricultural development in south-west Western Australia has led to extreme stream salinisation to the extent that 36 per cent of the divertible surface water resources is no longer potable and a further 16 per cent is of marginal quality.

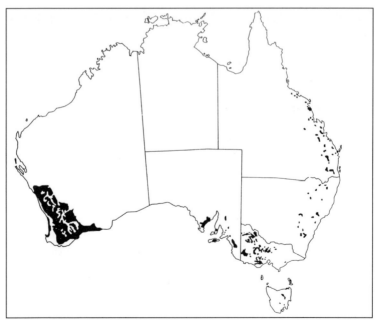

Figure 2-4: Schematic map of the areas susceptible to or containing dryland salinity (Schofield, 1992).

In the South Australian Gulf drainage division, of a mean annual runoff of 0.88×10^9 m^3, about 0.27×10^9 m^3 (30 per cent) are classified as divertible. And of these 0.16×10^9 m^3 (59 per cent) are fresh, 0.07×10^9 m^3 (26 per cent) are marginal, 0.03×10^9 m^3 (11 per cent) are brackish and 0.01×10^9 m^3 (4 per cent) are saline. Because of the shortage of freshwater in this drainage division, for the purpose of supplying water to major population centres like Adelaide, water has been transferred via a number of pipelines from the lower reaches of the River Murray (Engineering and Water Supply Department, 1987; Eastburn, 1990). On average about 35 per cent of the 200×10^6 m^3 yr^{-1} of water distributed through the Adelaide Metropolitan water supply system is pumped from the River Murray. This percentage rose to 90 per cent in the drought year of 1983 (Engineering and Water Supply Department, 1991).

In Australia, three regions have particularly severe human-induced salinity problems. These are: the south-west of Western Australia, South Australia and the Murray-Darling Basin in the south-east of the continent (Peck et al., 1983). Salinity problems in the south-west of Western Australia and in South Australia (State Dryland Salinity Committee, 1990a and b) are examples of dryland salinity, while the Murray-Darling Basin endures a complex combination of irrigated and dryland salinity. The following sections describe the south-west of Western Australia and the Murray-Darling Basin cases.

Dryland salinity in the south-west of Western Australia

Following European settlement in Western Australia in 1829 the population increased very slowly and reached only 180 000 by 1900. The growth of agricultural development was also slow. In 1900 there were only 30 000 ha sown to wheat in the whole of the state. The most rapid periods of development were 1900–1930 and 1955–1985; by 1985 the area sown to wheat exceeded 4.6 Mha (Schofield et al., 1988).

In the south-west of Western Australia dryland salinity associated with non-irrigated land, commonly caused by the removal of native vegetation, has been a threat to the land and water resources of the state, causing major social, economic and environmental problems. In this part of Australia, apart from a few observations recorded by the early settlers indicating that saline flows occurred east of the Darling Range, Bleazby (1917) and Wood (1924) were the first to discuss the impact of European settlement of Western Australia on stream water quality. Bleazby (1917) cited a number of cases where railway supplies had become saline. Wood (1924) gave the first clear reference to the association of agricultural clearing with stream salinity. The following is quoted from Wood's (1924) paper:

> For many years I have been interested in the fact that in certain districts in the southern portion of Australia where destruction of the native

vegetation has taken place rapidly, there has followed a very noticeable increase in salinity in the streams draining the area.

I first noticed this over 30 years ago in Yorke Peninsula, South Australia. Watercourses were rare but there were many depressions where soakage from higher ground gathered during the winter. In some of these depressions, gardens and orchards were planted with good results for a while, but in a few years, owing to a great increase of salt in the soil, many of them became quite useless and had to be abandoned.

Again some years later, about 1897, in the Northam-Toodyay district, I heard it suggested that destruction of the native vegetation turned the water in the creeks salty; and about 1904 I thought that I could see evidence of increase of salinity in the Goomalling Agricultural Area.

By 1905 a number of railway water supplies had become too salty for economical use in boilers, and various officers of the Department gave a lot of time and thought to the problem.

In the late years of the nineteenth and early years of the twentieth centuries Wood (1924) also observed an increase in the salinity of surface water resources in Cranbrook, Yarnening, Wagin, Tambellup and Blackwood Rivers, and provided some salinity data expressed in grain per gallon.

Wood (1924) put forward a hypothesis to account for the increases in stream salinity and the origin of the salt. Its main features are:

- The origin of the salt is oceanic, being continuously brought in from the sea via the atmosphere and deposited on the land as rain or dry fallout.

- The removal of native vegetation allowed more water to enter the deeper saline aquifers resulting in groundwater level rising to the soil surface near drainage lines and increased salinity of soil and streams.

According to Peck (1978) and Williamson et al. (1987), development of a salinity problem requires a source of salt, a source of water and a mechanism by which the salt may be moved to the soil surface or into a stream. In the south-west of Western Australia rainfall is the main source of the salt stored in the soil. Geographic distribution of the atmospheric salt shows that chloride (Cl^-) precipitation decreases with increasing distance from the coast. In 1973 chloride precipitation over the region was more than 100 kg ha^{-1} at coastal sites, decreasing to 50 kg ha^{-1} about 30 km inland and to 10 kg ha^{-1} at very inland sites (Hingston and Gailitis, 1976). The cyclic salt input in rainfall has accumulated in the soils over previous millennia, and there is no need to postulate that the salt in Western Australian soils has its origin in the Miocene sea (Teakle, 1937; Dimmock et al. 1974). Schofield et al. (1988) have reported the results of a number of investigations concerning the geographic distribution of soil salt content. Generally, research shows that soil salt content varies inversely with average annual rainfall. For example, with respect to the Darling Range, analysis of soil cores of some 40 laterite profiles to 40 m depth and covering a rainfall range of 560–1350 mm yr^{-1} reveals that salt storages were increasing systematically with decreasing annual rainfall, ranging from an average of 17 kg m^{-2} above 1000 mm yr^{-1} to 95 kg m^{-2} at 600 mm yr^{-1} (Dimmock et al., 1974).

Two basic types of soil salinity profiles (monotonic and bulge) were described by Johnston et al. (1980). For monotonic profiles salinity increases almost linearly from near the soil surface to some depth below, beyond which it is fairly constant. Bulge salinity profiles were defined as those having a maximum salinity at an intermediate depth in the profile.

Although soil salinity generally increases with decreasing rainfall, there is high local variability. Higher soil salinities accrue in valley floors and lower slopes than in high slopes and divide locations. There is also a good correlation between the clay content and salt content of the soil.

In major parts of south-west Western Australia, saline groundwater systems are usually found at a relatively shallow depth and provide a route for salt transport to streams. Groundwater is the major source of stream solute and its discharge is sensitive to both rainfall and land use.

A number of mechanisms, including transpiration, evaporation, overland flow, throughflow, stream flow, shallow and deep infiltration, groundwater movements, capillary action and salt leaching in salt-affected soils, contribute to land and water resources salinisation (Conacher et al., 1983a). In the case of stream salinisation, the discharge of salt to stream generally takes place by three mechanisms: direct run-off, groundwater discharge and throughflow discharge (Schofield et al., 1988).

Extent of dryland salinity

Dryland salinity has resulted in the loss of considerable areas of productive farmland. The Australian Bureau of Statistics has conducted farm surveys in Western Australia about every five years (1955, 1962, 1974, 1979, 1984 and 1989) to obtain estimates of the amount of cleared arable land that has become too saline for conventional crop and pasture species. In the first survey carried out in 1955, 73 436 ha or 0.5 per cent of cleared land was saline. The 1989 survey indicates that the salt-affected land has expanded to 443 441 ha or 2.83 per cent of 15.7 Mha of cleared arable land that was previously productive (George, 1990). The worst affected areas were in the 350 mm to 600 mm rainfall zone.

According to George (1990), the area of saltland increased at a rate of 11 000 ha per year between 1955 and 1989. The area cleared for agriculture has nearly doubled over the same period, although the rate of clearing has fallen since 1974 (Figure 2-5). It should be noted that the discrepancy caused by the results of the 1979 survey and the subsequent decline in 1984 is not generally supported by farmers' observations. Most report a rapid growth of salinity in the 1960s, followed by sustained growth throughout the 1970s and 1980s (Select Committee on Salinity, 1988).

Although a reduction in the rate of clearing may restrict the development of new areas of saltland in the future, it may take tens to hundreds of years to halt the increasing area of saltland resulting from past clearing. Estimates based on trends from the results of the six surveys conducted by the Australian Bureau of Statistics show that if appropriate land management systems are not implemented over large areas, about 7 per cent of the cleared farmland in the south-west of Western Australia could be salt-affected by the end of the next century (George, 1990).

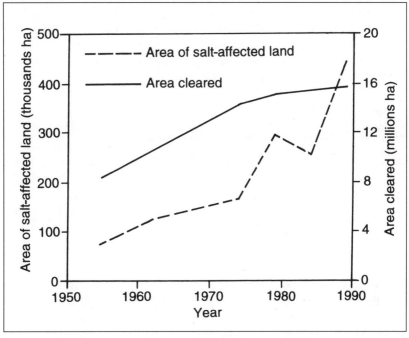

Figure 2-5: A comparison of the area affected by salinity in Western Australia and the area cleared for agriculture (George, 1990).

Extent of stream salinity

According to Schofield (1989), after the early observations and explanation of stream salinisation, the understanding and concern of water managers became surprisingly dormant for a long period and tended to decline with new generations. From 1930 to the late 1950s, salinity was largely regarded as an agricultural problem. There was little concern for water supplies because fresh sources were generally plentiful. However, the demand for water increased rapidly during the 1950s. The deterioration of water resources due to salinity became increasingly apparent through the 1960s and 1970s.

A regional distribution of the surface water resources, their salinity and rates of salinity change for south-west Western Australia is shown in Figure 2-6. It is clear from this figure that the major rivers of the region are already badly salinised and that their salinities are increasing. Streams arising in areas with rainfall greater than 1100 mm yr^{-1} are fresh whether forested or cleared. Fresh streams may also occur in lower rainfall areas where clearing is minor or absent. Below 900 mm yr^{-1} rainfall, streams are usually brackish or saline where clearing has been significant.

Rivers which show high rates of increase in stream salinity are given in Table 2-6. All these catchments have been partly cleared for agriculture. Two rivers are currently fresh (Capel and Preston) and six are marginal. Of the four catchments with periods of record greater than 40 years, the Frankland, Collie and Murray, but not the Warren, have shown substantially higher rates of salinity increase since 1965. Figure 2-7 shows the stream salinity trend of four rivers in the South-west Drainage Division.

Table 2-6: Stream salinity trends of major rivers.

Catchment	Period of record	Area cleared (per cent)	Average stream salinity over last 5 years of record (mg L^{-1})	Rate of stream salinity increase over period of record (mg L^{-1} yr^{-1})	Rate of stream salinity increase since 1965 (mg L^{-1} yr^{-1})
Denmark R	1960–86	17	890	25	26
Kent R	1956–86	40	1870	52	58
Frankland R	1940–86	35	2192	44	74
Warren R	1940–86	36	870	12	15
Perup R[a]	1961–86	19	3410	132	117
Wilgarup R[a]	1961–86	33	863	20	14
Blackwood R	1956–86	85	2192	52	58
Capel R	1959–76	50	423	15	14
Preston R	1955–75	50	354	8	11
Thomson R	1957–85	45	534	18	17
Collie R	1940–86	24	730	11	24
Murray R	1939–86	75	2792	39	93
Williams R[b]	1966–86	90	2425	95	95
Hotham R[b]	1966–86	85	3711	89	89
Woorollo Bk	1965–86	50	2092	44	39
Brockman R	1963–86	65	2040	76	72
Helena R	1966–85	10	1257	48	48

(a) Tributaries of the Warren River;
(b) Tributaries of the Murray River.

Source: Schofield and Ruprecht (1989).

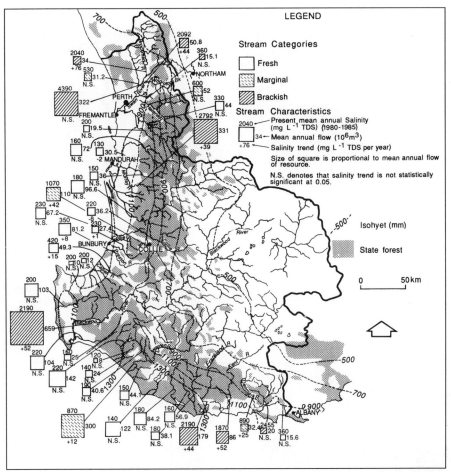

Figure 2-6: The distribution of surface water resources, salinities and rates of salinity change in relation to rainfall and forest cover in the south-west of Western Australia (Steering Committee for Research on Land Use and Water Supply, 1989).

The extent of salinisation is such that, in 1985, 36 per cent of the divertible surface water resources were no longer potable and a further 16 per cent were of marginal quality. For example, the largest river of the region is the Blackwood, with a mean annual flow of 0.66×10^9 m^3. About 85 per cent of the catchment was cleared in 1986 and the river reached an average salinity of about 2200 mg L^{-1} with an average annual increase rate of 52 mg L^{-1} (Schofield and Ruprecht, 1989). Additionally, considerable biological degradation of wetlands and rivers in low rainfall areas has occurred (Steering Committee for Research on Land Use and Water Supply, 1989).

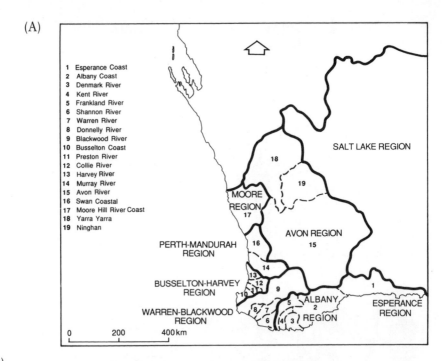

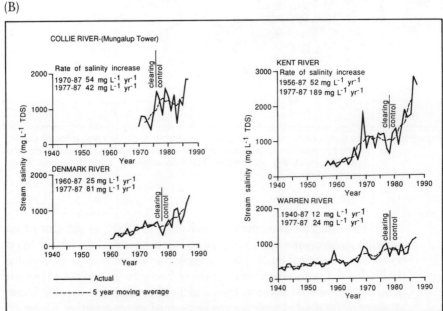

Figure 2-7: (A) River basins of the South-west Drainage Division (Schofield et al. 1988).
(B) Stream salinity trends of a number of catchments (Steering Committee for Research on Land Use and Water Supply, 1989).

Trends in stream salinity

Monitoring of flow and salinity data over the past 20 to 40 years has enabled a regional analysis of stream salinity trends. In fully forested catchments stream salinity has been declining over the past two decades (Table 2-7). The main reason for this is considered to be the general decline in groundwater level in response to generally lower rainfall conditions. A clear example is the case of the Bee Farm Road Catchment, where groundwater close to the gauging station fell 4.4 m from 1976 to 1981 and the annual stream salinity declined at an average rate of 128 mg L^{-1} yr^{-1}. The rainfall for the period was 10 per cent below the long-term average. The decline in groundwater level would mean a decrease in salt contribution to the stream both directly as groundwater flow and indirectly as salt movement to the surface soil layers during summer and subsequent leaching to streams during winter.

Table 2-7: Stream salinity trends of fully forested catchments in south-western Western Australia.

Catchment	Area (km²)	Period of record	Annual rainfall (mm)	Annual stream salinity (mg L^{-1})	Period of record salinity trend (mg L^{-1} yr^{-1})	Salinity trend for 1976–86 (mg L^{-1} yr^{-1})
Carey Bk	40.6	75–86	1420	116	+0.1	−0.1
Nth Dandalup R	153	39–86	1300	185	−1.5	−7.4
Little Dandalup R	396	67–86	1300	138	−3.7	−4.5
Harvey R	148	70–86	1250	114	−2.0	−2.2
Stones Bk	14.7	72–86	1220	181	−0.6	−3.7
Barlee Bk	164	62–86	1170	152	−0.2	−2.8
Yarragil Bk	72.5	51–86	1075	389	−7.7	−16.3
April Rd North	2.1	76–86	1070	122	−1.9	−1.9
Harris R	383	52–86	1000	224	−8.0	−5.4
Deep R	458	75–86	990	184	3.5	+4.9
Little Darkin R	40.1	67–86	900	390	−25.9	−19.7
Canning R	544	68–86	890	293	−17.0	−12.6
Pickering Bk	31.1	74–86	750	229	4.9	−12.1
Tunnel Rd	2.07	75–86	730	285	−52	−47.2
Bee Farm Rd	1.81	75–86	730	480	−106	−128[b]
Chalk Bk	104	58–86	700	293	−5.0	−11.7
Yarra Rd[a]	6.3	74–82	680	90	−8.3	−7.1

(a) Current salinity trend for Yarra Rd is from 1976 to 1982 due to closure of station in mid-1983.
(b) No flow recorded from 1984 onwards.

Source: Schofield and Ruprecht (1989).

In contrast to fully forested catchments, agricultural clearing almost invariably resulted in an increase of stream salinities. However, the degree of increase is highly variable between catchments, depending on such factors as annual rainfall, salt storage, groundwater hydrology, proportion of catchment cleared and clearing history (Schofield et al., 1988).

Annual flow-weighted mean salinities of minor rivers and all gauged tributaries of major rivers are given as a function of catchment average rainfall and percentage land cleared in Table 2-8. Stream salinities are low in all rainfall classes where there has been no clearing. The effect of clearing in areas above 1100 mm yr^{-1} rainfall is to marginally increase stream salinity but with average salinities remaining well in the fresh range. The average effect of clearing is to raise stream salinities from fresh to marginal in the 900–1100 mm yr^{-1} rainfall zone; to marginal or brackish in the 700–900 mm yr^{-1} rainfall zone; and to brackish in the 500–700 mm yr^{-1} rainfall zone. There were no uncleared catchments in the data set below 500 mm yr^{-1} rainfall but the streams were highly saline in catchments with clearing.

Table 2-8: The regional effects of agricultural clearing on stream salinity in annual rainfall classes.

Annual rainfall (mm)	Percentage of catchment cleared		
	0	0–50	50–100
>1100	144 (17)	176 (14)	233 (2)
900–1100	260 (13)	500 (4)	697 (2)
700–900	386 (5)	1095 (5)	756 (4)
500–700	70 (1)	1272 (1)	3488 (10)
<500		11988 (1)	19255 (5)

Stream salinity measurements in mg L^{-1}, averaged for the number of catchments shown in brackets.

Source: Schofield and Ruprecht (1989).

Changes in water and salt balance components

Agricultural development has modified the components of water and salt balance in south-west Western Australia. Several investigations have been carried out to clarify these changes. Peck and Hurle (1973) investigated the ratio of chloride input (atmospheric) to output (stream flow) for a range of forested and cleared catchments. They found that the output to input ratio (O/I) for forested catchments was in the range of 1.1 to 6, but in the range of 3 to 31 when more than 30 per cent of the catchment was used for agriculture. Recently, Williamson et al. (1987) investigated the effects of land clearing in the Collie River Basin (2830 km^2) on the O/I ratio of water and salt. Five small catchments of less than 4 km^2 were selected, two (Salmon and Wights) in a high rainfall zone and three (Dons, Ernies and Lemon) in low rainfall areas (Figure 2-8). Three of the catchments (Wights, Lemon and Dons) were cleared in 1977 and the other two (Salmon and Ernies) remained uncleared and have been used as controls. In the cleared catchments, forest vegetation had been removed and pasture or crop in the manner common to commercial agriculture in the region established.

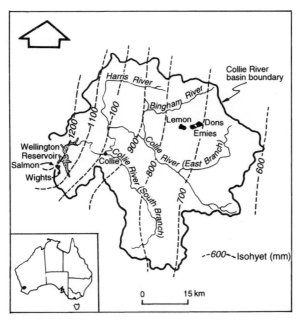

Figure 2-8: Location map for experimental catchments in the Collie River Basin. (Williamson et al. 1987).

The mean output to input (O/I) ratios for water over the ten-year period (1974 to 1983) in the Salmon and Ernies catchments were 0.10 and 0.01, respectively (Table 2-9). The mean O/I ratio for chloride in the Salmon catchment was 1.34, indicating a net loss of salt, for a mean annual salt fall of 7 gm^{-2}. In Ernies catchment the average chloride O/I ratio was 0.01 for a mean annual salt fall of 3.4 gm^{-2}.

Table 2-9: Characteristics of five experimental catchments and the water and chloride output to input (O/I) ratio over the period 1974–1983.

Catchment	Area (km²)	Area cleared (km²)	Mean annual rainfall (mm)	Water O/I			Chloride O/I		
				Min.	Max.	Mean	Min.	Max.	Mean
Western group:									
Salmon	0.82	Nil	1123	0.02	0.25	0.10	0.8	2.4	1.34
Wights	0.94	0.94	1027	0.02	0.48	0.24	1.5	14.6	7.48
Eastern group:									
Ernies	2.70	Nil	738	0	0.08	0.01	0	0.42	0.01
Dons	2.50	1.33	721	0.01	0.07	0.03	0.01	0.33	0.12
Lemon	3.44	1.84	737	0.01	0.07	0.03	0.01	0.37	0.18

Source: Williamson et al. (1987).

In Wights catchment clearing increased the O/I ratio of water from 0.07 in 1975 (preclearing) to 0.3 in 1980 and 0.48 in 1983 with a mean value of 0.24 over the period of ten years. The chloride O/I ratio also showed a substantial increase in export of salt with values changing from about 2 (in 1975) to 15 (in 1982) and a mean value of 7.48 over the period of ten years.

In the case of the Dons and Lemon catchments, which are in a lower rainfall area, the absolute values of O/I of water and chloride are about an order of magnitude lower than Wights. The reason is that these catchments are accumulating salt in response to clearing. Unlike Wights catchment, in Dons and Lemon catchments the deeper saline groundwater did not contribute to stream flow. The source of the additional salt load in these catchments was direct run-off and discharge from ephemeral perched groundwater systems in the lower valley section. It was predicted that the potentiometric surface of the aquifer would rise to the soil surface in the cleared valley of Lemon catchment during 1987 and in the partially cleared lower valley of Dons catchment by about 2010. It is predicted that the salt O/I ratio will increase subsequently to more than one.

Williamson and van der Wel (1991) investigated the O/I salt ratio in the Mt Lofty Ranges (South Australia), as a comparison between the O/I ratio of salt in Western Australia and South Australia. There, land use is also dominated by agricultural activities (with a lower degree of land clearing) and rainfall ranges from 500 to 1000 mm yr^{-1}. They found that the salt O/I of 21 catchments ranged from 0.5 to 8.2 with a mean value of 4.2, showing a more than fourfold increase in salt output compared to what it would have been had land remained under native vegetation.

Peck and Williamson (1987a) studied the groundwater systems and the effects of forest clearing in the five experimental catchments of the Collie Basin. Results of a number of investigations in these catchments have been published in a special issue of the *Journal of Hydrology* (Peck and Williamson, 1987b). More recent results are available in Schofield and Bari (1991), Bari and Schofield (1991 and 1992) and Ruprecht and Schofield (1991).

Management options

Appropriate management of salt-affected land and streams depends partly on the climatic, hydrologic and soil properties of the salt-affected area, as well as the economic constraints. In Western Australia, a range of rehabilitation methods have been investigated in the past. Those methods currently favoured include tree and shrub planting, agronomic manipulation and land drainage (Schofield et al. 1989). Evidently the appropriate rehabilitation technique for a particular catchment would in many cases be a mixture of these techniques, the extent of each depending on the local conditions and the management objectives. Tree planting strategies are widely regarded as potentially the most effective means of controlling stream and land salinity problems. This option is particularly attractive because only part of the agricultural land requires reforestation, allowing substantial agricultural production to continue, and because tree plantations could be economically viable in their own right (Schofield et al., 1989). Conacher et al. (1983a) argue that revegetation with indigenous eucalypts and woodland species in order to restore hydrological equilibrium is not a practicable economic proposition for the wheat belt farmers. Empirical evidence suggests that at least 50 per cent of salt-affected land or saline stream catchments would have to be revegetated. The alternative of using commercial crops (including tree) and pasture species which would simulate the hydrological behaviour of indigenous vegetation is an attractive and perhaps ideal alternative.

Schofield et al. (1989) have reviewed four agricultural strategies which may exert some level of salinity control. These are: perennial pastures; surface drainage; fodder crop trees in recharge zones; and halophytic fodder shrubs in discharge zones. The concept of integrated catchment management has been adopted for the Denmark River catchment, with the objective of reducing stream salinity by maximising land productivity (Schofield et al., 1989 and 1991). Reforestation options with commercial potential were seen as a means of achieving this objective. The following sections describe some of the aspects of land and stream rehabilitation.

Dryland salinity management

The rehabilitation of salt-affected land can rarely be achieved by working on the salt-affected area itself. It is often necessary to modify the land use on the catchment area above the saltland. Replanting trees and shrubs in the recharge areas of a catchment will increase the evapotranspiration, reduce the groundwater levels and subsequently will alleviate salinity problems. Planting trees and shrubs on recharge areas of agricultural catchments is advocated by the Western Australian Department of Agriculture. While it is not economic to indulge in large-scale planting, there are often areas in the catchments which can be identified as specific recharge areas which are small, make a major contribution to recharge and do not produce economic crops or pastures which could be planted to trees or shrubs (Western Australian Department of Agriculture, 1988a).

Treatments applied on or immediately around discharge areas can be either of engineering type, such as drains or banks, or biological, such as shrubs, grass or tree plantations. The Western Australian Department of Agriculture has devoted considerable resources over a number of years to select appropriate species and develop establishment techniques for revegetation of saltland. Malcolm (1982) has produced a guide to the selection of salt-tolerant shrubs for forage production from salt-affected sites in the south-west of Western Australia. Important criteria are site hydrology, severity of salinity and rainfall.

According to Malcolm (1986), forage production on saline soils is possible at economically acceptable levels. In addition, growing halophytes has other benefits such as erosion control, lowering watertables and soil amelioration. More recently, Malcolm (1990) described the establishment of 1200 ha of saltbush in the North Stirling District of Western Australia, where about 30 per cent of the land is salt-affected. In this district saltland is in the valley floor and hillside seepages. Land is highly saline and will not grow normal crops and pastures, and farmers regarded it as wasteland. According to Malcolm (1990), farmers have discovered this establishment of saltbush useful for grazing and reported a net return of A$125 (US$100) per hectare per annum, for land which was previously regarded as wasteland.

Aside from revegetation, a number of engineering methods for the control of salinisation and waterlogging have been tried. Conacher et al. (1983a and b) have evaluated the effectiveness of throughflow interception for controlling salinisation and waterlogging in the wheat belt. In this region, a large and influential group of farmers considered the causal mechanism of waterlogging and salinisation to be the development of perched soil water in low-lying areas, and not a rising watertable. Consequently these farmers have constructed graded and contour interceptors (shallow ditches and banks) to divert throughflow from salt-affected areas. An evaluation of this scheme has been undertaken, based on interviewing 16 farmers, farming on 22 018 ha of which 19 906 ha were cleared and 2399 ha salt-affected. It showed that interceptors had only partly achieved their objectives. All 16 farmers experienced problems with their interceptors. The most common difficulty, reported by 10 farmers, was leakage through imperfectly sealed banks. This investigation showed that following interceptor construction, the amelioration of salt-affected land was only partial. In diminishing order of success, improvements reported after at least two years are: (a) the reduced extent of waterlogging; (b) pasture quality; (c) crop yield; and (d) soil properties other than waterlogging. Such improvements have generally occurred on some but not all parts of individual farms, and in general there has been little improvement of severely salt-affected land.

According to Conacher et al. (1983a), deep draining has also been constructed to lower watertables in the wheat belt. However, the low hydraulic conductivity of most wheat belt subsoils would necessitate a close spacing (40–50m) of such drains. In general, deep drainage may not be even technically feasible in some locations, and it would certainly not appear to be an economic proposition for wheat belt farmers. The value of their production per unit area is considerably less than in irrigated areas where deep drainage is a common practice for the control of waterlogging and salinisation.

Tube-drainage has also been used throughout the agricultural areas, but has rarely proved cost-effective. In areas where groundwater levels have been lowered, tube-drains were installed at a shallow depth of about 1m and close spacing of 15 m (McFarlane, 1991). The pumping of saline water from aquifers has also been tried in agricultural areas (George and Nulsen, 1985).

A major problem with drainage and pumping is the disposal of saline effluents. A recent regulation requires landholders to notify the Commissioner of Soil Conservation at least 90 days before discharging saline effluents onto other lands or into water bodies or a watercourse. For the purpose of notification, saline water is defined as having a TDS of more than 2000 mg L^{-1} or being more saline than the water into which it is being discharged (McFarlane, 1991). Further information concerning the application of engineering methods for the control of salinity and waterlogging problems is available in a number of articles in the *Journal of Agriculture — Western Australia* (1985, Volume 6, No. 4. 106–135).

Recent investigations show that a combination of both engineering and biological treatments is most effective for combating dryland salinity problems. George (1991) reported the effectiveness of such a combined management strategy applied to a number of salt-affected sites in the wheat belt, where soil salinity is caused by the discharge of regional and perched perennial groundwater systems. In each reported case, the combined method was designed to allow restoration of the saline area to a condition capable of returning to the previous farming system, and at the same time to minimise costs, the time needed for reclamation and the amount of land taken out of production. According to George (1991), groundwater interception using drains has been successful in reclaiming saline soils within 1–2 years and plantations of eucalypts (*E. globulus, E. cladocalyx, E. camaldulensis*, etc.) were capable of intercepting groundwater and reclaiming seeps within five years. Evidence from one site suggested that 300 eucalypts were sufficient to intercept and transpire approximately 1000 m^3 of brackish to saline groundwater and to transpire 365 mm of rainfall on the site. Consequently, water-levels within the sandplain seep and soil salinities in the affected area were reduced. Then wheat was established on the 12 ha previously salt-affected and waterlogged area. Despite above-average rainfall, the crop established and produced a yield of approximately 1.2 tonnes ha^{-1} (district average). George (1991) concluded that sandplain seeps can be rehabilitated by drains and strategically placed plantations of trees. While these methods may be sufficient on their own for seep controls, it is considered that the inclusion of high water-use annual and perennial pasture and crop species should be encouraged.

Stream salinity management

For the purpose of stream salinity management, surface water catchments of the south-west have been classified into a number of groups depending on their salt hazard and extent of agricultural clearing (Schofield, 1989). The principal groups are: forested; marginally cleared; extensively cleared; and totally cleared.

The forested catchments have freshwater resources and are being managed with a policy of high protection of water quality. The main activities in these catchments are bauxite mining and forest operations. Research programs have been established to assess likely impacts and to develop appropriate operational guidelines so as to minimise any potential increase in stream salinity resulting from these operations.

The marginally cleared catchments have stream salinities within or close to the marginal water category (500 < TDS < 1500 mg L^{-1}) and often have an increasing salinity trend. These catchments are predominantly forested but extend inland to high salt hazard areas which have been partially cleared for agriculture. Despite the legislative actions to control land alienation during the 1960s and clearing in the late 1970s, stream salinities have continued to increase at an alarming rate (e.g.: Kent, 58 mg L^{-1} yr^{-1}; Collie, 42 mg L^{-1} yr^{-1}; Denmark, 26 mg L^{-1} yr^{-1}). Further supplementary action is required to halt stream salinity increase if these catchments are to remain viable potable water resources for future water supplies. Partial catchment reforestation has been the most promising approach of a range of possible options.

The extensively and totally cleared catchments include some of the major water resources of the south-west, namely the Murray, Blackwood and Avon Rivers. These rivers are now well into the brackish and saline categories and their reforestation is not an economic proposition at the present time. However, the tributaries of these basins which are forested and yield potable water could be developed for water supply.

Catchments suitable for partial reforestation have similar hydrological features. In their high rainfall parts they are characterised by high volume, low salinity yield, whereas the opposite holds in their low rainfall areas. Typically 80 per cent of the total salt load but less than 30 per cent of the total stream flow volume is generated from the agriculturally developed land below the 900 mm isohyet (Schofield, 1989). In these catchments the control of salinity can be achieved only by reducing salt discharge from the cleared land below the 900 mm isohyet. This objective could be achieved by partial reforestation of these catchments.

Research to develop appropriate strategies for partial reforestation has involved tree selection, tree establishment, and an evaluation of a range of tree planting strategies with respect to groundwater and stream salinity control.

The initial criterion for tree selection is adaptation to the environment, which in south-west Western Australia includes adaptation to climate, soils, pests, diseases, fire and in some locations waterlogging and salinity. A few species have been identified which are adapted to waterlogging and salinity (saline seep) conditions. These include *Casuarina obesa*, *Eucalyptus sargentii*, *E. camaldulensis* and *E. occidentalis*. These species can provide some secondary benefits in addition to salinity control but not sufficient to make them commercial propositions based on returns from their secondary benefits only.

For well-drained soils, away from saline seeps, a wide range of adapted species have been identified. Several are considered to have commercial potential. These include *Pinus radiata* for timber and other wood products and the fast growing *E. globulus*, *E. viminalis*, *E. botryoides* and *E. saligna* for pulpwood production.

Having satisfied the adaptation criterion, the other two main criteria are water use and commercial value. Knowledge of the water use capacity of various species for a range of conditions could allow trees to be selected to maximise the effectiveness of reforestation or minimise the area of agricultural land required for reforestation. Analysis of the results of a few studies on the transpiration of trees showed that the tree species *Pinus radiata*, *Eucalyptus cladocalyx*, *E. globulus*, *E. leucoxylon*, *E. maculata* and *E. wandoo* annually transpire significantly more than clover (*Trifolium subterraneum*), as probably do most other tree species (Schofield et al., 1989).

The commercial value of species has been easier to determine and there are good data on a number of eucalypt and pine species. An economic analysis of *P. radiata* agroforestry in the Manjimup region found that agroforestry could be substantially more profitable than grazing alone in the long term. To overcome the long wait for returns from the timber crop, the State Government has developed the Softwood Sharefarming Scheme whereby a farmer can receive from the State an annuity plus a share of the revenue at harvesting. Field trials with wide-spaced eucalypt plantations have shown that some eucalypts are also suitable for agroforestry. A number of eucalypt species also show promise for growing in dense plantations to produce pulpwood. Good quality farmland could give timber yields ranging from 15 cubic metres per hectare per year (m^3 ha^{-1} yr^{-1}) at 600 mm yr^{-1} rainfall to more than 25 m^3 ha^{-1} yr^{-1} at 900 mm yr^{-1} rainfall (Schofield, 1989).

Tree establishment requires site preparation, which includes ripping in late summer/autumn and herbicide spraying around May for weed control prior to hand planting of the seedling in June or July. In non-seep sites, reforestation may be achieved by direct seeding or the planting of seedlings. Direct seeding may be the cheaper method, but planting is generally more reliable. Fertilising is also required. The current practice is to fertilise all seedlings, either at planting time or within one month of planting (Schofield et al., 1989). In saline seep areas seedlings are planted into ridge mounds. Research indicates that a double ridge mound with seedlings planted in the trough between the ridges gives better establishment success than a single ridge mound.

In order to evaluate reforestation strategies a number of experimental sites were established in the late 1970s (Table 2-10 and Figure 2-9). Sites have been evaluated in terms of their effectiveness in controlling groundwater level and salinity, and in timber, pulpwood and pasture production. In the

Table 2-10: Characteristics of experimental reforestation sites.

Site	Planting year	Main species planted	Proportion of site cleared (per cent)	Proportion of cleared area replanted (per cent)	Initial planting density (stems ha⁻¹)	Estimated stem density in 1986 (stems ha⁻¹)	Mean reforestation crown cover at Dec 1987 (per cent)	Initial[a] depth to watertable (m)	Initial groundwater salinity (mg L⁻¹)	Decline in watertable (1979–88) (m)
Flynn's Farm										
Landscape	1977	E. wandoo, E. camaldulensis, P. pinaster, P. radiata	98	8	670	500	43	2.1	4600	0.9
Hillslope	1978/1979	E. camaldulensis, E. wandoo	100	54	1 200	1 000	29	3.3	7400	2.8
Agroforestry	1978	P. radiata, P. pinaster, E. camaldulensis	51	58	380, 760, 1 140	75, 150, 225	14	4.4	2400	1.2
Stene's Farm										
Strip plantings	1976 to 1978	E. camaldulensis, P. radiata, E. globulus, P. pinaster, E. wandoo	31	14	1 200	600	47	2.7	7700	0.5
Valley plantings	1979	E. wandoo, E. rudis, E. camaldulensis	44	35	625	500	41	6.3	5400	2.0
Agroforestry	1978	E. camaldulensis, E. sargentii, E. wandoo	25	57	1 250	150, 900	25	2.7	6600	1.8[b]
Arboretum	1979	63 eucalypt plus 2 pine species	35	70	625	0–600	39	7.1	5400	5.6

(a) Depth corresponds to the lowest groundwater level during the year;
(b) Decline in watertable (1982–88).

Source: Steering committee for Research on Land Use and Water Supply (1989).

following paragraphs, only the effects of reforestation on groundwater levels and salinity will be discussed. Other aspects are described in Schofield et al. (1989 and 1991).

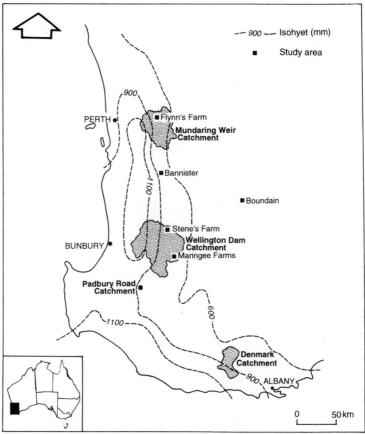

Figure 2-9: Location of experimental reforestation sites (Steering Committee for Research on Land Use and Water Supply, 1989).

Four partial reforestation strategies were tested:
- Dense plantation in lower slopes and groundwater discharge areas.
- Wide spaced plantation covering most of the cleared area in which trees are planted at wide spacing (low density) with pasture grazing between trees.
- Strip or small blocks strategically placed planting covering a small proportion of the cleared area allowing agriculture to be continued on the unplanted areas.
- Dense plantation covering more than 50 per cent of the cleared area.

In each strategy a component of lower slope and/or discharge area was included in the plantation. This was considered necessary if groundwater solute discharge to the stream was to be eliminated (Schofield, 1989).

Monitoring of groundwater levels on all reforested sites showed that watertables were lowered over a 7–10 year period. As shown in Table 2-10, watertables declined 0.5 to 5.6 m. In general the rate of decline was higher during the early years of plantation, reduced in the following years and even plateaued in the latest years of monitoring in some of the sites (Flynn's Farm Hillslope) while it continued to decline at Stene's Arboretum (see Schofield et al. (1989) for the graphs of the annual rainfall and groundwater level variations at the experimental sites).

According to Schofield et al. (1989, 1991) the effectiveness of the various reforestation strategies in reducing groundwater level beneath plantation has been found to depend primarily on the area and density of trees planted. A regression of average rate of watertable reduction against proportion of cleared area reforested is shown in Figure 2-10A. The four sites included in the regression have similar crown covers (39–47 per cent) whereas the two sites excluded have significantly less crown cover (Flynn's Hillslope 29 per cent, Flynn's Agroforestry 14 per cent). Over the measurement period the groundwater levels beneath pasture at Flynn's and Stene's sites were lowered on average by 6 mm yr^{-1}. Using this information and the regression in Figure 2-10A, it can be shown that reforestation of 22 per cent of the cleared area is required to lower the watertable at a rate of 200 mm yr^{-1} (i.e. 2 metres over 10 years) relative to the ground surface at these sites with about 700 mm yr^{-1} mean annual rainfall (Schofield et al., 1991).

Rainfall over the measurement period was 10 per cent less than the long-term (1926–88) average. If rainfall had been the long-term average, it is estimated that groundwater levels beneath pasture on Flynn's and Stene's sites would, on average, have risen at a rate of 360 mm yr^{-1}. In this case the regression of Figure 2-10A indicates that about 52 per cent of the cleared area would need to be reforested to lower the watertable at 200 mm yr^{-1} relative to the ground surface.

A second regression, which takes into account the crown cover of the reforestation, is shown in Figure 2-10B. In this example the average rate of watertable reduction was regressed against the 'total percentage tree cover' as represented by the product of proportion of cleared land reforested and reforestation crown cover. The quality of this regression, which includes all sites, implies that total percentage tree cover is the most important factor (for the given reforestation strategies) in the lowering of the watertable beneath reforested areas.

Groundwater salinity response to reforestation

In the past, concern has been expressed about the potential of salt concentrations to increase beneath reforestation stands located in valley floor

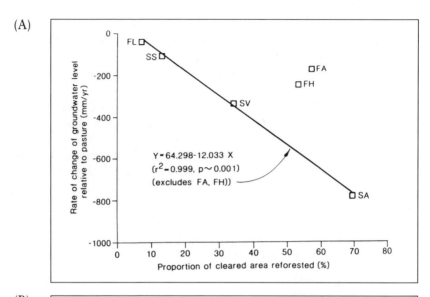
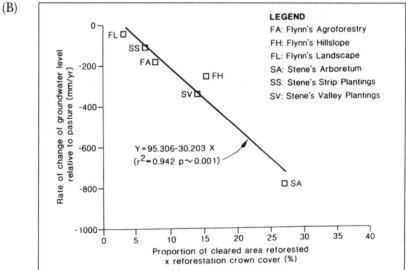

Figure 2-10: Dependence of rate of change of groundwater level under reforestation relative to pasture on: (A) proportion of cleared land reforested, and (B) product of the proportion of cleared land reforested and crown cover of reforestation (Schofield et al., 1991: 28).

and lower slope locations and affect their long-term viability. However, according to Schofield (1989), monitoring of groundwater salinity at all sites above 700 mm yr^{-1} of rainfall showed that salinities at the watertable have decreased. For example, at Flynn's Farm and Stene's Farm sites over the period 1979–86, the arithmetic mean of salinity data decreased 32 per cent for pasture and 12 per cent for the forested area. However, at one site (Boundain) with annual rainfall of about 500 mm yr^{-1}, groundwater salinities have increased below reforestation, while below pasture, groundwater salinity decreased or remained static. There is no obvious explanation why groundwater salinity beneath reforestation increased at this site, but not at the other sites, although rainfall may be a factor (Schofield et al., 1989). At this site (Boundain), measurement of soil salt content showed that under reforested areas, salt content decreased over the depth range of 0–0.75 m, increased over the depth range of 0.75–2.25 m and remained the same below 2.25 m. Under the pasture there was little change in the salt content of the profile (Steering Committee for Research on Land Use and Water Supply, 1989). According to Schofield et al. (1989), the general decrease of groundwater salinity beneath reforestation stands over the period 1976–86 shows that early concerns that lower slope plantation could be adversely affected by increasing groundwater salinity have not materialised.

Effect of reforestation on stream flow, salinity and salt load

The effect of reforestation on stream flow, salinity and salt load has been evaluated at Padbury Road catchment, located about 180 km south of Perth (Bell et al., 1987). This small catchment of 0.9 km^2 with a mean annual rainfall of 880 mm was progressively cleared from early this century and by 1977 only 34 per cent of the catchment remained under native forest. Due to clearing, the stream salinity increased from about 500 mg L^{-1} to 1000 mg L^{-1} TDS from 1960 to 1975. Between 1977 and 1983 the catchment was planted with *Pinus radiata*, *Eucalyptus globulus* and *E. resinifera*. After reforestation stream flow decreased and by 1986 was only 9 per cent of the predicted flow if the catchment had not been forested (a reduction of 91 per cent). Over the period 1981–86 stream salinity increased in four out of six years with the maximum being 531 mg L^{-1} in 1984 when the annual rainfall was 732 mm and much below the average. Salt discharge has decreased over time to 10 per cent of the predicted salt discharge without reforestation in 1986 with an annual rainfall of 607 mm. During 1986 predicted salt discharge was 0.14 t ha^{-1} while the observed value was 0.24 t ha^{-1}.

Operational reforestation for salinity control

In the late 1970s, the Water Authority of Western Australia declared an interest in a number of potentially high value water resource catchments which had marginal but deteriorating salinities. These catchments were the Mundaring, Wellington, Warren, Kent and Denmark. Clearing controls were placed on these catchments in 1976 and 1978 (Schofield et al., 1991). The Mundaring catchment is the source of water for the goldfields and agricultural water supply schemes. It required little reforestation, which is completed, and the inflow salinity to the reservoir has stabilised at about 500 mg L^{-1} TDS.

Wellington has the largest water yield of any individual reservoir in the south-west of Western Australia (approximately 100×10^6 m^3). The continued increase in its inflow salinity has necessitated an active program of catchment management and rehabilitation. The catchment spans a rainfall range from over 1200 mm yr^{-1} in the west to 600 mm yr^{-1} in the east of the catchment (Figure 2-8). Prior to agricultural development the salinity of the Collie River at Wellington Dam has been estimated to be between 200 to 250 mg L^{-1}. Agricultural development commenced at the turn of the century and expanded slowly over the following 30 years. Growth in agricultural development virtually ceased through the Depression years and it was not until the 1950s that agricultural development again expanded. In the late 1950s land clearing accelerated. This was reflected in increased salinity through the late 1960s and early 1970s (Figure 2-11). In 1961 the State Government prevented further alienation of Crown Land in the catchment. In November 1976 the State Government introduced legislation to control further clearing. Application of the legislation has effectively held the level of clearing to 64 000 ha or 23 per cent of the total catchment and has avoided the expansion of agriculture to a possible 100 000 ha or 35 per cent of the total catchment (Schofield et al., 1989).

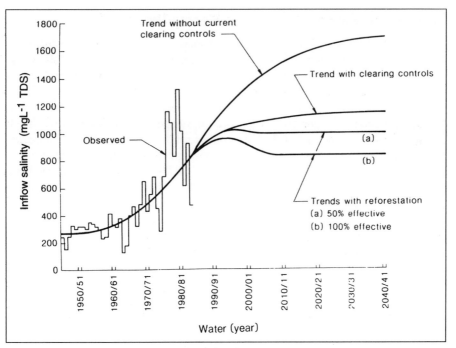

Figure 2-11: Observed and predicted inflow salinities to Wellington Dam for various salinity control measures (Schofield et al., 1989).

Following consideration of a number of options, reforestation of cleared farmland in the drier, high salt-yielding part of the catchment was commenced in 1979/80. The areas planted have been concentrated in the eastern and south-eastern portion of the catchment, where the annual rainfall is usually less than 750 mm. The reforestation strategy involved planting along the valley floors and lower side slopes. By 1990, 6200 ha had been reforested (Schofield et al., 1991).

Figure 2-11 shows the observed and predicted inflow salinities to Wellington Dam for various salinity control measures. It shows that without current clearing controls salinity would increase to about 1800 mg L^{-1} by the year 2040, while with the clearing control it would reach a maximum of 1150 mg L^{-1} in the same year. The effect of the current reforestation program on the inflow of salinity in a year of medium inflow has also been investigated. Two levels of control over salinity have been assumed. If the program is moderately effective (reducing groundwater discharge by 50 per cent), then average inflow salinities are likely to peak in the mid-1990s and return to about 1000 mg L^{-1} by the year 2010. If the reforestation is 100 per cent effective and reduces groundwater discharge to zero, then it is predicted that salinities could be returned to 850 mg L^{-1} TDS by 2010. Much higher salinities will of course occur in drought years.

Salinity and waterlogging in the Murray-Darling Basin

Major salinity problems permeate the south-east of the Australian continent in the Murray-Darling Basin (MDB). It consists of two drainage basins (the Murray and the Darling) separated by a ridge of low uplands and covers 1.06 million km^2 or one-seventh of the surface area of Australia (Figure 2-12). About 57 per cent of the basin is in New South Wales, 25 per cent in Queensland, 12 per cent in Victoria and 6 per cent in South Australia. It produces about one-third of Australia's total output from rural industries, supports 25 per cent of the nation's cattle and dairy farms, about 50 per cent of its sheep, lambs and cropland and almost 75 per cent of its irrigated land. The production is valued at some A$10 billion (US$8 billion) annually (Murray-Darling Basin Ministerial Council, 1989).

The Murray-Darling Basin has a mean annual rainfall of 430 mm, varying from 1500 mm in the east and south-east of the catchment to about 250 mm in the south-west. The basin covers a wide range of climatic zones: subtropical in the north, humid in the eastern Highlands, temperate in the south and hot and dry in the west. However, based on daily total solar radiation, air temperature and precipitation data at 397 stations, Nix and Kalma (1982) divided the MDB into three broad climatic zones and 13 subregions:

- **Climatic Zone A** extends inland from the southern and eastern coastline no more than 200 km and includes much hilly, mountainous and tableland terrain that bounds the MDB in the south and east. It is a relatively humid zone with an average annual rainfall of 953 mm.

- **Climatic Zone B** covers the Murray-Murrumbidgee lowlands and has a much reduced average rainfall of 437 mm yr^{-1}.
- **Climatic Zone C** covers that part of the Darling River Basin with an average annual rainfall of 318 mm.

Subdivisions within these zones reflect the general trends of increasing aridity from east to west and an increasing incidence of summer rainfall to the north.

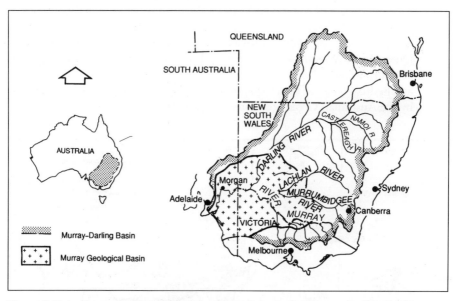

Figure 2-12: Murray-Darling Basin in south-east Australia (Murray-Darling Basin Ministerial Council, 1988).

The Murray-Darling Basin consists of 26 major catchments. Run-off within the basin is highly variable. About 60 per cent of the basin, mainly in the south-west and western regions, provides little or no run-off to rivers. The Darling River and its tributaries contribute 12 per cent to the River Murray. The Murrumbidgee contributes 13 per cent and River Murray tributaries upstream of the Murrumbidgee Junction contribute 75 per cent of the flow of the River Murray (Murray-Darling Basin Commission, 1988).

The annual run-off of the Murray-Darling system varies markedly from 3.7×10^9 m^3 (drought year) to 48×19^9 m^3 (flood year), with the average annual run-off being 24.3×10^9 m^3. These large variations in stream flow are characteristic of Australian rivers but not of large rivers elsewhere in the world. Surface water resources of the basin are intensively regulated, consisting of a complex system of locks and weirs, reservoirs and lakes (Figure 2-13). The two largest reservoirs are the Dartmouth and the Hume with storage capacities of 4×10^9 m^3 and 3.04×10^9 m^3, respectively (Jacobs, 1990). Total water storage capacity of the basin is 30×10^9 m^3 or more than 1.2 times the average annual run-off of the basin, and around 90 per cent of water used in the basin is used for irrigation (Simmons et al., 1991).

Geology

The Murray and the Darling drainage basins have a different geological history, and this has implications for the amount and source of soluble salts they now contain (Williams, 1991). The geology of the Darling Basin has been the subject of a number of studies. Detailed summaries of the geology in the New South Wales portion of the Basin can be found in Packham (1969), while Day et al. (1983) provide similar information for the Basin in Queensland. Wasson (1982) gives a brief summary of the geology and geomorphology of the entire Murray-Darling Basin. However, the geology and hydrogeology of the Darling Basin is presently subject to an investigation by the Australian Geological Survey Organisation.

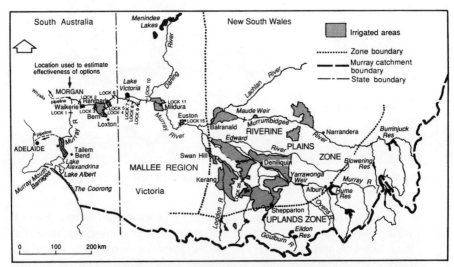

Figure 2-13: River Murray system, regulation works and irrigation regions (Murray-Darling Basin Ministerial Council, 1988).

The geology of the Murray Basin has been described by Brown and Stephenson (1991) and Brown (1989). It is a shallow sedimentary basin about 600 m in depth, filled with Tertiary and Quaternary sediments, and covers an area of about 300 000 km^2. Sedimentation commenced about 60 million years ago in the basin. Sediment is of fluvial, fluvio-lacustrine, shallow marine and aeolian origin.

The surface of the Murray Basin is almost entirely mantled by Quaternary sediments which provide a record of fluctuations in climate and groundwater levels. In the east, the flat landscape of the Riverine Plain is underlain by fluvio-lacustrine sediments, whereas in the west aeolian sediments dominate the landscapes of the Mallee Region. Within these landscapes active and fossil groundwater discharge complexes can be identified. The extent of fossil discharge complexes shows that salination of the landscape has been widespread in the past 0.5 million years (Brown, 1989).

Palaeoclimate and salinity

Climatic changes have had a major influence on the sedimentary history of the MDB. These changes, particularly over the past half-million years, have had profound effects on river flow, lake formation and accumulation of salt across the basin. While the climatic changes of the Darling Basin are not well understood, Bowler (1990) investigated the climatic changes of the Murray Basin. According to Bowler (1990), in the past six million years there have been a number of major environmental changes in the Murray Basin:

- Six million years ago the climate was relatively wet with high summer rainfall and rainforest extended across the landscape.
- About four million years ago the climate became drier and the sea began to retreat. In the western part of the Murray Basin, the change to a drier climate was accompanied by the development of a great lake named 'Lake Bungunnia'.
- Over the past million years Lake Bungunnia began to dry up and the climate underwent a series of climatic oscillations.
- In the past 500 000 years there have been several cyclic climatic changes. These changes have produced the modern landscape of the Murray Basin with its dunes, lakes and river channels. The crescent-shaped dunes or 'lunettes' occurred on the eastern margin of the lakes which are scattered throughout the entire Murray Basin.
- About 55 000 years ago the surface depressions of the Murray Basin were all brim full with freshwater.
- About 36 000 years ago a substantial change occurred. The availability of surface water diminished. The lakes became partially dry with increased salinity.
- About 18 000 years ago at the time of the glacial maximum, the dry phase intensified causing lakes to disappear and generally creating an inhospitable environment. At this stage salts which concentrated at the surface were blown across the landscape, producing a salinised environment. During the subsequent wetter period much of these salts was then available to be leached through the soil and into groundwater. This set up a recycling

system from west to east by aerial transport, and from east to west by leaching, streams and groundwater flow. The landscape of the Murray Basin at the last glacial phase was a largely saline wasteland and the salt recycled by these processes remains in the groundwater of the basin today.
- In the warming period since the glaciation, or about 13 000 years ago, recovery from the saline phase started and trees returned to the landscape, watertables fell and many saline lakes dried. However, processes involved in this recovery are not adequately understood.

Origin of salt

Salt in the Murray-Darling Basin can originate from the atmosphere as cyclic salts, blown in from the ocean; it can be released by chemical weathering of rock materials; and it can be leached out of the pore structure of sediments and rocks of marine origin that underlie and surround much of the basin (Williams, 1991). Redistribution of soluble salts accumulated in the Murray-Darling Basin is indicated by inland salt lakes, dry salinas and areas of saline soils. Aeolian movement of salts may include transfers from the ocean and from salinas in central Australia.

Blackburn and McLeod (1983) sampled the rainwater at 24 stations within the MDB and three stations near the South Australian coast. Samples were analysed for calcium, magnesium, potassium, sodium, chloride and sulphate, while carbonate and nitrate were estimated. Their analysis showed that:

- The chloride concentration ranged from 13 kg ha^{-1} (at 200 km from the coast) to 3.4 kg ha^{-1} (at 650 km from the coast) with a mean value of 7.5 kg ha^{-1}. For the coastal stations chloride concentration ranged from 77 kg ha^{-1} (at 35 km from the coast) to 40 kg ha^{-1} (at 100 km from the coast) with a mean value of 57 kg ha^{-1}.
- The aggregate concentration for eight ions in rainwater at the stations within the basin ranged from 3.3 mg L^{-1} to 35 mg L^{-1} with a mean value of 9 mg L^{-1}. Higher concentrations in the basin refer principally to stations in a semiarid district and are associated with relative abundance of calcium and carbonate derived from calcareous soils. For the three stations near the South Australian coast, concentrations ranged from 14 mg L^{-1} to 23 mg L^{-1} with a mean value of 18 mg L^{-1}.
- The mean annual deposition rate of salt was 34 kg ha^{-1} for stations within the basin, compared to 137 kg ha^{-1} for coastal stations.
- Comparison of the mean annual ionic concentration for stations within the basin with those of inland Western Australia indicates that ionic concentrations for the MDB are much lower.
- Values for the ratio of calcium ions to chloride ions are generally at least ten times the value for seawater, indicating that terrestrial sources were significant for the calcium ion concentration.

Blackburn and McLeod (1983) have estimated the oceanic input of chloride and compared these estimates with discharge of chloride by the Murray and Darling Rivers. They concluded that in 1974–75 the oceanic input of chloride to the Darling catchment was mostly retained there, but the accession to the Murray catchment was all flushed out, with an input-output ratio of 1:3. The sources of excess chloride are salts leached from the soil profile during irrigation, seepage of saline groundwater and discharge from saline aquifers (Williams, 1991).

Early records of salinity

The Murray-Darling Basin had salinity problems due to natural phenomena before European settlement. According to Evans (1989), the pre-European settlement flow of the River Murray at Morgan was 13.2×10^9 m^3 yr^{-1} at a salinity of 490 µS cm^{-1}. However, settlement in the basin has modified the hydrological balance of the basin and aggravated river salinity problems and caused land salinisation.

The River Murray was first seen by explorers Hamilton Hume and William Howell in 1824. Subsequent explorers linked the Murray-Darling system (Gutteridge Haskins & Davey, 1970). In February 1829, Captain Sturt reached the Darling River near the present site of Bourke and found the water too salty to drink (Close, 1990).

According to Powell (1989: 47), in 1853 a grazier in Victoria wrote to Governor La Trobe and said:

> Now that the only soil is getting trodden hard with stock, springs of salt water are bursting out in every hollow or watercourse, and as it trickles down the watercourses in summer, the strong tussocky grasses die before it, with all others.

Settlement in the basin gradually developed and the river system was increasingly used for purposes of navigation, irrigation and water supply. The first irrigation scheme in the Murray Basin was commenced in the 1870s in the Kerang area of Victoria. In 1887 irrigation schemes were initiated by the Chaffey brothers at Mildura in Victoria and at Renmark in South Australia (Alexander, 1928; Powell, 1989). Gutteridge Haskins & Davey (1970) give a brief account of the history of the development of navigation, irrigation and water supply in the Murray Basin.

Irrigated land salinisation began to emerge as a problem soon after irrigation commenced. In the Kerang region salt problems were first noticed in the 1890s, less than 20 years after the commencement of irrigation (Macumber and Fitzpatrick, 1987). Thomas (1939) recorded that salt troubles were encountered in Mildura a few years after its establishment, and the affected area steadily increased until the installation of a comprehensive drainage scheme in 1936–37. According to Mackay (1990), salinisation of irrigated soils appeared in the Cohuna Irrigation District (south-east of

Kerang) around 1900. In 1911 the salt-affected area was less than 80 ha; by 1913 it had grown to 400 ha. With the opening of Torrumbarry Weir in 1929, more water was available for irrigation and the salinisation of the Kerang-Cohuna area accelerated. By the early 1930s, salinisation had extended over much of the Kerang region. Drains were then constructed to remove saline groundwater and the saline drainage water was carried into the River Murray via Barr Creek. A detailed account of the early settlements in the irrigated areas of Victoria, development of irrigation, lack of adequate drainage facilities and expansion of salinity in these areas is available in Barr and Cary (1992). With respect to the development of dryland salinity, Morris (1991) reported that a state-wide survey in Victoria in 1952 (Cope, 1958) showed that the affected area was not less than 4000 ha.

Irrigated land salinity and waterlogging

European settlement in the basin has been characterised by rapid development which resulted in the degradation of the basin's natural resources. The degradation was largely due to lack of understanding of the climate, the fragile soils, and the way in which the basin's biological processes interrelate. For example, clearing of native vegetation has often resulted in soil erosion and soil salinity at sites remote from where the clearing took place. Installation of dams and weirs along the River Murray (Figure 2-13) has stabilised the river flow, thereby drastically altering the river's ecology and resulting in the reduction of the native fish population. Irrigation of land has resulted in rising groundwater salinising soil, streams and rivers (Murray Darling Basin Ministerial Council, 1989).

With respect to ecology, over 7000 wetlands with a total area of more than 200 000 ha have been identified in the Murray Basin (Pressey, 1990). As a result of the construction of dams and weirs many wetlands never dried out. This represents an enormous change to the wetland environment of the Murray Valley, because drying and refilling are natural processes to which flora and fauna are adapted. According to Pressey (1990), some of the ecological changes in the Murray wetlands because of the hydrological change are:

- The death of fringing river red gums (*Eucalyptus camaldulensis*) due to drowning.
- The invasion of wetland meadows by river red gums because of less frequent flooding.
- Replacement of species adapted to drying and refilling by species adapted to permanent flooding.
- Reduced breeding by waterfowl because permanent inundation reduces the numbers of important food organisms.

Changes in the quality of water in some of the Murray wetlands have also been significant. The water which drains from irrigation areas is often saline so that many wetlands have increased in salinity. An extreme example is Fletcher Lake, near the confluence of the Darling and Murray, the bed of which is now thickly encrusted with salt. Increases in salinity are associated with changes in the species present, and often the result is reduced species diversity (Pressey, 1990).

Chesterfield (1986) investigated changes in the vegetation of the red gum (*Eucalyptus camaldulensis*) forest at Barmah in Victoria. Factors influencing vegetation changes include the effect of river regulation, mainly from the construction of Hume Weir in 1934 and the Dartmouth Dam in 1980. Regulation has caused less frequent winter/spring flooding and the entry of water into the forest during summer as a result of high river levels managed for irrigation.

River salinity and land salinisation are particularly severe in the lower part of the MDB or the Murray geological basin (Figure 2-14). The Murray geological basin contains a number of major aquifers in Pliocene, Pleistocene and Quaternary formations. Hydrogeologic features and salinity problems in the basin are documented by a number of authors such as Lawrence (1975), Tickell and Humphrey (1985), Macumber and Fitzpatrick (1987), Evans (1988), Evans and Kellett (1989) and Macumber (1991).

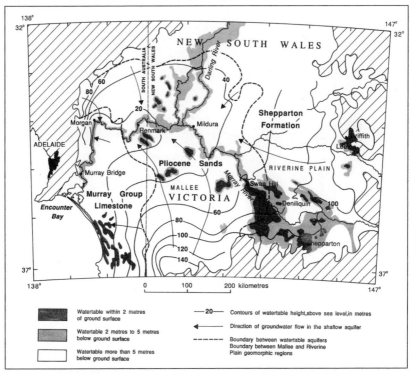

Figure 2-14: Piezometric map and depth to watertable in Murray Basin (Evans, 1988).

The Murray Basin has a saucer shape, without a major outlet to the ocean. Recharge zones for confined aquifers are generally around the margins of the basin. Shallow unconfined aquifers are recharged predominantly from the margins as well as from rainfall and stream flows. Discharge from the deep aquifers is by upward leakage through confining layers to the watertable aquifers. The major proportion of discharge from the watertable aquifers is either to the lower reaches of the River Murray and its tributaries (Figure 2-14) or by direct evaporation from the capillary zone.

European settlement in the basin has gradually changed the hydrological balance of the basin. The main change has been the removal of the previous existing forest cover and its replacement by crops and pastures, the modification of grasslands by heavy grazing and cultivations, and the introduction of major irrigation schemes. Approximately 0.5 million km^2 of the basin's pre-European settlement vegetation cover has been removed. This equates to approximately half the total area of the basin (Dryland Salinity Management Working Group, 1993). The removal of trees from highlands and the plains and the development of irrigation has increased recharge to aquifers. This has led to the gradual filling up of regional aquifer systems (Figure 2-15). Rising groundwater levels are bringing salt stored in the aquifers to the surface, polluting streams and causing destruction of soil and vegetation.

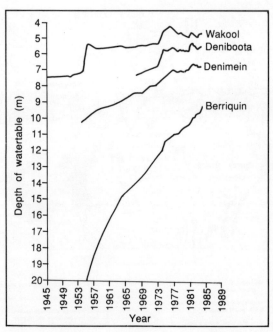

Figure 2-15:   Temporal change of watertable levels in the irrigated areas of the Murray Basin (Murray-Darling Basin Ministerial Council, 1988).

Evapotranspiration is the main mechanism for concentrating salts in groundwater. As a consequence the aquifer system of the basin shows a layering of salinity. The most saline water occurs towards the top of the sequence, the fresher water occurring at the bottom of the sequence. The salinity of groundwater is variable, ranging from fresh to 300 000 mg L^{-1} TDS. In general the better quality water is found around the basin margins. Total salt load in the aquifers and the aquitards of the basin exceeds 100 000 million tonnes. In an average year the River Murray exports about 5.5 million tonnes of salt to the sea and around 1.8 million tonnes of new salt are produced every year within the basin, mainly by evaporation. The net export of 3.7 million tonnes of salt represents less than 0.004 per cent of the total salt stored in the Murray Basin (Evans and Kellett, 1989).

Monitoring of groundwater levels shows that they have been rising rapidly. Rate of rise has been of the order of 10 to 20 cm per year with the most significant rise during periods of heavy rainfall. In fact, as a result of the mid-1970s high rainfall, the associated rise in regional groundwater pressures resulted in expansion of the regional groundwater discharge zones and caused major damage due to waterlogging and salinity. Unfortunately the present trend towards increasing salinity continues with little evidence that equilibrium is yet in sight (Macumber et al. 1988). Table 2-11 shows the past and predicted future expansion of the areas with shallow watertables in the Riverine Plain Zone of the Murray Basin, if nothing is done to arrest current trends.

Table 2-11: Expansion of the shallow watertable areas in the Riverine Plain.

Region	1985[a] (ha)	1990[b] (ha)	2040[b] (ha)
Kerang (Victoria)	240 000	300 000	310 000
Shepparton (Victoria)	120 000	170 000	310 000
Wakool (New South Wales)	34 000	80 000	230 000
Deniliquin (New South Wales)	22 000	70 000	270 000
Murrumbidgee (New South Wales)	143 000	100 000	180 000
Total	559 000	720 000	1 300 000

Source: (a) Murray-Darling Basin Ministerial Council (1988), (b) Gutteridge Haskins & Davey et al. (1990).

In the Kerang Region, the initial watertable was at a depth of 6–9 m in the north and up to 12 m in the south (Mabbutt, 1978), with groundwater salinity increasing to the north and west. By the 1930s, about 50 years after the introduction of irrigation, shallow watertables and surface salinisation were widespread (Chartres, 1987). In Victoria about 400 000 ha of land are already damaged by salt. Similar problems exist in irrigated and dryland areas of New South Wales and South Australia.

Some drainage facilities (surface drains, tile drains, tube-wells and well points) exist in the irrigated areas of the basin. Where they have been installed, they have been efficient in maintaining low watertables and preventing soil salinisation. However, their expansion would pose a serious disposal problem, considering the high salt content of the drained waters. Salt pick-up of the drainage water varies widely in the basin depending on the salinity of groundwater. Table 2-12 provides some estimates for the high and low groundwater salinity zones. As an example, the subsurface drainage of the Barr Creek (Kerang) catchment, with an area of 60 000 ha, would pick up 2.4 million tonnes of salt per year, assuming the minimum pick-up rate of 40 tonnes per hectare. It is evident that the amount of salt that needs to be exported from the high groundwater salinity areas is extremely high and this salt could not be economically disposed of to the River Murray (Blackmore, 1990).

Table 2-12: Estimates of the annual salt pick-up rate of the drainage water in a few irrigated areas of the Murray Basin.

Groundwater salinity zone	Region	Drainage system	Salt pick-up rate (tonnes ha^{-1} yr^{-1})
High	Kerang	Tile drainage	40–50
High	Wakool	Tube-wells	10–15
Low	Leeton	Tube-wells	2–3
Low	Shepparton	?	1

Source: Blackmore (1990).

Dryland salinity

Dryland salinity exists in the Victorian, South Australian and New South Wales part of the Murray-Darling Basin. Victoria has a total land area of 22.8 Mha, an agricultural land area of 9.2 Mha and forest cover of 7.9 Mha. Dryland salinity occurs in many locations in that state including the lower slopes of the Great Dividing Range, the Riverine Plain and the Mallee Region (Figure 2-16). A recent estimate shows that currently more than 100 000 ha is affected by dryland salinity. This area is expected to expand to 298 710 ha over the next 30 years to the year 2020 (Table 2-13).

According to Charman and Junor (1989), a recent survey of dryland salinity in New South Wales (Soil Conservation Service of New South Wales, 1989) shows that the main areas of concern are in the Southern Tablelands, Central Western Slopes and Lower Hunter Valley. The most intensely salted area is in the Yass-Boorowa-Crookwell-Gunning district.

In South Australia dryland salinity is also a major land degradation problem causing significant concern amongst affected landholders for the future sustainability of their agricultural enterprises. Nearly 225 000 ha of agricultural land is affected by dryland salinity in South Australia. A 50 per cent increase in dryland salinity has been predicted in the South Australian Mallee region alone within the next 30 years if current trends continue (Dooley, 1991).

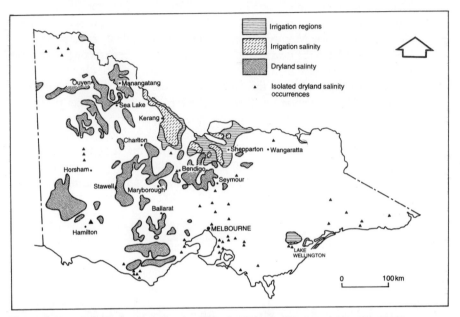

Figure 2-16: Irrigated and dryland salinity in Victoria (Hunter and David, 1992).

In eastern Australia and in particular in the Murray-Darling Basin, the impacts of land clearing and dryland salinity have yet to be fully felt. How far dryland salinity will spread, how fast and how to control the spread are questions of great concern and urgency (Landsberg, 1992). Further information concerning the dryland salinity in the Murray-Darling Basin and its management is available in Dryland Salinity Management Working Group (1993) and Unon (1993).

Table 2-13: Expansion of the irrigated and dryland salt-affected land in Victoria over the next 30 years.

Salt-affected land	Current extent (ha)	Future extent[a] (ha)	Total[a] (ha)
Irrigated	301 540	180 640	482 180
Dryland	100 110	198 600	298 710
Unspecified	200	n.a.[b]	200
Total	401 850	379 240	781 090

(a) Estimated for year 2020; *Source:* Salinity Planning Working Group, Victoria (1992).
(b) Not available.

River salinity

Isolated chemical analyses of Murray water are available from early this century but regular monitoring of water quality only commenced in 1934 at Murray Bridge in South Australia (Close, 1990). Since that time, the number of sites sampled and the frequency of sampling have increased. There are currently over 70 sites on the Murray and its tributaries where salinity is measured.

Salinities in the lower Murray at Morgan have been recorded for more than 45 years (Figure 2-17). It can be seen that the salinities have fluctuated widely over time. Most of the fluctuation in salinity can be explained by a corresponding variation in flow. However, salinities of the peaks are marginally higher now than they were in the 1940s.

Prior to the development of the dams and weirs along the river system, the River Murray would cease to flow. During the drought of 1914–15 salinity at Morgan rose to 10 000 µS cm^{-1} (6000 mg L^{-1}). In contrast, since the construction of dams the maximum salinity at Morgan during prolonged droughts has been around 1500 µS cm^{-1} or 900 mg L^{-1} TDS (Mackay et al., 1988).

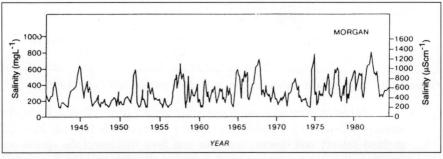

Figure 2-17: Average monthly salinities in the River Murray at Morgan in South Australia (Close, 1990).

Along the River Murray, the annual average salinity increases from less than 40 µS cm^{-1} in the headwaters to about 618 µS cm^{-1} at Morgan in South Australia (Figure 2-18). The salinity levels in the lower reaches of the River Murray have been increasing for many decades. These increases have resulted from:

- An increase in the diversion of freshwater from the rivers and streams of the Murray-Darling Basin. This diversion is currently averaging 9.5×10^9 m^3 per annum (Blackmore, 1991).
- An increase in drainage flows from irrigation areas which are becoming increasingly more salinised due to rising watertables.
- An increase in groundwater flows to the tributaries and directly to the River Murray, resulting from the removal of the deep-rooted native vegetation from much of the catchment.

Evans (1989) reported that approximately 30 per cent of the current salt load in the River Murray is human-induced, largely from groundwater inflow. A very large volume of saline water from both human-induced and natural saline groundwater sources ends up in the River Murray. This volume is thought to be increasing owing to both dryland and irrigation effects.

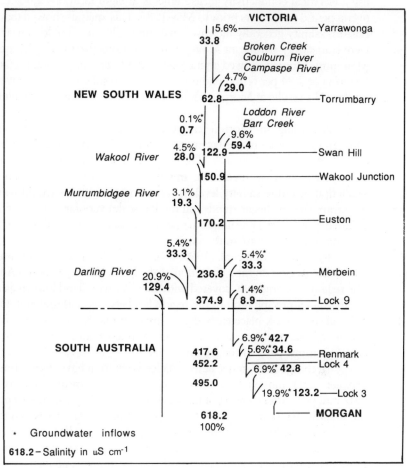

Figure 2-18: Contribution of each reach of the River Murray to the average salinity at Morgan in μS cm⁻¹ and in percentage (Blackmore, 1986; Close, 1990).

Trends in river salinity

Cunningham and Morton (1983) investigated the trend in salinity of the River Murray at Morgan in South Australia by analysing 517 monthly chloride concentration values over a 43-year period from January 1938 to January 1981. The data set was highly skewed, ranging from 20 mg L^{-1} to 365 mg L^{-1} chloride with most of the records in the lower part of the range, interspersed by sporadic runs of high values. Their analysis indicated an annual increase of 1.43 per cent with 95 per cent confidence limits of 0.04 and 2.84 per cent.

Additional statistical analyses to assess changes in salinity along the River Murray have been performed by Morton and Cunningham (1985), using electrical conductivity measurements at eight locations along the river over a period of 16 years prior to May 1983. This analysis showed that below Euston, salinity generally increased over time, although the log-linear trends were statistically significant at only three locations (Red Cliffs, Morgan and Mannum). Morgan showed an increase of 2.58 per cent per annum (adjusted for flow) or 2.34 per cent (unadjusted), and the results at Mannum were very similar. The results at Morgan confirmed those reported in Cunningham and Morton (1983).

According to Morton and Cunningham (1985), salinity levels at Morgan and Mannum are becoming serious. The maximum desirable limit of 800 $\mu S\ cm^{-1}$ (500 mg L^{-1} TDS) has been exceeded quite frequently and salinity will increase unless the trends are arrested. They conclude that prediction of future salinity levels is more complex, and it would be unwise to extrapolate the linear trends (on the log-scale) very far. Such a prediction would depend on the presumption that the factors affecting salinity in the past will continue to increase in the same way.

Recently Allison et al. (1990) investigated the effect of land clearing of the Mallee Region on the salinisation of the River Murray in South Australia. The region was originally covered with eucalypt woodland known as mallee. Mean annual precipitation of the region lies between 250 mm and 450 mm. Most clearing took place early this century, primarily for dryland cropping. The remaining mallee vegetation ranges from less than 20 per cent to 70 per cent, in various parts of the region.

Recharge rates under native mallee vegetation have been estimated at between 0.04 mm yr^{-1} to 0.09 mm yr^{-1} with a mean value of about 0.07 mm yr^{-1}. After clearing of native vegetation recharge increased owing to the shallow depth of rooting of the crops and pasture species which replaced the deep-rooted trees. Recharge rates after clearing have been estimated at six sites within the study area. Recharge rates range from less than 3 mm yr^{-1} to more than 50 mm yr^{-1} with a mean value of 17 mm yr^{-1}, an increase of about two orders of magnitude.

Depth to watertable in the study area ranges from more than 70 m to less than 10 m, with much of the area having a watertable deeper than 30 m. Deep watertables along with the low rate of recharge means that there is a considerable delay in the response of the aquifer to the increased recharge. Although much of the region has been cleared for more than 50 years, the watertable is rising currently only in areas where it is less than 40 m below the ground surface. The development of land salinisation has thus been slow and restricted to areas where the watertable was initially within only a few metres of ground surface. Because of the large depth to watertable over much of the study region there is considerable scope for increases in hydraulic gradients towards the river once the watertable begins to rise.

Using a 'pseudo-three-dimensional' model of the groundwater system of the region with a 25 km grid over the area, Allison et al. (1990) estimated the river salinity at Morgan and Tailem Bend (about 150 km south of Morgan). The model was calibrated in steady-state with the present watertable configuration, using the pre-clearing recharge rate of 0.1 mm yr^{-1}. The present groundwater recharge is largely a pre-clearing scenario, because the increased recharge resulting from clearing has not yet reached or is just reaching the watertable over much of the region. The effects of post-clearing recharge were simulated in transient mode. As the detailed land clearing history was not available, all clearing was assumed to have taken place in 1920. The model outputs, expressed in terms of river salinities at Morgan and Tailem Bend, are given in Figure 2-19. River salinities at these stations are expected to increase by about 0.75 µS cm^{-1} yr^{-1} and 1 µS cm^{-1} yr^{-1}, respectively, over the next 200 years.

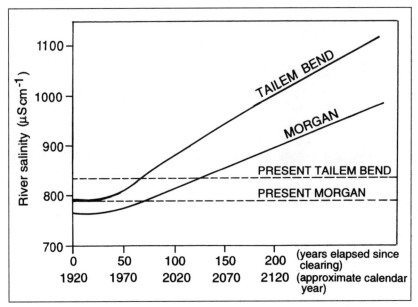

Figure 2-19: Predicted increase in river salinity at Morgan and Tailem Bend (Allison et al., 1990).

Although it has not suffered the land and stream salinity problems to the same extent as the Murray Basin, the Darling Basin contributes a significant amount of salt, 21 per cent, to the River Murray system. It has been suggested that the effect of land clearing in the Darling Basin is yet to be fully felt. Jolly (1989) investigated the potential for increased stream salinisation in the Darling Basin. He provided the mean electrical conductivity, mean annual stream flow and mean annual salt load of the major streams of the Darling Basin for 1970–79. These data show that there is a general increase in salinity of the tributaries moving south with the most saline water coming from the Namoi and Castlereagh Rivers (Figure 2-12). In the lower reaches of the Darling River, salinities increase markedly, presumably due to the contribution of more saline water from the Menindee Lakes system. In terms of total amount of salt load, the Namoi and MacIntyre Rivers are the largest contributors, with 321 000 tonnes yr^{-1} and 263 000 tonnes yr^{-1}, respectively.

According to Jolly (1989), there is a loss of salt from the lower reaches of several rivers of the Darling Basin. Although the reasons are not clear, possible explanations are: inadequate flow data; losses brought about by diversion of water and hence salt for irrigation; and recharge to the Tertiary aquifers which underlie the Darling River.

Management options

Under the Australian Constitution the ownership and hence management of land and water resources is vested in the states. Consequently each state has set up its own agencies and structures for the development of policy in relation to water and land use and their management. For this reason management of salinity problems in the Murray-Darling Basin was initially the responsibility of the New South Wales, Victorian and South Australian State Governments. But for a few decades there has been considerable friction between the states about who should be responsible for salinity problems in the River Murray. The river salinity gradually increased in South Australia and the South Australians consider that the problems have been caused to a large extent by the two upper states of New South Wales and Victoria. The situation became even more difficult in the mid-1970s when the Victorian Government proposed a system of subsurface drainage in the Shepparton Region that involved disposal of salty water to the River Murray (Close, 1990). Following many years of dispute, discussions and negotiation, the Murray-Darling Basin Agreement was signed by the Commonwealth and State Governments of New South Wales, Victoria and South Australia on 30 October 1987 and came into force on 1 January 1988 (Crabb, 1988 and 1991). The aim of the Agreement is to promote and coordinate planning and management of water, land and environmental resources of the Basin. Through this agreement and the latest one which became effective from 1 July 1992, institutions and procedures have been put in place to tackle the Basin's problems, including the management of salinity (for further information concerning the Murray-Darling Basin Agreement see Part One, Section E, page 97).

In the Murray-Darling Basin, the rise in watertables in response to increased aquifer recharge caused by the land clearing and excessive irrigation is the main cause of the salinity problems. Therefore the salinity problem is fundamentally a groundwater problem and its solution lies essentially in watertable control (Macumber, 1990).

A wide range of options is available for the management of the waterlogging, irrigated and dryland salinity and river salinity. These options include: measures designed to reduce accessions to the groundwater systems, including surface drainage and improved irrigation methods; measures to control watertables by subsurface drainage, including tile drainage and groundwater pumping; conjunctive use of surface and groundwater; salt interception options, and disposal; water management options, including water pricing and transferable water entitlements; a broad variety of agronomic and land management options, including drainage water reuse schemes and land retirement; measures designed to optimise farm productivity under saline conditions; measures to control dryland salinity; and reforestation and agroforestry. The following sections detail some of these options.

Engineering options

Traditionally, the State and Federal Governments have acted to reduce river and land salinity problems principally through engineering solutions. These solutions include interception of saline groundwater and the diversion of irrigation returns, both with disposal to evaporation basins and to the River Murray. Other engineering schemes consist of pumping groundwater in order to lower watertables. Table 2-14 shows that the operation of salinity mitigation prevents on average the discharge of about 162 000 tonnes per year of salt to the river. These schemes also have been successful in lowering the high groundwater levels and prevention of soil salinisation, for example in the case of Wakool-Tullakool in New South Wales.

After the establishment of the Murray-Darling Basin Ministerial Council, a Salinity and Drainage Strategy was adopted by the States. This has set up a framework for joint government action to manage salinity problems in the basin. The strategy incorporates an initial program of salinity mitigation works that will reduce river salinity at Morgan by 80 µS cm^{-1} (Table 2-15). These schemes will be funded equally by the three States and the Commonwealth. At the completion of these works the upper States will be entitled to salinity credits of 15 µS cm^{-1} each. They will then be able to construct drainage works or undertake other proposals that increase downstream salinity provided that the impact of salinity does not exceed their available salinity credits. In future, each State will have the option to contribute to additional salinity mitigation schemes and receive credits in proportion to the effectiveness of the scheme and the State's contribution to it (Blackmore, 1991).

Table 2-14: Annual average operational data of salinity mitigation schemes in the Murray Basin.

Scheme	State	Description	Completion date	Water diverted (10^6 m^3)	Salinity of water (mg L^{-1})	Salt load (tonnes)	Salt kept out of river (tonnes)
Barr Creek	VIC	Drainage diversion, evaporation	1968	8.920	3 260	29 000	29 100
Lake Hawthorn	VIC	Drainage diversion, evaporation	1968	5.090	3 020	15 400	12 700
Mildura-Merbein	VIC	Groundwater pumping, evaporation	1981	1.950	19 400	37 700	20 700
Buronga	NSW	Groundwater pumping, evaporation	1979	2.220	38 300	85 100	27 200
Curlwaa	NSW	Groundwater pumping, evaporation	1974	0.660	4 800	3 190	2 040
Rufus River	SA	Groundwater pumping, evaporation	1983	1.170	21 100	24 600	18 500
Noora	SA	Drainage diversion, evaporation	1983	4.910	10 100	49 700	49 700
Wakool-Tullakool[a]	NSW	Groundwater pumping, evaporation	1981	7.890	18 100	143 000	0
Wakool-Tullakool[b]	NSW	Groundwater pumping, evaporation	1983	4.780	12 500	59 500	0
Renmark Reservoir	SA	Drainage diversion, evaporation	1983	0.135	12 600	1 700	1 700
Total				37.725	11 900	448 990	161 640

Source: Murray-Darling Basin Commission (1988).

VIC: Victoria, NSW: New South Wales, SA: South Australia.
(a) Stage one; (b) Stage two.

Table 2-15: Impact of the interception schemes on the salinity of the River Murray at Morgan.

Interception scheme	State	Salinity impact at Morgan (µS cm^{-1})
Woolpunda	South Australia	−39.8
Waikerie	South Australia	−15.9
Mallee Cliffs	South Australia	−7.5
Chowilla	South Australia	−10.5
Mildura-Merbein	Victoria	−7.5
Total		−79.4

Source: Murray-Darling Basin Ministerial Council (1988).

According to Macumber (1990), the salinity credit of 15 µS cm^{-1} allows an additional salt load of only about 80 000 tonnes per year per State, which is clearly well below the amounts that can be generated from the available aquifers within the irrigated areas. This necessitates some degree of non-River Murray disposal. The intercepted saline groundwater would be pumped to evaporation basins some distance from the river.

Conjunctive use of surface and groundwater

Conjunctive use of surface and groundwater cannot be used everywhere in the basin due to the high salinity of groundwater resources in major parts of the basin and particularly in the shallow aquifers. However, this policy should be encouraged wherever it is efficient and cost-effective. Salinity Pilot Program Advisory Council (1989) recommended more groundwater extraction in the Shepparton region, and Evans and Nolan (1989) described a groundwater management strategy for the Victorian Riverine Plain. According to Evans and Nolan (1989), the strategy centres on the development and careful management of the good quality groundwater resources in northern Victoria.

The Riverine Plain to the east of the Murray Basin comprises predominantly good to moderate quality groundwater with a salinity of 500–3000 mg L^{-1} TDS. Within the Riverine Plain, groundwater levels are rising at the rate of 0.05 to 0.1 m yr^{-1}. Large tracts of both irrigated and dryland are threatened or already severely affected by salinity and/or waterlogging.

The major hydrogeological units of the Riverine Plain are as follows:
- The uppermost unit is the Shepparton Formation, which is a composite aquifer-aquitard complex. Its fine to coarse sand bodies form an essentially unconfined aquifer system.
- The Deep Lead consists of two identifiable units, The Renmark Group and the overlying Calivil Formation. Both of these consist of sand and gravel interbedded with less permeable river sediments. The Deep Lead may be

considered to be a single, highly transmissive aquifer containing good quality water.

The shallow watertable in the Shepparton Formation tends to be between 1–10 m below the surface and is influenced by leakage to or from underlying Deep Lead. In the eastern Riverine Plain, the Deep Lead generally has a potentiometric surface below that of the Shepparton Formation, hence leakage is downwards. However, the Deep Lead potentiometric surface is rising at a rate of up to 0.2 m yr^{-1} and thus will equal or exceed the shallow watertable within a few decades, if it is not managed through controlled pumping to maintain vertical drainage of the overlying Shepparton Formation. In the western part of the Riverine Plain the Deep Lead is generally artesian.

Extensive predominantly surface-water-based irrigation exists over much of the Riverine Plain. A complex channel system exists to distribute 1.6×10^9 m^3 yr^{-1} of water from dams located in the highlands. Within the Victorian Riverine Plain there are approximately 950 groundwater wells licensed to extract 255×10^6 m^3 yr^{-1} for irrigation purposes. The groundwater usage has increased from an estimated 4.5×10^6 m^3 in 1974 to 108×10^9 m^3 in 1986. Currently groundwater usage is approximately 8 per cent of the applied irrigation water across the region (Evans and Nolan, 1989).

In addition to private groundwater usage, there is substantial public-scale groundwater pumping for watertable and salinity control purposes. There are 79 pumping installations in the eastern region protecting high value horticultural and pasture (18 000 ha) areas and another six pumps directed at pasture protection. Almost all the groundwater pumped is discharged to surface water drains and channels where approximately 60 per cent is used for irrigation and 40 per cent is discharged to the River Murray. Groundwater salinity is generally about 2600 mg L^{-1}. In the order of 10.9×10^6 m^3 are pumped and hence 11 300 tonnes of salt enter the River Murray per year.

Significant scope exists for future development of groundwater in northern Victoria. In this area it is believed that groundwater could meet around 20 per cent of the total water demand. Groundwater development is being encouraged for its resources potential and also because of its role in salinity mitigation.

According to Evans and Nolan (1989), careful management is required for Deep Lead pumping bores. Such bores must be located to minimise migration of poor quality groundwater from either bedrock or overlying aquitards. The extraction rate for a region also must be carefully assessed to achieve optimum yield and minimal water quality decrease. Moreover, groundwater management costs should be borne by the whole irrigation community and not by the individual groundwater users. The need for a fully integrated groundwater and surface water allocation policy is essential to deal with the salinity issue. Such a policy would aim at having a consistent price for all water, where any specific subsidies to encourage groundwater use are identified and built into the total water pricing and salinity mitigation structure. Allocation criteria for new or transferred surface water and groundwater would consider the total water available to individual properties, and licences would be issued accordingly.

Water pricing

Both the States and the Commonwealth have subsidised Murray-Darling irrigation in the past to achieve regional development goals. This subsidisation persists to the present time. However, the emphasis has changed from support for new investment in infrastructure to partial coverage of operational and maintenance expenditures. Currently, the level of subsidisation is of the order of A$300 million (US$240 million) a year (Simmons et al., 1991). Some States have published plans for reducing subsidisation of irrigation and for applying the 'user pays' principle to water charging. In Victoria, gravity surface water supplies are currently at A$12.10 (US$9.70) per 1000 m^3 per year and rising at a rate of inflation plus 2 per cent per year towards the full cost of A$22.50 (US$18.0) at 1989 value (Evans and Nolan, 1989).

Transferable water entitlements

Transferability of water entitlements breaks the links between land and water and enables individual farmers to sell their surface water entitlement to other farmers. Considerable progress has been made by the States in increasing the transferability of water entitlements. All of the States involved in the basin have introduced transferability to varying extents. It is estimated that the potential return to increasing efficiency in water use through making water entitlements transferable is around A$40 million (US$32 million) a year for the basin as a whole (Simmons et al., 1991). Transferability of water entitlements would also facilitate solution to salinity and waterlogging problems of the basin by removing water application in the areas of high watertable and low productivity, and which are salt-affected, to the areas of high productivity which are not at risk from salinisation.

Saline water disposal

The cost-effective disposal of saline waters from drainage or mitigation schemes is an important issue in the Murray Basin. Evans (1989) has reviewed a wide range of disposal options, including reuse of marginal quality water, dilution and reuse of drainage water, disposal to evaporation basins, disposal to shallow and deep aquifers, dilution flow, desalination and pipeline transfer to the sea.

A great deal of attention is now being paid to the reuse options. In the New South Wales and Victorian irrigation regions, reuse of marginal quality water is probably the major disposal technique, where salinity is low. The practice of dilution and reuse is the major disposal method for almost all of the Murrumbidgee Irrigation Area.

Evaporation basins have been widely used. Most of the basins have been natural depressions and saline lakes. They vary in size from a few hectares serving individual farms to a few thousand hectares. Evans (1989) has provided an inventory of 35 significant evaporation basins in the Murray Basin. He has provided an area ratio of 1:30 for the subsurface drainage by pumping at Wakool-Tullakool in New South Wales. In protecting an area of 60 000 ha, the evaporation basin covers 2000 ha. He also reported an area ratio of 1:10 for tile drainage schemes in New South Wales.

Disposal to shallow aquifers is common in South Australia's irrigation regions and some occurs in Victoria and New South Wales. There are about 1000 shallow disposal bores in the South Australian part of the Murray Basin. Most bores dispose of tile drainage effluents with a salinity of 2000–3000 μS cm^{-1} into groundwater with a salinity of 38 000 μS cm^{-1}. There are 75 shallow disposal bores in the Victorian part of the basin and 25 in the New South Wales part of the basin. The disposal of saline water by injection into deep aquifers is not currently undertaken anywhere in the Murray Basin.

The practice of providing dilution flows specifically to reduce river salinity at various key times is now widely used. The feasibility of increased dilution flows has been evaluated. Although increased dilution flows in the lower river can have a significant impact on river salinity, the provision of large-scale dilution flows (by new storage development, reduction of irrigation water, etc.) is not economical at the moment.

Desalination processes produce freshwater and highly saline waste. With seaboard plants, disposal of saline waste to the ocean is usually acceptable, but for inland situations, some form of disposal (e.g. to evaporation basins, deep injection) should be considered. Studies show that even ignoring the cost of disposing of highly saline waste, the costs of desalination are far above the value of the water or the costs of evaporative disposal options.

Various options involving pipeline transfers to the sea have been considered since 1978. For this option, the principal factor influencing cost other than the length of pipeline is the degree of concentration that can be achieved before pumping. A smaller (more highly saline) volume costs much less to pump. Detailed studies of various pipelines showed that these projects are very expensive and not cost-effective. Table 2-16 shows the estimated capital costs and the total maintenance costs of various outfall schemes for the Riverine Plain Zone to the sea. The volumes of water that would require disposal by the year 2040 including effluents from existing drainage and interception schemes are estimated to be 608×10^6 m^3 and 335×10^6 $m^3 yr^{-1}$, respectively, under full and partial watertable control (Gutteridge Haskins & Davey et al., 1990).

Table 2-16: Total outfall schemes costs for the Riverine Plain Zone over a period of 50 years (1990–2040) in A$[a] million.

	Full watertable control		Partial watertable control	
	No conc.	1:8 conc.[b]	No conc.	1:8 conc.
Subsurface drainage	420	420	350	350
Collection	730	73	550	550
Concentration	–	590	–	320
Regional transfer works	1200	280	820	200
Outfall	3700	800	2500	550
Total capital costs	6050	2820	4220	1970
Total annual O&M[c] costs (2040)	66	17	41	11

(a) A$1 = US$0.80; (b) concentration; (c) operation and maintenance.

Source: Gutteridge Haskins & Davey et al. (1990).

The capital cost of a Mallee Region outfall would amount to approximately A$770 million with operating and maintenance costs of about A$6 million per year, considering the disposal of 78×10^6 m^3 yr^{-1} of saline effluents.

Evans (1989) provided the relative costs of saline water disposal options for the Murray Basin using January 1988 prices for all schemes, a 5 per cent internal rate of return and a 30 year lifetime for the project. Also, a marginal value for water of A$50 (US$40) per 1000 m^3 is assumed. As Figure 2-20 shows, evaporation basins are the most cost-effective options while the desalination and pipeline to the sea options are very expensive.

Dryland salinity management

In general, for groundwater pumping to be economically successful in dryland areas, the aquifers must be sufficiently permeable and fresh to permit successful extraction and re-use. However, in large dryland areas of the highlands and the mallee, aquifers commonly have relatively low permeabilities and are often too saline. Notable exceptions to this are the narrow highland valleys and parts of the Riverine Plain, where good quality aquifers are developed in fluvial sands and gravels. Elsewhere in the highland regions, where regional groundwater systems predominate, there is little technical or economic potential for subsurface draining and agronomic techniques are usually the only option (Macumber, 1990).

The Goulburn Dryland Region in Victoria covers an area of 1.9 Mha or about 10 per cent of the occupied area of Victoria. Already 3500 ha are obviously salt-affected and unproductive. A further 3500 ha are showing early signs of becoming salted. The total area of saline land is estimated to be increasing at a rate of 2 per cent to 5 per cent per annum across the catchment. At this rate the area affected by salt would be of the order of 20 000 to 80 000 ha at the end of 50 years (Garrett, 1992).

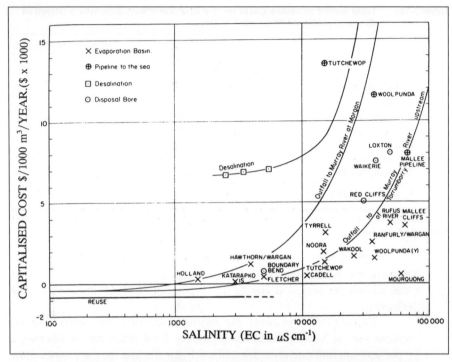

Figure 2-20: Relative costs of specific saline water disposal projects in Australian dollars (≈0.8 US$) and in January 1988 prices (Evans, 1989).

The Goulburn Dryland Salinity Management Plan is one of the plans prepared under the auspices of the Victoria State Salinity strategy, 'Salt Action: Joint Action'. The primary objective of the plan is to reduce rainfall accessions to the groundwater system by planting areas of high and moderate infiltration potential (recharge area) with high water using trees, pastures and crops. The secondary objective is to establish vegetation cover on denuded salted land and to control erosion, evaporation and salt accumulation from the areas. According to Garrett (1992) and Morris (1991), the management plan identified some 40 000 ha of high recharge land and a further 33 000 ha of moderate recharge land for which treatment to increase water use by vegetation was proposed, and included four basic work types, as follows:

- Tree planting at high density (200 trees per ha) on 4000 ha and low density (20 trees per ha) on 24 000 ha of recharge area.
- Perennial pasture establishment on high and moderate recharge areas.
- Lucerne planting on high and moderate recharge areas.
- Rehabilitation of discharge areas by the sowing of salt-tolerant species.

The Victorian Government accepted the recommendation of the Goulburn Dryland Salinity Management plan in June 1990 and implementation is now under way. The first year of the plan has seen achievements of better cooperation between government agencies and community, and the implementation of 700 ha of salinity control works (Garret, 1992). Incentives of up to 70 per cent of the cost of establishing high density trees and 50 per cent for low density trees are provided to encourage landowners to carry out the required work. Establishment of perennial pastures in the recharge areas also receives a 50 per cent subsidy (Morris, 1991). During review of the draft plan, the ability of trees at either 20 or 200 stems per hectare to control recharge was questioned. The available information on tree water use and recharge reduction is inconclusive, so the government undertook to provide funding for research and investigation aimed at defining more closely the density of planting required for recharge control in each land management unit. Similarly, the omission of tree planting from the recommended treatments for rehabilitation of saline discharge areas was questioned, and the government undertook to support investigations into the effectiveness and economics of tree growing in such areas.

Reforestation and agroforestry

In Australia approximately 20 per cent of the farmers plant trees. Most of these farmers are in Victoria where 60 per cent of the State farmers plant trees, followed by Western Australia (50 per cent). In contrast, in Queensland, New South Wales and South Australia only 4 per cent, 9 per cent and 22 per cent respectively of the farmers planted trees in the past two years (Prinsley, 1991). The purpose for which farmers plant trees varies widely, with only 9 per cent for salinity reclamation and 7 per cent for preventing salinity. According to Prinsley (1991), the main reasons given by farmers for not planting trees are as follows:

- Economical and financial reasons, including cost of establishment, noneconomic benefits, cost of maintenance (60.6 per cent).
- Technical reasons, including lack of knowledge, previous failures (14.4 per cent).
- Competition with agriculture, including loss of land, incompatibility with agriculture (14.0 per cent).
- Other reasons, including low priority and negative attitude (11.0 per cent).

These statistics show that for a successful program of tree plantation farmers need financial and technical assistance and that as far as possible planted trees should have a commercial value.

Dumsday et al. (1989) have analysed the economic aspects of broad-scale revegetation as a management tool in the basin. They have concluded that broad-scale revegetation is not economical or practical as a groundwater and salinity management tool and that the use of a tree-based system of land use should be evaluated in terms of local conditions. Two major impediments to the widespread adoption of tree-based systems are cited. One is the lack of commercial viability in low rainfall areas and the other is the risk associated with the long-term investment in forestry systems. They suggested that revegetation in the form of agroforestry or woodlot forestry provides economic benefits over traditional systems of agriculture in some areas.

Morris (1991) has reviewed the use of trees in salinity control in Victoria. As he reported, the Victorian salinity strategy incorporated a revegetation plan, particularly for dryland areas. This plan includes the use of deep-rooted perennial crops and pastures, as well as tree planting, retention of remnant native vegetation and encouragement of natural regeneration. The regional communities are largely responsible for undertaking revegetation projects, with government support through research, technical advice, financial assistance and community education programs. Although the plan is directed mainly at reducing groundwater recharge, discharge area revegetation is also acknowledged as useful for reducing erosion and improving the appearance of the area. Rehabilitation of selected salt-affected areas is specified as a strategic target of the revegetation plan. In irrigation regions, tree growing to assist with the control of channel seepage is encouraged.

Table 2-17 shows the number of trees planted by organisations and individuals in Victoria during 1989 for salinity control. The great majority of tree planting was by government agencies and almost two-third of this was located in the Goulburn, North-east and South-east salinity management regions. Further information concerning the revegetation of saline land in Victoria is available in West (1990).

Table 2-17: Number of trees planted for salinity control by organisations and individuals in Victoria during 1989.

Group	Trees planted	Survival (%)
Government agencies	1 726 784	94
Municipal bodies	7 525	87
Landcare and Farm Tree groups	57 342	76
Farmers	1 904	72
Total	1 793 555	93

Source: Morris (1991).

Major agroforestry research and demonstration plantings were established at six locations in Victoria between 1983 and 1985, with the aim of providing definitive cost-benefit information for a range of tree-agriculture combinations. Early results from these trials have provided valuable information to guide the selection of tree species and management techniques for agroforestry on a wide range of sites, including saline areas at one location.

At some sites, soil moisture and/or watertables have been monitored to provide data on the effectiveness of trees at different densities on groundwater recharge and discharge. More extensive agroforestry research and demonstration projects are now under way, including a multi-farm development in north-eastern Victoria funded by the Murray-Darling Basin Commission, and the Shepparton Irrigation Region Agroforestry Project (SIRAP) funded by the National Soil Conservation Program and State Timber Industry Strategy funds. SIRAP, which commenced in 1989, is the first large-scale irrigated agroforestry project in Victoria. It involves 28 landholders in five locations in the lower Goulburn Valley, establishing trees on 10 per cent to 20 per cent of each farm. With appropriate monitoring of watertables and salinity effects in addition to productivity of trees and agriculture, this project has the potential to resolve some of the important information deficiencies preventing direct government support for tree growing as a salinity control measure in irrigation areas (Morris, 1991).

The role of trees in controlling dryland salinity in South Australia has been reviewed by Lay (1990) and Dooley (1991). According to Dooley (1991), South Australia has in place a strategy for investigating and demonstrating the role of trees in controlling dryland salinity. The State is employing a cooperative approach involving government agencies, research organisations and community groups to address this serious land degradation problem. Research work, much of which is in its early stages, is based on the whole catchment approach, incorporates interstate results where appropriate, and is designed to have widespread practical applications.

The various on-farm uses of trees that potentially increase farm productivity while combating salinity have been promoted in South Australia. Government support for rural revegetation is demonstrated by the launch in 1990 of the Rural Tree Planting Strategy (managed by the Department of Agriculture), which includes the establishment of the State Tree Centre for coordination of activities and development and dissemination of information, and the appointment of rural revegetation officers to service landholder demand across the State. Many Landcare groups are currently establishing sites which demonstrate the role of trees in dryland salinity control. Similarly, State-sponsored extension programs at the district level complement those being funded by national agencies such as the National Soil Conservation Program (NSCP), the Natural Resources Management Strategy (NRMS), and the One Billion Trees Program.

The role that trees can play in controlling salinity is seen to include both recharge reduction and increased water use at discharge sites, and State programs are addressing the need for relevant technical and economic information. Extension of the information, to encourage significant landholder adoption of tree planting for salinity control, under conditions which are cost-effective for farming enterprises, will remain a high priority.

The Soil Conservation Service of New South Wales has had a

significant involvement in dryland salinity work over many years. Recent initiatives of the Service include the State Tree Policy, issued in 1987 to promote the conservation and increased planting of trees on rural land. The associated Trees on Farms Program is designed to encourage both farmers and community groups to undertake tree planting in strategic areas for soil and water conservation, improvement of silvicultural capability, conserving native flora and fauna and improving aesthetic qualities of the environment. The control of tree clearing and the planting of trees is seen to be particularly important in areas prone to salinisation, including irrigation districts (Charman and Junor, 1989).

Community involvement in salinity management

There is increasing community involvement in planning and management of salinity problems. This involvement is supported by the Federal and State Governments. With respect to salinity problems in the Murray-Darling Basin, in December 1986 the Murray-Darling Basin Ministerial Council formed a Community Advisory Committee. One of the roles of the Committee was to provide two-way channels of communication between the Ministerial Council and the Murray-Darling Basin Community (Stone, 1991; Department of Primary Industries and Energy, 1989).

In Victoria, the Government has adopted the coordinated strategy 'Salt Action: Joint Action' to manage the salinity problems in the State (Government of Victoria, 1988; Salinity Bureau, 1989). Coordination of the strategy is through the 'Salinity Bureau'. The salinity strategy 'Salt Action: Joint Action' is based mainly on the principle of community ownership of the salinity problem and joint action between the community and government to manage the problem. According to Hunter and David (1992), the basis of the strategy consists of:

- Catchment philosophies for addressing salinity control in an integrated and systematic manner.
- Community involvement at all stages of planning and implementation.
- Preparation of regional strategies and intensive subregional salinity management plans in key priority areas.
- Statewide inter-agency coordination, particularly through the preparation of a state salinity budget.

In order to manage the salinity problems, Victoria has been divided into nine salinity control regions covering the entire State and a number of subregions in the irrigated and dryland areas (Government of Victoria, 1988; Stone, 1991). Control regions are based on major catchments so that the causes and effects of each salinity problem generally lie within the same region.

Originally the central feature of 'Salt Action: Joint Action' was identification of areas with severe salinity problems and commitment to support the community to develop salinity management plans for those areas. New plans are being developed for the entire State. Each plan is prepared by a committee comprising representatives of the local community, farmers and conservation organisations. The committees are assisted by interdepartmental technical support groups and, where needed, by independent consultants. Government guidelines require the plan to consider a range of control programs and to outline the economic, environmental and social implications of the options considered, together with the proposed cost-sharing arrangements between the community and government. The management plans developed by community groups are then released for public comment. Following discussion and revision of the plan, government announces its support for the plan. Through this process the communities are involved in integration of information, setting targets in planning, reassessment of plans, measuring performance and adjusting activities accordingly within the context of catchment management, and in a coordinated way. It is important to note that plans that do not have community support are not funded, and the beneficiaries of any community supported salinity control works are required to share the cost of the works with the government.

Although the procedures and guidelines for the development of the community salinity management plans seem relatively clear, in practice, because of the conflicts of interest within community groups and between different groups of farmers (especially regarding the resolution of the complex issue of cost-sharing, which is one of the main concerns of the government), it takes a few years to develop a salinity management plan acceptable to both the community and the government. Barr and Cary (1992) and Wilkinson and Barr (1993) describe these dilemmas with reference to the development of salinity management plans for the Barr Creek, Shepparton, Campaspe West and a number of other regions in Victoria.

To date a number of plans have been prepared and supported by the government. The Victorian government announced its support of the plan for the Barr Creek catchment in 1987. In June 1990 the government announced its support for management plans for Tragowel Plains area, Campaspe West area (only very limited support for those parts of the plan which had community support), Shepparton Irrigation Region and Goulburn Dryland Catchment (Government of Victoria, 1990). In February 1991 a draft management plan was prepared for the Nangiloc-Colignan area, and was supported in November 1991 (Nangiloc-Colignan Community Salinity Working Group, 1991). A draft salinity management plan for Sunraysia was prepared in November 1991 (Sunraysia Community Salinity Working Group, 1991). It was expected that by the end of 1992 salinity plans would be completed for all areas in Victoria suffering from significant salinity problems (Hunter and David, 1992).

In New South Wales, 'Salt Action' is a strategy for joint action by the government and the community to manage and control salinity levels within the land and water resources of the state. Salt Action is part of the Total Catchment Management Policy and its key objectives are to: control existing land and water salinisation; prevent further degradation; rehabilitate affected areas; protect sensitive ecosystems from the impact of salinity; raise public awareness and understanding of the problem; and foster sustainable production methods. The New South Wales government provides funds for capital works, farm assistance, research, education and planning through the joint efforts of the community and government.

Economic damage and cost

In south-east Australia, Grieve et al. (1986) have estimated the economic losses for dairy and winter cereal production in the Berriquin and Wakool District (New South Wales) due to waterlogging and soil salinity as more than A$13 million (US$10 million) in 1984 prices. For the Murray-Darling Basin the agricultural loss amounts to A$260 million (US$208 million) per year (Murray-Darling Basin Ministerial Council, 1989) and the costs of the effects of salinity on quality of urban water supplies are estimated to be around A$65 million (US$52 million) a year (Simmons et al., 1991).

In Western Australia, a recent estimate by the Western Australian Department of Agriculture (1988b), reported by the Western Australia Legislative Assembly (1991), shows that the annual loss due to 443 000 ha of land affected by dryland salinity is about A$62 million (US$50 million) and the damage due to waterlogging of 500 000 ha of land under cropping is A$90 million (US$72 million). In addition, costs associated with stream salinity are estimated to be A$40 million (US$32 million) annually in 1990 prices (Water Authority of Western Australia, personal communication).

In both cases, the Murray-Darling Basin and the south-west of Western Australia, estimated economic damage underestimates the total loss to the community, as they do not include environmental damage such as the loss of natural habitat and landscape amenity.

Remedial actions for salinity control are expensive and each year the State and Federal Governments have to allocate considerable financial resources for this objective. For example, in Victoria, where salinity costs the Victorian economy more than A$60 million a year in lost agricultural production and reduced water quality, the government has spent A$177 million (in 1991/92 dollars) over the past eight years (Hunter and David, 1992). The 1991/92 Victorian Government salinity budget was A$26.9 million, showing an increase of A$17 million since the first coordinated budget on salinity in 1985/86. The salinity budget is made up of over 300 projects undertaken mainly by government agencies. According to Hunter and David (1992), plans already endorsed by the Victorian Government will cost about A$900 million over the next thirty years to implement with costs being shared almost equally by the landholders and government. In the 1989–90 financial year the Murray-Darling Basin Commission spent about A$12 million of its A$34 million budget on investigation, construction and operation of salinity mitigation schemes (Murray-Darling Basin Commission, 1991).

References

Alexander, J.A. 1928. *The Life of George Chaffey: A Story of Irrigation Beginnings in California and Australia.* Melbourne: Macmillan. 382 pp.

Allison, G.B., Cook, P.G., Barnett, S.R., Walker, G.R., Jolly, I.D. and Hughes, M.W. 1990. Land clearance and river salinisation in the western Murray Basin, Australia. *Journal of Hydrology.* 119(1/4): 120.

Australian Bureau of Statistics. 1988. *Agricultural Land Use and Selected Inputs, Australia 1986-87.* Canberra: Australian Government Publishing Service. (ABS Catalogue no. 7411.0). 15 pp.

Australian Bureau of Statistics. 1990. *Year Book Australia 1990.* Canberra: Australian Government Publishing Service. 846 pp.

Australian Geographic Society. 1988. *The Australian Encyclopaedia.* Fifth edition. Australian Geographic. 6: 2295.

Australian Water Resources Council. 1987. *1985 Review of Australia's Water Resources and Water Use.* Canberra: Australian Government Publishing Service. v.1: Water Resources Data Set, 158 pp; v.2: Water Use Data Set, 114 pp.

Bari, M.A. and Schofield, N.J. 1991. Effects of agroforestry-pasture association on groundwater level and salinity. *Agroforestry Systems.* 16(1): 1331.

Bari, M.A. and Schofield, N.J. 1992. Lowering of a shallow, saline watertable by extensive eucalypt reforestation. *Journal of Hydrology.* 133(3/4): 273–291.

Barr, N. and Cary, J. 1992. *Greening a Brown Land, The Australian Search for Sustainable Land Use.* Melbourne: Macmillan. 331 pp.

Bell, R.W., Loh, I.C. and Borg, H. 1987. *The Effect of Non-Valley Reforestation on Water Quality and Quantity in the Padbury Reservoir Catchment and its Regional Implication.* Leederville: Water Authority of Western Australia, Water Resources Directorate. (Report no. WS5). 74 pp.

Bergman, I.D. 1989. *A National Water Resource Inventory for the 1990s.* Canberra: The Institution of Engineers, Australia. (Civil College Technical Report, 5 May). 6 pp.

Blackburn, G. and McLeod, S. 1983. Salinity of atmospheric precipitation in the Murray-Darling Drainage Division, Australia. *Australian Journal of Soil Research.* 21(4): 411–434.

Blackmore, D.J. 1986. The River Murray option for salinity reduction. In. *Hydrology and Water Resources Symposium 1986, River Basin Management, Griffith University, Brisbane, 25–27 November 1986.* Canberra: The Institution of Engineers, Australia. (National Conference Publication no. 86/13). 312–316.

Blackmore, D.J. 1990. Water quality management in a saline environment. In. Humphreys, E., Muirhead, W.A. and van der Lelij, A. eds. *Management of Soil Salinity in South East Australia.* Proceedings of a symposium held at Albury, New South Wales, on 18–20 September 1989. Australian Society of Soil Sciences, Riverina Branch. Wagga Wagga: Charles Sturt University-Riverina. 109–114.

Blackmore, D.J. 1991. Murray-Darling Basin Initiative: A case study in integrated natural resources management. *Water and the Environment.* (Newsletter of the Water Research Foundation of Australia). 305: 2–8.

Bleazby, R. 1917. Railway water supplies in Western Australia — difficulties caused by salt in soil. *Institute of Civil Engineers London, Proceedings.* 203: 394–400.

Bowler, J. 1990. The last 500 000 years. In. Mackay, N. and Eastburn, D. eds. *The Murray.* Canberra: Murray-Darling Basin Commission. 95–109.

Brown, C.M. 1989. Structural and stratigraphic framework of groundwater occurrence and surface discharge in the Murray Basin, south-eastern Australia. *BMR Journal of Australian Geology & Geophysics.* 11(2/3): 127–146.

Brown, C.M. and Stephenson, A.E. 1991. *Geology of the Murray Basin, South-eastern Australia.* Canberra: Bureau of Mineral Resources. (Bulletin 235). 430 pp.

Brown, J.A.H. 1983. *Australia's Surface Water Resources.* (Water 2000 Consultants Report no. 1). Canberra: Australian Government Publishing Service. 177 pp.

Charman, P.E.V. and Junor, R.S. 1989. Saline seepage and land degradation — a New South Wales perspective. *BMR Journal of Australian Geology & Geophysics.* 11(2/3): 195–203.

Chartres, C. 1987. Australia's land resources at risk. In. Chisholm, A. and Dumsday, R. eds. *Land Degradation, Problems and Policies.* Cambridge: Cambridge University Press. 7–26.

Chesterfield, E.A. 1986. Changes in the vegetation of the river red gum forest at Barmah, Victoria. *Australian Forestry.* 49(1): 4–15.

Close, A. 1990. River salinity. In. Mackay, N. and Eastburn, D. eds. *The Murray*. Canberra: Murray-Darling Basin Commission. 127–144.

Conacher, A.J., Combes, P.L., Smith, P.A. and McLellan, R.C. 1983a. Evaluation of throughflow interceptors for controlling secondary soil and water salinity in dryland agricultural areas of south-western Australia: I. Questionnaire surveys. *Applied Geography*. 3(1): 29–44.

Conacher, A.J., Neville, S.D. and King, P.D. 1983b. Evaluation of throughflow interceptors for controlling secondary soil and water salinity in dryland agricultural areas of south-western Australia: II. Hydrological study. *Applied Geography*. 3(2): 115–132.

Consultative Committee for the Irrigation Strategy Study. 1990. *The Irrigation Strategy Study South-West Western Australia: Phase 1 Report*. Leederville: Water Authority of Western Australia. (Report no. WP 95). 41 pp.

Cope, F. 1958. *Catchment Salting in Victoria*. Melbourne: Soil Conservation Authority of Victoria. 88 pp.

Crabb, P. 1988. *The Murray-Darling Basin Agreement*. Canberra: Centre for Resource and Environmental Studies, The Australian National University. (CRES working paper 1988/6). 30 pp.

Crabb, P. 1991. Resolving conflicts in the Murray-Darling Basin. In. Handmer, J.W., Dorcey, A.H.J. and Smith, D.I. eds. *Negotiating Water: Conflict Resolution in Australian Water Management*. Canberra: Centre for Resource and Environmental Studies, The Australian National University. 147–159.

Cunningham, R.B. and Morton, R. 1983. A statistical method for the estimation of trend in salinity in the River Murray. *Australian Journal of Soil Research*. 21(2): 123–132.

Day, R.W., Whitaker, W.G., Murray, C.G., Wilson, I.H. and Grimes, K.G. 1983. *Queensland Geology*. Brisbane: Geological survey of Queensland. (Publication no. 383). 192 pp.

Department of Primary Industries and Energy. 1989. *Proceedings of the First Community Conference of the Murray-Darling Basin Ministerial Council's Community Advisory Committee, Albury 16–17 November 1988*. Canberra: Australian Government Publishing Service. (Various pagings, about 115 pp).

Dimmock, G.M., Bettenay, E. and Mulcahy, M.J. 1974. Salt content of lateritic profiles in the Darling Range, Western Australia. *Australian Journal of Soil Research*. 12(2): 63–69.

Dooley, T. 1991. The role of trees in controlling dryland salinity — a South Australian overview. In. *The Role of Trees in Sustainable Agriculture, A National Conference. Sept. 30–Oct. 3, 1991*. Albury, New South Wales (Sponsored by The Rural Industries Research and Development Corporation, Canberra). Salinity. 45–49.

Dryland Salinity Management Working Group. 1993. *Dryland Salinity Management in the Murray-Darling Basin*. Canberra: Murray-Darling Basin Ministerial Council. 191 pp.

Dumsday, R. G., Pegler, R. and Oram, D.A. 1989. Is broadscale revegetation economic and practical as a groundwater and salinity management tool in the Murray-Darling Basin? *BMR Journal of Australian Geology & Geophysics*. 11(2/3): 209–218.

Eastburn, D. 1990. The River. In. Mackay, N. and Eastburn, D. eds. *The Murray*. Canberra: Murray-Darling Basin Commission. 3–15.

Engineering and Water Supply Department. 1987. *Water Resources Inventory: Information on Water Availability and Use for the Water Resources Management Strategy, 1987*. Adelaide: Engineering and Water Supply Department. 86 pp.

Engineering and Water Supply Department. 1991. *Annual Report 1990-91*. Adelaide: Engineering and Water Supply Department. 78 pp.

Evans, R.S. 1989. Saline water disposal options in the Murray Basin. *BMR Journal of Australian Geology & Geophysics*. 11(2/3): 167–185.

Evans, R.S. and Nolan, J. 1989. A groundwater management strategy for salinity mitigation in Victorian riverine plain, Australia. *Groundwater Management: Quantity and Quality. Proceedings of the Benidorm Symposium, October 1989*. (IAHS Publication no. 188). 487–499.

Evans, W.R. compiler. 1988. *Preliminary Shallow Ground Water and Salinity Map of the Murray Basin (1:1 000 000 Scale Map)*. Canberra: Bureau of Mineral Resources, Geology & Geophysics.

Evans, W.R. and Kellett, J.R. 1989. The hydrogeology of the Murray Basin, southern Australia. *BMR Journal of Australian Geology & Geophysics*. 11(2/3): 147–166.

Food and Agriculture Organization of the United Nations (FAO). 1989. *Production Yearbook*. Rome: FAO. v.42. 350 pp.

Framji, K.K., Garg, B.C. and Luthra, S.D.L. 1981. *Irrigation and Drainage in the World: A Global Review*. New Delhi: International Commission on Irrigation and Drainage. v.I. 491 pp.

Garrett, B. 1992. The Goulburn dryland salinity management plan: A revegetation strategy. In. *Catchments of Green: A National Conference on Vegetation and Water Management*. Adelaide Convention Centre, 23–26 March 1992. Canberra: Greening Australia Ltd. Conference Proceedings. v. A. 59–66.

George, P.R. and Nulsen, R.A. 1985. Saltland drainage: Case studies. *Journal of Agriculture — Western Australia*. 26(4): 115–118.

George, R. 1990. The 1989 saltland survey. *Journal of Agriculture — Western Australia*. 31(4): 159–166.

George, R.J. 1991. Management of sand plain seeps in the wheat-belt of Western Australia. *Agricultural Water Management*. 19(2): 85–104.

Gordon, I.J. 1991. *A Survey of Dryland and Irrigation Salinity in Queensland*. Brisbane: Department of Primary Industries, Land Management Research Branch. 12 pp.

Government of Victoria. 1988. *Salt Action: Joint Action*. Melbourne: Salinity Bureau, Department of the Premier and Cabinet. 54 pp.

Government of Victoria. 1990. *Victorian Government Support for Salinity Management Plans: Tragowel Plains Area, Campaspe West Area, Shepparton Irrigation Region, Goulburn Dryland Catchment*. Melbourne: Salinity Bureau, Department of the Premier and Cabinet. 36 pp.

Grieve, A.M. 1990. Land management options to control waterlogging and salinity under irrigation. In. Humphreys, E., Muirhead, W.A. and van der Lelij, A. eds. *Management of Soil Salinity in South East Australia*. Proceedings of a symposium held at Albury, New South Wales on 18–20 September 1989. Australian Society of Soil Sciences, Riverina Branch. Wagga Wagga: Charles Sturt University — Riverina. 281–289.

Grieve, A.M., Dunford, E., Marston, D., Martin, R.E. and Slavich, P. 1986. Effects of waterlogging and soil salinity on irrigated agriculture in the Murray Valley: A review. *Australian Journal of Experimental Agriculture*. 26(6): 761–777.

Gutteridge Haskins & Davey. 1970. *Murray Valley Salinity Investigation*. Canberra: Murray-Darling Basin Commission. v.1 The Report, 476 pp; v.2 Maps, 24; v.3 Summary, 67 pp.

Gutteridge Haskins & Davey, ACIL Australia and Australian Groundwater Consultants. 1990. *A Pipeline to the Sea, Pre-Feasibility Study*. Canberra: Murray-Darling Basin Commission. 111 pp.

Hingston, F.J. and Gailitis, V. 1976. The geographic variation of salt precipitated over Western Australia. *Australian Journal of Soil Research*. 14: 319–335.

Hunter, G. and David, G. 1992. Victoria's catchment-based salinity program, achievements and challenges. In. *Catchments of Green: A National Conference on Vegetation and Water Management*. Adelaide, 23–26 March 1992. Canberra: Greening Australia Ltd. Conference Preceedings. v. A. 15–21.

Jacobs, T. 1990. River regulation. In. Mackay, N. and Eastburn, D. eds. *The Murray*. Canberra: Murray-Darling Basin Commission. 39–58.

Jacobson, G., Habermehl, M.A. and Lau, J.E. 1983. *Australia's Groundwater Resources*. (Water 2000 Consultants Report no. 2). Canberra: Australian Government Publishing Service. 65 pp.

Jacobson, G. and Lau, J.E. 1988. Australia's groundwater systems under stress. *Water Quality Bulletin*. 13(4): 107–116.

Johnston, C.D., McArthur, W.M. and Peck, A.J. 1980. *Distribution of Soluble Salts in Soils of the Manjimup Woodchip Licence Area, Western Australia*. CSIRO Division of Land Resources Management. (Technical paper. no. 5). 29 pp.

Jolly, I.D. 1989. *Investigation into the Potential for Increased Stream Salinisation in the Darling Basin*. Glen Osmond, South Australia: Centre for Research in Groundwater Processes. (Report no. 10). 103 pp.

Landsberg, J. 1992. Role of the Murray-Darling Basin Commission in integrated catchment management. In. *Catchment of Green: A National Conference on Vegetation and Water Management*. Adelaide Convention Centre, 23–26 March 1992. Canberra: Greening Australia Ltd. Conference Proceedings. v.A. 9–14.

Lawrence, C.R. 1975. *Geology, Hydrodynamics and Hydrochemistry of the Southern Murray Basin*. Melbourne: Geological Survey of Victoria. (Memoir 30). 359 pp.

Lay, B. 1990. Salt land revegetation: A South Australian overview. In. Myers, B.A. and West, D.W. eds. *Revegetation of Saline Land*.

Proceedings of a workshop held at the Institute for Irrigation and Salinity Research, 29–31 May 1990. Tatura, Victoria: Institute for Irrigation and Salinity Research. 15–20.

Mabbutt, J.A. 1978. *Desertification in Australia*. Kingsford, New South Wales: Water Research Foundation of Australia. (Report no. 54). 127 pp.

Mackay, N. 1990. Understanding the Murray. In. Mackay, N. and Eastburn, D. eds. *The Murray*. Canberra: Murray-Darling Basin Commission. ix-xix.

Mackay, N., Hillman, T. and Rolls, J. 1988. *Water Quality of the River Murray, Review of Monitoring 1978–1986*. Canberra: Murray-Darling Basin Commission. 62 pp.

Macumber, P. 1990. The salinity problem. In. Mackay, N. and Eastburn, D. eds. *The Murray*. Canberra: Murray-Darling Basin Commission. 111–125.

Macumber, P.G. 1991. *Interaction Between Ground Water and Surface Systems in Northern Victoria*. Melbourne: Department of Conservation and Environment. 345 pp.

Macumber, P.G., Dyson, P.R., Jenkin, J.J. and Moran, R.A.J. 1988. Possible impacts of the greenhouse effect on salinity in Victoria, Australia. In. Pearman, G.I., ed. *Greenhouse: Planning for Climate Change*. Melbourne: CSIRO Publications. 252–260.

Macumber, P.G. and Fitzpatrick, C.R. 1987. *Salinity Control in Victoria: Physical Option*. Melbourne: Department of Water Resources. (Report no. 15). 50 pp.

Malcolm, C.V. 1982. *Wheat-belt Salinity, a Review of the Salt Land Problem in South Western Australia*. Western Australia: Department of Agriculture. (Technical Bulletin no. 52). 65pp.

Malcolm, C.V. 1986. Production from salt-affected soils. *Reclamation and Revegetation Research*. 5: 343–361.

Malcolm, C.V. 1990. Saltland agronomy in Western Australia — an overview. In. Myers, B.A. and West, D.W. eds. *Revegetation of Saline Land*. Proceedings of a Workshop held at The Institute for Irrigation and Salinity Research, 29–31 May 1990. Tatura, Victoria: Institute for Irrigation and Salinity Research. 21–25.

McFarlane, D. 1991. A review of secondary salinity in agricultural areas of Western Australia. *Land and Water Research News*. (Western Australian Steering Committee for Research on Land Use and Water Supply). 11: 7–16.

Morris, J.D. 1991. The use of trees in salinity control in Victoria, Australia: Progress and potential. In. *The Role of Trees in Sustainable Agriculture, A National Conference. Sept. 30–Oct. 3, 1991*. Albury: New South Wales (Sponsored by the Rural Industries Research and Development Corporation, Canberra). Salinity. 63–75.

Morton, R. and Cunningham, R.B. 1985. Longitudinal profile of trends in salinity in the River Murray. *Australian Journal of Soil Research*. 23(1): 1–13.

Murray-Darling Basin Commission. 1988. *Murray-Darling Basin Commission, First Annual Report 1988*. Canberra: Murray-Darling Basin Commission. 129 pp.

Murray-Darling Basin Commission. 1991. *1990 Annual Report*. Canberra: Murray-Darling Basin Commission. 130 pp.

Murray-Darling Basin Ministerial Council. 1988. *Draft Salinity and Drainage Strategy*. Canberra: Murray-Darling Basin Commission. 12 pp.

Murray-Darling Basin Ministerial Council. 1989. *Draft Murray-Darling Basin Natural Resources Management Strategy*. Canberra: Murray-Darling Basin Ministerial Council. 19 pp.

Nangiloc-Colignan Community Salinity Working Group. 1991. *Nangiloc-Colignan Draft Salinity Management Plan*. Red Cliffs, Victoria: Rural Water Commission. 128 pp.

Nix, H. 1988. Australia's renewable resources. In. Day, L.H. and Rowland, D.T. eds. *How Many More Australians*. Melbourne: Longman Cheshire. 65–76.

Nix, H.A. and Kalma, J.D. 1982. The climate of the Murray-Darling Basin. In. *Murray-Darling Basin Project Development Study, Stage 1: Working Papers*. Canberra: CSIRO Division of Water and Land Resources. 22–38.

Packham, G.H. ed. 1969. The geology of New South Wales. *Journal of Geological Society of Australia*. 16 (1): 654 pp.

Parkinson, G. ed. 1986. *Atlas of Australian Resources*. v.4. Climate. Canberra: Division of National Mapping. 60 pp.

Peck, A.J. 1978. Salinization of non-irrigated soils and associated streams: A review. *Australian Journal of Soil Research*. 16(2): 157–168.

Peck, A.J. and Hurle, D.H. 1973. Chloride balances of some farmed and forested catchments in south-western Australia. *Water Resources Research*. 9(3): 648–657.

Peck, A.J., Thomas, J.F. and Williamson, D.R. 1983. *Salinity Issues, Effect of Man on Salinity in Australia*. (Water 2000 Consultants Report no. 8). Canberra: Australian Government Publishing Service. 78 pp.

Peck, A.J. and Williamson, D.R. 1987a. Effects of forest clearing on groundwater. *Journal of Hydrology*. 94 (1/2): 47–65.

Peck, A.J. and Williamson, D.R. eds. 1987b. Hydrology and salinity in the Collie River Basin, Western Australia. *Journal of Hydrology*. 94 (1/2): 198 pp.

Powell, J.M. 1989. *Watering the Garden State: Water, Land and Community in Victoria 1834-1988*. Sydney: Allen & Unwin. 319 pp.

Pressey, B. 1990. Wetlands. In. Mackay, N. and Eastburn, D. eds. *The Murray*. Canberra: Murray-Darling Basin Commission. 167–181.

Prinsley, R.T. 1991. *Australian Agroforestry: Setting the Scene for Future Research*. Canberra: Rural Industries Research and Development Corporation. 90 pp.

Ruprecht, J.K. and Schofield, N.J. 1991. Effects of partial deforestation on hydrology and salinity in high salt storage landscapes. I: Extensive block clearing. II: Strip, soil and parkland clearing. *Journal of Hydrology*. 129: 19–38 and 39–55.

Salinity Bureau. 1989. *Victorian Salinity Program: Annual Review 1988-89*. Melbourne: Salinity Bureau, Department of the Premier and Cabinet. 84 pp.

Salinity Pilot Program Advisory Council. 1989. *Shepparton Land and Water Salinity Management Plan (A Draft)*. Melbourne: Salinity Bureau, Department of the Premier and Cabinet. 151 pp.

Salinity Planning Working Group, Victoria. 1992. *Regional Salinity Impacts*. Melbourne: Salinity Bureau, Department of the Premier and Cabinet. (Draft Report). 42 pp.

Schofield, N.J. 1989. Stream salinisation and its amelioration in south-west Western Australia. In. *Proceedings of the Baltimore Symposium, May 1989*. (IAHS Publication no. 182). 211–220.

Schofield, N.J. 1992. Tree planting for dryland salinity control in Australia. *Agroforestry Systems*. 20(1): 1–23.

Schofield, N.J. and Bari, M.A. 1991. Valley reforestation to lower saline groundwater tables: Results from Stene's Farm, Western Australia. *Australian Journal of Soil Research*. 29(5): 635–650.

Schofield, N.J., Bari, M.A., Bell, D.T., Boddington, W.J., George, R.J. and Pettit, N.E. 1991. The role of trees in land and stream salinity control in Western Australia. In. *The Role of Trees in Sustainable Agriculture, A National Conference. Sept. 30 – Oct. 3 1991*. Albury: New South Wales. (Sponsored by The Rural Industries Research and Development Corporation, Canberra). Salinity. 21–43.

Schofield, N.J., Loh, I.C., Scott, P.R., Bartle, J.R., Ritson, P., Bell, R.W., Borg, H., Anson, B. and Moore, R. 1989. *Vegetation Strategies to Reduce Stream Salinities of Water Resources Catchments in South-West Western Australia*. Leederville: Water Authority of Western Australia, Water Resources Directorate. (Report no. WS 33). 81 pp.

Schofield, N.J. and Ruprecht, J.K. 1989. Regional analysis of stream salinisation in south-west Western Australia. *Journal of Hydrology*. 112: 19–39.

Schofield, N.J., Ruprecht, J.K. and Loh, I.C. 1988. *The Impact of Agricultural Development on the Salinity of Surface Water Resources of South-West Western Australia*. Leederville: Water Authority of Western Australia, Water Resources Directorate. (Report no. WS 27). 69 pp.

Select Committee on Salinity. 1988. *Report on Salinity in Western Australia*. Perth: Parliament of Western Australia, Legislative Council. 72 pp.

Simmons, P., Poulter, D. and Hall, N.H. 1991. *Management of Irrigation Water in the Murray-Darling Basin*. Canberra: Australian Bureau of Agricultural and Resource Economics. (Discussion Paper 91.6). 42 pp.

Smith, R.C.G., Mason, W.K., Meyer, W.S. and Barrs, D. 1983. Irrigation in Australia: Development and prospects. In. Hillel, D. ed. *Advances in Irrigation*. New York: Academic Press. v.2. 99–153.

Soil Conservation Service of New South Wales. 1989. *Land Degradation Survey of New South Wales 1987/88*. Sydney: Soil Conservation Service of New South Wales. 32 pp.

State Dryland Salinity Committee. 1990a. *Technical Strategy to Address Dryland Salinity in South Australia*. Adelaide: Department of Agriculture. 18 pp.

State Dryland Salinity Committee. 1990b. *Dryland Salinity in South Australia: Annual Report 1989/90*. Adelaide: Department of Agriculture. 23 pp.

Steering Committee for Research on Land Use and Water Supply. 1989. *Stream Salinity and its Reclamation in South-West Western Australia*. Leederville: Water Authority of

Western Australia, Water Resources Directorate. (Report no. WS 52). 23 pp.

Stone, S. 1991. Community-led land and water use planning: Some Victorian experiences. In. Handmer, J.W., Dorcey, A.H.J. and Smith, D.I. eds. *Negotiating Water: Conflict Resolution in Australian Water Management*. Canberra: Centre for Resource and Environmental Studies, The Australian National University. 219–236.

Sunraysia Community Salinity Working Group. 1991. *Sunraysia Draft Salinity Management Plan*. Mildura, Victoria. 94 pp.

Teakle, L.J.H. 1937. The salt (sodium chloride) content of rainwater. *Journal of Agriculture-Western Australia*. 14(2): 115–123.

Thomas, J.E. 1939. *An Investigation of the Problem of Salt Accumulation in a Mallee Soil in the Murray Valley Irrigation Area*. Melbourne: Council for Scientific and Industrial Research. (Bulletin no. 128). 88 pp.

Tickell, S.J. and Humphrey, W.G. 1985. *Ground Water Resources and Associated Salinity Problems of the Victorian Part of the Riverina Plain*. Melbourne: Department of Industry, Technology and Resources. (Geological Survey of Victoria Report no. 84). 197 pp.

Unon. 1993. *A National Conference: Land Management for Dryland Salinity Control*. LaTrobe University, Bendigo, 28 September – 1st October 1993. 276 pp.

Wasson, R.J. 1982. Geology, geomorphology and mineral resources of the Murray–Darling Basin. In. *Murray-Darling Basin Project Development Study, Stage 1: Working Papers*. Canberra: CSIRO Division of Water and Land Resources. 1–9.

West, D.W. 1990. Revegetation of saline-lands: Perspectives from Victoria. In. Myers, B.A. and West, D.W. eds. *Revegetation of Saline Land*. Proceedings of a Workshop held at the Institute for Irrigation and Salinity Research, 29–31 May 1990. Tatura, Victoria: Institute for Irrigation and Salinity Research. 5–11.

Western Australian Department of Agriculture. 1988a. *Salinity in Western Australia — A Situation Statement*. Perth: Western Australian Department of Agriculture, Division of Resource Management. (Technical Report no. 81). 116 pp.

Western Australian Department of Agriculture. 1988b. *Situation Statement — Soil and Land Conservation Program in Western Australia*. Perth: Western Australian Department of Agriculture, Division of Resource Management. (Technical Report).

Western Australia Legislative Assembly. 1991. *Select Committee into Land Conservation, Final Report*. Perth: Legislative Assembly. 171 pp.

Wilkinson, R. and Barr, N. 1993. *Community Involvement In Catchment Management: An Evaluation of Community Planning and Consultation in the Victorian Salinity Program*. Melbourne: Department of Agriculture. 197 pp.

Williams, B.G. 1991. Salinity and waterlogging in the Murray-Darling Basin. In. *Environmental Research in Australia: Case studies*. Canberra: Australian Government Publishing Service. 198 pp.

Williamson, D.R. 1990. Salinity — an old environmental problem. In. *Year Book Australia 1990*. Canberra: Australian Government Publishing Service. 202–211.

Williamson, D.R., Stokes, R.A. and Rupprecht, J.K. 1987. Response of input and output of water and chloride to clearing for agriculture. *Journal of Hydrology*. 94: 1–28.

Williamson, D.R. and van der Wel, B. 1991. Quantification of the impact of dryland salinity on water resources in the Mt Lofty Ranges, S.A. In. *International Hydrology and Water Resources Symposium 1991: Challenges for Sustainable Development, 2–4 October 1991, Perth*. Canberra: The Institution of Engineers, Australia. (Publication no. 91/22). v.1. 48–52.

Wood, W.E. 1924. Increase of salt in soil and streams following the destruction of the native vegetation. *Journal and Proceedings of the Royal Society of Western Australia*. 10: 35–47.

Chapter 3: China

Introduction

The People's Republic of China (PRC) is the world's third largest country by area (after the Commonwealth of Independent States and Canada) and the most populated. It covers an area of 9 560 980 km² and has a varied topography with highlands in the west and plains in the east. This physical feature of gradual descent from the west to the east causes all the major rivers to run eastwards. China comprises the following physiographic features (China Handbook Editorial Committee, 1983):

- **Mountain ranges** run in different directions across the length and breadth of the country (Figure 3-1). These are: the west-east ranges consisting of the Tienshan, Yanshan, Kunlun, Qinling and Nanling Mountains; the north-east–south-west ranges consisting of the Greater Hinggan, Taihang, Changbai, Wuyi and Taiwan Mountains; the north-west–south-east ranges consisting of the Qilian and Altai Mountains; and the north-south ranges consisting of the Helan, Liupan and Hengduan Mountains.

- **Plateaus** vary in both height and physical features. They are located mainly in the western and central parts of the country. The major plateaus are: the Qinghai-Tibet Plateau in western China, which is the highest plateau in the world with an elevation of 4000–5000 m; the Inner Mongolia Plateau in northern China, which is the second largest plateau in the country and stands at an average elevation of 1000–2000 m above sea-level; the Loess Plateau, bounded by the Qinling Mountains in the south and the Taihang Mountains in the east which rises 800–2000 m above sea-level; and the Yunnan-Guizhou Plateau in south-western China, which is 2000 m high in Yunnan and 1000 m high in Guizhou.

- **Basins** as large as hundreds of thousands of square kilometres are found in the west, notably the Tarim, Junggar, Turpan, Qaidam and Sichuan basins. Medium and small basins are found mainly in the east.

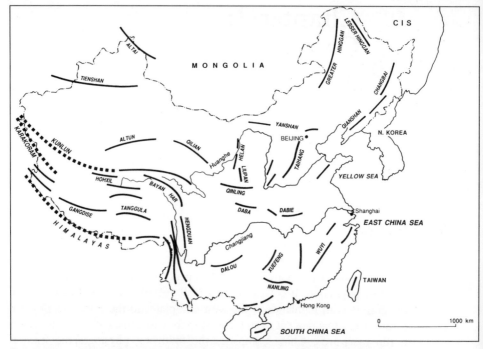

Figure 3-1: Principal mountain ranges of China (China Handbook Editorial Committee, 1983: 18).

- **Plains** lie mainly in the north-east and eastern seaboard regions, covering 1.12 million km² or a little more than 10 per cent of the country's total area. The main plains together constitute the bulk of the country's plain area, extending in one stretch to form a north-south plain belt. They have gentle terrain, fertile soil, mild climate and provide a base for China's major agriculture and industry. The main plains are: the North-east Plain, which is the largest plain in China, covers 350 000 km² in north-east China and measures 1000 km from north to south and 400 km from west to east in its widest part; the North China Plain, also called the Huang-Huai-Hai Plain, which is the second largest plain, covering an area of about 300 000 km², and is the product of the alluvial deposits from the Huanghe, Huaihe and Haihe Rivers and is an important agricultural region, with the largest cultivated acreage in China; and the Middle-Lower Changjiang Plain, which consists of a series of plains of varying width on both sides of the Changjiang River, covering 200 000 km² with an elevation mostly less than 50 m above sea-level.
- **Deserts** cover 1 095 000 km² or 11.4 per cent of the country's total area, of which sand deserts account for 637 000 km² and the Gobi and stone deserts 458 000 km². The major deserts are: the Taklimakan, Gurbantunggut,

Badinjaran, Tenger and Muus Deserts. Most of the deserts are found in Xinjiang, Qinghai, Gansu, Inner Mongolia and Shaanxi in north-western and northern China. The desert lands in Xinjiang constitute about 60 per cent of the country's total.

Figure 3-2: Administrative divisions of China (China Handbook Editorial Committee, 1983).

China supported a population of 1134 million in 1990 with an average growth rate of 1.4 per cent during the 1980s (World Bank, 1992: 268). China's population is unevenly distributed. The population density ranges from more than 2000 persons per km^2 in Shanghai to one person per km^2 in Tibet (James, 1989: 46). Generally eastern China, especially the eastern seaboard, has a population density of 300–400 persons per km^2, while western China is more sparsely populated. The average population density of the nation is about 100 persons per km^2. China is divided into 22 provinces (Figure 3-2), five autonomous regions and three independent municipalities, the latter being Beijing, Shanghai and Tianjin.

Rainfall and climate

Because of its immense size, climate and precipitation vary greatly in different parts of China. The mean annual precipitation of the country is about 630 mm. Some 32 per cent of the country is considered humid, 15 per cent semihumid, 22 per cent semiarid and 31 per cent arid (Editorial Department of the PRC Year Book in Beijing, 1987: 5). At one extreme, the coastal area in the south-east enjoys plentiful rainfall and a humid climate with an annual precipitation of 1000–2000 mm, while in some parts of north-west China, the annual rainfall is below 50 mm. If the isohyet of 400 mm is taken as the boundary line, China can be divided into two parts (Figure 3-3). East of the line is a humid area and to the west is the arid section of Central Asia. In the major part of the arid region, annual precipitation is under 200 mm. It reduces rapidly toward China's driest centres, such as southern Xinjiang, western Gansu and north-western Qaidam Basin. These dry centres all have less than 50 mm of rainfall per year. The interior of the Tarim Basin has less than 30 mm and Qiemo has an average annual precipitation of only 9.2 mm and no rain in some years (Ren et al., 1985).

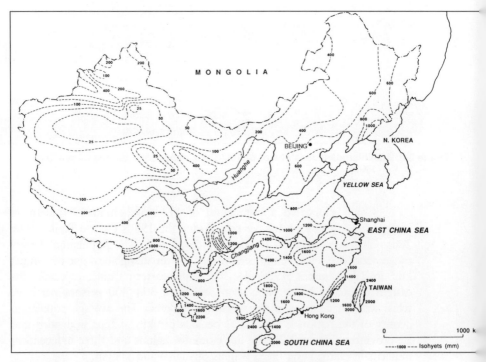

Figure 3-3: China's annual precipitation (Ren et al., 1985: 66).

The precipitation is strongly seasonal over most parts of the country. Taking China as a whole, the amount of precipitation is scant from November to February. June through August is a wet season for the whole country. In between are transition periods from dry to wet or from wet to dry. In the north-east, north China and Inner Mongolia, precipitation is concentrated in summer, comprising 60 to 70 per cent of the annual rainfall. The rainy season in these areas is short and the dry season exceptionally long (Ren et al., 1985). The precipitation in China is not only unevenly distributed spatially but varies considerably from year to year, causing major drought and flood problems.

Water resources

Surface water

China has a particularly large number of rivers (Figure 3.4). There are more than 50 000 rivers, each with a drainage area of over 100 km^2 including 1580 rivers with over 1000 km^2 and 79 with over 10 000 km^2. Among them the Changjiang River and the Huanghe River are the longest and the second longest rivers in Asia (James, 1989: 73).

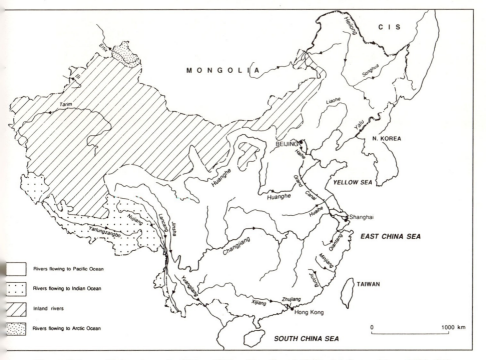

Figure 3-4: Major rivers in China (China Handbook Editorial Committee, 1983: 91).

All the major rivers of China flow in a generally west to east direction to the Pacific Ocean. River pattern is dense in the east and the south but sparse in the west and the north. Of all the rivers, 56.7 per cent run into the Pacific, 6.5 per cent into the Indian and 0.5 per cent into the Arctic Ocean (Figure 3-4). In northern and western China, there are many rivers without outlet to the sea, the largest being the Tarim River, which meanders for 2179 km through southern Xinjiang Province.

The long-term average annual precipitation in China totals approximately 6000×10^9 m^3 with an average annual depth of 630 mm. This is less than the world continental average of 800 mm and the Asian average of 740 mm (Yao and Chen, 1983). The average annual volume of run-off from China's rivers is 2600×10^9 m^3 with an average annual depth of 272 mm, or about 43 per cent of the precipitation. This means that the total amount of evaporation is about 3430×10^9 m^3, which is equivalent to a nationwide water depth of 360 mm (Ren et al., 1985). China's river flow is 6.5 per cent of the world total and 13.9 per cent of Asia (James, 1989), ranking sixth in the world after Brazil, the former Soviet Union, Canada, the United States and Indonesia (Chen and Wu, 1987).

Table 3-1: Characteristics of China's major rivers.

River	Drainage area (km^2)	Length (km)	Average flow (m^3 s^{-1})	Annual flow (10^9 m^3)	Depth of run-off (mm)
Changjiang	1 807 199	6 380	31 055	979.35	542
Ertix	50 862	442	348	10.98	216
Haihe	264 617	1 090	717	22.60	85
Hanjiang	34 314	325	942	29.71	866
Heilong	1 620 170	3 420	8 600	270.90	167
Huaihe	185 700	1 000	1 113	35.10	189
Huanghe	752 443	5 464	1 820	57.45	76
Ili	56 700	375	374	11.79	208
Jiulong	14 689	258	445	14.02	954
Lancang	164 799	1 612	2 354	74.25	412
Liaohe	164 104	1 430	302	9.53	58
Minjiang	60 924	577	1 978	62.37	1 024
Nujiang	142 681	1 540	2 222	70.09	469
Qiantang	54 349	494	1 484	46.80	861
Yalu	62 630	773	1 040	32.76	541
Yarlungzangbo	246 000	1 940	3 699	116.65	474
Yuanjiang	34 917	772	410	12.92	370
Zhujiang	452 616	2 197	11 075	349.20	772

Source: Ren et al. (1985: 89).

Table 3-1 provides basic data on the drainage area, length and annual flow of the major rivers in China. The Huanghe River, the fifth longest in the world with a length of 5464 km, flows across arid and semiarid areas where there is no ample surface run-off. Its flow is not heavy, averaging 1820 m^3 s^{-1}, even less than that of the Minjiang, with a length of only 577 km.

In addition to natural rivers, China has many canals. The Grand Canal, connecting Hangzhou in the south to Beijing in the north, is the world's largest canal system. It extends in a generally north–south direction for about 1794 km. The other outstanding examples are the spider web of canals in the Changjiang Delta and the new Red Flag Irrigation Canal built through the mountains of Henan Province.

China is a land of many lakes, covering an area of over 80 000 km^2. Lakes fall into two main categories: freshwater and salt lakes. Freshwater lakes cover an area of 36 000 km^2. The middle-lower Changjiang region has the most lakes because it has plenty of rainfall and a low gentle terrain. Salt lakes are found mainly in inland regions like the Qinghai-Tibet Plateau, Xinjiang and Inner Mongolia. In the Qaidam Basin the hundred or more salt lakes covering a total area of 8500 km^2 have a salt reserve of 100 000 million tonnes (China Handbook Editorial Committee, 1983).

China has an undulating terrain, concentrated rainfall and heavy rainstorms, and because of the destruction of natural vegetation cover in the past, the surface run-off results in a great deal of erosion. According to statistics, the total mass of silt transported into the ocean by China's rivers reaches 2600 million tonnes per year. Of this amount, silt discharge of the Huanghe is about 1600 million tonnes or more than 60 per cent of the national total (Ren et al., 1985).

In 1988, the total water supply capacity of all water projects had reached 498.6×10^9 m^3, of which surface water and groundwater accounted for 89.4 per cent and 10.6 per cent, respectively (Zhang et al., 1992). The annual water use in China is about 444×10^9 m^3 (Ministry of Water Conservancy, 1990). Agricultural use, industrial use and domestic use accounted for 388.5×10^9 m^3 (87.5 per cent), 47×10^9 m^3 (10.6 per cent) and 8.5×10^9 m^3 (1.9 per cent), respectively (Zhang et al., 1992). Moreover, it is estimated that by the end of this century the total water demand will increase by 100 to 150×10^9 m^3.

The rate of utilisation of water resources in the Huang-Huai-Hai Plain is relatively high and will increase substantially in the future. Despite further increases in the utilisation rate of the region's run-off and adoption of water-saving measures, the long-term plans of the pertinent provinces and municipalities estimate that by the turn of the century the normal dry year water deficit will be around 70×10^9 m^3 (Yao and Chen, 1983). Therefore, transfer of water from the Changjiang is required to address this problem.

Groundwater

The east-west trending Qinling-Kunlun structural zone (Figure 3-1) separates the whole territory of China into northern and southern parts both geologically and physiographically, bringing about remarkable differences in the distribution of groundwater. Further, owing to the differences in geological structure, topography, climate and surface water in an east-west direction, the distribution of groundwater also varies in an east-west direction. According to the Department of Technical Co-operation for Development (1986) and the Institute of Hydrogeology and Engineering Geology (1988), the groundwater resources of China occur mainly in the following geological formations:

- **Groundwater in unconsolidated sediments.** In the vast plains in the eastern part of China and in the interior basins, groundwater occurs in the unconsolidated sediments. In the plains, unconsolidated Quaternary alluvio-diluvial, alluvio-lacustrine and marine deposits have accumulated in varying thickness. The aquifers consist mainly of sands and gravel and often appear in multilayered form. In the Huang-Huai-Hai Plain, Quaternary sediments are quite well developed. The thickness of sediments varies from place to place from 200 to 600 m and reaches a maximum of more than 1000 m in the depressions.

- **Groundwater in karstic formation.** There are appreciable differences in the distribution of karstic formations between the areas north and south of the Qinling-Kunlun Mountains. In the northern area they occur mainly in dolomitised Cambrian and Ordovician rocks which are moderately karstified. In the southern areas, karstic formations occur abundantly in the Upper Palaeozoic and Lower Mesozoic carbonate rocks. These rocks are younger in age, pure in carbonate composition and rather intensely karstified, resulting in a series of underground rivers and huge solution caves, and also typical karst landscapes.

- **Groundwater in fractured rocks.** The exposed bedrock in mountainous areas makes up about two-thirds of the total territory of China. Rocks are of various ages and kinds. Among them granitic rocks are rather widespread, accounting for about one-third of the total exposed area of bedrock over the whole country. The main rock types include magmatic, metamorphic and clastic rocks. Aquifers are mainly of the fracture type and locally of the porous type.

Rough estimates indicate the long-term safe yield to be approximately 700×10^9 m^3. Of this, 480×10^9 m^3 are distributed in the Changjiang Basin and regions to the south of it (Yao and Chen, 1983). This reserve has been utilised only in extremely rare instances, one case being in the Nanyang Basin. Only 70×10^9 m^3 or about 10 per cent of China's groundwater lie under one-quarter of the cultivated land which is located in north China in the municipalities of Beijing and Tianjin and the provinces of Hebei, Shandong and Henan. The most extensive extraction of groundwater is in the Huang-Huai-Hai Plain, which was about 27×10^9 m^3 in 1978 (Zhu and Zheng, 1983). A large portion of the area north of the Huaihe is irrigated by wells. In recent years, excessive extraction of groundwater in some places has produced cones of depression, particularly in the deep aquifers. Some of the cones have overlapped over broad areas. The groundwater in the coastal plains is highly salinised and some of it is unusable (Yao and Chen, 1983). According to Xu (1987) and Zou (1987), about 2.2 to 2.4 million wells exist in the northern region of China for the extraction of groundwater. The annual groundwater extraction is about 53×10^9 m^3.

Land cover

According to FAO (1989), in 1987, out of a total land area of 932.64 Mha, about 96.98 Mha were under cultivation. Of these lands 44.83 Mha were irrigated. Permanent pasture covered 319.08 Mha, forest and woodlands 116.57 Mha and other land 400.02 Mha. Another statistic (Editorial Department of the PRC Year Book in Beijing, 1987: 4) shows that China has over 100 Mha of cultivated land, 319.08 Mha of grassland and 115.25 Mha of forest. Comparing the two sets of statistics indicates that, in the second case, the cultivated area is slightly (3 Mha) larger, grasslands have the same coverage and the area covered by forest is 1.32 Mha smaller. Therefore the two sets of statistics could be considered as consistent. Kaplan and Sobin (1982: 193) provided a relatively detailed distribution of the cultivated and irrigated area of China for 1978.

Irrigation

Framji et al. (1981: 222–237) have documented the historical development of irrigation and drainage in China since 4000 years ago, and have described the main features of recent projects.

About 100 Mha of the land surface in China is arable. These arable lands are located mostly in eastern China and nearly all are under cultivation. Since the foundation of the republic in 1949, the country has engaged in a vigorous water conservation and irrigation development program. Table 3-2 shows the capacity of reservoirs and their numbers. As shown in Table 3-2, the total number of reservoirs is more than 83 000, with a total storage of 430×10^9 m^3. The irrigated land in 1949 was about 16 Mha. It has increased greatly since then and is now about 48 Mha, out of which 17 Mha is in reservoir irrigation districts, 11 Mha is in well irrigated districts and about

20 Mha is irrigated by other means (Yao and Chen, 1983). China has more irrigated land than any other nation. In terms of the number of irrigation systems, Xu (1987) reported that there are 204 large irrigation systems (each one over 20 000 ha), more than 5000 medium-sized (670 to 20 000 ha) and numerous small projects with areas under 670 ha.

Table 3-2: Number and capacity of reservoirs in China (1985).

Item	Small (0.1–10 x 10^6 m^3)	Medium (10–100 x 10^6 m^3)	Large (>100 x 10^6 m^3)	Total
Number	80 4780	2401	340	83 219
Capacity (10^9 m^3)	56.4	66.2	307.5	430.1

Source: James (1989: 1470, Table 49).

Because of continuing population growth, the loss of cultivated land to non-agricultural uses constitutes a major problem. Per capita cultivated land has reduced from 0.18 ha since 1949 to 0.11 ha in 1979. Also, severe competition exists for the allocation of water resources between agricultural and non-agricultural uses.

About 80 per cent of the sown area of China is devoted to food crops, the most important being rice and wheat. Under the combined action of water conservation and agricultural measures, China's agricultural production has developed rapidly despite the reduction in cultivated land and the competition for water supply. In 1984, the total national grain and cotton yield reached 407 million tonnes and 6.08 million tonnes, respectively, or 3.6 and 13.7 times that in 1949 (Zou, 1987).

Salinity

China has about 20 Mha of salt-affected soils, of which about one-fifth is alkaline (Yu and Hseung, 1983). The salt-affected soils are widely distributed and can be found almost all over China, from the tropical to the temperate zones, from the coastal areas in the east to those inland in the west, from the lowland to the plateau. More than 14 Mha of salt-affected lands are lying mainly in Xinjiang, Qinghai, Gansu, Ningxia, Inner Mongolia and Heilongjiang Provinces or Regions (Hseung and Liu, 1990). The formation processes of salt-affected soils are roughly divided into: recent processes of salt accumulation induced by seawater, shallow groundwater and surface run-off; the process of residual accumulation of salts formed during geological periods; and the process of alkalisation (Wang and Li, 1987; Wang et al., 1990).

According to the natural conditions of salt-affected soil formation and the characteristics of salt migration and accumulation in the soils, as well as the geochemical features of the soils, China is divided into eight salinisation regions and 27 subregions (Wang and Li, 1987; Zhao and Zhu, 1992). The eight salinisation regions are as follows:

- **The coastal humid and semihumid regions with seawater submergence salinisation.** In these regions, the salt in the soil and in groundwater comes mainly from seawater. Soils with different salinities are distributed in the form of continuous or discontinuous ribbon parallel to the coastal line; the nearer to the seashore, the higher the salt content of the soil and groundwater.
- **The north-eastern semihumid and semiarid region.** This region is dominated by the accumulation of chloride-bicarbonate and sulphate-bicarbonate and there occur alkaline soils.
- **The Huang-Huai-Hai Plain.** This region is dominated by the accumulation of chloride-sulphate and sulphate-chloride. Various types of salt-affected soils appear in patches scattered in the farming lands.
- **The Mongolian Plateau, arid and semidesert.** This region is dominated by the accumulation of chloride-bicarbonate. There occur saline and alkaline soils.
- **The middle and upper Huanghe River semiarid and semidesert region.** This region is known as the principal salt source of the Huang-Huai-Hai Plain, with a varied type of salinisation and the extensive accumulation of sulphate.
- **The Gansu-Xinjiang desert salinised region.** In this region, with the exception of the Ertix River, all the surface water either disappears in the Gobi Desert or converges into inland lakes. Soils of this region contain a large amount of salt and are characterised by significant accumulation of gypsum and magnesium carbonate.
- **The Qinghai-Xinjiang extremely arid desert region.** In this region, there is a variety of saline soils. With the exception of chloride, sulphate and carbonate-bicarbonate, there exists predominantly the accumulation of nitrate in soils of some areas, and in certain regions, salts of boron, lithium and potassium are accumulated.
- **The Xizang (Tibet) Plateau alpine frozen desert salinised region.** This region is known as the highest salinised region above sea-level in the world and has numerous salt lakes. The types of salinisation vary with different areas; they are mainly characterised by the accumulation of bicarbonate-chloride or bicarbonate-sulphate.

In China, the alkaline soils are widely distributed from Inner Mongolia in the north to the Huang-Huai-Hai Plain and from the coastal region in the east to Xinjiang in the west. They are even spread in the Tibet Plateau. The alkaline soils are dispersed as spots in the soils, and often coexist with saline soils (Yu and Hseung, 1983).

In terms of secondary salinisation, no data is available for the whole of China. However, Vermeer (1977) has reported that in 1965 about 6.7 Mha of cultivated lands were salinised. A recent publication (James, 1989: 502) reports that since the early 1950s, due to the implementation of water and soil conservation policies, 4.57 Mha of saline-alkaline land in the country have been reclaimed. In the remainder of this chapter, discussion is based mainly on the Huang-Huai-Hai Plain because it is a major salt-affected area and a reasonable amount of information is available on this region. Other areas are covered together in a separate section.

Salinity in the Huang-Huai-Hai Plain

The Huang-Huai-Hai Plain (Figure 3-5) was formed by the alluvial deposits of the Huanghe (Yellow), Huaihe and Haihe Rivers. It is the largest plain of China, covering 314 400 km^2 and spanning seven provinces and municipalities including Hebei, Shandong, Henan, Jiangsu, Anhui, Beijing and Tianjin. It has a population of 150 million, accounting for about 15 per cent of the national total. This plain, with 20 Mha of cultivated land, of which 12.7 Mha are irrigated (Lou, 1987), is an important agricultural region, producing rice, wheat and cotton. It is characterised by its favourable conditions of heat and light. Its accumulated temperature over 10°C is 3800°C–4900°C, with the mean temperature of the hottest months being 24°C–28°C; sunlight is abundant, with annual sunshine time of 2800 hours (Z. Zhao, 1989); and there is a frost-free period of 170–220 days (Liu and Hseung, 1990). Therefore, cropping systems of three crops every two years or two crops a year can be adopted (Liu and Hseung, 1990). However, the development of agricultural production in the Huang-Huai-Hai Plain is greatly affected by drought, flood, salinisation and alkalisation.

The Huang-Huai-Hai Plain has flat relief and is part of the semiarid and semihumid region. It is subject to a monsoon climate and has an annual precipitation of 600–1000 mm which is unevenly distributed, with about 60–70 per cent in summer (You et al., 1984; Hseung et al., 1981). The rainfall is usually concentrated in July and August while for the rest of the year the region is in a drier climatic condition. In the rainy season, there are mostly rainstorms and the maximum daily rainfall may amount to 100–300 mm, which often results in flooding and waterlogging (Liu and Hseung, 1990). Potential evaporation of the region is very high and is several times more than precipitation. The climatic, geomorphologic, hydrologic and hydrogeologic condition of the Plain leads to drought in spring and autumn and flooding in summer, which aggravates the salinisation and waterlogging of soils. At the same time, because the upper reaches of the Huanghe and Haihe Rivers flow through Loess Plateau, a large amount of clay and silt is carried downward. According to Mosely (1985), the Huanghe River has an annual silt load of 1640×10^6 tonnes and an average silt concentration of 37.6 kg m^{-3}. These settle in the river course due to the decrease of river flow velocity over the plain. Owing to the long-term settling of clay and silt, the river bed has continued to rise to the extent that it is 3–10 m higher than the land surface in the surrounding area (Song, 1989) and has formed a so-called 'river over the land'. Consequently the watertable is raised due to intensified lateral seepage of river water (Hseung et al., 1981). The salt content of rivers is low but it exceeds 500 mg L^{-1} TDS in the lower reaches of the Haihe River (Wang and Liu, 1983).

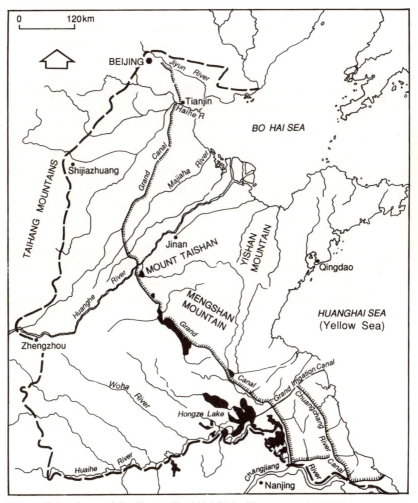

Figure 3-5:   Major rivers of the Huang-Huai-Hai Plain (Hseung et al., 1981).

In attempting to solve the drought problem, water was diverted blindly from the Huanghe River in the late 1950s for irrigation. Many reservoirs were built in the Plain and irrigation was undertaken without construction of reasonable drainage systems. These factors caused the watertable to rise, which inevitably induced extensive secondary salinisation, swamping of soils and deterioration of ecosystems in the Plain (Hseung et al., 1981). Salinisation usually occurs in spring and autumn and desalinisation occurs in summer, which is the rainy season. The time of accumulating salts in the soils amounts to 5–6 months, while that of desalinisation is only about 3 months per year, so the crops always suffer from salt injury (Liu and Hseung, 1990).

The formation of salt-affected soils in the Huang-Huai-Hai Plain is mainly affected by factors such as climate, topography, hydrology, hydrogeology, soil texture and human activities (Liu and Hseung, 1990; Yu, 1992). Some of these aspects have been briefly described in this section. The following section describes in more detail the hydrogeological conditions.

Hydrogeologic conditions

According to Wei and Zhao (1983), Quaternary deposits have settled on the Huang-Huai-Hai Plain in great thickness (200–600 m). From the piedmont to the coastal plains, hydrogeological conditions change with marked regularity. Aquifer materials become finer, individual aquifers become thinner and more numerous and the water quality worsens. The water-bearing strata, composed of Quaternary unconsolidated deposits, can be divided into shallow watertable aquifers and deep aquifers.

Since the Quaternary Period, the sedimentary environments on either side of the Huanghe River have differed, resulting in a great disparity in hydrogeological conditions. The piedmont diluvial-alluvial fans are well developed, with aquifers thicker than 100 m and composed of relatively coarse materials. This area has abundant pure freshwater. South of the Huanghe River, no piedmont diluvial-alluvial fans have developed. In the flat plain area, the alluvial and alluvio-lacustrine deposits are dominant. The total thickness of the Quaternary sediments is thinner than that in the north. The aquifers are composed mainly of layers of medium and fine sands. The water quality is good with TDS concentration of less than 1000 mg L^{-1} in most parts of the area, 100 to 2000 mg L^{-1} in some parts and rarely exceeding 2000 mg L^{-1} locally (Department of Technical Co-operation for Development, 1986).

Through the alluvial plain, variations in the lithological characteristics, structure and thickness of the deposits are extremely complicated. The aquifers are numerous but not thick, with water yielding decreasing capacity eastward. The watertable is 2–4 m deep in the alluvial plain and less than 2 m deep in coastal areas and in some depressions (Figure 3-6). Confined aquifers are shallow in the west (40–50 m) and deep (200–300 m) in the east, with a confined water level lying mostly within 3–5 m of the ground surface.

The general direction of groundwater flow in the shallow aquifers is from the west to the east.

Figure 3-6: Sketch-map of the depth (in m) of shallow aquifers in the Huang-Huai-Hai Plain (Zhu and Zheng, 1983).

There is a marked variation in the salinity of groundwater. The piedmont contains freshwater with a salinity of less than 1000 mg L^{-1}. Most parts of the alluvial plain have slightly saline water (1000–3000 mg L^{-1}). Highly saline water predominates in the coastal areas, generally greater than

5000 mg L^{-1} and sometimes even in excess of 30 000 mg L^{-1}. Saline groundwater covers 100 000 km^2 of the plain, one-third of the region's total area (Wei and Zhao, 1983).

Estimates provided by Zhu and Zheng (1983) show that the annual volume of shallow groundwater in the plain is 56.7×10^9 m^3, of which 47.6×10^9 m^3 is fresh (salinity of < 2000 mg L^{-1}), 5.4×10^9 m^3 is brackish (2000–5000 mg L^{-1}) and 3.7×10^9 m^3 is saline (> 5000 mg L^{-1}). The Beijing alluvial fan is very rich in groundwater and its annual recharge rate is 300 000–400 000 m^3 per square kilometre. Statistics show that the total shallow groundwater extraction from the plain was 27×10^9 m^3 in 1978. Hebei Province extracted 7.8×10^9 m^3 per year during the period 1974–78. Henan Province extracted 9.3×10^9 m^3 over the same period. The Beijing area, with the heaviest rate of extraction, pumps about 2.5×10^9 m^3 per year (Zhu and Zheng, 1983). According to statistics, about 80 per cent of the shallow groundwater resources is recharged by the infiltration of rainfall, 7 per cent by the seepage of rivers and canals, 7 per cent by irrigation and 6 per cent by the lateral inflow from mountains.

Extent of salinisation

Salt-affected soils exist in the littoral and inland areas of the Huang-Huai-Hai Plain (Figure 3-7). In the littoral areas such as around the Bo Hai Sea, the main cause of salinisation is the inundation of lands by seawater due to transgression and tidal activities. In inland areas, primary salt-affected soils are scattered over the paleo-depressions, while secondary salt-affected soils occur in the cultivated areas.

Recent investigation shows that 3.06 Mha of saline-alkaline soils exist in the region. These account for 15.28 per cent of the total 20 Mha of the cultivated area (Jia, 1989). Table 3-3 shows that 29.38 per cent (898 000 ha) of the cultivated area is seriously affected and 13.76 per cent (421 000 ha) is moderately salt-affected. Affected soils are widespread over Hebei Province and the northern part of Shandong Province.

Table 3-3: Salt-and alkali-affected soils in cultivated areas of the Huang-Huai-Hai Plain.

Type	Affected area (ha)	Percentage of the total affected area (%)
Seriously salt-affected	897 827	29.38
Moderately salt-affected	420 673	13.76
Slightly salt-affected	1 139 200	37.27
Alkali-affected	408 060	13.35
Undivided	190 840	6.24
Total	3 056 600	100.00

Source: Jia (1989).

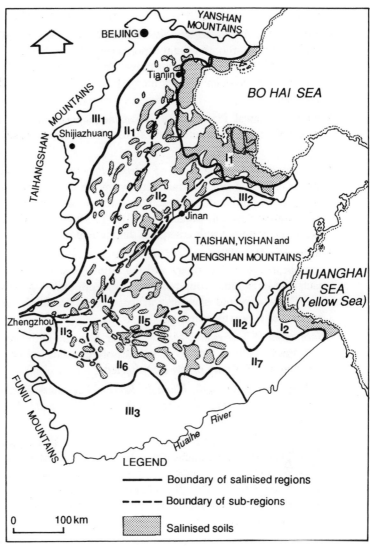

Figure 3-7: Salinised soil in the Huang-Huai-Hai Plain (Project Group of the Ministry of Geology and Mineral Resources, 1987).

As Figure 3-7 shows, soils of the Plain have been classified into three regions and twelve subregions. The main features of these regions are as follows:

Region I Littoral Plain: Alluvio-marine plain with smooth terrain and wide distribution of salt-affected soils. It is divided into two subregions.

- *Subregion I-I (Bo Hai Sea Littoral):* Has shallow aquifers consisting of silt and fine sand. Watertable is at a depth of 0–3 m and the TDS content ranges from 5000 to 10 000 mg L^{-1}. Soils are mainly salinised.

- *Subregion I-2 (Huanghai Sea Littoral)*: Consists mainly of marine sediments with smooth terrain topography. Watertable is at a depth of less than 1 m and the TDS content ranges from 3000 to 20 000 mg L^{-1} from west to east. Also, soil salinity increases from low to high along the same direction.

Region II Central Plain: Fluvio-lacustrine deposits with smooth terrain, impeded run-off and multiaquifer system. There are seven subregions in this area:

- *Subregion II-1 (Mid-Plain of Hebei Province)*: Has smooth terrain topography. Aquifer is formed chiefly by the fluvio-lacustrine medium to fine grained sand and silt sand. Watertable is at a depth of 3–5 m and the TDS content generally ranges from 3000 to 50 000 mg L^{-1}. Light, medium and heavy salinised soils emerge in this subregion.

- *Subregion II-2 (Western and north-western Shandong Province)*: Has a complicated geomorphology and the aquifer system is multi-layered, but each aquifer is thin. Groundwater is at a depth of 2–4 m. Salt-affected soils are fragmentary and island-like in shape. They are lightly salinised but the fringe area is heavily salinised.

- *Subregion II-3 (Eastern Henan Plain)*: Has an undulated topography and thick shallow aquifer. Total dissolved solids content is less than 1000 mg L^{-1}. Soils are lightly salt-affected.

- *Subregion II-4 (Along the Huanghe River)*: Has watertable about 2 m deep. Groundwater is brackish with a TDS of 2000 mg L^{-1}. It contains some heavily salinised soils.

- *Subregion II-5 (South-western Shandong Plain)*: This area belongs to the flood plain of the Huanghe River. Groundwater system consists of three aquifers. The upper aquifer contains freshwater, the middle saline water and the lower freshwater. The saline aquifer is at a depth of 10–30 m, with a TDS of 3000–5000 mg L^{-1}. It contains salt-affected soils.

- *Subregion II-6 (Jiangsu and Henan Provinces)*: This area consists of smooth slope and shallow interdepression. It contains a large distribution of saline groundwater in western parts and fragmentary in the east. Soils are lightly to moderately salinised.

- *Subregion II-7 (Flood plain of the Huanghe River)*: Consists of silt, sand, clayey soil and sandy soil. Aquifers have low productivity. The riverbed is higher than the land surface in the surrounding area and seepage from the river raises the groundwater level, causing salinisation.

Region III Piedmont Area: Non-salinised soils with abundant fresh groundwater resources. There are three subregions in this area:

- *Subregion III-1 (Piedmont in Yanshan and Taihangshan Mountains)*: In the

slope plain of the piedmont, groundwater level is deep and its quality is good.
- *Subregion III-2 (Piedmont in Shandong hilly area)*: Has a complicated topography and groundwater depth changes substantially. Groundwater is abundant and its quality is good. Local depressions are easily subject to flooding during the rainy season and form small areas of salinised soil after the rainy season.
- *Subregion III-3 (Northern Anhui Plain)*: This area is the vast flood plain of the Huanghe River with smooth topography. Salinised soil occurs in local areas.

Salinisation in other regions of China

Secondary salinisation occurs throughout the arid and semiarid regions of China. However, no published information is available describing the situation in these regions. The following is a short description of the problem in a few regions.

In the north-west arid region of Xinjiang, precipitation ranges from less than 50 mm a year in the east to 50–100 mm in the south and about 200 mm in the north, while evaporation is about 2000 mm per annum. The total area under cultivation in 1949 was about 1.2 Mha but had expanded to 3 Mha by 1961. This increase was made possible by the development of surface water storage capacity to 2.9×10^9 m^3 and construction of irrigation facilities (Vermeer, 1977). In 1964, the total annual volume of surface water in Xinjiang was estimated at 85.2×10^9 m^3 and of groundwater at 18.5×10^9 m^3. At that time, 40 per cent of the surface water resources had been used for farming, while utilisation of groundwater was yet to commence. Ten years later it was reported that 55 per cent of the annual volume of water was channelled into areas under irrigation.

Initially a loss of up to 70 per cent of water during conveyance was estimated. Such a large amount of seepage and evaporation, in combination with other factors, contributed to soil salinisation and alkalisation. In 1953 and again in 1954 about 700 ha had to be abandoned. In 1960, due to backward irrigation methods, some 60 per cent of the cultivated land in the lower reaches of the rivers in South Xinjiang was salt-affected. Subsequently, the watertable was lowered and the extent of salinisation reduced by lining half of the canals with pebbles or cement, and by improved irrigation and cultivation techniques. However, there is little data on the effects of these efforts. In 1972 it was reported that during the preceding 20 years the People's Liberation Army had created nearly 170 000 ha of high-yield farmland and improved 533 000 ha of alkaline land in Xinjiang (Vermeer, 1977). According to Li and Wang (1989), currently there are 3.2 Mha of arable land in Xinjiang. However, soil salinisation has been aggravated and 1.3 Mha are affected by secondary salinisation. Other statistics provided by

Gu and Hseung (1990) indicate that 2.67 Mha of arable land exists in Xinjiang and secondary salt-affected soils are estimated to constitute about 30 per cent of the total cultivated area.

The Hotao irrigation area in the Huanghe River bend in Inner Mongolia has a precipitation of 100–300 mm and evaporation of 2000 mm per year. The total irrigated area was about 133 000 ha before 1946 and increased to 333 000 ha in 1976. At the end of the 1950s and in the early 1960s, as the amount of water used for irrigation was increasing and insufficient precautions were taken, the watertable rose considerably, causing severe salinisation. By 1962 about 34 per cent of the irrigated area had developed secondary salinisation (Vermeer, 1977). According to Yu (1990), secondary salinisation is still expanding in this irrigated area. By the end of the 1980s salinised soils occupied half of the irrigated area.

Shanxi Province has a semiarid to semihumid climate, with an annual precipitation of 400–500 mm and potential evaporation of 1400–2200 mm. The distribution of the rainfall is uneven over the year, with 70–75 per cent of the total annual precipitation in June to September. According to a mid-1980s survey, salt-affected soils in the province covered an area of 300 000 ha. These are mainly formed under the influence of a shallow watertable. In areas where the watertable is more than 3 m deep, there is no soil salinisation. In the areas where the watertable is less than 2 m deep, salt accumulates at the soil surface. In well-irrigated areas where there is a significant drop in the watertable, soils are not salinised (Z. Zhao, 1989).

Management options

The history of prevention and control of flooding, drought, waterlogging, salinisation and alkalisation of soils in the Huang-Huai-Hai Plain can be divided into four stages (Liu and Hseung, 1990):

- From 1949 to the end of the 1950s, large-scale projects for the harnessing of the Huaihe River valley were conducted and put into effect in order to solve the problems of flooding and waterlogging. At the same time a long-term program for harnessing the Huanghe River valley and the Haihe River valley was devised. The harnessing of the Huaihe River valley consisted mainly of dredging canals, heightening dykes and raising drainage ability for flood. Also, many large and medium-sized reservoirs were built on the upper reaches, as well as the regulating reservoirs on the middle reaches of the river.

- From the end of the 1950s to the beginning of the 1960s the problem of drought became more urgent. Subsequently during 1958–1962 large-scale water diversions from the rivers and reservoir construction in the plain for irrigation were carried out. However, these measures, accompanied by intensive irrigation without drainage facilities, resulted in rising watertables and accelerated development of secondary salinisation.

- At the beginning of the 1960s, irrigation had to be stopped temporarily to combat salinisation, and attention was devoted towards the development of drainage facilities. From 1965 pumped-well irrigation and drainage were adopted in the Huang-Huai-Hai Plain, which had a marked effect on the control of salinisation and alkalisation.
- From the beginning of the 1970s, combinations of well and surface (ditch) irrigation-drainage were extensively adopted in this plain. This effectively controlled and regulated the surface and groundwater, alleviating drought, flooding, soil salinisation and alkalisation problems in the region.

According to Q. Zhao (1989), alleviating the disasters of drought, flood, salinisation and alkalisation which frequently affect the Huang-Huai-Hai Plain requires management of the water resources. Management options consist of irrigating when it is dry; draining when it is flooded; setting up a combined system of wells, ditches and canals for both irrigation and drainage; regulating and storing water resources; pumping out saline groundwater and replenishment with freshwater in some areas; cultivating rice and developing agriculture in depressions; ameliorating soils to raise soil fertility in coordination with development of agriculture, forestry and animal husbandry (Q. Zhao, 1989). It is important to note that the overall management of the Huang-Huai-Hai Plain's problems requires appropriate measures in each particular part of the plain, taking into account such factors as topography, hydrology, hydrogeology, soil characteristics, irrigation and drainage facilities, agricultural practices and the nature of the ecological systems (Hseung et al., 1981).

The following is a brief description of the some of the measures adopted for the management of the salt-affected soils of the plain.

Management of saline soils
Well-irrigation and drainage

In order to manage secondary salinisation, the watertable should be kept below the critical depth of about 3 m from the land surface. In this respect, well-irrigation (irrigation with groundwater) and well-drainage (drainage by pumping groundwater) have proved to be very effective. The well-irrigated area has increased over the years to reach 40 per cent of cultivated land (Q. Zhao, 1989). According to You et al. (1984), design characteristics of the pumped-wells should be adapted to local conditions because conditons of soil improvement are quite different. You et al. (1984) describe the following areas:

- In the upper and middle parts of the plain, groundwater in both shallow and deep aquifers is mostly fresh and suitable for irrigation. In these parts of the plain, soil salinisation can be eliminated by the application of well-irrigation, which provides water for irrigation and simultaneously lowers the watertable. Moreover, it is necessary to build the main drainage system in combination with shallow field drains (ditches) for preventing soil waterlogging. By adopting these measures good results have been achieved in Fengqiu County, Henan Province, where 4800 pumped wells were built during 1965–1974 but the salt-affected area had decreased by 80 per cent in 1974 compared with that in 1964 (You et al., 1984). This was due to the drop of watertable from 1.5 m to 3.5 m below the soil surface and downward movement of the salt in the soil profile (Yu, 1992).

- In the lower parts of the plain and shallow-flat or trough depressions, the shallow groundwater has a salt content of 3000 to more than 10 000 mg L^{-1} TDS, although the deep groundwater is fresh and can be used for irrigation. In these areas shallow wells should be adopted in combination with deep wells. The shallow wells are used for pumping and drainage of saline groundwater and controlling the watertable, and the deep wells for pumping freshwater in deep aquifers for irrigation and leaching salt from the soil.
- In the salt-affected coastal areas with a low and flat topography, shallow groundwater is highly saline and fresh groundwater is available in deep aquifers (generally below 400 m from the land surface). Because of the high salt content and severe waterlogging of the soils in these coastal areas, it is necessary to direct river water for irrigation, leaching the salts, desalting the groundwater and eliminating the waterlogging. Therefore, it is necessary to build an irrigation system and to use the surface drainage system for draining water and salt.
- In the areas with serious collapse of drainage ditches, pumped-wells combined with surface drainage or pumped-wells combined with subsurface pipe drainage should be adopted. The pumped-well is mainly used for pumping and drainage of highly saline groundwater and promoting the establishment of a fresh groundwater layer with the infiltration of irrigation water and rainfall.

Experiments in many areas of the Huang-Huai-Hai Plain have shown the beneficial effects of the withdrawal of saline groundwater and recharging the aquifer with freshwater (Liu and Hseung, 1990). Generally, this measure has the following benefits: withdrawal of saline groundwater in early spring may lower the watertable and prevent the resalinisation of surface soil; withdrawal of saline groundwater before the rainy season can create favourable conditions for absorption and storage of rainwater and prevention of waterlogging; withdrawal of saline groundwater in the rainy season may accelerate the desalinisation of groundwater; and the withdrawal of saline groundwater after the rainy season may lower the watertable rapidly and prevent both the waterlogging and resalinisation of soils.

Surface drainage

Surface drains play an important role in controlling watertables and management of salinisation, flooding and waterlogging. Both deep and shallow drains have been developed to achieve these objectives (Hseung and Liu, 1990). According to Yu (1992), it is difficult to implement deep drains in the coastal area. Therefore, a shallow and dense drainage network with a depth of 1.2 m–1.5 m and spacing of 50 m between the drains should be implemented.

Rice cropping

Cropping rice is a traditional method for improvement of salt-affected lands in China. According to Hseung and Liu (1990), as early as about 600 BC, there was an account in an ancient book of soil improvement on salt-affected soils by means of cropping rice. The authors said: 'The strongly saline and alkaline soils taste very salty and bitter. They are low-yielding land, but suitable for cropping rice'.

In the growing season of rice, owing to constantly maintaining a water layer on the soil surface, the process of salt-leaching can proceed continuously, and the desalinisation is gradually deepened to the lower part of the soil profile. The degree of desalinisation is also intensified in line with the years of cropping rice (Hseung and Liu, 1990).

Warping of salt-affected lands

Along the Huanghe River salt-affected soils are formed by a shallow watertable induced by lateral seepage of the river. Watertable depth is in the range of 1 m–1.5 m below the land surface (Yu, 1992) and the fine texture of the soils makes drainage by surface drains or wells very difficult. Warping salt-affected land and growing rice on it proved very successful. According to Song (1989), warping involves the channelling of water from the river to fields. When the silt content is high in the Huanghe River, warping deposits 30 cm or more of silt. For the average silt content of the river, a warping area of 30–37 ha and warping time of 50–60 days are required. The warping field should be provided with intake gate, drainage gate, dykes and other facilities. This practice would reduce capillary rise considerably and would facilitate salt leaching during the rainy season. Then rice can be cultivated on the warped area. Paddy fields without warping would be resalinised two years after being transferred into dryland. A land warped with 15 cm silt would be resalinised 4 to 6 years after transforming to dryland, while a field deposited with over 30 cm silt would basically no longer be resalinised (Song, 1989).

Afforestation

Afforestation can control soil salinisation by lowering the watertable through the process of taking up water from the soil and its transpiration. The transpiration rate varies greatly with tree species. In general the fast-growing tree species with a large canopy have a high rate of transpiration which can play a major role in lowering watertables. These trees include *Salix* sp., *Marous* sp., *Fraxinus* sp., *Populus diversifolia* and *Populus nigra* (Hseung and Liu, 1990). In Shouguang Forest Farm of Shandong Province, the watertable dropped to 1.2–1.5 m below the land surface in a *Robinia pseudoacacia* plantation of five years old. The salt content in a soil profile of 1 m depth was reduced to 0.02–0.05 per cent, while the watertable depth was 1 m in the rainy season and the salt content was 0.2 per cent, before afforestation (Hseung and Liu, 1990).

Afforestation in salt-affected land can improve soil physical properties and increase the organic matter content in the soil. Some shrub species such as *Amorpha fruticosa* and others are also good resources of green manure (Hseung and Liu, 1990). In addition, forest belts have a protective range within which wind speed can be reduced, air temperature and evaporation rate of soil are decreased and air humidity is increased. Accordingly, the rate and intensity of salt accumulation in soil are reduced. According to Q. Zhao (1989), the current forest coverage of the plain is 5–8 per cent and it is possible to increase this to 12–18 per cent.

Other measures

Other measures for the management of salt-affected soils and improvement of their productivity include: land levelling; deep ploughing; tilling and harrowing at the proper time of the year; cultivation of deep-rooted and high water use crops such as alfalfa; green manure cropping; cropping legume crops mixed with forage grasses; and rotation of rice crop with other crops (see Hseung and Liu, 1990 for further information).

Management of alkaline soils

There are many measures for management of alkaline soils, but each measure has its specific effect and limitations (Yu and Hseung, 1983). Chemical reclamation and hydraulic engineering measures are mainly adopted to eliminate harmful sodium ions. These measures require specified technical standards and greater investment. The agricultural measures are generally lower in cost but take more time to be efficient. Moreover, the reclamation measures adopted are dependent on the degree of soil alkalisation. For soils with a low or medium degree of alkalisation, it is only needed to improve the soil by deep ploughing combined with applying organic manure or planting green manures. For strongly alkaline soils, it is necessary to apply chemical amendments such as gypsum, phosphogypsum and calcium sulphate and to have adequate irrigation and drainage facilities. Where water is abundant, soil may be reclaimed by planting rice (Yu and Hseung, 1983).

Yu (1992) describes the characteristics of the alkaline soils of the Huang-Huai-Hai Plains and the measures to be undertaken for their improvement. These soils are of a light texture and are low in organic matter.

The low and moderately alkaline soils of the plain can be improved by using deep tillage together with the application of organic manure. To improve strongly alkaline soils chemical amendments such as gypsum have to be applied and relatively alkali-tolerant crops (cotton, barley, etc.) should be cultivated. As a result a good harvest can be obtained and yield increases of 10 to 200 per cent can be achieved.

According to Yu (1989), secondary alkalisation in the Huang-Huai-Hai Plain is caused by desalination of salt-affected soils and irrigation with alkaline water. This problem can be avoided as long as attention is paid to the close coordination of irrigation water quality, improvement of irrigation and drainage facilities and agricultural measures.

References

Chen, J. and Wu, G. 1987. Water resources development in China. In. Ali, M., Radosevich, G.E. and Khan, A.A. eds. *Water Resources Policy for Asia*. Proceedings of the regional symposium on water resources policy in agro-socio-economic development. Dhaka, Bangladesh, 4–8 August 1985. Rotterdam: A.A. Balkema. 51–60.

China Handbook Editorial Committee. 1983. *Geography*. Beijing: Foreign Languages Press. (China Handbook Series). 260 pp.

Department of Technical Co-operation for Development. 1986. *Ground Water in Continental Asia (Central, Eastern, Southern, South-Eastern Asia)*. New York: United Nations. (Natural Resources/Water Series no. 15). 391 pp.

Editorial Department of the PRC Year Book in Beijing. 1987. *People's Republic of China Year Book 1987*. Beijing. Xinhua Publishing House. 646 pp.

Food and Agriculture Organization of the United Nations (FAO). 1989. *Production Yearbook*. Rome: FAO. v.42. 350 pp.

Framji, K.K., Garg, B.C. and Luthra, S.D.L. 1981. *Irrigation and Drainage in the World: A Global Review*. Third edition. New Delhi: International Commission on Irrigation and Drainage. v.1. 491 pp.

Gu, G. and Hseung, Y. 1990. Utilization and improvement of desert soils. In. Institute of Soil Science, Academia Sinica. ed. *Soils of China*. Beijing: Science Press. 806–815.

Hseung, Y. and Liu, W. 1990. Improvement and utilization of salt-affected soils. In. Institute of Soil Science, Academia Sinica. ed. *Soils of China*. Beijing: Science Press. 734–761.

Hseung, Y., Zhu, S. and Wang, Z. 1981. *Ecological Regionalization of the Huang-Huai-Hai Plain in China*. Nanjing: Institute of Soil Science, Academia Sinica. (Soil Research Report no. 4). 16 pp.

Institute of Hydrogeology and Engineering Geology. 1988. *Hydrogeologic Map of China (1:4 000 000)*. Beijing: China Cartographic Publishing House.

James, C.V. ed. 1989. *Information China: The Comprehensive and Authoritative Reference Source of New China*. (Organised by the Chinese Academy of Social Sciences). Oxford: Pergamon Press. v.1. 1–440, v.2. 441–929, v.3. 930–1621.

Jia, Y. 1989. Institute of Hydrogeology and Engineering Geology, Chinese Academy of Geological Sciences. Shijiazhuang, Hebei. (Personal communication).

Kaplan, F.M. and Sobin, J.M. 1982. *Encyclopedia of China Today*. Third edition. London: Macmillan Reference Books. 446 pp.

Li, S. and Wang, Z. 1989. Dynamics of soil salinization in new irrigation area of northern foot of Mount Tianshan. In. *Proceedings of the International Symposium on Dynamics of Salt-Affected Soils*. Nanjing, 4–10 October 1989. Nanjing: Institute of Soil Science, Academia Sinica. 134–135.

Liu, W. and Hseung, Y. 1990. The comprehensive improvement of the soils in the Huang-Huai-Hai Plain. In. Institute of Soil Science, Academia Sinica. ed. *Soils of China*. Beijing: Science Press. 724–733.

Lou, P. 1987. The economic effectiveness of variable irrigation supplies in North China

Plain. In. *Transactions, Thirteenth International Congress on Irrigation and Drainage.* Rabat, Morocco, September 1987. New Delhi: International Commission on Irrigation and Drainage. v.I-D (Symposium). 115–129.

Ministry of Water Conservancy. 1990. *Handbook of Natural Resources of China.* Beijing: Science Press. (In Chinese).

Mosely, P. 1985. Upstream-downstream interactions as natural constraints to basin-wide planning for China's River Huang. In. Lundquist, J., Lohm, U. and Falkenmark, M. eds. *Strategies for River Basin Management: Environmental Integration of Land and Water in a River Basin.* Dordrecht: D. Reidel. 131–140.

Project Group of the Ministry of Geology and Mineral Resources (MGMR). 1987. *Study on Groundwater Resources in the Huang-Huai-Hai Plain.* Beijing: MGMR. (Science Report in Chinese).

Ren, M., Yang, R. and Bao, H. 1985. *An Outline of China's Physical Geography.* Beijing: Foreign Languages Press. 471 pp.

Song, R. 1989. The effects of warping with Yellow River water on improvement of salt-affected soil. In. *Proceedings of the International Symposium on Dynamics of Salt-Affected Soils.* Nanjing, 4–10 October 1989. Nanjing: Institute of Soil Science, Academia Sinica. 159–161.

Vermeer, E.B. 1977. *Water Conservancy and Irrigation in China: Social, Economic and Agrotechnical Aspects.* The Hague: Leiden University Press. 350 pp.

Wang, J. and Liu, Y. 1983. An investigation of the water quality and pollution in rivers of the proposed water transfer region. In. Biswas, A.K., Dakang, Z., Nickum, J.E. and Changming. L. eds. *Long-Distance Water Transfer: A Chinese Case Study and International Experiences.* Dublin: Tycooly International (for the United Nations University). 361–371.

Wang, Z. and Li, L. 1987. *The Processes and Regionalization of Soil Salinization and Alkalization in China.* Nanjing: Institute of Soil Science, Academia Sinica. (Soil Research Report no. 17). 13 pp.

Wang, Z., Hseung, Y. and Yu, R. 1990. Saline and alkali soils. In. Institute of Soil Science, Academia Sinica. ed. *Soils of China.* Beijing: Science Press. 261–290.

Wei, Z., and Zhao, C. 1983. Natural condition in the proposed water transfer region. In. Biswas, A.K., Dakang, Z., Nickum, J.E. and Changming, L. eds. *Long-Distance Water Transfer: A Chinese Case Study and International Experiences.* Dublin: Tycooly International (for the United Nations University). 97–114.

World Bank, 1992. *World Development Report 1992: Development and the Environment.* New York: Oxford University Press. 308 pp.

Xu, Z. 1987. Improving water management through training in China. In. *Transactions, Thirteenth International Congress on Irrigation and Drainage.* Rabat, Morocco, September 1987. New Delhi: International Commission on Irrigation and Drainage. v.I–C. 367–373.

Yao, B. and Chen, Q. 1983. South-north water transfer project plans. In. Biswas, A.K., Dakang, Z., Nickum, J.E. and Changming, L. eds. *Long-Distance Water Transfer: A Chinese Case Study and International Experience.* Dublin: Tycooly International (for the United Nations University). 127–149.

You, W., Hseung, Y. and Liu, W. 1984. *Effect of Pumped-Well Drainage on the Prevention and Control of Drought, Waterlogging, Salinization and Alkalization in Huang-Huai-Hai Plain and its Utilization.* Nanjing: Institute of Soil Science, Academia Sinica. (Soil Research Report no. 11). 7 pp.

Yu, R. 1989. Secondary alkalization of soil in Huang-Huai-Hai Plain. In. *Proceedings of the International Symposium on Dynamics of Salt-Affected Soils.* Nanjing, 4–10 October 1989. Nanjing: Institute of Soil Science, Academia Sinica. 96–97.

Yu, R. 1990. Soil salinization and alkalization are the significant issues in soil degradation. In. *Proceedings of National Symposium on Prevention of Land Degradation.* Beijing: Science and Technology Press. (In Chinese).

Yu, R. 1992. Salt-affected soils and their improvement and utilization in the Huang-Huai-Hai Plain. In. *Proceedings of the International Symposium on Strategies for Utilizing Salt Affected Lands.* Bangkok, 17–25 February 1992. Bangkok: Department of Land Development, Ministry of Agriculture and Cooperatives. 220–225.

Yu, R. and Hseung, Y. 1983. *Alkaline Soils in China.* Nanjing: Institute of Soil Science, Academia Sinica. (Soil Research Report no. 9). 16 pp.

Zhang, Z., Chen, B., Chen, Z. and Xu, X. 1992. Challenges to and opportunities for development of China's water resources in the 21st Century. *Water International.* 17(1): 21–27.

Zhao, Q. 1989. Comprehensive management and development of agriculture in the Huang-Huai-Hai Plain. In. *Proceedings of the International Symposium on Dynamics of Salt-Affected Soils*. Nanjing, 4–10 October 1989. Nanjing: Institute of Soil Science, Academia Sinica. 17–24.

Zhao, Q. and Zhu, S. 1992. Improvement of salt affected soils in China. In. *Proceedings of the International Symposium on Strategies for Utilizing Salt Affected Lands*. Bangkok, 17–25 February 1992. Bangkok: Department of Land Development, Ministry of Agriculture and Cooperatives. 226–231.

Zhao, Z. 1989. Type of water-salt movement in soil of Shanxi Basin and improvement of salt-affected soils. In. *Proceedings of the International Symposium on Dynamics of Salt-Affected Soils*. Nanjing, 4–10 October 1989. Nanjing: Institute of Soil Science, Academia Sinica. 147–149.

Zhu, Y. and Zheng, X. 1983. Shallow groundwater resources of the Huang-Huai-Hai Plain. In. Biswas, A.K., Dakang, Z., Nickum, J.E. and Changming, L. eds. *Long-Distance Water Transfer: A Chinese Case Study and International Experience*. Dublin: Tycooly International (for the United Nations University). 257–270.

Zou, G. 1987. The investment distribution for irrigation and drainage projects in China. In. *Transactions, Thirteenth International Congress on Irrigation and Drainage*. Rabat, Morocco, September 1987. New Delhi: International Commission on Irrigation and Drainage. v.I–D. 79–84.

Chapter 4: Commonwealth of Independent States

Introduction

This chapter covers the former territory of the Soviet Union, which currently consists of the Commonwealth of Independent States, three Baltic Republics (Estonia, Latvia and Lithuania) and Georgia. Although not strictly correct, for the present it will be referred to as the Commonwealth of Independent States (CIS).

The Commonwealth of Independent States occupies more than 22 402 200 km² of continuous landmass. From east to west, the continental territory of the CIS stretches for nearly 10 000 km and from north to south, it measures approximately 5000 km (Figure 4-1). In addition, the CIS possesses a number of large and small islands in the Arctic and Pacific Oceans. The perimeter of the continental CIS is close to 60 000 km (Medish, 1990).

The major part of the Commonwealth of Independent States is relatively flat and has the following principal physiographic regions (Gregory, 1975):

- **The European Plain** is primarily a vast plain with an elevation ranging from 100 m to 400 m and is the most densely populated part of the country.
- **The Ural Mountains** form the eastern edge of the European Plain and stretch more than 2000 km from north to south. The Ural Mountains, which separate European Russia from Asian Russia, are topographically unimpressive with elevations ranging from 300 m to 800 m in their middle part. The highest peak of the Ural Mountains is about 1894 m above sea-level in northern Ural.
- **The West Siberian Lowland** is located to the east of the Ural Mountains and extends north to south from the shores of the Arctic Ocean to the steppes of Kazakhstan and eastward to the Yenisei River. It is very flat, poorly drained and generally marshy or swampy.
- **The Central Siberian Plateau** lies between the Yenisei and Lena Rivers with an average elevation of about 450 to 900 m above sea-level.

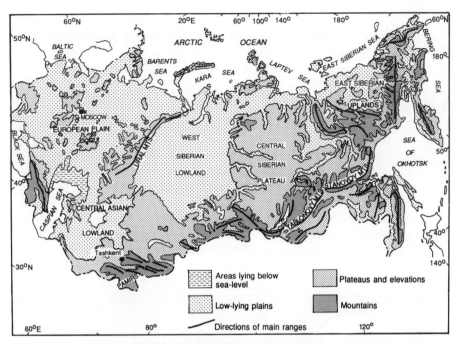

Figure 4-1: Principal physiographic features of the Commonwealth of Independent States (Pokshishevsky, 1974 and Medish, 1990).

- **The East Siberian Uplands** are located to the east of the Lena River. The topography consists of a series of mountains and basins. The highest ranges in the region reach maximum elevations of about 2300 to 3200 m. To the east, toward the Pacific Ocean the mountains are higher and volcanic activity becomes prevalent.
- **The Central Asian Lowland** is a broad lowland region. In the west is the Caspian depression which lies about 27.2 m below mean sea-level. A large portion of the Central Asian Lowland is occupied by the Kara Kum and Kyzyl Kum Deserts.
- **The Southern Mountain Systems** lie in the south and east of the country. The western part of these mountain systems includes the Carpathian Mountains, the Crimean Mountains and the Caucasus Mountains, which run between the Caspian and Black Sea. On the east of the south Caspian Sea, the Pamirs, the Tien Shan and the Altai Mountains form the frontier with Afghanistan and western China. These are the most rugged mountains in the Commonwealth of Independent States and their highest peak is at 7498 m in the Pamirs. Other mountain ranges continue north-easterly along the southern border of the Commonwealth of Independent States all the way to the Pacific Ocean.

The Commonwealth of Independent States had a population of 290.9 million in July 1990 with a growth rate of 0.7 per cent (Central Intelligence Agency, 1990).

Rainfall and climate

According to Kalinin (1971), 11.7×10^{12} m^3 of water falls annually over the Commonwealth of Independent States as rain and snow. This volume is equivalent to an average annual precipitation of about 531 mm. The amount of precipitation is unevenly distributed throughout the CIS (Figure 4-2). Average annual precipitation is 400–600 mm over the greater part of the plains. In the westernmost districts, especially in the Baltic regions, the amount of precipitation is 600–800 mm because of proximity to the Atlantic. The Pacific monsoon brings 600–800 mm of precipitation to the Far East (Gregory, 1975).

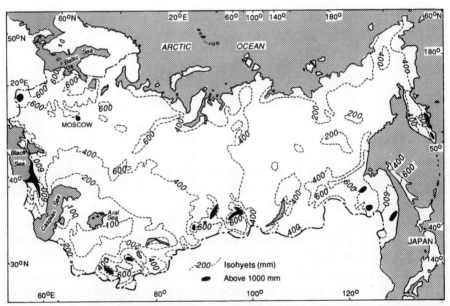

Figure 4-2: Average annual precipitation of the Commonwealth of Independent States (Pokshishevsky, 1974).

In the high ranges of the mountains such as the Caucasus, the Tien Shan, the Altai and others, the amount of precipitation is between 600 and 1600 mm. The southern districts of the East European and the West Siberian Plains, as well as the mountains of Central and Eastern Siberia and the North East, are dry. They have a total annual precipitation of 200 to 400 mm. In

Central Asia, the annual precipitation is below 200 mm, while in the Kara Kum and Kyzyl Kum deserts it is only 100 mm. The highest annual precipitation in the CIS is 2500 mm in the mountain districts of the Caucasus.

All the main world climatic types, with the exception of tropical forest and savanna climates, are found in the Commonwealth of Independent States (Gregory, 1975). About 80 per cent of the CIS is in the temperate zone, 18 per cent in the Arctic and 2 per cent in the subtropical zone. In general the climate is continental with short hot summers and cold winters, although there are exceptions because of the size of the country and the variety of local conditions. The climate is harshest in the Arctic where the polar night lasts for several months in the winter. The European areas of the CIS have a milder climate. The mean January temperature does not fall below −15°C and the mean July temperature is between 10°C and 20°C. The climate becomes increasingly continental eastward from the Baltic and Black Seas, the winters becoming colder and the summers hotter.

Western Siberia and northern Kazakhstan have a markedly continental climate. Eastern Siberia is a region with an extremely continental climate. The difference between mean summer and winter temperatures reaches 50°C–60°C. The East Siberian summer is short but relatively warm. The Far East, which faces the Pacific Ocean, has a distinctive monsoon climate. In the Black Sea coastal region, the summer is moderately hot and the winter is mild, and the only region in the CIS with a humid subtropical climate is found here. Central Asia is the hottest and driest region of the country. Temperatures as high as 50°C have been recorded in the town of Termez in Uzbekistan.

Water resources

Surface water

In the Commonwealth of Independent States there are some 200 rivers longer than 500 km (Gregory, 1975). The average annual surface run-off of the Commonwealth is about 4740×10^9 m^3 (Khublarian, 1990a), which is equivalent to 215 mm over the land area of the country. As shown in Table 4-1 and Figure 4-3, the CIS can be subdivided into five major river basins. About 63 per cent of the CIS annual flow originates in the Arctic Ocean Basin where the major rivers are the Yenisei, Lena and Ob. It is followed by the Pacific Ocean Basin which has a share of 21 per cent of annual flow. Table 4-2 provides characteristics of the major rivers of the CIS.

Many inland seas and lakes occur in the territory of the Commonwealth of Independent States. The largest are the Caspian Sea and the Aral Sea (Table 4-3). The Aral Sea has shrunk considerably during the past three decades and its salinity increased following diversion of major

Table 4-1: Major river basins of the Commonwealth of Independent States.

Basin	Share in area (%)	Share in annual flow (%)
Arctic Ocean	54	63
Pacific Ocean	15	21
Caspian and Aral Seas	23	9
Black Sea	6	3
Baltic Sea	2	4
Total	100	100

Source: Micklin (1991a).

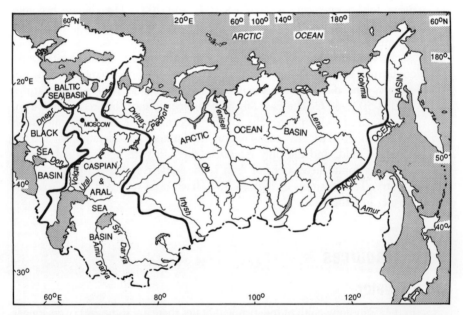

Figure 4-3: Major river basins of the Commonwealth of Independent States (Micklin, 1991a).

rivers for the development of irrigation in Central Asia. Lake Baikal accounts for over 80 per cent of the freshwater lake resources of the CIS and about 20 per cent of the world (Khublarian, 1992). Salinity of Lake Baikal water is about 120 mg L^{-1} and over 2600 species of animal and plants exist in this lake. Lake Balkhash is a large and slightly saline lake in the south-west of Kazakhstan. It is divided into two basins, a south-western and an eastern basin. These have somewhat different limnological conditions, but the most obvious one is that the south-western basin has a salinity of less than 2000 mg L^{-1} whereas the salinity of the eastern basin is about 4000 mg L^{-1} TDS (Peter, 1992).

Table 4-2: Characteristics of the major rivers of the Commonwealth of Independent States.

River	Drainage basin	Drainage area (1000 km^2)	Length[a] (km)	Mean annual flow (10^9 m^3)	Period of record
Yenisei	Arctic Ocean	2580	3490	630	1936–1980
Lena	Arctic Ocean	2490	4400	532	1934–1981
Ob	Arctic Ocean	2990	3650	404	1930–1980
Amur	Pacific Ocean	1855	2820	344	1933–1980
Volga	Caspian Sea	1380	3350	254	1982–1980
Pechora	Arctic Ocean	322	1810	130	1930–1980
Kolyma	Arctic Ocean	647	2130	128	1927–1981
Amu Darya	Aral Sea	309	1415	70	1930–1980
Dnepr	Black Sea	505	2200	54	1930–1980
Syr Darya	Aral Sea	219	2210	37	1930–1980

Source: Khublarian (1990a), (a) Korzun et al. (1978).

Urvantsev (1985) reported that over 2200 reservoirs are under operation in the CIS with a total volume of 870×19^9 m^3. Major CIS reservoirs are multi-purpose and operate in cascade (e.g. see Kubijovyč, 1984: 684–686). Capacity of some of the CIS reservoirs is given in Table 4-4.

According to Urvantsev (1985), the total withdrawal from water bodies in 1982 was about 344×10^9 m^3. From the total water withdrawal, 30×10^9 m^3 was from groundwater and the main water consumer was irrigation (195×10^9 m^3).

Table 4-3: Characteristics of selected major lakes of the Commonwealth of Independent States.

Lake/Sea	Area (km^2)	Maximum depth (m)	Volume (km^3)	Water quality (mg L^{-1} TDS)
Caspian Sea	374 000	1025	78 200	11 000
Aral Sea[a]: – In 1960	68 000	68	1 090	10 000
– In 1989	40 400	54	374	30 000
Lake Baikal	31 500	1741	23 000	120[b]
Lake Balkhash	18 200	26	112	2 000[c]
Lake Issyk-Kul	6 200	702	1 730	5 800[d]
Laka Ladoga	17 700	230	908	Fresh
Lake Onega	9 630	127	295	Fresh

Source: Korzun et al. (1978: 34); (a) Micklin (1991a); (b) Khublarian (1992); (c) Peter (1992); (d) Hammer (1986: 553).

Table 4-4: Capacity of some of the major reservoirs of the Commonwealth of Indepdent States.

Reservoir	River	Capacity (10^9 m^3)
Kuibyshev[a]	Volga	34.6
Toktogul[b]	Naryn	19.5
Nurek[b]	Wahsh	10.5
Kapchagay[a]	Ili	6.6
Chardarya[b]	Syr Darya	5.7

Source: (a) Framji et al. (1983: 1408); (b) Raskin et al. (1992).

Groundwater

The Commonwealth of Independent States has a wide variety of physical, climatic and geographical conditions, as well as a very complex geological and tectonic structure (Nalivkin, 1960; Khain, 1985; Zonenshain et al., 1990). These factors explain the extremely uneven distribution of groundwater resources and their diverse hydrodynamic and hydrochemical characteristics. Within the drainage basin of the Arctic Ocean, particularly in its northern part, a permanent layer of permafrost with a thickness of 100 m to more than 500 m (Pokshishevsky, 1974) substantially reduces the recharge area of the artesian basins, impedes the infiltration of precipitation and transforms a large part of the fresh groundwater from a liquid to a solid phase.

Groundwater resources of the country occur in unconsolidated alluvial deposits, consolidated sediments, karstic formations and fractured igneous and metamorphic rocks (Kudelin, 1977). Many artesian basins containing fresh groundwater exist in the country. The Western-Siberian Basin is one of the earth's largest artesian basins. It occupies an area of about 3.2 million square kilometres. The basin includes Precambrian, Palaeozoic, Mesozoic and Cainozoic deposits over a thickness of about 6000 m. Several aquifer systems have been identified in the Jurassic, Cretaceous, Oligocene and Quaternary deposits of the basin (Department of Technical Co-operation for Development, 1986). Most of the exploitable groundwater resources of the country are formed in large artesian basins and in large intermontane depressions such as the Fergana and Chu Basins. The thickness of freshwater bearing strata varies from 100 m to 700 m and in some areas like Baikal may exceed 2000 m (Department of Technical Co-operation for Development, 1986).

The yield of karstic springs is variable. In the Chu-Sarysu Basin yields attain 30 L s^{-1}, while those in the Naryn Basin reach 2000 to 3000 L s^{-1}. Yields from springs located in fractured zones amount to 5 to 20 L s^{-1} (Department of Technical Co-operation for Development, 1986).

As shown in Table 4-5, the potential exploitable fresh groundwater resources of the country (the quantity that could be obtained by the rational

Table 4-5: Groundwater resources of the Commonwealth of Independent States and their utilisation (in 10^9 m^3 per year).

Republic	Potential exploitable resources		Proven reserves at 1.1.1985	Extraction in 1984	Utilisation in 1984			Predicted extraction for year 2000
	Fresh	With salinity of 1 to 3 g L^{-1}			Water supply	Irrigation	Waste	
Russia	175.10	8.0	19.00	13.85	11.70	0.25	1.90	29.68
Ukraine	23.30	1.6	5.50	5.95	4.05	0.45	1.45	7.90
Belorussia	16.20	-	1.55	1.35	1.05	-	0.30	1.84
Moldavia	0.92	0.3	0.50	0.26	0.26	-	-	0.58
Estonia	2.20	-	0.20	0.44	0.18	-	0.26	0.65
Latvia	2.20	-	0.45	0.20	0.20	-	-	0.46
Lithuania	0.95	-	0.50	0.46	0.46	-	-	0.65
Georgia	3.20	-	2.40	1.02	1.02	-	-	1.36
Armenia	5.00	-	4.05	3.90	0.86	2.28	0.76	1.75
Azerbaijan	5.18	0.5	2.20	2.85	0.33	2.52	-	2.40
Kazakhstan	47.30	13.6	13.10	2.60	1.88	0.62	0.10	6.80
Uzbekistan	30.00	7.5	5.20	7.90	3.44	2.28	2.18	10.80
Kyrgyzstan	14.00	1.4	3.40	1.65	0.93	0.70	0.02	3.50
Tadjikistan	6.40	0.5	2.10	2.10	0.75	1.07	0.28	3.84
Turkmenistan	1.20	0.8	1.00	0.60	0.20	0.24	0.16	0.89
Total	333.15	34.2	61.15	45.13	27.31	10.41	7.41	73.10

Source: Khublarian (1990a).

capture systems in terms of technical and economic feasibility) have been estimated to be about 333×10^9 m^3 per annum. Of this, 45 per cent is located in the European part of the country, 30 per cent in Central Asia and Kazakhstan, while the remaining 25 per cent is in Siberia and the Far East (Khublarian, 1990a). The proven fresh groundwater resources of the CIS at January 1985 were estimated to be about 61.5×10^9 m^3 per year, with Russia containing 19×10^9 m^3 yr^{-1}, followed by Kazakhstan, 13.1×10^9 m^3 yr^{-1}, Ukraine, 5.5×10^9 m^3 yr^{-1} and Uzbekistan, 5.2×10^9 m^3 yr^{-1} (Table 4-5).

In the Commonwealth of Independent States groundwater is being used increasingly for public water supply and in a number of southern republics for irrigation. A few decades ago water supply for towns and cities was mainly based on surface water, but now about 60 per cent of town and cities are using groundwater for municipal water supply, 20 per cent are using both groundwater and surface water and about 20 per cent rely on surface water only (Kovalevsky, 1985).

Groundwater in the Commonwealth of Independent States is also widely used for irrigation. The main republics which use groundwater for irrigation are Azerbaijan, Uzbekistan, Armenia, Tadjikistan, Kyrgyzstan, some areas of Russia, Kazakhstan and Ukraine. Large aquifers consisting of alluvial fans and intermontane sediments, and to a lesser extent confined aquifers, are mainly developed for irrigation (Vartanyan et al., 1985).

Table 4-5 also shows groundwater extraction during 1984 for each republic. The total was about 45.13×10^9 m^3, of which 27.31×10^9 m^3 was used for domestic water supply, 10.41×10^9 m^3 for irrigation and 7.41×10^9 m^3 was wasted. Predictions show that groundwater extraction will increase to 73.1×10^9 m^3 by the year 2000.

The groundwater monitoring network of the CIS consists of more than 30 000 wells. About 20 000 are first-class and some 10 000 are second-class observation wells (Zaltsberg, 1988). In the first-class observation wells all components of the groundwater regime are monitored. Second-class observation wells are used to provide site specific data, such as the groundwater level in the irrigated area or in the vicinity of engineering structures (dams, canals, reservoirs, etc.).

Land cover

According to the FAO (1989), of a total land area of 2227.2 Mha, about 232.57 Mha were under cultivation in 1987. This consisted of 228.20 Mha of arable land and 4.37 Mha of permanent crops. Irrigated land constituted 20.48 Mha and the remaining 212.09 Mha of the cultivated land was not irrigated. Permanent pasture covered 371.60 Mha, forest and woodlands 944.0 Mha and other land 679.03 Mha.

Irrigation

Irrigation has been a mainstay of agriculture in Central Asia for thousands of years. Archaeological researchers have confirmed the existence of some irrigation systems, along and between the Syr Darya and Amu Darya, which are 3500 years old (Micklin, 1987; French, 1983a). However, technical means of irrigation were rather primitive, consisting of manual water lifted from the river and flood irrigation, without any regulatory structures. Only towards the end of the nineteenth century did there begin to appear engineering irrigation structures and pumping stations (Framji et al., 1983).

Before 1913 there were less than 4 Mha of irrigated land in the Commonwealth of Independent States. In the following years reconstruction and development projects were undertaken in Central Asia, Trans-Caucasus, Northern Caucasus, Volga Basin, Dnepr Basin and Crimea. These activities resulted in the expansion of the irrigated area of the country, attaining about 9.9 Mha in 1965 and 20.78 Mha in 1988 (Table 4-6).

Table 4-6: Expansion of the irrigated area in the Commonwealth of Independent States since 1913.

Year	Irrigation area (Mha)	Reference
1913	3.97	Dukhovny (1985)
1940	6.09	Dukhovny
1950	7.38	Dukhovny
1960	9.84	Dukhovny
1965	9.90	Dukhovny
1970	11.10	FAO (1987)
1975	14.49	FAO
1980	17.49	FAO
1985	19.95	FAO
1987	20.48	FAO (1989)
1988	20.78	FAO (1990)

According to Framji et al. (1983), the existing water resources of the CIS make it possible to irrigate 36 Mha, and with inter-basin water transfer this figure could reach 100 Mha. The most promising areas for future water and land development are the steppe zone and the vast territory which is crossed by big rivers such as the Volga, Don, Dnepr and Ural.

Many canals existed in the country before 1917, but these canals were mainly used for fluvial transport (French, 1983b). Since 1917 numerous huge reservoirs and a great number of large and long canals have been constructed for the development of irrigation. The total length of the permanent irrigation network in the country exceeds 530 000 km. The length of the collectors and drainage network in the irrigated zone is about 240 000 km (Framji et al., 1983).

At the first stage of land development, the drainage systems comprised a network of open drains, but presently this practice is being replaced by more advanced methods of subsurface drainage and under suitable conditions by vertical drainage. In 1970 the extent of subsurface drainage on irrigated land was only 16 per cent of the entire length of the drainage network, but by 1977 it had increased to 52 per cent. Table 4-7 shows the expansion of the drained area in the Commonwealth of Independent States between 1967 and 1978.

Table 4-7: Expansion of drained area in the Commonwealth of Independent States.

Year	Drained area (Mha)	Area with subsurface draining (Per cent of drained area)
1967	5.8	37
1970	7.4	45
1975	10.1	57
1978	11.6	63

Source: Framji et al. (1983: 1419).

Further information relating to irrigation and drainage in the CIS is available in Central Intelligence Agency (1974: 22–24), Micklin (1978), Framji et al. (1983: 1404–1433) and Tochenov (1986: 154), while Kubijovyč (1988: 355–56) provides similar information for Ukraine.

Salinity

The Commonwealth of Independent States has about 218.04 Mha of saline and alkaline soils or approximately one-quarter of the 954.8 Mha of the world's saline and alkaline soils (Szabolcs, 1989). These soils cover 170.72 Mha in the Asian part of the CIS and 47.32 Mha in the European part. The large extent of salt-affected soils is mainly due to natural factors, and to a much lesser extent, human activities.

More than 90 per cent of the 50.80 Mha of salt-affected soils in Europe is found in the European part of the Commonwealth of Independent States. These soils are located in the following regions:

- Pre-Caspian lowlands contain a massive area of salt-affected soils. This area is influenced by the Volga, Ural and Emba Rivers and in part by the Atrek River. The salt content differs from place to place but is generally increasing towards the shores of the Caspian Sea.
- The Trans-Caucasian republics of the CIS, namely Georgia, Armenia and Azerbaijan, have considerable areas of different types of salt-affected soils.

- Ukraine has a large area of salt-affected soils, partly in the valley of Dnepr, the southern part of the republic, and Crimean Peninsula.
- Moldavia is another republic with a massive area of salt-affected soils, mainly in the Danube Delta and near Odessa.

In the Asian part of the CIS saline soils cover about 57 per cent of the 299.29 Mha of Asia's salt-affected soils (Szabolcs, 1989). Saline soils are widely distributed in the Central Asian republics, namely in Kazakhstan, Uzbekistan, Turkmenistan, Tadjikistan and to some extent in the Kyrgyzstan, as well as in some parts of Russia.

In the Asian part of the CIS, alkalinity is much less widespread than salinity. Alkalinity sometimes occurs jointly with salinity in desert and semidesert areas, but soil alkalinity becomes dominant towards the north. One of the largest alkaline areas can be found in Western Siberia along the upper valleys of the Ob and Irtysh Rivers. Alkaline soils also exist in Eastern Siberia in the basin of the Lena River. The vast semidesert areas east of the lower reaches of the Volga River are very rich in saline and alkaline soils (Szabolcs, 1989).

Salt-affected soils, arising from human activities and particularly irrigation, exist in Central Asia, the Ukraine, Northern Caucasus and Trans-Caucasus. In these areas groundwater levels rose as a result of the large-scale construction of reservoirs and irrigation systems (Khublarian and Yushmanov, 1985).

In the late 1960s, because of secondary salinisation in Central Asia, land abandonment equalled the area of newly irrigated land. The situation evidently has improved since then, leading for example to a doubling of cotton yield in Khorezm Oblast, in Uzbekistan and in the Murgab oasis in Turkmenistan. For the whole country, salinised land decreased from 40 per cent of the irrigated area in 1964 to 20 per cent in 1975 (Micklin, 1978: 15). In other words, if these estimates are accurate, the area of salt-affected lands in the irrigated areas of the Commonwealth of Independent States has decreased from 4 Mha to 2.9 Mha, if it is considered that there were 9.9 Mha and 14.49 Mha of irrigated land for 1964 and 1975 (see Table 4-6). This was achieved mainly by improving drainage facilities and lining irrigation canals. Originally, drainage facilities in the more important irrigated areas consisted predominantly of open drains that became choked with weeds and silted and required annual cleaning. This was a particularly serious problem in southern Kazakhstan and Central Asia and modern subsurface drainage facilities were constructed in these areas. However, the current extent of human-induced salt-affected soils of the country is estimated to be about 3.7 Mha (Khublarian, 1990b) or 18 per cent of the irrigated lands. As a very large proportion of the CIS irrigated and human-induced salt-affected lands exist in Central Asia, this area will be given extended treatment in the following sections.

Salinity and environmental problems in Central Asia

Secondary salinisation of land and water resources in Central Asia is mainly due to excessive and inefficient water use for irrigation, the climatic and hydrogeologic conditions of the region, lack of attention to drainage problems and mistakes in management of water resources which allow return flows with high salt concentration to be discharged to the rivers. Therefore, some brief background information about the region and the development of its water resources will be provided prior to a description of the extent of salinisation and its management.

Central Asia is mainly a lowland desert with low precipitation (100–200 mm per year) and high potential evapotranspiration ranging from 1000 mm in the north to over 2250 mm in the south of the desert zone, resulting in severely arid conditions (Micklin, 1991a).

Although most of Central Asia is desert, it has substantial water resources. These resources are highly regulated with a regulation degree of 94 per cent in Amu Darya and 86 per cent in Syr Darya. Current withdrawals exceed the available water resources of the region and it is mainly through reuse of surface, subsurface and other water resources that requirements are met (Dukhovny, 1989). The average annual flows of the Amu Darya and Syr Darya Rivers are 70×10^9 m^3 and 37×10^9 m^3, respectively (see Table 4-2). Discharges are maximum where the rivers exit mountains but decrease rapidly as they cross the deserts and approach the Aral Sea.

During the period from 1911 to 1960, 56×10^9 m^3 of river water, originating annually in the region, reached the Aral Sea. In the mid-1970s, due to the withdrawal of water for irrigation, it dropped to $7-11 \times 10^9$ m^3 per annum, while in the 1980s the inflow of river run-off to the Aral Sea ceased almost completely in a number of years (Glazovskiy, 1991).

Groundwater resources of the region are highly saline and the largest reserves of fresh groundwater tend to lie at great depth, necessitating expensive deep drilling and high operational costs (Micklin, 1987). The extractable groundwater resources of the five republics of Central Asia are about 25×10^9 m^3 per year (see Table 4-5). According to Micklin (1991a), groundwater resources that are not hydraulically connected with river flow are about 18×10^9 m^3 per annum, consisting of 9×10^9 m^3 in the Amu Darya Basin and 7.9×10^9 m^3 in the Syr Darya Basin.

Central Asia is a region with very high potential for cultivation of crops, such as cotton, grapes, fruits and vegetables. Land resources suitable for irrigation in the basins of the Amu Darya and Syr Darya comprise about 30 Mha, consisting of 16.3 Mha of first-class, 8.7 Mha of second-class and 5 Mha of third-class fertility (Tsurikov, 1989). Currently less than 8 Mha of these lands are irrigated. Because of the shortage of water resources in the region, further expansion of irrigated lands requires a restricted water saving policy.

In the 1950s to early 1960s a decision was made for the massive expansion of irrigation in Central Asia. It was proposed that the following objectives be met (Glazovskiy, 1991): increase raw cotton production in the country to provide the population with cotton fabrics and articles of clothing made from them; increase the export of raw cotton to augment foreign currency earnings; increase the production of vegetables and fruits; transform the region into a national garden and orchard; supply the population with meat and rice; and provide employment for the local population. The development project consisted of the construction of large reservoirs (Table 4-8), hydraulic structures and irrigation and drainage networks. One of the major components of the scheme is the Kara Kum Canal. Construction of the canal started in 1953 (Micklin, 1978; 8–9). The canal originates at Kerki, on the left bank of the Amu Darya River, and extends 1300 km westward. In May 1962 Kara Kum Canal brought Amu Darya water to Ashkhabad, which had a water supply problem (Parker, 1983: 143).

Table 4-8: Characteristics of the major reservoirs in Central Asia.

Reservoir	River	Year of construction	Surface area (km^2)	Storage capacity ($10^9 \, m^3$)
Amu Darya Basin:				
Nurek	Vahsh	1975	98.0	10.50
Tuyamuyun	Amu Darya	1985	650.0	7.23
Kattakurgan	Zerafshan	1952/68	84.5	0.90
South-Surhan	Surhan Darya	1964	64.6	0.80
Syr Darya Basin:				
Toktogul	Naryn	1974	284.0	19.50
Chardarya	Syr Darya	1965	900.0	5.70
Kayrakkum	Syr Darya	1956	513.0	4.03
Chakir	Chakir	?	69.1	2.43
Charvak	Chirchik	1970	40.3	1.99
Andijan	Kara Darya	1980	59.0	1.79

Source: Raskin et al. (1992).

The resolution of the Central Committee of the Communist Party and the then Soviet Council of Ministers of 1956 on the Golodnaya Steppe development marked a turning point in the development of irrigation projects in Central Asia. The first large-scale irrigation and development project was accomplished in the Golodnaya Steppe on an area of 300 000 ha. The experiences and learning from the mistakes of the Golodnaya project were widely applied in design and construction of other irrigation projects such as Karshi Steppe, Bukhara and Amu Darya Delta (Tursunov, 1989). The main irrigation complexes of the region are shown in Figure 4-4.

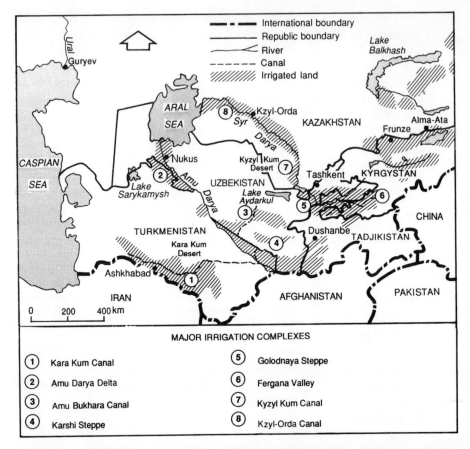

Figure 4-4: Irrigation development in Central Asia (Micklin 1991a: 90 and Kotlyakov, 1991).

Table 4-9 shows the irrigated area in the five republics of Central Asia, which has increased from 2.98 Mha in 1913 to 7.86 Mha in 1980, representing an increase of 264 per cent. During the same period the irrigated area of the country increased from 3.97 Mha to 17.49 Mha (Table 4-6), representing an increase of 440 per cent.

Irrigated agriculture in Central Asia is important to both the national and regional economy. It produces 95 per cent of national cotton, 40 per cent of rice, 25 per cent of vegetables and 32 per cent of fruits and grapes (Razakov, 1989). In terms of irrigated land allocated to different crops, of an irrigated area of 7.61 Mha in 1987, about 3.85 Mha was allocated to cotton, 1.70 Mha to fodder, 0.43 Mha to rice and 0.42 Mha to orchards (Raskin et al., 1992).

Table 4-9: Increase in irrigated areas in the republics of Central Asia (in 1000 ha).

Republics	1913	1940	1950	1960	1970	1980
Uzbekistan	1339	2008	2122	2665	2751	3407
Kazakhstan	696	994	1194	1482	1451	1930
Turkmenistan	307	373	385	434	643	942
Kyrgyzstan	452	794	797	834	883	975
Tadjikistan	211	325	336	391	518	605
Total	2978	4494	4864	5806	6246	7861

Source: Dukhovny (1985).

Irrigation is by far the dominant user of water in the region. The net water demand in 1987, or water requirement by the final users for crop growth, industrial processes and households, was about 97.32×10^9 m^3 (Raskin et al., 1992). This consisted of 79.38×10^9 m^3 for irrigation (81.6 per cent), 8.02×10^9 m^3 for industry (8.2 per cent), 6.3×10^9 m^3 for domestic water supply (6.5 per cent) and the remaining 3.62×10^9 m^3 for livestock and fishery (3.7 per cent). To overcome the demand, 127.44×10^9 m^3 was withdrawn from surface and groundwater resources. Groundwater contributed 12.3×10^9 m^3 (4×10^9 m^3 in the Amu Darya Basin and 8.3×10^9 m^3 in the Syr Darya Basin) and was extracted from the confined aquifers of the region.

Modern irrigation techniques are not widely used in the region. According to Glazovsky (1990a), 89.5 per cent of irrigated areas in the Uzbekistan were watered by furrows and only 1.5 per cent with sprinklers in 1977. On-farm water application rates are high in the region. For example, according to Raskin et al. (1992), application rates for the three main crops in 1987 were as follows, with the maximum application rates corresponding to the Kara Kum irrigation areas: cotton, from 7400 to 10 600 m^3 ha^{-1} yr^{-1}; fodder, from 8800 to 13 300 m^3 ha^{-1} yr^{-1}; and rice, from 22 700 to 33 400 m^3 ha^{-1} yr^{-1}.

Research has indicated that lower application rates would be sufficient for crop production. According to the research of the Central Asian Research Institute of Irrigation (SANIIRI) in Tashkent, the water requirement (evapotranspiration less precipitation) of cotton in Uzbekistan ranges from 4800 to 7200 m^3 ha^{-1}, and the use of drip irrigation allows an irrigation rate of 3300 m^3 ha^{-1} (Dukhovny, 1992). These figures are much less than the current application rate in the region.

Efficiency of the water used for irrigation, measured as the ratio of water arriving at the field to the withdrawal, is low for Central Asia. According to Micklin (1991a), a commonly cited figure for the Aral Basin is 60 per cent and for the irrigated systems in Uzbekistan, it has been said to range from 52 to 61 per cent. These figures, characteristic of the early 1980s, indicate that at least 40 per cent of the water withdrawn was lost before it reached the field, primarily due to filtration from earthen canals.

Irrigated land salinisation

Excessive and inefficient water use in the irrigated area of Central Asia is the major cause of secondary salinisation of land and water resources of the region. Irrigation and drainage systems were inadequately constructed as a consequence of hasty development. Also, research findings, especially relating to soil properties, appropriate crops and irrigation management, were inadequately used (Walker et al., 1989).

Due to erroneous irrigation practices, secondary salinisation exists in 2.1 Mha of the irrigated areas of the region (Rozanov, 1991) and land losses have reached 1 Mha (Glazovskiy, 1991). Estimates show that in 1985 moderately and strongly salinised soils in the irrigated region of Central Asia ranged from 35 to 80 per cent, with 15 per cent of the irrigated land (about 1.2 Mha) in an extremely unsatisfactory condition (Glazovskiy, 1991). Production losses were considerable and estimated to be between 18 and 40 per cent (Table 4-10). According to a recent estimate, in 1990 about 4.2 Mha were irrigated in Uzbekistan, out of which 1.2 Mha were salt-affected (Dukhovny, 1992). This represents a ratio of salt-affected to irrigated land of 29 per cent which is considerably lower than the previous estimate of 60 per cent, for 1985 (Table 4-10). Also, the regional movement of salts has been altered because of the development of irrigation. Previously the Aral Sea received 20 million tonnes of salt annually from the major rivers flowing to the sea (Rozanov, 1991). Now these salts are accumulating on the plain, in soils, sediments, groundwater (under the irrigated areas and along canals), and in numerous small and large lakes such as Sarykamysh and Aydarkul (Arnasay).

Table 4-10: Estimates of moderate and strongly salt-affected soils and production losses in the irrigated regions of the five Central Asian republics for 1985.

Republic	Salt-affected soils (%)	Production loss (%)
Uzbekistan	60	30
Kazakhstan	60–70	30–35
Turkmenistan	80	40
Kyrgyzstan	40	20
Tadjikistan	35	18[a]

Source: Glazovskiy (1991); (a) Glazovsky (1990a).

Of the 2.02 Mha of land used for cotton production in Uzbekistan, approximately 50 per cent suffers from secondary salinisation. The best water used to irrigate cotton has a salt content of less than 1000 mg L^{-1}, but mostly the water contains 1000 to 2000 mg L^{-1} of salt. The watertable is generally 1 to 3 m from the surface. Salt accumulated in the surface layers is removed by leaching. More than 1 Mha of cropped land in Uzbekistan is leached every year (Walker et al., 1989). The 150 000 ha used for rice production in Uzbekistan is mainly salt-affected, including the rice fields in Karakalpak and Khorezm. The salinity of irrigation water used for rice production is generally less than 2500 mg L^{-1}. Some state farms use drainage water for irrigation containing up to 4000 mg L^{-1} of salt (Walker et al., 1989).

In Kazakhstan salinity problems exist in irrigated regions along the Syr Darya and near the north shore of the Caspian Sea. Problems along the Syr Darya are situated near Kzyl-Orda. Problems in the Caspian Sea area of Kazakhstan are centred around Guryev (Figure 4-4). As in the Kzyl-Orda region, these problems are associated with saline irrigation with water of up to 4000 mg L^{-1}, restricting application to only the more salt-tolerant crops such as lucerne, maize and sorghum.

In Turkmenistan, in the irrigated areas of the Kara Kum Canal, large areas have been waterlogged and salinised along the entire canal mainline and up to 50 km either side of the canal (Rozanov, 1991). The landscape around Ashkhabad is abundant with spots of salt efflorescence. Groundwater here has risen too high as a result of seepage from the canal. The groundwater has a high dissolved solids content and with intensive evaporation salt accumulates on the surface. In eastern Kara Kum, the area of irrigated land is 184 400 ha, of which only 140 000 ha has drainage and 26 300 ha are saline and in an unsatisfactory condition. Overall, 90 per cent of the irrigated land in Turkmenistan is now in an unsatisfactory state due to salinisation (Rozanov, 1991). However, this estimate is 10 per cent higher than the estimate provided in Table 4-10.

Dryland salinisation

It is assumed by scientists in the CIS that secondary salinity problems only occur as a result of irrigation (Walker et al., 1989). In northern Kazakhstan, starting 35 years ago, 25 Mha of perennial grassland were converted to agricultural use in areas receiving about 200–400 mm precipitation per annum. Soils are relatively sandy and the area is apparently underlain by saline groundwater at depths ranging from about 2–100 m. During agricultural development the perennial native pastures (dominated by *Stipa* sp.) were ploughed in and replaced with a rotation of winter wheat, barley, fallow, winter wheat followed by five years of volunteer pasture grazed by sheep.

Prior to agricultural development there were areas of primary dryland salinity. Measures referred to as 'dry amelioration' were developed to improve these areas to the point where cropping was possible. About 0.5 per cent of the developed area has been treated with dry amelioration. Dry amelioration involves various phases. The first and most critical phase is profile mixing by deep ploughing. The second phase is chemical amendment, which includes addition of organic fertilisers, gypsum and phosphogypsum. The third phase is establishment of salt-tolerant plants such as *Melilotus* spp.

Unfortunately no work has been done on the effect of agricultural development on groundwater levels. It is well known that in Canada and the United States salinity problems followed conversion of perennial grasslands to wheat with fallow rotation and that in Australia salinity developed following agricultural developments of native bushland. The CIS example is of considerable interest for two reasons. Firstly, the CIS scientists should be alerted to the possibility that a salinity problem may be developing and, secondly, it would be valuable to know why salinity problems have not yet developed (Walker et al., 1989).

Surface water salinisation

During the first half of the 1980s, drainage run-off, that is, the run-off of drainage and wastewaters, over the entire Aral region ranged from 29 to 46×10^9 m^3 per year (according to data from different authors). Recently, because of the expansion of irrigated areas, drainage run-off has increased even more, and amounts to at least $46-47 \times 10^9$ m^3 per year. From this total volume of drainage run-off, $25-26 \times 10^9$ m^3 is discharged into rivers, $11-12 \times 10^9$ m^3 into lakes and $14-15 \times 10^9$ m^3 into the desert (Glazovskiy, 1991: 82). It must be emphasised, however, that the existing accounting system does not make possible a precise determination of the volume of drainage run-off. Accordingly, all the cited figures on drainage run-off require refinement.

The discharge of sizeable drainage run-off to the rivers has sharply increased the level of salinity and polluted water by pesticides, herbicides, agricultural chemicals and remains of mineral fertilisers. As a result an extremely unfavourable ecological situation has been observed downstream of both Amu Darya and Syr Darya Rivers (Tsurikov, 1989). According to Razakov (1989), water salinity in the Upper Syr Darya at the confluence of the Narin and Kara Darya averages 300 mg L^{-1}. At the outflow from the Fergana Valley it grows up to 1200–1400 mg L^{-1}, in the Chardara region up to 1600–1700 mg L^{-1}, and in the downstream reaches at Kzyl-Orda it is 1800–2000 mg L^{-1} (Figure 4-5). Other estimates for the level of salinity in the lower reaches of the two major river systems are: 1500–3000 mg L^{-1} (Glazovskiy, 1991: 75), 1500–2500 mg L^{-1} (Tsurikov: 1989, 42), and finally 1500 mg L^{-1} for the Amu Darya and 2500 mg L^{-1} for the Syr Darya during winter when water levels are low (Orechkine, 1990: 1383).

Irrigation drainage waters are partly diverted to two large lakes, Aydarkul (also called Arnasay) and Sarykamysh. The Aydarkul has an area of 2300 km^2 and a volume of 20×10^9 m^3 (Micklin, 1991a), while Lake Sarykamysh has a surface area of 3000 km^2, volume of 30×10^9 m^3 and maximum depth of over 40 m. In the 1950s, there were only several seasonal lakes in Sarykamysh region with an aggregated area near to 100 km² (Micklin, 1991b). Since their origin in the 1960s these two lakes have developed considerable fishery and wildlife importance. Sarykamysh is a noted haven for migratory waterfowl, some of which are rare. Fish catches in Aydarkul and Sarykamysh were 10 000 and 3000 tonnes, respectively. A nature reserve has been proposed to protect part of Aydarkul and a protected area has already been established for Lake Sarykamysh (Micklin, 1991b). However, the fishery and ecological value of both lakes is threatened by rising salinities and contamination from herbicides, pesticides and fertiliser contained in irrigation drainage. In 1987 the average salinity for Sarykamysh was 12 000 mg L^{-1} and this could rise to 15 000–17 000 mg L^{-1} by the year 2000, adversely affecting the fish. However, commercial fishing was halted in Sarykamysh in 1987 because of pesticide contamination. The full cut-off of irrigation drainage water now sent to the lake (so that it could be delivered to the lower reach of the Amu Darya River and the Aral Sea) would drop Sarykamysh's level 15–17 metres and raise average salinity to 40 000 and 50 000 mg L^{-1}, wiping out all fish species (Micklin, 1991a; 57–58).

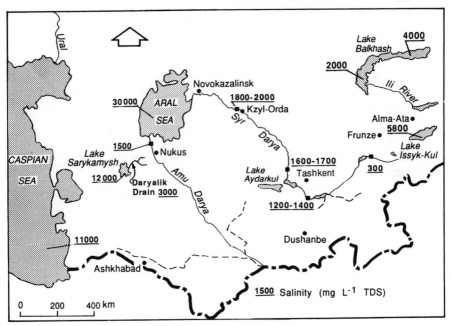

Figure 4-5: Salinity of surface water bodies in Central Asia (Data from various sources).

Like the Aral Sea, Lake Balkhash also suffers from a reduction in water input. The reduction is the result of the construction of Kapchagay Reservoir on the Ili River in 1969, which is the major river flowing to the lake. Prior to the construction of the reservoir, Ili River had an average annual flow of 12×10^9 m^3, carrying about 6 million tonnes of salt to the lake (Peter, 1992). The reservoir was designed to have a total capacity of 28.14×10^9 m^3 and useful capacity of 6.6×10^9 m^3. However, it has not been permitted to reach its full supply capacity because it was recognised that complete filling would affect the water-level of the lake. In 1986 the reservoir had been allowed to fill to only half of its capacity. Salinity varied slightly during the years but for 1976–1979 was between 318 and 411 mg L^{-1}.

The reservoir is multipurpose and, among others, serves for hydroelectricity production, irrigation and flood control. According to Peter (1992), while there is no doubt that the reservoir has successfully achieved many of its objectives, this has been at the cost of downstream environmental degradation. In general, impacts include: wetland degradation; land salinisation; rising salinities in Lake Balkhash; decline in the value of the Lake's fishery; and alteration of the hydrological patterns. A primary cause of these impacts was a fall in the lake's water-level between 1970 and 1989. This fall was about 2 m and was associated with a decrease in the lake's surface area of 261 km^2. Associated with the fall in the water-level of the lake has been an overall increase in the mean salinity of the lake. During the period 1970–1980 it rose from 2230 to 2750 mg L^{-1} (Peter, 1992). This increase has been correlated with a reduction in aquatic productivity. Additionally, water from the lake has become increasingly less useful for domestic, agricultural and industrial supplies. Adding to the deteriorative effects of the reservoir, there are more widespread catchment phenomena. Of particular importance since the 1970s has been an increase in the pollution of the Ili and other rivers discharging to the lake by industrial and urban waste water and irrigation drainage water. As a result, Lake Balkhash has become seriously polluted with toxic chemicals, including heavy metals.

In the Ili Delta, soils are salinised and profound changes occurred in the vegetation. To reverse these processes several measures have been proposed. Peter (1992) describes these measures, which include: maintaining the water-level in the reservoir at 10 m below its full supply level, at which evaporation losses would exceed 1.5×10^9 m^3 yr^{-1}; greater use of regional groundwater resources, which have a potential of 3.5 to 3.7×10^9 m^3 yr^{-1}; more effective usage of water for irrigation; and restriction on further increase in the area irrigated (originally it was expected to irrigate 360 000 ha with water from the reservoir, now an area of only 100 000 ha is regarded as a reasonable value).

Forecasts for the future water-level and salinity of the lake are difficult to make. But if continued anthropogenic losses occurred with a prolonged natural period of aridity, causing the lake water-level to drop from 341 m (in 1990) to 339 m, salinity in the south-western basin could exceed 3000 mg L^{-1} and in the eastern basin might reach 6000 mg L^{-1} TDS (Peter, 1992).

Management options

Management of salinity problems in the Aral Basin is part of an overall policy to increase the efficiency of irrigation water use. Efforts to improve irrigation through reconstruction of older facilities, lining of earthen canals, automation, telemechanisation and water pricing are now being stressed not only to save water but to alleviate salinisation and waterlogging and raise yield on irrigated lands (Micklin, 1987). Table 4-11 summarises the estimates for 1983 of the required measures in the region and in Uzbekistan.

Table 4-11: Estimates for 1983 of required measures for the improvement of the irrigation and drainage systems in the Aral region and in Uzbekistan.

Required measures	Aral region	Uzbekistan
Reconstruction of irrigation systems	4.2 Mha	2.2 Mha
Advanced methods of drainage	2.8 Mha	1.0 Mha
Land levelling and the introduction of advanced irrigation methods	2.0 Mha	1.3 Mha
Lining of main and interfarming canals	84 000 km	48 000 km

Source: Dukhovny (1992).

The irrigation efficiency improvement program has produced some positive results. In Uzbekistan, by 1986 irrigation canals on 1.3 Mha had been rebuilt, the levelling of fields had been completed on 0.63 Mha, and 2.2 Mha had received lesser reclamation improvements (Micklin, 1991a: 18). As a result, the efficiency of the irrigation system (expressed as a ratio of water arriving at the field to water withdrawal) in Uzbekistan rose from 48 per cent (at an unspecified earlier date) to 62 per cent. The reconstruction of old irrigation systems between 1962 and 1988 in some 4 Mha in the Aral Basin raised their average efficiency from 48 per cent to 64 per cent and the average withdrawal in the Central Asian region dropped from 18 700 m^3 ha^{-1} in 1980 to 16 000 m^3 ha^{-1} in 1985. By achieving the target of 8000 m^3 ha^{-1} and an efficiency of 80 per cent, it might be possible to irrigate 7.8 Mha with a withdrawal of 62×10^9 m^3 instead of 107×10^9 m^3, achieving a reduction of 45×10^9 m^3 (Micklin, 1991a: 19).

In Central Asia about 80 per cent of existing and prospective irrigation land is located in areas of inadequate natural drainability (Umarov, 1989). Therefore irrigation without adequate drainage facilities would be accompanied by watertable rise and soil salinisation.

General technical means used for land drainage are surface and subsurface horizontal drainage, vertical drainage and combined drainage. Subsurface drainage systems widely used in the region comprise pipes (tile, plastic, asbestos, cement) with a diameter of 7 to 30 cm. These are placed underground at a depth of 2 to 4 m and surrounded by protective filtering, and at a spacing of at least 50 m. Filters consist of sand or sandy gravel with a thickness of 15 to 18 cm. Recently use has also been made of synthetic filtering materials. Vertical drainage wells are usually made with a diameter of 0.9–2.3 m and operational column of 0.4 m. Combined drainage comprises a system of horizontal drains connected to a well. The wells of combined drainage have a diameter of 0.5 m, casing of 0.1 m and a maximum depth of 30 m (Umarov, 1989). The conditions of application for each of the above mentioned type of drainage evidently are based on the hydrogeological characteristics of the area. According to Framji et al. (1983: 1418), most of the more than 5000 wells used for vertical drainage in the CIS are installed in the republics of Central Asia and Kazakhstan.

The Golodnaya Steppe irrigated area is an example of an irrigation development in a very arid region with adequate drainage facilities. In this area 350 000 ha of land is irrigated and the length of the drainage system is about 20 000 km. This system permits very successful cotton production with common irrigation water application of 8500 to 9500 m^3 ha^{-1} (Dukhovny, 1992).

Other policies for saving water and alleviating salinity problems consist of (Glazovskiy, 1991):

- Removal of saline soils with low productivity which use enormous amounts of water without appreciable return. For example, withdrawal of at least 0.5 Mha of these lands would save 7×10^9 m^3 of water. Even considering that 15 per cent of the irrigated lands in the Aral Basin is in an extremely unsatisfactory condition, it will be desirable to remove an even larger area.
- Reduction of the area allocated to rice and cotton production. It would be desirable to reduce cotton-producing irrigated land by 1 to 1.3 Mha and save 10 to 15×10^9 m^3. Rice is the crop requiring the most water. A reduction of the area planted to rice in the region by at least 100 000 ha would be desirable.

In areas where soils are heavy and seepage low, tree plantation would reduce the groundwater mound and improve the local microclimate (Walker et al., 1989).

Aral crisis

The Aral Basin is facing a huge environmental crisis. The main reasons for this crisis are: an erroneous strategy of development; mistakes in the strategy of agricultural development (introduction of cotton as a monoculture, an excessive expansion of irrigated agriculture, a wide use of agricultural chemicals and development of low productivity lands); and low quality of design, construction and exploitation of irrigation systems.

Problems of the Aral Basin were neglected and hidden from the public for many years. Their resolution became a priority by the late 1980s. Following is a brief description of this crisis extracted from Micklin (1988 and 1991a), Glazovsky (1990a) and Glazovskiy (1991). Information about the Aral crisis is also available in Ellis (1990), Glazovsky (1990b), Orechkine (1990), Kotlyakov (1991), Williams and Aladin (1991) and Golubev (1993).

Because of major river diversions, the Aral Sea is disappearing. Between 1960 and 1987 its level dropped 14.34 m (from 53.41 to 39.07 m), its area decreased by 40.6 per cent (from 68 000 to 40 393 km^2), its volume diminished by 66 per cent (from 1090 to 370 km^3) and its average salinity rose from 10 000 to 30 000 mg L^{-1}. One serious problem is the blowing of dust and salt from the 28 000 km^2 of exposed seabed. Between 40 and 150 million tonnes are blown off annually, reaching 9.5 tonnes per ha in some areas of Karakalpak. Dust and salt are deposited in a vast area of 150 000 to 200 000 km^2.

The fishery resources of the Aral Sea have suffered as well. Once, many fish species lived in the Aral Sea. Fishing was a major industry in the region, employing 60 000 people, and the annual catch was about 40 000 tonnes. Due to the increase of salinity and the introduction of exotic fishes, all species of fish have now disappeared (a detailed description of the biological features of the Aral Sea is provided by Williams and Aladin, 1991). The major fishing ports of Muynak and Aralsk are now tens of kilometres from the shore. For some time fishermen were using a canal dug to provide them access to the sea. Eventually, even this became useless and fishing ships were subsequently left abandoned in the canal.

The problems of the Aral Sea are not limited to its desiccation, the disappearance of fish from its remaining salty water, the death of the fishing industry and blowing of dust and salt from its dried bed. A number of other problems have developed. These include toxification problems resulting from the excessive use of fertilisers and chemicals for cotton and other crop production. There are also health problems, lack of medical facilities and inadequate drinking water and sewage treatment facilities.

For decades, huge dosages of chemicals, fertilisers, pesticides and defoliants have been used for cotton and other agricultural production. Consequently, the spread of chemicals in the soil, drainage water and the water supply decreased land productivity and caused poisoning of tens of thousands of inhabitants of the region. Excessive use of chemicals and environmental degradation, particularly in the lower part of the Aral Basin and close to the sea, resulted in an increase in respiratory, eye, gastrointestinal, kidney and cardiovascular ailments, tuberculosis, typhoid, throat cancer and even mental diseases. Apart from health problems, the divorce rate, social tensions and emigration of the population are also increasing.

The deltas of the Amu Darya and Syr Darya have been severely degraded and lost much of their former economic and ecological importance. The climate of the region around the sea has grown more extreme as the water body has shrunk and lost its moderating influence. Living conditions around the Aral have become much more difficult as clean water supplies have disappeared and become polluted, medical and health conditions deteriorated and employment opportunities vanished.

Any solution of the Aral crisis should consider the two aspects of the crisis: to solve health, social, ecological and economic problems resulting from mistakes in agricultural development and unsatisfactory design, building and exploitation of the irrigation system; and to preserve the Aral Sea at its 1989 level, which requires an inflow of about $32-35 \times 10^9$ m^3 of water per year. Although the two aspects of the problem are interrelated, the problems of the Aral Sea preservation are secondary compared to the general problems of the Aral Basin. Proposals have been put forward for transfer of Caspian Sea water and transfer of part of the Volga or Siberian river run-off. Apart from the prohibitive costs of such diversion, this would only save the Aral Sea and would not solve the overall problems of the basin.

Proposals relating to changes in the agricultural structure and irrigation system are capable of solving these overall problems and would save enough water to preserve the Aral Sea. These changes include: retiring of salt-affected land, decreasing cotton production and the area used for rice production, crop rotation, introduction of new varieties of plants, reconstruction of the irrigation system, more extraction from groundwater resources, and rational use of drainage water.

In brief, priorities in the region of the Aral Sea can be defined as follows:

- **Short-term measures.** Improvement of living and health conditions of the population, providing healthy drinking water, construction of sewage systems, optimising the use of fertilisers and other agricultural chemicals, drastic improvement of medical services and providing the population with high quality foodstuffs.
- **Medium-term measures (2–10 years).** Reconstruction of irrigation systems and water saving, use of drainage water and further use of groundwater, introduction of new varieties of plants, using new technologies in irrigation and agriculture, introduction of new farm and economic management including private farming and water pricing.
- **Long-term measures (more than 10 years).** Solution of demographic problems and diversifying the economic activities in the region.

Salinity problems in other areas of the CIS

Information concerning secondary salinity problems in other areas of the CIS is very limited. Following is a brief description of the situation in some of the salt-affected areas.

In Ukraine, before 1917 only 17 400 ha of land were irrigated (Kubijovyč, 1988: 355–6). The irrigated area reached 89 700 ha by 1940 and in the 1950s the construction of large-scale irrigation systems was begun. By 1968, about 748 000 ha of Ukraine's arable land in many areas including Crimea, Kherson and Kiev were irrigated. In 1966 another large-scale irrigation program was initiated in the southern Ukraine where the land was fertile but precipitation was scarce. Since that time, the irrigated area has increased considerably and reached 2.48 Mha in 1984. Irrigation has been the most important water consumer in Ukraine since 1965. It was estimated that in 1980 about 10.4×10^9 m^3 of Dnepr (Dnieper) water, or 42.8 per cent of its annual run-off, was used for irrigation. However, irrigation efficiency was low and the use of unlined canals contributed to excessive losses, rising watertables and salinisation.

More recently Ukraine has become the object of a major water diversion scheme that would involve the transfer of $16-23 \times 10^9$ m^3 per year of water northward from the Danube to irrigate 2.6–4.6 Mha in the Dnepr Basin. The first step through Sasyk Lake was completed in 1983, but because of uncoordinated planning and insufficient research the irrigation of the Budzhak Steppe resulted in soil salinisation and alkalisation, making the area infertile. The continuing controversy over large-scale diversion schemes has put the completion of this project in doubt. According to Kreyda et al. (1990), completion of the project largely depends on the solution of the irrigation water problem. As part of the project, water from the former marine Sasyk Lake (lagoon) should have become suitable for irrigation by 1981 as a result of planned water exchange with the Danube. The factors responsible for widespread degradation (salinisation, alkalisation, increased bulk density) include significant alkalinity of irrigation water (pH of 8.2 to 9.2), high salt content (1100 to 3000 mg L^{-1} TDS) and an unfavourable combination of ions (Na$^+$/Ca^{++} ratio of 4 to 10 and Cl$^-$/SO$_4^-$ of more than 1). Results of a three-year study showed that acidification of irrigation water and application of gypsum and manure, as well as different tillage practices, had no significant improvement effects on the land irrigated with Sasyk Lake water. Kreyda et al. (1990) conclude that without improving the quality of irrigation water in the lake it will be impossible to prevent land degradation processes by the above mentioned traditional practices.

Fedorishchak et al. (1990) describe the secondary salinisation of soils of the Dniester Delta flood plain. Here reduced run-off of the Dniester, in connection with water diversion for irrigation and urban water supply, has significantly affected the groundwater salinity and land salinisation. The main causes of the salinisation are the intrusion of Black Sea water into the Dnestrovskiy Lagoon as a result of regulation and diversion of flow from the Dniester River. This has led to an increase in salinity of not only lagoon waters but also of groundwater of coastal areas, including the estuarian portion of the river valley. Under the dry climatic conditions of the area, an increase in salinisation of shallow groundwater led to secondary salinisation of the delta flood plain. The major areas of saline soils are confined to the central portion of the flood plain.

According to Ryskov and Gurov (1988), extensive irrigation of Chernozems in the Lower Don Basin has increased the water supply to the landscape considerably. Consequently salinisation, alkalisation and waterlogging have developed in the irrigated area. Ryskov and Gurev (1988) describe the results of their study within the Bagayevskiy-Sadkovskiy and Azov Irrigation Systems (BSIS and AIS, respectively). Before the introduction of irrigation, groundwater depth was about 10 to 15 m. Groundwater recharge occurred only in wet years at an average rate of 30–50 mm yr^{-1}. Irrigation on the flood plain of the Lower Don has been practised since the 1950s. Calcium bicarbonate water of the Tsimlyansk Reservoir, with a salt content of 300–700 mg L^{-1}, is delivered to the BSIS and sodium-sulphate water of the Veselovskiy Reservoir, with a salt content of about 2000 mg L^{-1}, is supplied to the AIS. After the beginning of irrigation, because of intense infiltration from canals and irrigated fields amounting to 45–50 per cent of the water intake, the watertable rose by 3–12 m. Lands with a depth to groundwater less than 2 m occupy up to 20 per cent of the irrigated area and lands with depth to groundwater of 2 to 3 m cover as much as 30 per cent of the area under irrigation. The salinity of the groundwater in the BSIS varies mainly within 3000 to 1000 mg L^{-1} TDS and that in the AIS 30 000 to 40 000 mg L^{-1} TDS in some places. Prolonged irrigation has drastically changed the hydrologic and hydrogeologic conditions of the area. Groundwater recharge increased fourfold from 150 m^3 ha^{-1} to 600 m^3 ha^{-1} per year. The average annual drainage water is about 200 m^3 ha^{-1} and the load of the chemicals from the landscape into the regional drains (Don, Sal and Manych Rivers) has increased sharply.

Salt balance calculations for the BSIS show that 63 000 tonnes of salt are delivered annually to the system with irrigation water. The amount of salt removed with drainage and subsurface flow is 104 000 tonnes. The difference is about 41 000 tonnes or 0.8 tonnes per ha. Thus, on the whole, irrigation with freshwater results in the desalination of the landscape. However, its scale is insignificant compared with salt reserves, which reach 300–500 tonnes ha^{-1} in the upper 10 m. Complete desalination of the landscape at the present rate may take several hundred years. In the Azov Irrigation System the salinity of the landscape has not decreased because of the high salt content of the irrigation water. Here the salt has been redistributed in the landscape.

In general, Chernozems, irrigated with freshwater in well-drained areas of the BSIS, will evolve in the direction of desalination. However, signs of salt accumulation are evident almost everywhere in BSIS despite the general desalination of the landscape and the aquifer. The reason is the widespread shallow groundwater which brings water and soluble salts to the land surface via capillary rise.

Based on the results of a three-year field experiment, Orlov et al. (1989) recommend the application of a mixture of phosphogypsum and copperas (green ferrous sulphate heptahydrate) instead of traditional gypsum to reclaim the sodic soils of the lower Don flood plain. The recommended chemicals are the wastes of a titanium dioxide production plant.

The Kura-Araks lowland is an intermontane depression and one of the largest irrigation regions of the Trans-Caucasus. Flooding ceased after the construction of the Mingechaur Reservoir on the Kura River. According to Babayev (1982), salinity of irrigation water has on average doubled from 400–500 mg L^{-1} to 800–1000 mg L^{-1} over a period of 30 years since 1950. The alkalinity of water has also increased. The net result has been a worsening quality of the soils. Salt-affected soils run parallel with the Kura-Araks rivers down to the southern shores of the Caspian Sea.

In Kalmykia (Lower Volga Region) about 100 000 ha of land are irrigated. However, 100 per cent of the irrigated land is saline to some degree. This not only means lost yield but more importantly potential growth of the saline desert and loss of productive land (Rozanov, 1991).

References

Babayev, M.P. 1982. Classification and diagnosis of irrigated soils of the dry subtropics of the eastern Transcaucasus region. *Soviet Soil Science*. 14(2): 50–59.

Central Intelligence Agency. 1974. *USSR Agriculture Atlas*. Washington DC: US Government Printing Office. 59 pp.

Central Intelligence Agency. 1990. *The World Factbook 1990*. Washington DC: US Government Printing Office. 382 pp.

Department of Technical Co-operation for Development. 1986. *Ground Water in Continental Asia (Central, Eastern, Southern, South-Eastern Asia)*. New York: United Nations. (Natural Resources/Water Series no. 15). 391 pp.

Dukhovny, V.A. 1985. Irrigation farming in Central Asia and its effectiveness. In. *Proceedings of the Tenth Session of the Committee on Natural Resources*. New York: United Nations. (Water Resources Series no. 59). 304–310.

Dukhovny, V.A. 1989. Management of river basins' water resources in conditions of water shortage (at present and in future). In. *Seminar-Cum-Study Tour on Water Resources Management and Use: Lectures of the USSR Experts, Tashkent*. United Nations: Economic Commission for Latin America and the Caribbean (ECLAC). 16–33.

Dukhovny, V.A. 1992. Central Asian Research Institute of Irrigation (SANIIRI), Tashkent. (Personal communication).

Ellis, W.A. 1990. The Aral: A Soviet sea lies dying. *National Geographic*. 177(2): 73–93.

Fedorishchak, M.R.P., Tsar, V.V. and Shishchenko, P.G. 1990. Secondary salinization of soils of the Dniester Delta flood plain. *Soviet Soil Science*. 22(4): 101–112.

Food and Agriculture Organization of the United Nations (FAO). 1987. *Production Yearbook*. Rome: FAO. v.40. 306 pp.

Food and Agriculture Organization of the United Nations (FAO). 1989. *Production Yearbook*. Rome: FAO. v.42. 350 pp.

Food and Agriculture Organization of the United Nations (FAO). 1990. *Production Yearbook*. Rome: FAO. v.43. 346 pp.

Framji, K.K., Garg, B.C. and Luthra, S.D.L. 1983. *Irrigation and Drainage in the World: A Global Review*. Third edition. New Delhi: International Commission on Irrigation and Drainage. v.III. 1161–1667.

French, R.A. 1983a. Introduction. In. Bater, J.H. and French, R.A. eds. *Studies in Russian Historical Geography*. London: Academic Press. v.1. 13–21.

French, R.A. 1983b. Canals in pre-revolutionary Russia. In. Bater, J.H. and French, R.A. eds. *Studies in Russian Historical Geography*. London: Academic Press. v.2. 451–481.

Glazovsky, N.F. 1990a. The Aral crisis: The source, the current situation, and the ways to solving it. Paper presented at The International Conference 'The Aral crisis: Causes, consequences and ways of solution'. Sponsored by the Special Research and Coordination Center 'Aral' (Institute of Geography, Moscow), Nukus, Uzbekistan, 2–5 October 1990. 17 pp.

Glazovsky, N.F. 1990b. *Aral Crisis: The Reasons of Emergence and the Ways for Solution*. Moscow: Nauka. 134 pp. (in Russian).

Glazovskiy, N.F. 1991. Ideas on an escape from the Aral crisis. *Soviet Geography*. XXXII (2): 73–89.

Golubev, G.N. 1993. State and perspectives of Aral Sea problem. In. Biswas, A.K., Jellali, M. and Stout, G.E. eds. *Water for Sustainable Development in the 21st Century*. Delhi: Oxford University Press. 245–254.

Gregory, J.S. ed. 1975. *The Geography of the USSR: An Introductory Survey*. Moscow: Novosti Press Agency Publishing House. 143 pp.

Hammer, U.T. 1986. *Saline Lake Ecosystems of the World*. Dordrecht, The Netherlands: Dr W. Junk Publishers. 616 pp.

Kalinin, G.P. 1971. *Global Hydrology* (Translated from Russian). Springfield, Virginia: US Department of Commerce, National Technical Information Service. 311 pp.

Khain, V.E. 1985. *Geology of the USSR: First Part, Old Cratons and Paleozoic Fold Belts*. Berlin: Gebrüder Borntraeger. 272 pp.

Khublarian, M.G. 1990a. *Water Resources: Rationalized Utilization and Protection*. Moscow: Knowledge Publisher. 38 pp. (In Russian).

Khublarian, M.G. 1990b. Water Problems Institute, Moscow. (Personal communication).

Khublarian, M.G. 1992. Water pollution and its consequences. In. Coles, J.N. and Drew, J.M. eds. *Australia and the Global Environmental Crisis*. (A publication of the Professors World Peace Academy of Australia). Canberra: Academy Press. 95–113.

Khublarian, M.G. and Yushmanov, I.O. 1985. Groundwater quality appraisal. In. *Hydrogeology in the Service of Man*. Memoires of the 18th Congress of the International Association of Hydrogeologists, Cambridge, 8–13 September 1985. v. XVIII, Part 3 (Groundwater quality management). 119–126.

Korzun, V.I., Sokolov, A.A., Budyko, M.I., Voskresensky, K.P., Kalinin, G.P., Konoplyantsev, A.A., Korotkevich, E.S., Kuzin, P.S. and Lvovich, M.I. eds. 1978. *World Water Balance and Water Resources of the Earth*. Paris: UNESCO. 633 pp.

Kotlyakov, V.M. 1991. The Aral Sea basin: A critical environmental zone. *Environment*. 33(1): 4–9 and 36–38.

Kovalevskiy, V.S. 1985. Monitoring of groundwater in the course of its development in the USSR. In. *Hydrogeology in the Service of Man*. Memoires of the 18th Congress of the International Association of Hydrogeologists, Cambridge, 8–13 September 1985. v.XVIII, Part 2 (Economic and social influence). 12–19.

Kreyda, N.A., Mikhaylyuk, V.I. and Kichuk, I.D. 1990. Degradation of Chernozems on the Danube-Dniester irrigation system. *Soviet Soil Science*. 22(1): 81–86.

Kubijovyč, V. ed 1984. *Encyclopaedia of Ukraine*. Toronto: University of Toronto Press. v.I. 952 pp.

Kubijovyč, V. ed. 1988. *Encyclopedia of Ukraine*. Toronto: University of Toronto Press. v.II. 737 pp.

Kudelin, B.I. Chief editor. 1977. *Ground-Water Flow Map of USSR Area*. (Scale: 1:2 500 000, 16 sheets). Moscow: Department of Geodsy and Cartography. (In Russian with brief descriptions in English).

Medish, V. 1990. *The Soviet Union*. Fourth edition. Englewood Cliffs, New Jersey: Prentice Hall. 411 pp.

Micklin, P.P. 1978. Irrigation development in the USSR during the 10th Five-Year Plan (1976–1980). *Soviet Geography*. XIX(1): 1–24.

Micklin, P.P. 1987. Irrigation and its future in Soviet Central Asia: A preliminary analysis. In. Holzner, L. and Knapp, J.M. eds. *Soviet Geography Studies in our Time*. Milwaukee: The University of Wisconsin. 229–261.

Micklin, P.P. 1988. Desiccation of the Aral Sea: A water management disaster in the Soviet Union. *Science*. 241:1170–1176.

Micklin, P.P. 1991a. *The Water Management Crisis in Soviet Central Asia*. Pittsburgh: University of Pittsburgh, Center for Russian and East European Studies. (The Carl Beck Papers in Russian and East European Studies no. 905). 120 pp.

Micklin, P.P. 1991b. Touring the Aral: Visit to an ecologic disaster zone. *Soviet Geography*. XXXII (2): 90–105.

Nalivkin, D.V. 1960. *The Geology of the USSR: A Short Outline*. (Including a 1:7 500 000 scale geological map of the USSR in full colour). Oxford: Pergamon Press. 170 pp.

Orechkine, D. 1990. La mer d'Aral menacée de disparition. *La Recherch*. 21: 1380–1388.

Orlov, D.S., Luganskaya, I.A. and Lozanovskaya, I.N. 1989. Chemical reclamation of saline-sodic soils of the Lower Don flood plain by some industrial wastes. *Soviet Soil Science*. 21(3): 78–89.

Parker, W.H. 1983. *The Soviet Union*. Second edition. London: Longman. 207 pp.

Peter, I. 1992. Lake Balkhash, Kszakhstan. *International Journal of Salt Lake Research.* 1(1): 21–46.

Pokshishevsky, V. 1974. *Geography of the Soviet Union.* Moscow: Progress Publishers. 279 pp.

Raskin, P., Hansen, E., Zhu, Z. and Stavisky, D. 1992. Simulation of water supply and demand in the Aral Sea region. *Water International.* 17(2): 55–67.

Razakov, R.M. 1989. Problems of integrated use and protection of water resources. In. *Seminar-Cum-Study Tour on Water Resources Management and Use: Lectures of the USSR Experts, Tashkent.* United Nations: Economic Commission for Latin America and the Caribbean (ECLAC). 52–64.

Rozanov, B.G. 1991. Once again on desertification. *Soviet Soil Science.* 23(7): 22–31.

Ryskov, Ya. G. and Gurov, A.F. 1988. Role of irrigation in the present evolution of terrace Chernozems of the Lower Don. *Soviet Soil Science.* 20(2): 40–48.

Szabolcs, I. 1989. *Salt-Affected Soils.* Boca Raton, Florida: CRC Press. 274 pp.

Tochenov, V.V. Chief editor. 1986. *USSR Atlas.* Moscow: Department of Geodsy and Cartography. 259 pp. (In Russian).

Tsurikov, G.S. 1989. Formulation and correction of master plans for irrigation development in river basins. In. *Seminar-Cum-Study Tour on Water Resources Management and Use: Lectures of the USSR Experts, Tashkent.* United Nations: Economic Commission for Latin America and the Caribbean (ECLAC). 34–43.

Tursunov, E.T. 1989. Development of irrigated agriculture under conditions of insufficient water resources (the Uzbek SSR case study). In. *Seminar-Cum-Study Tour on Water Resources Management and Use: Lectures of the USSR Experts, Tashkent.* United Nations: Economic Commission for Latin America and the Caribbean (ECLAC). 5–15.

Umarov, P.D. 1989. Advanced type of drainage and methods of its construction. In. *Seminar-Cum-Study Tour on Water Resources Management and Use: Lectures of the USSR Experts, Tashkent.* United Nations: Economic Commission for Latin America and the Caribbean (ECLAC). 108–123.

Urvantsev, G.S. 1985. Fundamentals of the multipurpose use and conservation of water resources in the USSR. In. *Proceedings of the Tenth Session of the Committee on Natural Resources.* New York: United Nations. (Water Resources Series no. 59). 302–304.

Vartanyan, G.S., Kulikov, G.V., Plotnikova, R.I., Shpak, A.A. and Yazvin, L.S. 1985. Significance of groundwater in the economic and social development of the USSR. In. *Hydrogeology in the Service of Man.* Memoires of the 18th Congress of the International Association of Hydrogeologists, Cambridge, 8–13 September 1985. v. XVIII, Part 2 (Economic and social influence). 20–24.

Walker, R.R., Malcolm, C.V. and Lyle, C.W. 1989. *Report on Australian Mission to USSR on Management of Salt Affected Soil and Crop Salt Tolerance, 8–31 August 1989.* Canberra: Department of Primary Industries and Energy. 30 pp.

Williams, W.D. and Aladin, N.V. 1991. The Aral Sea: Recent limnological changes and their conservation significance. *Aquatic conservation: Marine and Freshwater Ecosystems.* 1(1): 3–23.

Zaltsberg, E. 1988. Ground water monitoring in the USSR. *Ground Water Monitoring Review.* Winter 1988. 98–103.

Zonenshain, L.P., Kuzmin, M.I., Natapov, L.M. and Page, B.M. 1990. *Geology of the USSR: A Plate-Tectonic Synthesis.* Washington DC: American Geophysical Union (Geodynamics Series, v.21). 242 pp.

Chapter 5: Egypt

Introduction

The Arab Republic of Egypt occupies the north-east corner of Africa and the Sinai Peninsula of south-west Asia. The country has a total area of about 1 001 450 km². Its maximum length from north to south is about 1085 km and its maximum width near the southern border is about 1255 km. Egypt has 2900 km of coastline, two-thirds of which is on the Red Sea. Less than 4 per cent or about 35 600 km² of the land area of Egypt is settled or is under cultivation. This territory consists of the valley and delta of the Nile and a number of oases. The remaining 96 per cent of the country consists of desert. Egypt has the following physiographic regions (Figure 5-1):

- **The Nile Valley and Delta.** The Nile enters Egypt from Sudan and flows north for about 1545 km to the Mediterranean Sea. The width of the Nile Valley is not uniform, and varies from 0.3 to 24 km from south to north. From Cairo northward, the Nile divides into separate channels that fan out to form the Delta. The two main branches are the Damietta and Rosseta. The Delta is about 160 km long, 250 km wide and has an area of about 22 000 km². A series of four shallow brackish lakes exists along the seaward extremity of the Delta. From east to west these lakes are Manzala, Burullus, Idku and Marut. The Nile Valley and Delta cover an area of about 30 000 km² and have fertile soils, formed by deposits carried down by the Nile. In the Nile Valley the cultivated area mostly consists of a narrow strip of land surrounded by desert on both sides.

- **The Western Desert or Libyan Desert** is part of the Sahara and includes a vast sandy expanse called the Great Sand Sea. This desert makes up more than two-thirds of Egypt. It has an average elevation of 180 m above mean sea-level, with the highest point at 1082 m, and several depressions. The Qattara Depression has an area of about 18 100 km² and reaches a depth of 133 m below sea-level, the lowest point in Africa. The Oases of Siwa,

Kharga, Bahariya, Dakhla and the large lake of Qarun to the north of the town of Fayum are located in this region.

- **The Eastern or Arabian Desert** rises gradually from the Nile Valley to a mountain range bordering the Red Sea. The highest peak of this range is about 2187 m above mean sea-level.
- **The Sinai Peninsula** consists of sandy desert in the north and rugged mountains in the south with a peak of 2642 m, which is the highest point in Egypt.

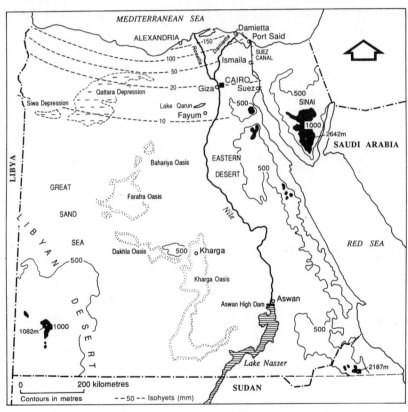

Figure 5-1: Main physiographic features and rainfall map of Egypt (physiography is based on: Department of Technical Co-operation for Development and Economic Commission for Africa, 1988, and isohyets prepared from data provided in FAO, 1984).

In 1800 Egypt had a population of about 2.5 million (Fisher, 1985); by 1987 it had risen to 50 million (Abu-Zeid, 1989a) and to 52 million in 1990, with a growth rate of 2.4 per cent in the 1980s (World Bank, 1992: 268). Nearly 97 per cent of the population lives along the Nile Valley and its Delta (Abu-Zeid, 1989b). Projections suggest that the population will be

62 million in the year 2000 and 86 million by 2025. Rapid population growth is a critical problem in Egypt because of the relatively small area of cultivated land. Indeed, in 1900 there were 1900 m^2 per person, whereas there were about 500 m^2 per person in 1980, indicating an approximate fourfold decrease over a period of 80 years (Kishk, 1986).

About 55 per cent of the country's population lives in rural areas and 35 per cent of the labour force is employed by the agricultural sector. Yet that sector contributes only 23 per cent of the gross products and 21 per cent of the total value of the national product (Amer and Abu-Zeid, 1989). Almost 90 per cent of the country's export earnings are derived directly from agriculture (Abu-Zeid, 1989b).

The sea-level at the Nile Delta coastal zone was approximately 41 m below the present level around 8500 years ago and with melting of glaciers there was a rapid rise in sea-level (Frihy, 1992). However, in anticipation of the rise in sea-level induced by the greenhouse effect and subsidence of the Nile Delta, Milliman et al. (1989) argue that, in the worst case, the sea-level in Egypt could be as much as 3.3 m higher by the year 2100 than it is at present. Subsequently, the country could lose up to 26 per cent of its limited habitable land. However, according to Abu-Zeid (1991), there is no evidence that this will happen. The Water Research Centre in Cairo is carrying out a study to come up with a more accurate estimate.

Rainfall and climate

The main feature of the Egyptian climate is its almost uniform aridity. With the exception of the Nile Valley, its Delta and a few oases, some 96 per cent of the country is desert. It rains in Egypt only from October to May, the summer months of June to September being dry. The most humid area of the country is along the Mediterranean coast (Figure 5-1). Alexandria, which is the wettest part of Egypt, receives only 191 mm of rain per year (FAO, 1984). Precipitation decreases rapidly to the south. Giza receives only 19 mm of rain a year from November to March and on average the number of rainy days is 5 per year. Further south, Aswan has an average annual rainfall of 1 mm per year. In many desert locations it may rain only once in two or three years.

Summer temperatures are extremely high, reaching 38–43°C in Cairo and even 49°C in the southern deserts and the Western Desert. The Mediterranean coast has cooler conditions with 32°C as a maximum. Winter temperatures average 13–21°C, but cold spells occur occasionally. Owing to the large extent of desert, hot dry sand-winds are fairly frequent, particularly in spring, and much damage can be caused to crops. Temperatures have been observed to rise by 20°C in two hours and wind speeds have reached 150 km per hour (Fisher, 1985). Potential evapotranspiration is high in Egypt. It is about 1400 mm in coastal areas of the Mediterranean Sea, and increases toward the south where it reaches about 2120 mm in Aswan (FAO, 1984).

Water resources

Surface water

The surface water resources of Egypt are limited to its share of the Nile River flow. Being the only river in Egypt and providing 97 per cent of its water requirements, the Nile River is the nation's lifeline. The balance of water supply, only 3 per cent, comes from groundwater and rainfall. The Nile is an international river shared by nine countries, Egypt, Sudan, Ethiopia, Uganda, Kenya, Tanzania, Rwanda, Burundi and Zaire (Figure 5-2). Its catchment area covers 2 900 000 km^2, which represents nearly 10 per cent of the land area of Africa (Gasser and Abdou, 1989). The total length of the river and its tributaries amount to 37 205 km, its main lake areas total 81 550 km^2 and its swampy areas stretch over 69 720 km^2 (Mageed, 1985).

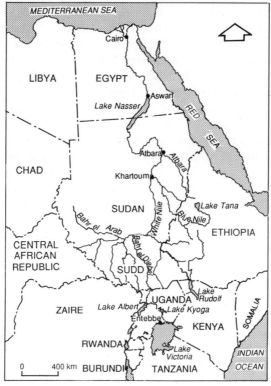

Figure 5-2:   The Nile Basin (Mageed, 1985).

The Nile River which flows northward to the sea is the longest in the world, being about 6700 km long. Its basin can be divided into the following main sub-basins (Abu-Zeid, 1983):

- **The Blue Nile,** which rises in north Ethiopia, has a run-off equal to 57 per cent of the Nile's total. It exhibits great seasonal flow variations and its flow is muddy during rainy seasons.
- **The White Nile,** whose headwater rises south of the equator, has a run-off comprising 29 per cent of the Nile run-off. It has a relatively steady monthly flow and its water is clear.
- **The Atbara River,** which, like the Blue Nile, rises in north Ethiopia, is a flashy river. It is dry for half the year. Its run-off is muddy and constitutes 14 per cent of the Nile run-off.

The discharge of the Nile River is subject to wide seasonal and annual variations. About 80 per cent of the total annual discharge is received during the three months of the flood season (August to October). The average annual flow of the Nile estimated at Aswan is about 84×10^9 m^3 (Abu-Zeid, 1989b). Its annual discharge is highly variable and may reach 151×10^9 m^3 as in 1978 or may drop to 42×10^9 m^3 as in 1913 (Abu-Zeid, 1983). During the flood period the maximum mean monthly discharge ranges from 10 000 to 11 000 m^3 s^{-1}, while in the low seasons it ranges from 400 to 500 m^3 s^{-1} (Shalash, 1990).

According to Abu-Zeid (1983), engineering efforts to control the Nile started during the Pharaonic epoch when King Mina, who ruled Egypt in the First Dynasty (3100–2905 BC) constructed the left bank to protect urban areas. Then he went on constructing canals to carry the Nile waters to lower lands. The concept of annual storage of part of the flood water was first introduced to the Nile system by Egyptians in 1898. The construction of the old Aswan Dam was completed in 1902. Its storage was 1×10^9 m^3. In 1912 the Aswan Dam was heightened to increase its annual storage capacity to 2.5×10^9 m^3. A second heightening of the dam took place in the year 1912, then another in 1933 to increase its capacity to 5.3×10^9 m^3. After the 1952 revolution in Egypt, the idea of constructing the Aswan High Dam (AHD), 6 km south of the Aswan reservoir, received great attention. The construction of the AHD started in 1960, with the financial and technical assistance of the former Soviet Union, and was completed in 1968 (Fisher, 1985).

The AHD is a rockfill dam with a total length of 3600 m and a height of 111 m above the riverbed. Its width is 980 m at the bottom and 40 m at the top. The resulting reservoir (Lake Nasser) is one of the largest artificial lakes in the world. It is about 500 km long and its surface area at normal supply level is 5760 km^2. The 162×10^9 m^3 capacity of Lake Nasser is distributed as follows (Abu-Zeid, 1983; Gasser and Abdou, 1989): 31×10^9 m^3 capacity for sediment deposits over 500 years; 90×10^9 m^3 for live storage; and 41×10^9 m^3 as flood room for protecting against high flood.

On the basis of an agreement reached between Egypt and Sudan in

1959 (Mageed, 1985), the average annual flow of the Nile (84×10^9 m^3) is divided into three parts: 55.5×10^9 m^3 for use by Egypt, 18.5×10^9 m^3 for use by Sudan and the remaining 10×10^9 m^3 is accounted for by evaporation and seepage losses from Lake Nasser. According to Abu-Zeid (1989a), of about 55.5×10^9 m^3 of the Nile water available to Egypt, some 14×10^9 m^3 constitute the net drainage to the sea, 2×10^9 m^3 are lost through evaporation, 32×10^9 m^3 comprise the evapotranspiration requirement and the remainder is used by other water users.

The Nile River carries 134 million tonnes of silt annually, mostly from Atbara and the Blue Nile catchments, which bring down the eroded surface soil of the Ethiopian mountains during the three rainy months of July to September. The maximum silt concentration varies from 3000 to 4000 mg L^{-1} in August and September. The minimum concentration of 25–30 mg L^{-1} is recorded in April and May. This silt built the fertile soils in the flood plain and shaped the hydraulic regime of the river. Currently, 98.5 per cent of the suspended load of the Nile is deposited in Lake Nasser and only 1.5 per cent of very fine matter is released below the dam (Shalash, 1990).

As irrigation drainage water is returned to the Nile River its salinity gradually increases toward lower Egypt. However, El-Guindy (1989) reports that the water quality of the Nile River is good overall for any type of water use with a total salt content generally not exceeding 350 mg L^{-1} TDS.

Construction of the AHD and its environmental and social impacts have been very controversial issues since its completion more than 25 years ago. A series of publications made the dam the most popular environmental problem in the world. It has been condemned for many reasons, including the loss of Mediterranean fishing, and rising salinity and reduction in the fertility of the Nile Valley (Lavergne, 1986). On the other hand, based on 20 years' operational data, Abu-Zeid (1989b) argues that even though the AHD contributed to some environmental problems, it has been clearly beneficial to Egypt's overall development.

Egypt's water resources are now under severe stress and its needs will increase for the foreseeable future. This is also the case in Sudan. Therefore both countries are turning to Ethiopia for their water requirements. However, Ethiopia is not bound by any agreement with Egypt over water supply and intends diverting some 4×10^9 m^3 of Blue Nile waters into its own irrigation projects despite opposition from Egypt and Sudan (Clarke, 1991: 104). The potential for conflict in sharing the Nile's water is also discussed by Smith and Al-Rawahy (1990) and Bulloch and Darwish (1993: 79–123).

Groundwater

According to the Department of Technical Co-operation for Development and Economic Commission for Africa (1988), Egypt has three main aquifer complexes corresponding to the sedimentary system which overlies the

Precambrian crystalline basement rocks. These are:

- **The Nubian Aquifer,** which extends into Libya, Chad and Sudan to form the artesian basin of north-eastern Africa. It consists mainly of sandstone and its thickness ranges from 100 m to 2500 m. This aquifer is artesian and within Egypt's borders its piezometric level declines from an elevation of 500 m in the south-west to approximately sea-level in the north. Groundwater quality of this vast system varies from place to place with a TDS generally below 500 mg L^{-1}. Its age is between 20 000 and 40 000 years (see Margat and Saad, 1984; Alam, 1989; and Thorweihe, 1991 for further information about the Nubian Aquifer).

- **The Middle Limestones** of Cenomanian to Upper Eocene age are composed of fissured limestone which is well karstified in many cases throughout Egypt. This limestone is the source of almost all natural springs in Kharga, Farafra, Siwa, Sinai and almost all the wells in the north of the Qattara Depression.

- **The Upper Oligocene to recent age formation**, which is mainly detrital. These deposits of marine or continental origin form an aquifer along the whole length of the 1000 km of coastline. This aquifer is highly affected by the intrusion of seawater.

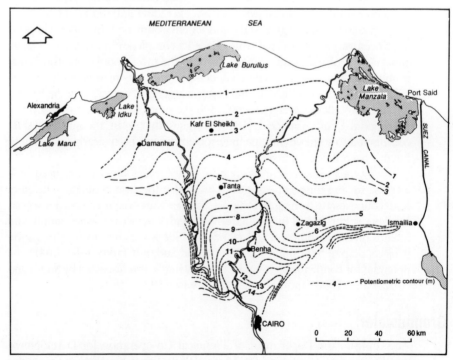

Figure 5-3:   Potentiometric contours of the main aquifer of the Nile Delta (Kashef, 1983).

One of Egypt's very important aquifers is found in the sand and gravel formation deposited in the depression of the Nile and Delta (Figure 5-3). The Nile Valley and Delta constitute one of the world's largest groundwater reservoirs with a capacity of 400×10^9 m^3 (120×10^9 m^3 in Upper Egypt and 280×10^9 m^3 in the Delta). The annual recharge of the aquifer is estimated at 8.6×10^9 m^3 (5.6×10^9 m^3 in the Delta and 3×10^9 m^3 in Upper Egypt). The thickness of the aquifer in the Delta zone increases towards the north, from 100 m in Cairo to 900 m to the Mediterranean coast. The aquifer is recharged by the waters of the Nile and the network of irrigation channels.

In some areas, the piezometric head is so high that groundwater flows in an upward direction. The upward flux depends on the difference between the piezometric head in the aquifer and the watertable level, the thickness of the clay cap and its vertical hydraulic conductivity. Areas with potential occurrence of upward leakage are located mainly in the northern part of the Delta, and in some parts along the Ismailia Canal and along the fringes of the Nile Valley and Delta (Amer et al., 1989a).

The withdrawal from the aquifer is for domestic, agricultural and other purposes. According to Attia and Tuinhof (1989), about 2.6×10^9 m^3 per year of groundwater are extracted at present for domestic, industrial and agricultural uses (Table 5-1). A plan has been drawn up to extract an additional 2.3×10^9 m^3 per year by the year 2000 from the Nile Valley and Delta aquifer.

Table 5-1: Present annual groundwater extraction (in 10^9 m^3).

Area	Domestic and industrial use	Agricultural use	Total
Nile Valley	0.4	0.6	1.0
Nile Delta	1.0	0.6	1.6
Total	1.4	1.2	2.6

Source: Attia and Tuinhof (1989).

Groundwater quality in the Delta varies from place to place and from year to year. Moreover, intrusion of seawater into the Delta aquifer is a major problem, reducing its reserves of freshwater considerably (Kashef, 1983). Investigations have proved that seawater intrusion has extended to a distance of about 130 km from the Mediterranean coast. This would leave only a relatively small triangular zone of freshwater in the south of the Delta. Figure 5-4 shows the salinity map of the Delta aquifer for 1978. According to Abu-Zeid (1989b), the isosalinity contour line of 250 mg L^{-1} (not shown on Figure 5-4) defines the locus of the upper point of the seawater wedge intrusion into the aquifer. When this was compared with the 1968 data it was found that within 10 years this line had moved seaward by approximately

2.5 km, indicating an increase in groundwater gradient toward the sea. If the present water balance of the Nile Delta is unchanged, the isosalinity line of 1000 mg L^{-1} is likely to continue moving seaward, improving the quality of the groundwater in the south of the delta (Abu-Zeid, 1989b). Apart from this improvement, the water quality of the aquifer has been deteriorating since the completion of the Aswan High Dam. This is due to the rise of the piezometric level and the exposure of groundwater to increased evaporation with a consequent increase in its salinity following development of irrigation in the Nile Valley and Delta. Since the construction of the Aswan High Dam the piezometric level has risen by an average of 2 m in Upper Egypt and by 1.5 m in the Delta. In some areas bordering the western part of the Delta, the watertable has risen by 28 m in the north and 20 m in the south, with an increase of 50×10^6 m^3 per year in the volume of water stored in the aquifer (Department of Technical Co-operation for Development and Economic Commission for Africa, 1988).

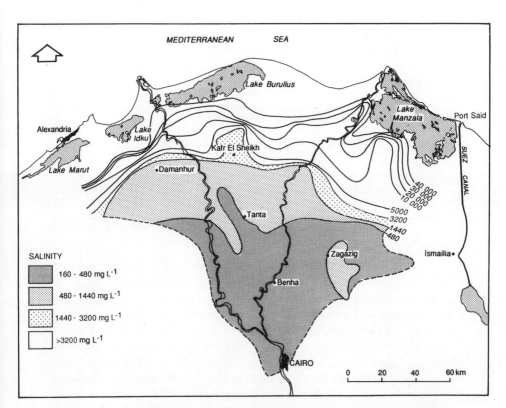

Figure 5-4: Salinity map of the Nile Delta aquifer for 1978 (Abu-Zeid, 1989b).

Land cover

In Egypt, where the major part of the country is covered by desert, arable land is very limited. Of a land area of 99.5 Mha, only 2.56 Mha are arable and under cultivation, 31 000 ha are covered by woodland and forest and the remaining are mainly covered by desert and by infrastructures such as cities and roads (FAO, 1989). However, according to Amer and Abu-Zeid (1989: 47), the current land resources for agricultural production in Egypt consist of about 2.31 Mha of old lands and 0.38 Mha of reclaimed lands, yielding a total of 2.69 Mha. This estimate is slightly (0.13 Mha) higher than the figure provided by FAO (1989).

Irrigation

Egypt has practically no rain and its agriculture depends on irrigation from the Nile River. For thousands of years, Egypt's agriculture was essentially confined to a narrow strip along the Nile and to its Delta. Irrigation has been practised throughout the Nile Valley for more than 5000 years. Until the nineteenth century, the basin system was the main irrigation system. Under the basin system the land was divided into basins of 4200 to 16 800 ha by the construction of dykes. Flood water was let into the compartments to a depth of one or two metres. After 40 to 60 days, when the river level had fallen, these lands were drained and a crop was grown from moisture uptake. The basin system continued to be the only method of irrigation in the country until about 1820, when the cultivation of cotton and sugarcane, requiring perennial irrigation, was first introduced. In 1826 a system of deep canals was developed for the irrigation of Lower Egypt with a view to growing cotton in the region. The deep canals silted up yearly and had to be cleared. To alleviate the situation the Delta Barrage, at the head of the Rosetta and Damietta branches of the Nile about 23 km north of Cairo, was constructed. Towards the end of the last century and during this century, many projects were undertaken and hydraulic structures were constructed for a better control of water supply, conversion of basin irrigation schemes into perennial irrigation systems and the reclamation of swamps and lakes in the north of Lower Egypt (Abu-Zeid, 1983 and 1989a).

The existing irrigation system consists of two dams at Aswan, and seven major barrages on the Nile and its branches (Figure 5-5) which divert water into 31 000 km of main canals. These canals deliver water into smaller canals serving 42 to 210 ha. Seepage from the extensive network of canals causes watertable rise and soil salinity problems (Amer and Abu-Zeid, 1989).

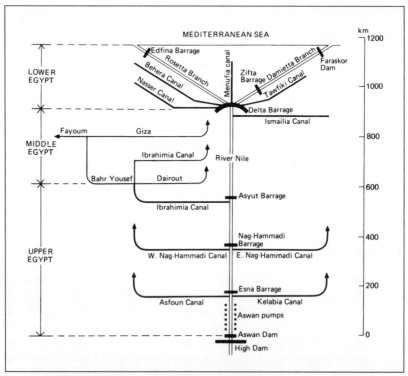

Figure 5-5: Schematic diagram of hydraulic works in the Nile Valley and its Delta (Abu-Zeid, 1989a, courtesy of the author).

In 1879 the area under irrigation was about 2.07 Mha. It increased to 2.26 Mha in 1907 and to 2.48 Mha in 1959 before the construction of the Aswan High Dam (Abu-Zeid, 1989a). Currently all Egyptian agricultural land (2.69 Mha) is irrigated and, since completion of the Aswan High Dam, all irrigation is perennial (Amer and Abu-Zeid, 1989). Irrigation is the main user of water in Egypt. About 31.6×10^9 m³ of surface water (Abu-Zeid, 1989a) and 1.2×10^9 m³ of groundwater (Attia and Tuinhof, 1989) are used for this purpose every year.

Perennial irrigation has made it possible to grow two or three crops each year, where only one grew before. The current cropping intensity is about 200 per cent (two crops per year on average) and may reach a higher level of 300 per cent if the additional quantities of water can be secured (El-Din El-Quosy, 1989). Agricultural products of Egypt consist of wheat, cotton, rice, corn, millet, vegetables and clover.

Abul-Ata (1977) reported that during the period of construction of the Aswan High Dam and of conversion of basin irrigation to perennial irrigation, the Ministry of Irrigation devised a system for installing drainage facilities, but its application was not seen to be feasible. However, soon after perennial irrigation was started in the former basins, the rise of watertables made the drainage problem and soil degradation priority issues.

Salinity

Irrigated land salinity

Soil salinity and alkalinity are major problems for Egyptian agriculture. Egypt's civilisation flourished for thousands of years, thanks to the annual flooding of the River Nile. The reason that these annual floodings did not cause rising watertables and accompanying soil salinity problems, as happened in Mesopotamia, is the high capacity for natural drainage in Egypt. The huge body of coarse sands underlying the Delta is capable of transmitting large quantities of groundwater (Amer and Abu-Zeid, 1989). Also, under the basin system of irrigation, salts were washed away in solution from the land each year by the flood waters.

The introduction of perennial irrigation resulted in a gradual but general rise in the watertable and accumulation of salt in the soil, because the natural drainage cannot cope any more with the increased groundwater recharge due to the increased canal and field percolation losses. Due to inadequate drainage, the cotton crop failed in 1909 and this led to strictly reducing the amount of water put on land to the minimum required for cropping (Framji et al., 1981: 342).

Abu-Zeid (1989b) reported that inadequate drainage has been a serious problem since 1938 and the need for farm drainage became more serious because of the following factors: the horizontal expansion of agriculture into sandy or light soils, and the fact that much of this land lies on the Nile Valley fringes of higher elevation; seepage from new irrigation systems; increased cropping intensities from one to two crops a year; and an increase in rice and sugarcane growing areas requiring high water application.

El-Gabaly (1977) estimated that about 33 per cent or 0.8 Mha is affected by salinity and poor drainage to varying degrees, with a loss in crop production estimated at 30 per cent of the potential production. The area affected by waterlogging and salinity increased at an alarming rate during the 1970s following the completion of the Aswan High Dam (Abu-Zeid, 1983). According to Kishk (1986), in 1982 almost all irrigated areas in Egypt (2.4 Mha) were found to be potentially salt-affected and at least 1.2 Mha of this area was more or less affected by salinity. The latest estimate shows that 0.88 Mha or about 33 per cent of the 2.69 Mha irrigated land is salt-affected (Abu-Zeid, 1991).

Figure 5-6 shows the distribution of soil salinity in the Nile Delta. It shows that:

- In the southern part of the Delta the soils are generally non-saline and the salinity varies between 1000 and 2000 µS cm^{-1}, while the soil salinity has been in equilibrium following the construction of the tile drainage system.
- The soils located between the elevation contour lines of 3 and 7 m above mean sea-level (MSL) have salinities ranging from 2000 to 4000 µS cm^{-1}. Patches of different size and different degree of salinity are scattered throughout this part of the Nile Delta, especially close to irrigation canals and where drainage is either non-existent or inadequate.
- The major portion of the salt-affected soils is found in the regions which extend from the east to west parallel to the sea coast and up to the contour line of 3 m above MSL. Salt accumulation has resulted in these soils from saline water intrusion from the sea, the north lakes and tidal marshes. This has been accelerated by the flat topography and low land level in that area, the closeness of its watertable to the soil surface and the absence of an adequate drainage system.

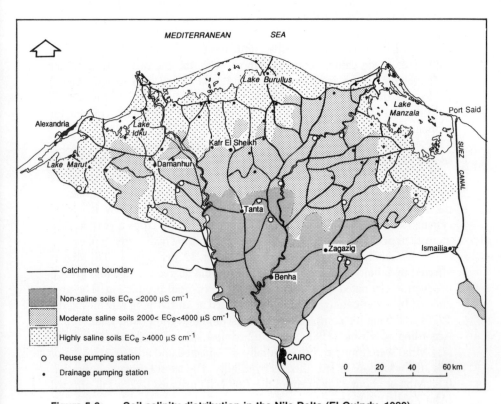

Figure 5-6:   Soil salinity distribution in the Nile Delta (El-Guindy, 1989).

According to El-Lakany et al. (1986), the reduction in agricultural production on salt-affected soils is about 30 per cent.

River salinity

The Nile River water is of high quality for irrigation, industrial and domestic water supply. According to Meybeck et al. (1989: 243–252), salinity is generally low in the headwaters of the Nile. Lake Victora at Entebbe (Figure 5-2) has a salinity of 48.5 mg L^{-1} and EC value of 95 µS cm^{-1}. Lake Kyoga has a salinity of 61.2 mg L^{-1} and EC value of 97 µS cm^{-1}. At Lake Albert salinity is about 393 mg L^{-1} with an EC value of 613 µS cm^{-1}. Since Lake Albert produces 15.1 per cent of the water of the White Nile, the salinity of the river at Sudd is about 111 mg L^{-1}. When the Nile reaches Khartoum its salinity increases to 172 mg L^{-1}. The salinity of the Blue Nile at Khartoum is about 147 mg L^{-1} with an EC value of 255 µS cm^{-1}.

Khalil and Hanna (1984) studied variations in the chemical constituents of Nile water at Giza through the 25 years from 1954 to 1979. Water samples were collected at Giza during 1954, 1964, 1973 and 1979. Then samples were analysed for pH, TDS, chemical constituents and suspended matter. Analysis of the results shows that before the construction of the AHD, TDS varied from an average of 184 mg L^{-1} in 1954 to 195 mg L^{-1} in 1964. After the construction of the AHD, the TDS continued to increase from an average of 209 mg L^{-1} in 1973 to 239 mg L^{-1} in 1979. These data show that the total salt content of Nile water increased by 29 per cent during the entire 25 years of study. The increase has been mainly due to the seepage of drainage water from the cultivated land of the Nile Valley.

According to Meybeck et al. (1989), the pattern of salinity of the Nile River has reversed since the construction of the AHD. Initially, the highest concentrations occurred during summer, but following construction of the AHD the lowest concentration occurs during this period (the summer), which coincides with the highest discharge period from the AHD.

The salt load of the Nile River increases sharply between the Delta Barrage and its outlets to the sea. By considering the quantity of water passing the AHD (55 × 10^9 m^3) and its salinity (200 mg L^{-1}), El-Din El-Quosy (1989) calculates the salt load of the Nile at this point to be about 11 million tonnes per year. The quantity of water passing the Delta Barrage is approximately 35 × 10^9 m^3 per year with a salinity of about 300 mg L^{-1}. Therefore the salt load of this water would be 10.5 million tonnes per year. This means that no salt accumulation takes place between the AHD and the Delta Barrage. The quantity of water leaving the Delta is 14 × 10^9 m^3 per year at an average salinity of 2300 mg L^{-1}, bringing the salt load of this water to some 32 million tonnes per year. The 21 million tonnes annual increase in salt load of the Nile River between the Delta Barrage and the Mediterranean Sea is too big to be attributed to the leaching of agricultural land. This difference might be caused by other factors such as saline upward leakage or seawater intrusion. No firm estimates of such phenomena have been obtained so far (El-Din El-Quosy, 1989).

Management options

Surface and subsurface drainage

In order to manage the problems of waterlogging, salinisation and the subsequent production loss of agricultural lands, the Egyptian government has given priority to land drainage. According to Amer and Abu-Zeid (1989), open drains have been excavated and pumping stations have been constructed since 1933. The drained area was about 0.8 Mha in 1952 and had increased to 1.34 Mha by 1966. The system consists of: open and covered drains (collectors and laterals); open collector drains that comprise both branch and sub-branch drains; a carrier network of main drains that receives water from the collector system and conveys it to the Nile or discharges it into the coastal lakes or the Mediterranean Sea; and pumping stations which transfer drainage water to the coastal lakes or to the Mediterranean Sea.

An arrangement was made to provide open drains such that the water level in the drains be at 1.5 m below ground surface, in regions having perennial irrigation systems. In 1942 it was observed that this level of 1.5 m was not enough. It was decided to keep the water-level in the open drains at 2.5 m below ground surface. This policy was enforced together with a new policy for drainage in 1958. The length of open drains increased from 12 200 km in 1952 to about 13 300 km in 1960 and 16 200 km in 1965. This main system was almost completed and since then its length has not increased much. Figure 5-7 shows a general view of the open drainage network in the Delta. Following the implementation of the 1958 drainage policy in several successive Five Year Plans, a comprehensive drainage program was initiated in the early 1970s. Table 5-2 shows the areas provided with open drains.

Table 5-2: Areas provided with open drains in the Delta and Upper Egypt (areas in 1000 ha).

Total area	Executed up to 1982	Remaining areas	First Five Year Plan 82/83–86/87	Second Five Year Plan 87/88–91/92
2768.64	1814.82	953.82	600.60	353.22

Source: Amer and Abu-Zeid (1989).

Drainage by gravity to the Nile, the coastal lakes and the Mediterranean Sea was the method prevailing during the first quarter of this century. This method worked well in high-lying lands, but not in low-lying areas. For such areas, pumping stations were envisaged. In 1987 the number of pumping stations was 129 with a capacity of 2459 m³ s⁻¹ and serving a total area of 2.77 Mha (Amer and Abu-Zeid, 1989).

Apart from the development of surface drainage, field drainage has been developed significantly. Since 1922 the introduction of field drainage was foreseen and studies carried out. The 1938 studies in 15 fields scattered all over the country resulted in the initiation of tile drainage networks. From 1942 to 1948 an area of 8000 ha was tile drained. The tile drains were of the clay type, 50 cm long and were manually installed. About 1952-53 the total area provided with tile drainage reached about 20 000 ha. During the period 1960-65, about 105 000 ha were provided with tile drains (Amer and Abu-Zeid, 1989). According to Abu-Zeid (1989b), the total area requiring subsurface drainage is estimated at 2.14 Mha. By the end of 1988 pipe drains had been installed in some 1.34 Mha and in the years to come, an additional 0.8 Mha will be provided with pipe drainage systems (Amer and de Ridder, 1989).

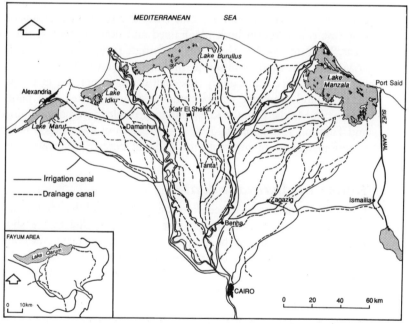

Figure 5-7:   Open drainage network in the Nile Delta (Amer and Abu-Zeid, 1989).

The installation of subsurface drainage systems in the Nile Delta is primarily meant to ensure proper drainage and soil conditions for such crops as cotton, maize and wheat. In large parts of the Delta, however, rice is grown in rotation with these crops. As the rice fields are commonly submerged throughout the growing season, the introduction of a subsurface drainage system in these areas gives rise to severe problems (El-Guindy and Risseeuw, 1987) such as:

- High percolation losses in rice fields to these drainage systems. If the system is left open, considerable amounts of water are lost through the tile drains, resulting in extremely high irrigation water applications;
- Farmers block the drainage systems (collectors) to reduce such losses, and this often causes severe clogging of the collectors;
- Over-pressure developing in the upstream parts of the laterals/collectors, due to high discharges from downstream rice growing areas. This seriously limits watertable control in the area served by the collector, and results in poor drainage of the crops grown together with rice in the same collector area. The situation becomes worse if the rice fields are located in the downstream part of the collector area, and even more serious when the collector is blocked in order to reduce irrigation water applications for rice crops.

El-Guindy and Risseeuw (1987) describe the problem of water management of rice fields in the Nile Delta and present the results of a study that was conducted in both tile-drained and non-tile-drained rice growing areas. Their findings have been used to formulate recommendations for better design and operation of subsurface drainage systems in the rice growing areas.

Subsurface drainage by tube-well is not in practice in Egypt, but its technical and economic feasibility has been investigated in a pilot area of 18 km^2 in Middle Egypt, and encouraging results have been obtained (Attia and Tuinhof, 1989; Shahin, 1987). Application of this method will alleviate waterlogging and salinity problems, and will increase freshwater supplies for irrigation.

Reuse of drainage water

Egypt is fortunate to have access to coastal lakes and the Mediterranean Sea for the disposal of its drainage water. However, due to the shortage of irrigation water, reuse of the drainage water is in practice in the country.

The monthly flow of drainage water to the sea is highest during the period of July–October, with a monthly rate of $1.2–1.6 \times 10^9$ m^3 (Abu-Zeid and Abdel-Dayem, 1991). At this time more than 400 000 ha of rice fields in the Nile Delta remain ponded and are continuously replenished with irrigation water to compensate for the higher summer evapotranspiration and excessive deep percolation. During the same season the rest of the Nile Delta is mainly cultivated with maize and cotton. The quantity of drainage water decreases in November after harvesting of the summer crops. Part of the agricultural area remains fallow (uncultivated) during November before farmers start land preparation for the next crop. The pre-irrigation of winter crops is mainly responsible for a drainage water flow increase in December. The drainage water flow during the rest of winter (from January till May) is generally less than in summer. It becomes too low during February, during which irrigation is not practised for three weeks and the irrigation system remains shut down for maintenance. The drainage flow during February (0.5–0.7×10^9 m^3) is mainly due to subsurface flow of groundwater to the drains (Abu-Zeid and Abdel-Dayem, 1991).

The salinity of drainage water varies from location to location depending on soil type, texture, irrigation practices, cropping patterns and other special factors like seawater intrusion. Even at each individual location, salinity of drainage water varies every month, with the highest observed values in February, when the irrigation system is closed for annual maintenance, and the lowest during the period July–October, influenced by the intensive irrigation and significant increase in the volume of drainage water. The average salinity of drainage water in the Middle Delta for 1986 was mainly between 750 and 2000 mg L^{-1}, but in the area close to Lake Burullus and the Mediterranean Sea it was in the range of 3000 to 8000 mg L^{-1} (Abu-Zeid and Biswas, 1990). Comparison of these data with those provided by El-Guindy (1989) for 1981 shows an increase in the salinity of drainage water during the period 1981–86, which could be due to the development of drainage facilities in the Delta.

The amount of leached salts is a function of the quantity of leaching water and the concentration of salts in the soil. Abdel-Dayem et al. (1989) have estimated the annual leaching of salts by the drainage water in Mashtul Pilot area located at 70 km north-east of Cairo to be about 4.6 to 7.6 tonnes per hectare.

At present about 13.5×10^9 m^3 of drainage water of reasonable quality flows unused every year to the Mediterranean Sea and the coastal lakes (El-Din El-Quosy, 1989). About 65 per cent of drainage water released to the sea or lakes (9×10^9 m^3 yr^{-1}) has a salinity of below 2000 mg L^{-1}. Currently an annual quantity of about 2.9×10^9 m^3 of drainage water with an overall average (weighted) salinity of 930 mg L^{-1} is used in existing reuse projects. Drainage water is mixed with the low salinity canal water to make a mixture with a salinity range of 500–600 mg L^{-1} before being used for irrigation. Apart from the official reuse of drainage water, it is believed that a quantity of 1.5 to 2×10^9 m^3 of drainage water is being reused every year by the farmers on a non-official basis. With the completion of the reuse projects under construction an extra quantity of 2.75×10^9 m^3 of drainage water will be used for irrigation, which brings the total to 5.75×10^9 m^3 (about 3×10^9 m^3 are already reused every year). Also, several plans are under development for the short-term and long-term reuse of the remaining drainage water.

Revegetation

Revegetation of the salt-affected land has been tested in Egypt. Growing native and exotic salt-tolerant shrubs and trees has been tested and found to be advisable (El-Lakany et al., 1986). One of the most palatable halophytes is the saltbush *Atriplex nummularia*. This species, introduced from Australia during the 1950s, has been grown successfully on the north-western coastal belt of Egypt, where it was found to be superior to *Atriplex halimus*, a native of the Mediterranean region. Exotic tree species, such as *Acacia saligna*, *Casuarina glauca* and *Eucalyptus camaldulensis*, have also been tried successfully. Provenance in some species can be critical and the recently introduced *Eucalyptus camaldulensis* has been found in a trial near Alexandria to be superior to species already established locally.

Economic damage

No estimates of the damage to the Egyptian economy from waterlogging, salinisation and production loss are available. However, according to Amer et al. (1989b), the Egyptian Government spends more than LE70 million (US$30 million) annually on drainage projects. This is a sizeable proportion of the country's investment effort in agriculture. Most of this effort is being financed from domestic funds, and the remainder from external sources, principally the World Bank. However, economic analysis shows that such investment is justified and for the most common crops, the annual average net income is definitely higher than for those grown without drainage. The additional net return due to drainage, after deducting the annual tile drainage charges to be paid by the farmers, is estimated to be in the range of: 11–14 per cent in the Nile Delta; 9–16 per cent in the cotton growing areas; and 28–44 per cent in sugarcane growing areas. It should be noted that farmers pay the cost of drainage facilities implemented by the government on their land, interest free over a period of 20 years, beginning 3 years after the installation of the tile drains.

References

Abdel-Dayem, S., El-Atfy, H.E., Ritzema, H.P. and Amer, M.H. 1989. Pilot areas and drainage technology. In. Amer, M.H. and de Ridder, N.A. eds. *Land Drainage in Egypt*. Cairo: Drainage Research Institute. 103–161.

Abul-Ata, A.A. 1977. The conversion of basin irrigation to perennial systems in Egypt. In. Worthington, E.B. ed. *Arid Land Irrigation in Developing Countries: Environmental Problems and Effects*. Oxford: Pergamon Press. 99–105.

Abu-Zeid, M. 1983. The River Nile: Main water transfer projects in Egypt and impacts on Egyptian agriculture. In. Biswas, A.K., Dakang, Z., Nickum, J.E. and Changming, L. eds. *Long-Distance Water Transfer: A Chinese Case Study and International Experiences*. Dublin: Tycooly International. (Published for the United Nations University). 15–34.

Abu-Zeid, M. 1989a. History and future role of water development and management in Egypt. In. Amer, M.H. and de Ridder, N.A. eds. *Land Drainage in Egypt*. Cairo: Drainage Research Institute. 23–42.

Abu-Zeid, M. 1989b. Environmental impacts of the Aswan High Dam: A case study. *Water Resources Development*. 5(3): 147–157.

Abu-Zeid, M. 1991. Cairo: Water Research Centre. (Personal communication).

Abu-Zeid, M. and Abdel-Dayem, S. 1991. Variation and trends in agricultural drainage water reuse in Egypt. *Water International*. 16(4): 247–253.

Abu-Zeid, M. and Biswas, A.K. 1990. Impacts of agriculture on water quality. *Water International*. 15(3): 160–167.

Alam, M. 1989. Water resources of the Middle East and North Africa with particular reference to deep artesian ground water resources of the area. *Water International.* 14(3): 122–127.

Amer, M.H. and Abu-Zeid, M. 1989. History of land drainage in Egypt. In. Amer, M.H. and de Ridder, N.A. eds. *Land Drainage in Egypt.* Cairo: Drainage Research Institute. 43–66.

Amer, M.H. and de Ridder, N.A. eds. 1989. *Land Drainage in Egypt.* Cairo: Drainage Research Institute. 377 pp.

Amer, M.H., Osman, M.A., Abdel Dayem, S., and Makhlouf, M.A. 1989a. Recent developments of land drainage in Egypt. In. Amer, M.H. and de Ridder, N.A. eds. *Land Drainage in Egypt.* Cairo: Drainage Research Institute. 67–93.

Amer, M.H., El-Guindy, S. and Rafla, W. 1989b. Economic justification of drainage projects in Egypt. In. Amer, M.H. and de Ridder, N.A. eds. *Land Drainage in Egypt.* Cairo: Drainage Research Institute. 327–339.

Attia, F.A.R. and Tuinhof, A. 1989. Feasibility of tube-well drainage in the Nile Valley. In. Amer, M.H. and de Ridder, N.A. eds. *Land Drainage in Egypt.* Cairo: Drainage Research Institute. 303–325.

Bulloch, J. and Darwish, A. 1993. *Water Wars: Coming Conflicts in the Middle East.* London: Victor Gollancz. 224 pp.

Clarke, R. 1991. *Water: The International Crisis.* London. Earthscan Publications Ltd. 193 pp.

Department of Technical Co-operation for Development and Economic Commission for Africa. 1988. *Ground Water in North and West Africa.* (Natural Resources/Water Series no. 18). New York: United Nations. 405 pp.

El-Din El-Quosy, D. 1989. Drainage water re-use projects in the Nile Delta: The past, the present and the future. In. Amer, M.H. and de Ridder, N.A. eds. *Land Drainage in Egypt.* Cairo: Drainage Research Institute. 163–174.

El-Gabaly, M.M. 1977. Water in arid agriculture: Salinity and waterlogging in the Near-East region. *Ambio.* 6(1): 36–39.

El-Guindy, S. 1989. Quality of drainage water in the Nile Delta. In. Amer, M.H. and de Ridder, N.A. eds. *Land Drainage in Egypt.* Cairo: Drainage Research Institute. 189–206.

El-Guindy, S. and Risseeuw, I.A. 1987. *Research on Water Management of Rice Fields in the Nile Delta, Egypt.* Wageningen: International Institute for Land Reclamation and Improvement. (ILRI Publication 41). 72 pp.

El-Lakany, M.H., Hassan, M.N., Ahmed, A.M. and Mounir, M. 1986. Salt affected soils and salt marshes in Egypt; their possible use for forage and fuel production. *Reclamation and Revegetation Research.* 5: 49–58.

Fisher, W.B. 1985. Egypt: Physical and social geography. In. *The Middle East and North Africa 1986.* Thirty-second edition. London: Europa Publications Limited. 337–379.

Food and Agriculture Organization of the United Nations (FAO). 1984. *Agroclimatological Data: Africa.* (Plant Production and Protection Series no. 22). Rome: FAO. v.1. (Countries north of the equator.) About 350 pp.

Food and Agriculture Organization of the United Nations (FAO). 1989. *Production Yearbook.* Rome: FAO. v.42. 350 pp.

Framji, K.K., Garg, B.C. and Luthra, S.D.L. 1981. *Irrigation and Drainage in the World: A Global Review.* Third edition. New Delhi: International Commission on Irrigation and Drainage. v.I. 491 pp.

Frihy, O.E. 1992. Holocene sea-level changes at the Nile Delta coastal zone of Egypt. *GeoJournal.* 26(3): 389–394.

Gasser, M.M. and Abdou, M.I. 1989. Nile water management and the Aswan High Dam. *Water Resources Development.* 5(1): 45–49.

Kashef, A.A.I. 1983. Salt-water intrusion in the Nile Delta. *Ground Water.* 21(2): 160–167.

Khalil, J.B. and Hanna, F.S. 1984. Changes in the quality of Nile water in Egypt during the twenty-five years, 1954–1979. *Irrigation Science.* 5: 1–13.

Kishk, M.A. 1986. Land degradation in the Nile Valley. *Ambio.* 15(4): 226–230.

Lavergne, M. 1986. The seven deadly sins of Egypt's Aswan High Dam. In. Goldsmith, E. and Hildyard, N. eds. *The Social and Environmental Effects of Large Dams.* Camelford: Wadebridge Ecological Centre. v.2 (Case studies). 181–183.

Mageed, Y.A. 1985. The integrated river basin development: The challenges to the Nile Basin countries. In. Lundqvist, J., Lohm, U. and Falkenmark, M. eds. *Strategies for River Basin Management: Environmental Integration of Land and Water in a River Basin.* Dordrecht: D. Reidel Publishing Company. 151–160.

Margat, J. and Saad, K.F. 1984. Deep-lying aquifers: Water mines under the desert? *Nature and Resources.* 20(2): 7–13.

Meybeck, M., Chapman, D.V. and Helmer, R. eds. 1989. *Global Freshwater Quality: A First Assessment.* Oxford: Blackwell Reference. (Published on behalf of the World Health

Organization and the United Nations Environment Programme). 306 pp.

Milliman, J.D., Broadus, J.M. and Gable, F. 1989. Environmental and economic implication of rising sea-level and subsiding deltas: The Nile and Bengal examples. *Ambio*. 18(6): 340–345.

Shahin, M. 1987. Technical feasibility of vertical drainage of irrigated lands in Egypt. In. *Proceedings of the International Groundwater Conference 1987: Groundwater and the Environment*. Malaysia: Universiti Kebangsaan. B16–B25.

Shalash, S. 1990. Effect of long-term capacity reservoir on large alluvial river with special reference to Aswan High Dam of Egypt. In. *The Impact of Large Water Projects on the Environment*. Paris: UNESCO. 63–77.

Smith, S.E. and Al-Rawahy, H.M. 1990. The Blue Nile: Potential for conflict and alternatives for meeting future demands. *Water International*. 15(4): 217–222.

Thorweihe, U. 1991. Hydraulic characteristics in the Nubian Aquifer system of the eastern Sahara. In. *Proceedings of the International Conference on Groundwater in Large Sedimentary Basins*. Perth, Western Australia, 9-13 July 1990. (Australian Water Resources Council Conference Series no. 20). Canberra: Australian Government Publishing Service. 278–287.

World Bank. 1992. *World Development Report 1992: Development and the Environment*. New York: Oxford University Press. 308 pp.

Chapter 6: India

Introduction

India is a vast country covering an area of about 3 287 800 km². It measures about 2980 km from east to west and about 3220 km from north to south with a coastline of 5690 km. India has the following principal physiographic regions (Figure 6-1).

Figure 6-1: Major physiographic features of mainland India.

- **The Himalaya Mountains** run nearly from north-west to south-east along the northern frontiers of India.
- **The Indo-Gangetic Plains,** in the south and parallel to the Himalaya Mountains, comprise a belt of flat alluvial lowlands. The average elevation is less than 150 m above mean sea-level (MSL). These plains are watered by three distinct river systems — the Indus, the Ganga and the Brahmaputra. The Indus River drains into the Arabian Sea, while the other two have their outfall in the Bay of Bengal.
- **The Great Indian Desert** is an arid region in the north-western part of the country and contains large tracts of fertile soils. The Indira Gandhi Canal Project diverts the Indus River tributaries to these areas. It transforms the desert into a land of prosperity by insured irrigation supplies.
- **The Deccan Plateau,** in the south of the Indo-Gangetic Plains, is a vast triangular tableland occupying most of the Indian Peninsula. It is a generally rocky and uneven plateau, divided into natural regions by low mountain ranges and deep valleys. The elevation of the plateau ranges mainly from about 305 to 915 m, with a highest peak of 2134 m.

Figure 6-2:   Political division in India (Muthiah, 1987).

- **The Coastal Mountain Belts** on the eastern and western sides of the triangular Deccan Plateau converge in south India. These ranges are named the Eastern and Western Ghats and their elevations above MSL are about 457 and 914 m, respectively. The Western Ghats rise sharply to elevations of 1000 to 1300 m, but some peaks are as high as 2438 m. The Eastern Ghats are less prominent and rise to elevations of 800 to 1000 m. The deltas along the Eastern Ghats are fairly extensive in area. Being flat and fertile, they are irrigated by some of the largest irrigation systems in the country. The deltas of the west-flowing rivers are not so extensive and the western coastal belt is comparatively narrow.

India supported a population of 850 million in 1990, with a growth rate of 2.1 per cent over the period 1980–90 (World Bank, 1992: 268). Politically, it consists of 25 states and seven union territories (Figure 6-2).

Rainfall and climate

Because of its size, peninsularity, topography and geographical position, climatic conditions in India are diverse on both a seasonal and regional basis. The diversity ranges from tropical wet to arid and semiarid. A tropical wet climate is characteristic of the west coast, especially in Kerala, Karnataka and Goa, where both temperature and rainfall are high. A tropical wet and dry climate is mainly found along much of the east coast and in the interior of the northern peninsula. A humid subtropical type of climate is found in the entire Indo-Gangetic Plains. There are humid subtropical regions in the extreme north-east also, where the rainfall is highest. Areas with arid and semiarid climate are mainly located in the west and north-west of India.

The Indian economy depends almost entirely on the monsoons which bring the seasonal rains, essentially for agriculture, general irrigation needs and power production. Although considerable areas of the country are provided for by irrigation, nearly 80 per cent of the cropped area still depends on seasonal rainfall. India has two monsoons, the south-west (June to September) and the north-east (October to December).

The average annual rainfall of India is 1250 mm, the highest for a land of such size anywhere in the world (Muthiah, 1987). However, the rainfall is highly variable in space and time, with the heaviest rains occurring over the north-eastern states and along the west coast (Figure 6-3). Cherrapunji, in the state of Meghalaya, receives a record rainfall of nearly 9000 mm a year. In contrast, some districts in south-west Rajasthan receive hardly 150 mm a year. According to Singh (1979), 30 per cent of the country receives an average annual rainfall of 0–750 mm, 42 per cent receives 750–1250 mm, 20 per cent receives 1250–2000 mm and 8 per cent receives more than 2000 mm.

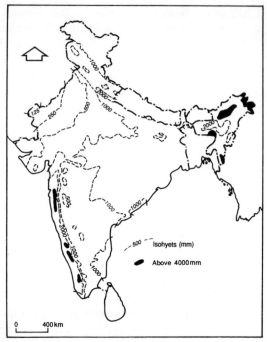

Figure 6-3: Average annual rainfall of India (Johnson, 1983).

Table 6-1 shows the distribution of arid lands in India. West Rajasthan is the principal arid zone in the country, constituting 62 per cent of arid zones. This is followed by Gujarat, Punjab and Haryana. These are all high temperature arid zones. The only cold arid zone of the country is in the extreme northern portions of the Jammu and Kashmir State, covering 70 300 km^2. The climatic conditions over this arid zone are partly due to its very high altitude (2400 to 4300 m). This cold arid zone can be considered a continuation of the cold deserts of China.

Table 6-1: Distribution of arid lands in India.

State	Area of the arid zones (km^2)	Percentage of the total arid zone in India
Rajasthan	196 150	61.9
Gujarat	62 180	19.6
Punjab	14 510	4.6
Haryana	12 840	4.0
Maharashtra	1 290	0.4
Karnataka	8 570	2.7
Andhra Pradesh	21 550	6.8
Total	317 090	100.0

Source: Krishnan (1971) and Mann (1986).

Water resources

Surface water

India is divided into 14 major river systems (Figure 6-4, Table 6-2 and Figure 6-14). The average annual flow of all Indian rivers is about 1897×10^9 m^3. The most prominent rivers are the Brahmaputra, Ganga and the Indus with average annual flows of 591, 557 and 73 billion m^3, respectively. The Brahmaputra rises in Tibet and runs eastward parallel to the Himalaya. In India it flows westward for about 720 km before entering Bangladesh. Its catchment in India is about 258 000 km^2, which is smaller than the catchments of the Ganga and the Indus (Mukherjee, 1985). The Ganga River is 2525 km long and has a catchment area of 760 000 km^2 which covers virtually the entire area of northern India. The Ganga River has a large number of tributaries. Some of them are of Himalaya origin, while the other tributaries such as Chambal and Banas originate in the Deccan Plateau. The Indus River has its origin in the Tibet Plateau and has a catchment area of more than 468 000 km^2 spread over China, India (320 180 km^2) and Pakistan. The main tributaries of the Indus River are the Jhelum, Chenab, Ravi, Beas and Sutlej. Other Indian river systems include the Mahanadi,

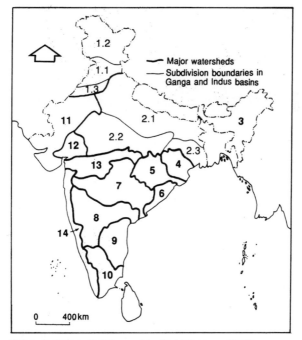

Figure 6-4: Major drainage divisions of India (Johnson, 1983).

Godavari, Krishna, Sabarmati and Narmada. Table 6-2 provides more information on the surface water resources of India and the available storage capacities on the various river systems. Although Table 6-2 shows that the total storage capacity of India was 139×10^9 in 1983, recent statistics (Murty, 1990) indicate that over 600 storage dams of various sizes, accounting for a storage capacity of 160×10^9 m^3, have been constructed since 1950 for irrigation, flood control and power generation.

Table 6-2: Surface water resources of India.

River system[a]	Annual flow (10^9 m^3)	Percentage of total flow (%)	Storage capacity (10^9 m^3)
1. Indus tributaries in India	73	3.8	14
1.1 Ravi, Beas, Sutlej	40	2.1	-
1.2 Jhelum, Chenab	31	1.6	-
1.3 Ghaggar	2	-	-
2. Ganga	557	29.2	31
2.1 Himalayan rivers	424	22.4	7
2.2 Right bank tributaries	88	4.6	20
2.3 Damodar and others	45	2.4	4
3. Brahmaputra	591	31.2	-
4. Between Ganga and Mahanadi	44	2.3	4
5. Mahanadi	71	3.7	8
6. Between Mahanadi and Godavari	17	0.9	-
7. Godavari	118	6.2	15
8. Krishna	63	3.3	30
9. Between Krishna and Cauvery	25	1.3	2
10. Cauvery and southwards	28	1.5	5
11. Rivers of Saurashtra and Kachchh	12	0.6	-
12. Mahi, Sabarmati, and others	16	0.8	5
13. Narmada and Tapi	64	3.4	11
14. West Coast rivers	218	11.5	14
Total	1 897	100.0	139

(a): See Figures 6-4 and 6-14 for the location of these river systems.   *Source:* Johnson (1983).

In terms of water use, irrigation is the major water user in the country. In 1990, out of a total water use of 510×10^9 m^3, about 439×10^9 m^3 or 86 per cent was used for irrigation. Surface water contributed 71 per cent (360×10^9 m^3) to the water supply, while the contribution of groundwater was about 29 per cent (150×10^9 m^3). Estimates show that water demand will increase almost twofold to a level of 1050×10^9 m^3 yr^{-1} from 1990 to 2025 (Table 6-3).

Table 6-3: The current and projected water demand and source of supply for India from 1900 to 2025 (in 10^9 m^3 yr^{-1}).

	1990	2000	2025
Water demand:			
Irrigation	439	630	770
Domestic	25	33	52
Energy	5	27	71
Industry	15	30	120
Others	26	30	37
Total	510	750	1050
Source of supply:			
Surface water	360	500	700
Groundwater	150	250	350

Source: Navalawala (1992).

Groundwater

Groundwater has been utilised for irrigation in India for many centuries, but until recently its exploitation was limited mainly to shallow dug-wells. In the mid 1930s, tube-well schemes tapping deep aquifers were taken up in Uttar Pradesh by the state government. Since then, similar developments have taken place in other states. Groundwater resources exist in a wide variety of formations from Precambrian crystalline and sedimentary rocks to Quaternary alluvium. Major hydrogeological features of these formations are as follows (Department of Technical Co-operation for Development, 1986):

- **Precambrian crystalline and sedimentary rocks.** The major part of peninsular India is occupied by igneous, metamorphic and sedimentary rocks of Precambrian age. These rocks are devoid of primary porosity, but have been rendered porous by weathering and fracturing. The weathered zone is usually extensive to a depth of 10 to 20 m. Groundwater occurs in weathered zones, joints and fissures. Limestones are subject to solution activity and are productive sources of water supply in some areas.
- **Palaeozoic group.** Palaezoic formations are mainly developed at high altitudes. Except in localised areas, groundwater development from this group is not very promising.
- **Mesozoic group.** Studies carried out in Rajasthan and Gujarat indicate that important aquifers exist in sandstones of Jurassic age.
- **Deccan Traps.** These consist of basalts and do not possess any primary porosity except in certain areas. Groundwater occurs in the weathered and jointed portions of upper horizons.

- **Tertiary formations.** Tertiary rocks comprise semi-consolidated to poorly consolidated conglomerates, sandstones and shales. Where poorly cemented, the conglomerates and sandstones are moderately to highly permeable and productive.
- **Quaternary alluvium.** Quaternary alluvium occupies a wide stretch in the Ganga, Brahmaputra and Indus valleys. The thickness of sediments exceeds 600 m. Such formations also occur along the east coast, in the Gujarat Basin and in narrow strips along the west coast. In Rajasthan much of the Quaternary deposits consist of blown sand and loess. The Sindhu-Ganga-Brahmaputra alluvium, adjacent to the Himalayan foothills, is very coarse-grained becoming fine-grained to the south.

Systematic investigations for assessment of groundwater resources have been in progress for the past 45 years. Various estimates have been made of the country's groundwater potential. As reported by Ghosh and Phadtare (1990), the Central Groundwater Board (CGWB) of India has estimated the annual groundwater recharge and the extractable groundwater resources of the country for 1988 as about 436.5×10^9 m^3 and 383.4×10^9 m^3, respectively. According to the CGWB estimate, the level of groundwater development was about 30 per cent for the whole country, with the exception of Punjab and Haryana undergoing 99.4 and 70.2 per cent development. These figures show that groundwater development in India is not high and that there is plenty of scope for further development.

Groundwater resources play a major role in irrigation, domestic and industrial water supply by providing about one-third of India's water requirements (see Table 6-3). Reddy (1989) reports that in 1982/83, of a total irrigated area of 39.9 Mha, 19.1 Mha or 48 per cent of irrigated area was fed by groundwater. Tube-wells irrigated 10.7 Mha and dug-wells the remaining 8.4 Mha. Observations show that groundwater levels have dropped 5–10 m in arid areas of India. This decline is usually explained as a result of the increased number of dug-wells and tube-wells, the appearance of powerful energised pumpsets and a decline in average annual rainfall. However, Reddy (1989) argues that the decline in groundwater levels in the arid zones of India is due to destruction of tanks which had played a significant role in recharging the aquifers.

At the end of the VIth Five Year Plan (1980–85), there were 8.7 million irrigation dug-wells, 3.36 million private tube-wells and 46 000 state tube-wells in India. The target for the VIIth Five Year Plan (1986–90) was an additional 1.25 million irrigation dug-wells, 1.41 million private tube-wells and 25 000 state tube-wells (Ghosh and Phadtare, 1990). According to Pathak (1985), utilisation of groundwater resources in canal-command areas offers a means of correcting the imbalance in the groundwater system which causes waterlogging problems.

Land cover

In India nearly 183 Mha are potentially arable (Abrol, 1986). In 1987 about 168.99 Mha were under cultivation, out of which 42.1 Mha were irrigated, 12.0 Mha were under permanent pasture, 67.1 Mha were covered by forest and woodland and 49.23 Mha were covered by other lands (FAO, 1989).

Irrigation

Framji et al. (1982: 516–597) have provided a comprehensive report on the development of irrigation in India. The following is a summary of their report. Further information is also available in Stone (1984).

While irrigation in India has been practised for thousands of years, the total area irrigated was relatively small until British influence in the nineteenth century. During the period 1800–1836 some of the pre-existing canals in the north (Yamuna Canals) and the south (Cauvery Delta System originally built in AD 200) were remodelled. After 1836 a number of large-scale irrigation projects were designed and constructed by the British Army engineers. These projects included the Upper Ganga Canal in Uttar Pradesh, the Upper Bari Doab Canal in Punjab and the Godavari Delta System in Andhra Pradesh. The Upper Ganga Canal, with a discharge of 269 m^3 per second, was opened in 1854. The system comprised 914 km of main canals and 3846 km of distributaries, irrigating an area of 639 000 ha. The Upper Bari Boab Canal was completed in 1859. Because of the partitioning of the country in 1947, a substantial part of the area irrigated by this canal was included in the newly created State of Pakistan. The area irrigated in the Indian part was 384 856 ha in 1960–61. The Godavari Delta System was completed in 1890. It consisted of 805 km of canals and 3219 km of distributaries. In 1961–62 the area irrigated by this system was 558 466 ha.

Irrigation in India has undergone continuous development. The total irrigated area, which was about one million ha in 1850, reached 11.7 Mha in 1900. Table 6-4 provides statistics related to the irrigated and cultivated area from 1900 to 1987. It shows a fourfold increase (from 11.7 Mha to 42.1 Mha) in the irrigated area and more than a twofold increase (from about 82 Mha to 169 Mha) in the cultivated area.

As reported by Yadav (1987), the Indian Government is planning to attain gradually its ultimate irrigation potential of 113 Mha. The significant increase in irrigated and cultivated land in India has successfully transformed the country from a food deficient state to one which is self-sufficient in the production of food grains. Total food grain production, about 50 million tonnes in 1950–51, increased to 108.4 million tonnes in 1970–71 and to 172 million tonnes in 1988–89 (Abrol, 1990). This has been achieved in spite of a continuous decline in per capita availability of land. In 1901 the per capita availability of land was 1.37 ha. It was reduced to 0.48 ha in 1950–51 and to 0.2 ha in 1980–81, and will further reduce to 0.15 ha towards the end of this century (Abrol, 1990).

Table 6-4: Irrigated and cultivated area in India from 1900 to 1987.

Year	Area irrigated (net) (Mha)	Area irrigated (gross) (Mha)	Area cultivated (net) (Mha)
1900	11.7	N.A.	81.95
1951	20.9	22.6	118.75
1956	22.8	25.6	129.16
1961	24.7	28.0	133.20
1966	26.3	30.9	136.20
1971	31.1	38.2	140.78
1976	34.5	43.2	142.22
1977	34.8	43.1	140.23
1978	36.7	45.3	142.80
1987	42.1	N.A.	168.99

Source: Framji et al. (1982: 542), except for 1987 which is from FAO (1989).

Salinity

India has a record of salinity research which goes back more than 100 years (Gupta and Pahwa, 1978). However, major detailed scientific investigations were initiated only a few decades ago. Considerable information is now available on different aspects of salt-affected soils in India, especially their nature, extent, characteristics, classification, reclamation and management. Gupta and Pahwa (1978) in their annotated bibliography provide about 1100 abstracts, covering original research articles, review books, bulletins and monographs related to soil salinity and alkalinity in India, published over a period of 114 years from 1863 to 1976.

In India the extent of salt-affected soils is considerable. There is a wide divergence in the figures reported by different researchers, ranging from about 7 to 26 Mha. Estimates compiled by the Central Soil Salinity Research Institute in Karnal show that about 7 Mha of salt-affected soils occur in India (Table 6-5). According to Yadav (1987), the extent of salt-affected soils in different states of India is as follows: Uttar Pradesh (1.30 Mha); Gujarat (1.21 Mha); West Bengal (0.85 Mha); Rajasthan (0.73 Mha); Punjab (0.69 Mha); Haryana (0.53 Mha); Maharashtra (0.53 Mha); Orissa (0.40 Mha);

Karnataka (0.40 Mha); Madhya Pradesh (0.22 Mha); Andhra Pradesh (0.04 Mha); and other states (0.04 Mha).

The nature and characteristics of the salt-affected soils vary considerably from one region to another depending on climate, topography, geology, soil texture, drainability, hydrology, groundwater depth, groundwater salt content and management practices. They occur in wide belts or as patches of varying expanse. In a fraction of the salt-affected soils, salt accumulation in the soil profile can be attributed to natural processes. In a majority of cases, however, soil degradation through salinisation and alkalisation can be attributed to the changes brought due to human intervention by way of introduction of irrigation, use of saline water or other developmental works, leading ultimately to accumulation of salt in a region (Foreword by I.P. Abrol in Bhargava, 1989). The main salt-affected areas in India are as follows (Szabolcs, 1989):

- **Salt-affected soils of the Indo-Gangetic Plains.** This is the largest salt-affected area of the country. The soils of this region are generally sodic, except for a few pockets which are only saline. The soluble salts mainly comprise carbonate and bicarbonates of sodium.
- **Salt-affected soils of the arid and semiarid regions.** The main soluble anions are chlorides and sulphates. Because of low rainfall, scarcity of irrigation water and poor quality of groundwater, reclamation of these soils is difficult.

Table 6-5: The geographic distribution of salt-affected soils in India.

Category	States in which the soils occur	Approximate area (Mha)
Salt-affected soils of the Indo-Gangetic Plains	Bihar, Haryana, Madhya Pradesh, Punjab, Rajasthan, Uttar Pradesh	2.50
Salt-affected soils of the arid and semiarid regions	Gujarat, Haryana, Punjab, Rajasthan, Uttar Pradesh	1.00
Salt-affected soils of the medium and deep black soil regions	Andhra Pradesh, Gujarat, Karnataka, Madhya Pradesh, Maharashtra	1.42
Coastal salt-affected soils: — Coastal salt-affected soils of the arid regions	Gujarat	0.71
— Deltaic coastal salt-affected soils of the humid regions	Andhra Pradesh, Orissa, Tamil Nadu and West Bengal	1.39
— Acid salt-affected soils	Kerala	0.02
Total		7.04

Source: Abrol (1990).

- **Salt-affected soils of the black soil regions.** These soils are mainly found in the states of Andhra Pradesh, Gujarat, Karnataka, Madhya Pradesh and Maharashtra. They are characterised by black clay dominated by montmorillonite clay minerals. They swell on wetting and crack on drying and show a very poor infiltration rate, thereby rendering the leaching process rather difficult, unless the physical conditions of the soils are first improved through the addition of a suitable amendment.
- **Coastal saline soils.** These soils are found along the coastline of the country. The most important tracts of coastal saline soils are in West Bengal, the deltas of the Krishna and Godavari Rivers and the coastal areas of Kerala, Maharashtra and Gujarat. While chlorides and sulphates are the dominant anions, sodium is the dominant cation, though magnesium is also present in appreciable amounts in some cases.

Alkaline (sodic) soils are also widespread under varying climatic conditions. They occupy about 2.5 (Abrol, 1990) to 3.6 Mha (Bhargava, 1989: 6) and are predominantly found in the states of Uttar Pradesh, Punjab, Haryana, Rajasthan, Gujarat and in other states to a lesser extent. The important characteristics of these soils, which are responsible for very low yields, are as follows (Szabolcs, 1989): excessive amounts of exchangeable and soluble sodium; a high soil pH, often exceeding 10; a deficiency of available calcium; an extremely low water intake rate and hydraulic conductivity; an imbalance of plant nutrients; and poor physical conditions leading to the hindrance of root penetration.

Irrigated land salinity

Like many other similarly placed countries which have introduced intensive irrigation, India has experienced waterlogging and salinisation of its soils. The major factors responsible for causing serious waterlogging and salinity problems (Rao et al., 1990) are: seepage losses from canal networks; excessive deep percolation losses from the applied irrigation water due to poor land development; inappropriate irrigation practices; irrigation with groundwater of high salinity; unsuitable cropping patterns; insufficient natural drainage; and hindrance to natural surface and subsurface drainage by canals and other development works. Watertables which were originally several metres deep rose after the introduction of irrigation. Unfortunately, adoption of large-scale subsurface drainage measures has always been postponed for technical, socioeconomic or other considerations.

Historically, the ill-effects of waterlogging received initial attention in the western Yamuna Canal Zone (in Uttar Pradesh) around 1850 (Framji et al., 1982: 555). The problems surfaced on the Deccan Plateau when the Nira Irrigation Project (in Maharashtra) was commissioned in 1884 and in Punjab by 1907. It caused serious waterlogging and salting in the deep black soil of the command area. The situation assumed alarming proportions with 6–7 per cent of the area being damaged annually. The Chakkanwali Reclamation Farm and the Punjab Irrigation Research Institute at Lahore were established in the pre-independence (pre-1947) era for investigating all the problems associated with irrigation, drainage and salinity.

Among the post-independence projects, where serious waterlogging and consequent salt problems have arisen, are the Chambal Project Command in Madhya Pradesh and Rajasthan, the Gandak Project Command in Bihar, the Tungabhadra Project in Karnataka, the Sriramsagar Project in Andhra Pradesh and the Ukai-Kakrapar Project in Gujarat (Table 6-6). In the case of the Chambal Project, completed in 1960, soils waterlogged within a few years of the introduction of irrigation. By 1968 groundwater was within 3 m of the surface over 405 000 ha (Johnson, 1983).

Table 6-6: Extent of waterlogging and soil salinity in some irrigation projects of India.

Irrigation project	State	Waterlogging		Soil salinity	
		Area (1000 ha)	% of I.P.[a]	Area (1000 ha)	% of I.P.[a]
Sriramsagar	Andhra Pradesh	60.0	47.6	1.0	0.8
Tungabhadra	Andhra Pradesh and Karnataka	4.6	1.3	24.5	6.7
Gandak	Bihar and Uttar Pradesh	211.0	21.1	400.0	40.0
Sarda Sahayak	Uttar Pradesh	303.0	28.3	50.0	4.7
Ramganga	Uttar Pradesh	195.0	33.0	352.4	59.6
Ukai-Kakrapar	Gujarat	16.2	4.3	8.3	2.2
Mahi-Kadana	Gujarat and Rajasthan	82.0	16.8	35.8	7.3
Tawa	Madhya Pradesh	–	–	6.6	3.8
Chambal	Madhya Pradesh and Rajasthan	98.7	20.3	40.0	8.2
Rajasthan Canal	Rajasthan	43.1	8.0	29.1	5.4
Total		1013.6		947.7	

(a): Irrigation potential created.   Source: Yadav (1987).

Salinity and alkalinity in India decreases or severely impairs the productivity of an estimated 7 million ha of otherwise productive land (Abrol, 1986). The severity of this salt problem is increasing. Increasing pressure on existing land resources requires protracted efforts to reclaim or revegetate the land that is already unproductive and to prevent future degradation through salinisation. In particular, if the government wants to augment the irrigated area of the country and realise its ultimate potential of 113 Mha, this objective can be achieved only if suitable measures are taken to prevent secondary salinisation (Yadav, 1987). The following sections provide more detailed information on the salinity and waterlogging problems in the states of Haryana, Punjab, Rajasthan and Gujarat.

State of Haryana

The state of Haryana covers an area of 44 222 km² (Tanwar, 1979) and lies in the north-western part of India (Figure 6-2). It supported a population of about 16.3 million in 1991. About 80 per cent (3.5 Mha) of the state area is agricultural land which is important for food grain production. Of 3.5 Mha of agricultural land, 57 per cent is irrigated, 22.8 per cent with groundwater and 34.2 per cent with canal water (Kulkarni et al., 1989). According to Tanwar and Kruseman (1985), some 2.4 Mha of the state can be irrigated with available water resources.

The average annual rainfall ranges from 2000 mm in the north-east in the Himalayan foothills to 200 mm in the arid areas of the west (Figure 6-5). About 80 per cent of the annual rainfall occurs in heavy monsoons between June and October and 15 per cent in the winter from December through March. The Yamuna River flows along the eastern border of the state and the Ghaggar flows from the north and disappears in the sand-dunes of Rajasthan. A large part of Haryana (7207 km²) has no proper surface water drainage. Only the zones adjacent to the Yamuna River (16 330 km²) and Ghaggar River (10 675 km²) have natural drainage to the river (Tanwar and Kruseman, 1985).

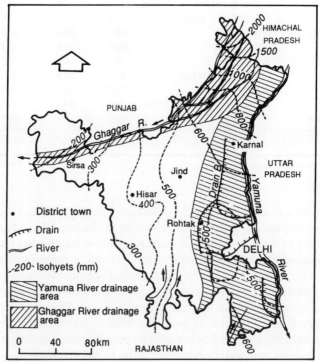

Figure 6-5:   Isohyetal and surface drainage pattern of Haryana (Tanwar, 1979).

Hydrogeology and salinity

Haryana State lies almost entirely in the Indo-Gangetic Plains of northern India, which constitute the surface expression of a structural depression. In front of the Siwalik Hills this structural depression is more than 3000 m deep, but in the south of the Narwana-Sonipat line (Figure 6-6) the basement occurs at a depth of 1000 m, gradually decreasing to 200 m (Tanwar and Kruseman, 1985). This depression is filled with alluvial materials. In the north, clay layers are intercalated with medium sand and gravel, while in the south, the clay and silt layers are predominant. The aquifer system is considered as a single unconfined aquifer, with transmissivity values ranging from 170 to 2600 m^2 d^{-1} and a specific yield of about 10–15 per cent.

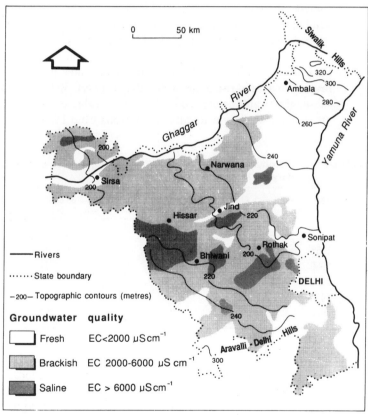

Figure 6-6:   Shallow groundwater quality map of Haryana (Tanwar and Kruseman, 1985).

A century ago the central and western parts of the state were mostly covered by semiarid woodland and scrubs, and the groundwater level was at a depth of 20–50 m. Canal irrigation was introduced about 100 years ago by diverting water from the Yamuna River. Now a network of canals more than 10 000 km in length is spread over the state. Currently the groundwater regime of the state can be broadly divided into two areas. The first area with fresh groundwater is in the north and the east covering an area of about 12 000 km². In this area fresh groundwater is pumped for irrigation from nearly 3000 deep state-owned tube-wells and more than 300 000 shallow private tube-wells, resulting in declining groundwater levels. The second area, covering 28 000 km² in the south and west, is underlain by brackish to very saline groundwater (Figure 6-6) and mainly depends on canal water for irrigation. According to Kulkarni et al. (1989), high salinity of most of the deeper aquifers and some of the shallow aquifers of south-western Haryana is due to connate waters. These saline waters were formed during a drier climatic phase in the Pleistocene. However, brackish waters of the shallower aquifers owe their origin to two principal causes: high evaporation of meteoric waters and mixing with older saline waters.

Losses through deep percolation in the conveyance system and in the field, and the absence of pumping groundwater in the south and west, resulted in a rapid rise of the watertable (between 30 to 100 cm per year) and subsequent waterlogging and salinisation. Table 6-7 shows the increase in areas with a shallow watertable in Haryana from 1955 to 1977.

Table 6-7: Distribution of areas under different ranges of watertable depth in Haryana from 1955 to 1977.

Year	Area in canal irrigation (Mha)	Per cent area under various depth to watertable ranges in June		
		0–3m	3–10m	>10 m[a]
1955	2.4	4	28	68
1960	2.3	9	31	60
1965	2.4	18	30	52
1970	3.2	12	46	42
1977	3.2	14	52	32

[a]: Watertable depth up to 50 m.
Source: Tanwar (1979).

At present the watertable is within 3 m of the surface, over an area of 400 000 ha which is critically affected by saline watertable rise. This affected area may ultimately increase fivefold to about 2 000 000 ha in the next 30 years (Kulkarni et al., 1989), threatening the livelihood of one million farmers and their families, and having a significant influence on the food grain production of Haryana and in turn of India as a whole (Tanwar and Kruseman, 1985). Apart from rising watertables, canal irrigation water (about 15×10^9 m³ yr^{-1}) brings nearly 2 million tonnes of salt to the area, thereby adding substantially to existing salinity problems (Rao et al., 1990).

Management options

The long run consequences of the expansion of areas affected by shallow watertable and salinisation threaten the economy of Haryana. Local communities would be disrupted, lands and villages would be deserted and both rural and urban unemployment would increase. According to Gangwar and van der Toorn (1987), these effects are already visible in parts of the Hisar and the Sirsa districts (Figure 6-6). Moreover, estimates show that the agricultural production loss due to waterlogging and salinisation was about Rs 270 million (US$9 million), and is expected to rise to about Rs 860 million (US$28.5 million) by the year 2000.

In 1982 the Haryana State embarked on a massive investigation into the cause and effects of this salinity problem and into potential solutions. These investigations were entrusted to Haryana State Minor Irrigation (Tubewells) Corporation (HSMITC) as the executive agency, with the assistance of UNDP/FAO (Gangwar and van der Toorn, 1987).

Tanwar and Kruseman (1985) describe various aspects of salinity management in Haryana. The following description is based mainly on their account. To achieve control of the watertable and salinisation in Haryana, three systems must be integrated. First, the surface water drainage system must function to prevent inundation of soils by monsoon rainfall (which otherwise would result in excessive groundwater recharge) and to transport groundwater drainage water to its place of reuse or final disposal. Second, the irrigation water supply system must provide the soil moisture needed by crops and leach salt left behind in the topsoil and the root zone. Third, the groundwater drainage system has to evacuate excess groundwater to prevent watertable rise and evacuate salt that has been leached from the root zone by excess irrigation water and rainfall.

Rao et al. (1990) describe the performance of an experimental subsurface drainage system to reclaim a highly salt-affected soil in a 10 ha area at Sampla (Rohtak District). They experimented with drainage spacings of 25 m, 50 m and 75 m at a depth of 1.75 m over a length of 200 m. Cement concrete tiles of 10 cm internal diameter and 30 cm long were used with gravel envelope in their experiments. According to their results, even subsurface drains at a spacing of 75 m provided sufficient drainage and facilitated the growing of crops within two to three years on lands that had previously remained barren for a considerable length of time due to high salinity.

Watertable control requires groundwater drainage and disposal of effluent. However, disposal of effluent is one of the major constraints in Haryana. In much of the state there is no possibility of discharging surface water through any streams. The Yamuna River is a potential receiver of saline drainage effluent, but its water should not exceed the EC value of 750 µS cm^{-1} because it is used for the Delhi water supply. This restricts the period when it can receive limited amounts of saline water to July and August, when river flow is high.

Two options for disposal of groundwater drainage effluents are unattractive in Haryana. Disposal into areas with deep groundwater levels is unattractive for two reasons. It is technically difficult and will only be possible for a limited period, and it is expensive, with disposal to evaporation ponds requiring large tracts of land equal to one-tenth the size of the drained area. The most attractive solution is the reuse of drainage water for irrigation, which will also alleviate the shortage of irrigation water. Under this option, drained water, with an EC value not exceeding 4000–6000 µS cm^{-1}, can be used in undiluted form for irrigation. This causes an accepted decline in crop yield of less than 30 per cent. Water with an EC value between 4000–6000, and even up to 10 000 µS cm^{-1}, can be reused after mixing with fresh canal water (EC value of 200 to 400 µS cm^{-1}). However, mixing effluent with an EC value above 10 000 µS cm^{-1} is not recommended.

The limited possibilities for disposing of drainage water make it imperative that the salinity of the drainage effluent be reduced as much as possible. As groundwater salinity increases rapidly with depth to unacceptable values, vertical drainage using deep tube-wells is not possible. Therefore, the solutions involve vertical drainage by shallow skimming wells and horizontal pipe drainage systems. Drainage by shallow skimming wells has a number of advantages compared to drainage by horizontal pipe drainage. These advantages are: the watertable can be controlled at any depth; they are relatively simple to operate; the investment costs are approximately half those of the horizontal pipe drainage system; and the groundwater quality will gradually improve because recharge water has a quality better than the original groundwater that is being pumped out. However, the groundwater drainage system can only be fully effective if the surface water disposal system and the irrigation water supply are also improved.

State of Punjab

The state of Punjab in the north-western corner of India (Figures 6-2 and 6-7) covers an area of 50 376 km^2, and in 1991 supported a population of about 20.2 million. The Plains of Punjab are bordered in the north-east by the Siwalik Hills. The Central Plains are 200 to 260 m above MSL. The elevation of the south-west semiarid region is less than 200 m above MSL. The average annual rainfall varies from 1200 mm in the north-east to 225 mm in the south-west (Singh, 1987).

There are two major geological units in the state. The Siwalik Hills, exposed in the north-east part and composed of Tertiary sedimentary formations, and the alluvial plain of Quaternary Period. Singh (1990) describes the application of LANDSAT and TM data to delineate the state into different hydromorphological units consisting of: hilly areas; piedmont areas; flat plains; flood plains; old river courses; and sand-dunes. More than 70 per cent of the state is occupied by flat alluvial plain and this area is mostly cultivated.

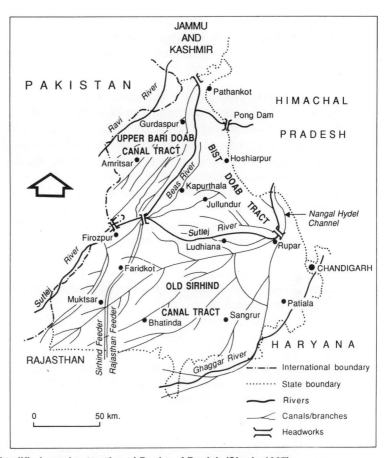

Figure 6-7: Simplified canal network and Doabs of Punjab (Singh, 1987).

The alluvial plain is dissected into three Doabs (area between two rivers) by three perennial rivers — the Sutlej, the Beas and the Ravi. The non-perennial Ghaggar River flows almost along the south-eastern border with the state of Haryana (Figure 6-7). The area between the Ravi and Beas Rivers is known as the Upper Bari Doab Tract. A triangular portion, bordered by the Beas and Sutlej Rivers and Siwalik Hills, is known as the Bist Doab Tract, and the area south of Sutlej River is known as the Old Sirhind Canal Tract.

The first modern perennial canal irrigation was initiated in 1851 in Upper Bari Doab and was completed in 1859 from the Ravi River. This was followed in 1882 by the Sirhind Canal from the Sutlej River (Framji et al., 1982). Subsequently, many canals were constructed to form a network over much of the Punjab Plains. Canal irrigation expands as far as southern Haryana and northern Rajasthan. However, canal irrigation was rapidly developed without considering the interaction between groundwater and surface water. The hydrologic regime was consequently disrupted and the watertable rose rapidly in irrigated tracts soon after the start of canal irrigation. This caused serious problems of waterlogging and salinisation.

Hydrogeology and salinity

As the Bist Doab Tract has very little canal irrigation compared to the other two tracts, the hydrogeologic behaviour of the Bari Doab and Old Sirhind Canal Tracts, with respect to canal irrigation, will be discussed.

In the Upper Bari Doab Tract, four aquifers have been identified. The top aquifer, with a thickness of 13 m to 50 m, consists of sand, silt and clay. In piedmont areas, pebbles, boulder and coarse sand are encountered. The second aquifer varies from 44 m to 63 m in thickness. The third and fourth aquifers have varying thickness and are composed mainly of sand and gravel (Singh, 1987). Groundwater flow direction is from the north-east to south-west. Depth to watertable ranges between 3 m and 20 m. Groundwater quality is good with EC values less than 2000 µS cm^{-1}. The tract started receiving canal irrigation water in 1860 when the Upper Bari Doab canal started flowing. During 1949–59 the canal was remodelled and the irrigation canal network expanded.

A substantial rise in groundwater levels took place during 1941–63 and large areas became waterlogged. After 1963 the watertable declined at a rate of 0.25 m to 0.50 m per year due to extensive horizontal and vertical drainage. As the groundwater quality is good and aquifers are productive, the number of irrigation tube-wells increased from about 7500 in 1960 to more than 88 900 in 1987. As a result groundwater levels have fallen at a rate of 0.3 m per year. It is suggested that artificial recharge can be carried out in the piedmont areas to arrest the drop of water-levels.

In the Old Sirhind Canal Tract aquifers consist of piedmont, alluvial and aeolian deposits. Depth of groundwater varies from 2 m to 35 m. Critical high watertable areas occur around Muktsar and Faridkot (Figure 6-7), where the watertable rests within 2 m of the ground surface. The general groundwater flow direction is from north-east to south-west. Groundwater quality of the shallow aquifer is highly variable with EC values ranging between 500 and 20 000 µS cm^{-1} (Singh, 1987). However, around unlined canals or distributors, groundwater is less saline (500–2000 µS cm^{-1}) and is suitable for irrigation. The aquifers deeper than 50 m are highly saline.

All over the south-west of the tract, the watertable has been rising at an average rate of 0.25 m per year from 1895 to 1950–60. From 1950–60 onwards, in the deep watertable area, the rate of rise was of the order of 0.5 m per year, and in the shallow watertable and waterlogged area (Muktsar) the rate of rise has nearly stabilised (Figure 6-8), indicating an increasing evapotranspiration loss as the watertable approached land surface. The following major causes for the rise in water levels, waterlogging and salinisation have been identified (Singh, 1989): excessive recharge due to seepage from irrigation canals and the distribution system; recharge due to blockage of surface run-off on account of the inadequate drainage system; and less withdrawal from the groundwater resources due to salinity and alkalinity of groundwater.

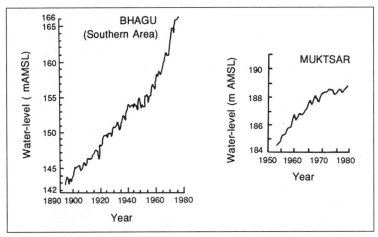

Figure 6-8: Representative hydrographs of south-western parts of the Old Sirhind Canal Tract (Singh, 1987).

In order to evaluate the size of excess recharge to the aquifer, a water balance study of the largest area affected by waterlogging was undertaken. The study area in Firozpur, Faridkot and Muktsar covered 173 345 ha of waterlogged land and showed that the annual recharge was 5.33×10^9 m^3. The discharge was 3.3×10^9 m^3, indicating an excess recharge of 2.03×10^9 m^3. The study was based on the assumption that the major canals and feeders are lined. However, because of serious damage to the lining of canals, the excess recharge could be as much as 2.83×10^9 m^3 yr^{-1} (Singh, 1989).

With the exception of some variations due to precipitation, there has been a considerable increase in the waterlogged areas in south-west districts of Punjab (Firozpur, Faridkot and Bhatinda). The waterlogged area has increased from 67 090 ha in 1978 to 199 235 ha in 1984, indicating an alarming increase. The waterlogged area decreased to 67 500 ha in 1987 because of the drought but increased sharply again to 176 500 ha in 1988. Estimates show that, if appropriate remedial measures are not taken, the waterlogged area will expand to 600 000 ha before the year 2000 (Singh, 1989).

The waterlogged areas in south-west Punjab are also affected by salinity and alkalinity problems. The estimated area affected by salinity and alkalinity is of the order of 60 000 ha, covering parts of Firozpur, Faridkot, Bhatinda, Sungrur and Amritsar Districts (Figure 6-7). The affected soils have EC values of 4000 to 40 000 μS cm^{-1} and high exchangeable sodium, and thus fall under the saline-alkaline soil category (Singh, 1989).

Management options

In order to control the groundwater regime, rise in water levels and salinisation caused by surplus groundwater balance, the following remedial measures have been suggested (Singh, 1987 and 1989):

- **Construction of surface drains, deepening of existing drains and weed control.** Construction of surface drains should be given top priority. They should be constructed deep enough below the watertable to increase the groundwater outflow from the area. Deepening of existing drains is also required for lowering the watertable. Weed growth should be checked and regular cleaning is required for making the drains more efficient.
- **Lining of canals and distributors.** This will reduce the seepage to groundwater reservoirs. It is expected that reduction of losses after lining canals and distributors will be about 75 per cent and 50 per cent, respectively.
- **Groundwater extraction.** Groundwater extraction can be increased by installing shallow tube-wells in areas where vertical drainage is feasible without causing upconing of saline waters. Shallow tube-wells should be installed where water is fit for irrigation. LANDSAT images have shown that old river courses of Sutlej River exist in the south-west of Punjab. Tube-wells can be drilled over these areas to increase the groundwater outflow component.
- **Blending of fresh and saline water.** Pumped low quality groundwater could be blended with canal water and used for irrigation. Experiments mixing canal water with saline water with an EC value of 17 000 $\mu S\ cm^{-1}$ show that a mixing ratio of 5:1 is required to obtain water with an EC value of 3000 $\mu S\ cm^{-1}$.
- **Construction of subsurface drainage systems.** In areas with shallow aquifers and low transmissivities of less than 100 $m^2\ d^{-1}$, vertical drainage or skimming by means of shallow tube-wells will result in upconing of saline water into the freshwater zone. In these cases horizontal subsurface drainage is the best method. PVC drains with a diameter of 100–150 mm can be installed at a depth of 2.5 m with the help of trench digging machines.
- **Afforestation.** Deep rooted vegetation can contribute to increased groundwater abstraction. A close spaced plantation of trees can remove more than 1000 mm of water per year from the soil profile, and up to 600 mm during the dry season.
- **Water and crop management practices.** Improving irrigation practices (such as the use of sprinklers and drip irrigation) and the plantation of high salt-tolerant crops will restrict the rise in water-levels.
- **Monitoring of the waterlogged and salt-affected areas.** Monitoring has a major role in management of waterlogged and salt-affected lands. In this

regard remote sensing technology can be used efficiently instead of conventional methods (see Sehgal and Sharma, 1988).

Development of surface and subsurface drainage systems generates low quality drainage water requiring disposal. The permanent solution to this problem requires detailed investigations. However, the following options are available as interim solutions:

- **Disposal to the Sutlej River.** Disposal by gravity could be possible during the monsoon period, but in the lower reach near the Rajasthan border the land surface gradient is not sufficient, and therefore pumping to the natural drains may be essential.
- **Blending option.** Drainage water could be blended with fresh canal water and used for irrigation.
- **Conjunctive use of fresh and saline water.** Waters with different quality could be used for irrigation during suitable periods of plant growth, that is, high quality water during germination and lower quality water after establishment of crop.
- **Disposal to evaporation ponds.** Saline water can be stored in ponds with sealed bottoms. This option can promote the development of fisheries because some varieties of fish can be developed in saline waters.

State of Rajasthan

Rajasthan is the second largest state in India after Madhya Pradesh and is mainly agricultural. The state is about 342 274 km^2 in area and in 1991 had a population of 43.9 million. Physically, Rajasthan can be placed into three natural divisions (Pal, 1977) — the arid west, subarid east and subhumid south.

- **Arid west.** About 75 per cent of north and western Rajasthan is a vast desert called The Great Indian Desert (Figure 6-9). It is the eastern part of the Thar Desert in Pakistan. The climate of this region is characterised by extremes of temperature ranging from below freezing point in winter to 52°C in summer. Rainfall is precarious and erratic, varying from less than 130 mm in the western part to over 500 mm in the eastern part. The mean annual value based on records since 1875 is 310 mm (Dhar and Rakhecha, 1979). Luni is the only river worth naming.
- **Subarid east.** Eastern Rajasthan for the most part consists of a highly fertile plateau. The principal rivers are the Chambal and Banas which drain almost all the north-east of the state. Rainfall over the area varies from 500 to 750 mm.
- **Subhumid south.** In contrast to western Rajasthan, the south is extremely rugged. The important rivers of the region are the West Banas, Sabarmati and Mahi, which flow towards the south-west and enter the Arabian Sea. The rainfall ranges from 700 to 1400 mm.

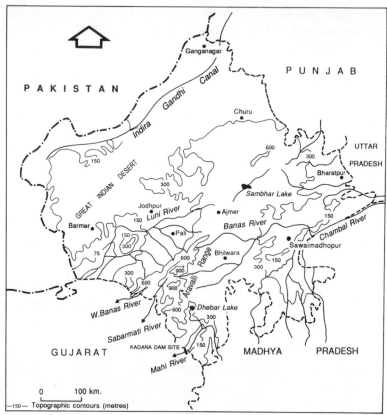

Figure 6-9: Physiographic map of Rajasthan (National Council of Applied Economic Research, 1963; and Pal, 1977).

There are 15 important rivers in Rajasthan. With the exception of the Chambal they are non-perennial. These rivers have a potential of 22.37×10^9 m^3 yr^{-1}. Of this only about 14.16×10^9 m^3 yr^{-1} could be effectively utilised (Pal, 1977). Because of the limitation of surface water resources, the state is dependent on its groundwater resources and the water transfer from Punjab rivers, mainly via the Indira Gandhi Canal (Rajasthan Canal). This 645 km canal with a capacity of 524 m^3 s^{-1} basically irrigates the western side along its length and some limited areas on its east side.

Hydrogeology and salinity

Over 80 per cent of western Rajasthan is covered by a blanket of sand and alluvium varying from a few centimetres to over 200 m thick. According to Chatterji (1980), based on 1975 statistics, of the total 3.95×10^9 m^3 of recharge, 2.61×10^9 m^3 were utilised for the following purposes: domestic (0.41×10^9 m^3 or 16 per cent); livestock (0.29×10^9 m^3 or 11 per cent); and irrigation (1.91×10^9 m^3 or 73 per cent). Although the difference between recharge and withdrawal shows a surplus of 1.34×10^9 m^3, in some districts (Barmer, Churu and Ganganagar) aquifers were overexploited. In these three districts, groundwater overdrafts were 17.4×10^6 m^3, 6.9×10^6 m^3 and 725.8×10^6 m^3, respectively. If the present rate of groundwater mining continues, the situation may become critical. In general, use of groundwater presents the following major problems: a major proportion of the groundwater reserves of this area (65 per cent) has a total salt content of over 3000 mg L^{-1}; watertables are deep, of the order of 60–100 m, except in the southern region where watertables are 15 to 30 m deep; and generally yield from wells is poor, ranging between 2 to 6 L s^{-1} with high discharge wells confined to very restricted areas.

With the introduction of canal irrigation, groundwater levels are rising at an alarming rate due to excess recharge (Table 6-8). Although the watertable is at a depth of 20–27 m in the Gang Canal area, and 34 m under the Indira Gandhi Canal (Rajasthan Canal) area, acute drainage problems could occur over the next few decades. However, scattered occurrences of perched watertables have also developed and as a result the land is becoming saline. In order to avoid the development of salinity, it would be desirable to lower the water-levels in these areas by mining it.

Table 6-8: Average rise in watertables in canal irrigated areas of Rajasthan.

Canal system	Year of start	Average rise in watertable (m yr^{-1})
Gang Canal	1927	0.3
Rajasthan Canal	1961	1.5
Bhakra Canal	1954	1.5
Ghaggar Flood Water Irrigation	1926	0.6

Source: Chatterji (1980).

In Rajasthan, where about 20 Mha are subject to arid climatic conditions, soil salinity and alkalinity pose major problems. There are two types of salinity in Rajasthan: natural and human-induced. Natural salinity has developed by weathering of rocks, release of soluble salt and its precipitation in depressions or playas of western Rajasthan. Many of these playas extend over vast areas. Salt concentration in many of these areas is very high and the land is devoid of any vegetation. The total area affected by natural salinity hazard is about 348 600 ha (Singh and Ghose, 1980).

Human-induced salinity is a recent phenomenon due to the development of the irrigation network, causing waterlogging and salinisation. A total of 613 500 ha is affected by this type of salinity hazard (Singh and Ghose, 1980). In western Rajasthan, canal irrigation is practised on a large scale, with the Rajasthan Canal having an irrigation potential of 1.15 Mha

(Murty, 1990). In other areas, of 320 000 ha of irrigated land, nearly 95 per cent is irrigated from shallow to deep wells, with groundwater quality ranging from low to medium (Roy and Kolarkar, 1977). The lands under irrigation with such saline water (1500 to 7000 µS cm^{-1}) are generally cultivated every alternate year or after two to three years, to allow the accumulated salt to leach out by rain before the next crop is sown. In spite of this practice, continuous irrigation has rendered many such lands saline and unproductive. In fact irrigation with saline groundwater has been the main cause for the development of salinity in the cultivated lands of the five districts of Jodhpur, Pali, Ajmer, Bhilwara and Bharatpur (Roy and Kolarkar, 1977). Mathur et al. (1977) investigated the groundwater quality of these five districts by taking 10 representative samples in each district. Groundwater samples were analysed for pH, EC, cations, anions and SAR. The study revealed that most groundwater samples contained high amounts of sodium salts, particularly sodium chloride, sodium sulphate and sodium carbonate and bicarbonate. Table 6-9 shows the range of EC values in 50 groundwater samples.

Table 6-9: Frequency distribution of groundwater electrical conductivity in five districts of Rajasthan.

Electrical conductivity (µS cm^{-1})	Number of samples in districts:					Total number of samples
	Jodhpur	Pali	Ajmer	Bhilwara	Bharatpur	
0–1000	1	–	2	1	2	6
1000–2000	1	1	3	3	1	9
2000–3000	–	1	–	1	1	3
3000–4000	1	–	1	–	1	3
4000–10 000	4	2	1	4	3	14
10 000–15 000	1	4	2	–	–	7
15 000–20 000	2	1	–	1	1	5
20 000–25 000	–	1	1	–	1	3
Total	10	10	10	10	10	50

Source: Mathur et al. (1977).

Another cause of salinity is the poor drainage condition, due to physical characteristics of the soil, such as heavy texture and low permeability, accompanied by a high watertable. This is the main cause of salinisation in the Chambal command area and in the Pali, Bhilwara, Kota and other districts. A further cause of salinisation is seepage from the main canals or reservoirs and the low-lying topographical situation accompanied by high evaporation during the summer months.

Management options

Roy and Kolarkar (1977) describe the management of saline and sodic soils. In Rajasthan increased salinity of soil has rendered many lands unsuitable for cultivation. However, salt-tolerant wheat is grown in light textured soils under irrigation with moderately saline water. After harvesting, the land is left fallow for two to three years, during which time the salt that has accumulated due to irrigation is leached down by rainwater and the land is again ready for cultivation.

Some saline lands provide fairly good pasture unless the salinity is excessively high. The succession of grass species suitable for highly saline to low saline conditions consists of *Sporobolus* sp., *Cynodon dactylon* (Bermuda grass), *Echinochloa* spp., *Aristida* spp., *Eleusine compressa, Dactyloctenium sindicum, Eragrestis tremula,* and *Dichanthium annulatum*. Among the more important salt-tolerant trees in Rajasthan, particularly in arid western Rajasthan, are *Prosopis juliflora, Salvadora persica* and *Tamarix articulata*.

In lands where the salinity is low and the soil is in the initial stage of salinisation, reclamation is achieved by growing a rice crop for one or two years with a prior leaching. Where salinity is somewhat high, prolonged leaching may be necessary before transplanting a coarse variety of rice. When the land is strongly saline or sodic, excessive flooding with water is necessary, followed by the addition of amendments like gypsum. A dressing of farmyard manure improves the structure of the soil. Growing *Sesbania aculeata* preceding the plantation of rice is very helpful in increasing the permeability of the soil.

For successful management and reclamation of salt-affected and alkaline soils, water must be available in sufficient quantity to leach down the salt below the root zone, and drainage should also be available. In arid western Rajasthan, water is scarce. Moreover, salt-affected lands are usually situated in low lying areas where drainage is impeded and there is little possibility of draining out the leached salt. In such areas reclamation by the usual methods is difficult and uneconomical. The only way to make use of these lands is to grow, wherever possible, salt-tolerant grasses (for grazing) and trees (for fuel and other uses).

State of Gujarat

The state of Gujarat (Figure 6-10) has an area of 195 984 km^2, which is about 6 per cent of the country. It supported a population of 41.2 million in 1991. The following brief description of the physiographic features, climate and rainfall of the state is based on Patel (1977).

The main physiographic feature of the state is the large central alluvial plain extending from Banaskhanta district in the north to Bulsar (Valsad) district in the south and the Saurashtra Peninsula, which is separated from it by the Gulf of Khambhat. The eastern border of the state is mountainous with ranges up to 1000 m high. Another feature is the large desert in the north called Rann of Kachchh, covering the northern part of the Kachchh district and the western part of the Banaskhanta district.

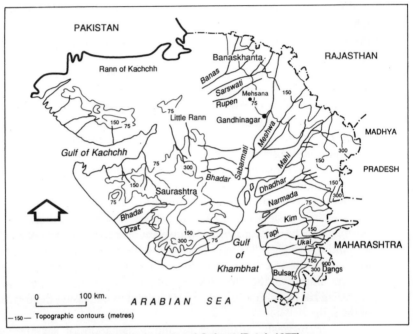

Figure 6-10: Main physiographic features of Gujarat (Patel, 1977).

A large number of rivers flow westward across the plain. Those in the north of Narmada, like Banas, Sabarmati and Mahi, flow only during the monsoon season. The Narmada, Tapi and some smaller rivers from the south have perennial flow. On the Tapi River is the state's largest irrigation dam at Ukai. Numerous rivers originate in the central plateau of the Saurashtra Peninsula and flow radially. They are all shallow and experience occasional flash floods in the monsoon season, and are dry for the rest of the year. The state has three types of climate. The climate is arid in the extreme north, which comprises the Kachchh district, the western part of Banaskhanta and Mehsana, and the northern fringe of the Saurashtra Peninsula. The extreme south districts of Bulsar and Dangs have a subhumid climate. The rest of the state experiences semiarid climate.

The monsoon season is of very short duration, about three months. During the remainder of the year there is no significant precipitation. Usually the monsoons commence by the middle of June and withdraw by the end of September. The precipitation in the first half of June and later half of September is very limited compared to the intervening period. July and

August are the months of high rainfall. The annual average rainfall of the state is 821 mm. However, the rainfall decreases from the south-east toward the north and north-west. The district of Dangs in the south-east has an annual rainfall of 2000 mm, while the district of Kachchh has an annual rainfall of 322 mm, which is the lowest in the state. South-east Gujarat experiences high daily rainfall of over 250 mm d^{-1}. The high rainfall intensity makes this part of the state prone to floods. Flood frequency is particularly high in Narmada, Tapi and other river catchments in south-east Gujarat (Patel, 1977). Overall, the low and uncertain rainfall with high variability in certain parts of the state, combined with limited irrigation facilities, has made Gujarat extremely susceptible to droughts and famine. It has experienced drought conditions of various magnitudes about 25 times since 1900. The 1985–86 drought was one of the most severe, affecting 14 000 of 18 000 villages, with a population of about 18–33 million (Patel and Shankara Iyer, 1987). To overcome this problem, 76 major and medium irrigation projects had been completed and 92 major and medium irrigation projects were in progress (up to March 1986). Also, under the major and medium irrigation projects, an irrigation potential of 1.096 Mha has been created.

Gujarat is one of the most agriculturally productive states of India and cultivation extends over much of the state. About 80 per cent of the cultivated area is dependent on rainfall. However, surface water and groundwater are used for irrigation. According to Kolvalli and Chicoine (1989), of the 1 715 000 ha irrigated in Gujarat in 1978–79, about 1 346 000 ha were irrigated by wells. Nearly 26 000 wells are estimated to be in the Mahi-Kadana command area alone.

In the Gujarat alluvial plain, groundwater in the eastern part adjoining the hilly areas is comparatively low in salinity, with EC values of less than 2000 µS cm^{-1}. The western half of the plain contains groundwater that is appreciably more saline than that in the east. There is a progressive deterioration in the water quality at greater depth, with EC values ranging from 2000 to 40 000 µS cm^{-1} or more (Department of Technical Co-operation for Development, 1986: 120).

In the Mehsana area, insufficient rain and frequent droughts have resulted in large-scale dependence on groundwater resources. Moreover, insufficient annual recharge and large-scale exploitation of groundwater resources for irrigation over the past few decades have resulted in rapid depletion of groundwater reserves. This is reflected in progressive declines in watertable and piezometric levels. It is estimated that the annual groundwater overdraft, from the phreatic aquifer and shallow confined aquifers down to 125 m depth, is of the order of 115×10^6 m^3, whereas cumulative overdraft since 1961 is of the order of 3.41×10^9 m^3. The cumulative declines in water levels in the phreatic aquifer between 1961 and 1984 ranged from 1 to 21 m (Bradley and Phadtare, 1989).

In the coastal plain of Saurashtra, farmers are heavily dependent on groundwater due to the low reliability of rainfall and lack of surface water for irrigation. This has resulted in intrusion of saline seawater to the aquifers. This issue and its effect on salinisation of coastal lands will be developed later.

Salinity

Problems of waterlogging and salinity in irrigated areas were not severe, since the major rivers were harnessed by weir schemes only. But with the construction of reservoirs like the Mahi, Shetrunj and Ukai, enormous quantities of water have been made available perennially. Some projects, like the Kakrapar and Mahi (Figure 6-11), have shown indications of waterlogging and/or salinity in quite a sizeable area for about a decade.

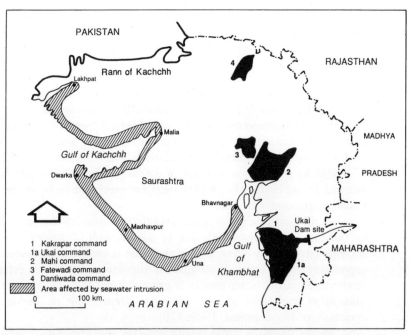

Figure 6-11: Major irrigation projects in Gujarat (Patel and Shankara Iyer, 1987) and the coastal areas affected by seawater intrusion (Mistry, 1989).

In Gujarat the following measures are adopted to assess the extent of waterlogging and/or salinity in the command areas: observation of the watertable; pre-irrigation soil surveying; and remote sensing (Bapat and Shah, 1984). Observation of groundwater levels has been carried out over the past two decades in almost all major irrigation schemes, during pre-monsoon (May) and post-monsoon (November) periods. The latest data available for such major schemes is given in Table 6-10. The data reveals that problems of waterlogging and salinity are increasing annually. The rate of rise in the watertable in the Kakrapar and Mahi areas has been recorded as 0.3 m per year.

Table 6-10: High watertable area in major irrigation schemes in Gujarat State for the pre-monsoon period (May).

Name of Project	Area (ha)	Area under various depth to watertable ranges		Year
		0–1.5 m (ha)	1.5–3 m (ha)	
Mahie R B C	293 860	3 820	43 785	1983
Kakrapar	406 008	7 534	74 735	1982
Shetrunji R B C	29 150	–	1 908	1981
Shetrunji L B C	30 161	21	615	1981
Ukai R B C	85 873	40	3 275	1983
Ukai L B C	121 392	1 838	1 134	1982
Dantiwada	81 000	–	1 870	1981
Fatewadi	129 555	–	6 628	1981
Panam	62 750	–	175	1981
Karjan	79 724	–	285	1983
Damanganga	56 070	–	32	1983
Total	1 405 543	13 253	144 457	

R B C: Right Bank Canal, L B C: Left Bank Canal   Source: Bapat and Shah (1984).

All irrigation projects are being covered by a systematic reconnaissance and detailed pre-irrigation soil survey. This survey gives an appreciation of the existing soils and their interpretation for various land uses. It also helps in identifying areas which are likely to have been affected by waterlogging and salinisation after irrigation. Remote sensing techniques were used to identify areas which are waterlogged and/or salt-affected. The technique was used for part of the areas in Kakrapar (in 1982), Mahi RBC (in 1982) and Fatewadi (in 1984). Gore and Bhagwat (1991) describe the use of 1977, 1983 and 1987 LANDSAT images in an irrigated area of the Gujarat State. In the following sections three cases of land salinisation in Gujarat will be described.

Salinity in Kakrapar Irrigation Project

With the construction of a weir at Kakrapar on the Tapi River, a gross area of 406 008 ha has been brought under irrigation since 1957–58. Since 1972, with the construction of the Ukai Reservoir, a large area under the command of the Kakrapar Project has been receiving water perennially. The majority of the soils of the command area have been derived from older alluvium and are commonly known as black cotton soils. These soils have montmorillonite as a predominant clay mineral. They have a very high potential fertility, high moisture holding capacity, high cation exchange capacity and can sustain increased crop production under irrigation. At the same time these soils have low to very low vertical and horizontal permeability, which makes their drainage difficult.

In view of the soil fertility, climatic conditions and the availability of a source of water, there was a sudden shift from the traditionally unirrigated crops to irrigated crops. Among the new crops, sugarcane gained a lot of momentum and in 1984 was grown in 11–12 per cent of the command area. Sugarcane cultivation is practised in 40 per cent of the cultivated area in Kamrej and Palsana, and 30 per cent in Navsari and Bardoli (see Figure 6-12 for locations). In addition to this, the area under banana and vegetable crops increased (Raman, 1984).

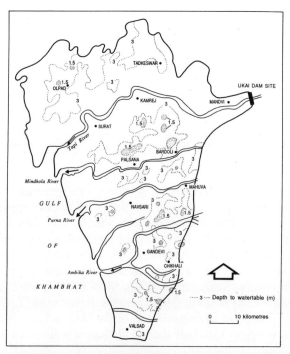

Figure 6-12: Watertable depth in the Kakrapar irrigation project for the period of pre-irrigation 1957–58 (Bapat and Shah, 1984).

Analysis of watertable levels in observation wells in the command area under the Kakrapar Weir shows clearly that there has been rapid rise in the watertable (Table 6-11). While in 1957 there was no area with watertable level up to 1.5 m, the area rose to 1283 ha during 1975 and 3213 ha during 1980. Another alarming fact is that, while 156 712 ha had watertables deeper than 9 m during the inception of the canal system in 1957, the area in this category was almost nil in 1980. In this command area the rate of rise in the watertable over a period of 20 years has been reported to be about 0.3 m per year. Figure 6-12 shows the depth of watertable for the pre-irrigation period of 1957–58, while Figure 6-13 shows the watertable depth for the pre-monsoon period 1982 in the Kakrapar irrigation project.

As a consequence of canal water recharging the aquifer, a considerable amount of salts is also added. With the rise in watertable, salt is brought to the surface by capillary rise. Thus soils gradually become salt-affected. Salt-affected soils are on the increase and thousands of hectares have become almost barren (Raman, 1984).

Table 6-11: Watertable depths in the Kakrapar Weir system in pre-monsoon periods (areas in ha).

Year	Depth to watertable (m)				
	0–1.5	1.5–3	3–6	6–9	>9
1957	–	1 037	40 920	204 910	156 712
1970	25	2 078	104 077	133 380	166 390
1975	1 283	21 557	209 848	88 339	34 992
1980	3 213	54 017	183 527	84 940	–

Source: Raman (1984).

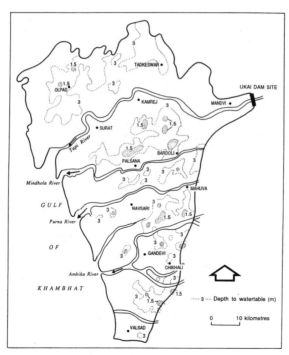

Figure 6-13: Watertable depth in Kakrapar irrigation project for the period of pre-monsoon 1982 (Bapat and Shah, 1984).

Major causes of waterlogging and land salinisation are as follows (Bapat and Shah, 1984): losses due to conveyance of water through unlined canals; change in cropping patterns and heavy concentration on sugarcane cultivation; too frequent irrigation resulting in excessive losses through fields (for example, while sugarcane needs to be irrigated 14–15 times during its growth period with 8 cm of water each time, farmers generally apply 20–22 irrigations); clayey soils with low natural drainage; and low topographic gradient of the command area.

Raman (1984) and Bapat and Shah (1984) describe the following remedial actions to overcome the problem of waterlogging and salinisation in the area:

- **Seepage control**. Seepage is one of the main factors contributing to watertable rise and salinisation. The main canals and distributors should therefore be lined.
- **Drainage improvement**. A pilot research project was undertaken in a village near the seacoast. The pilot experiment included different sizes of drains, depths of drains and spacing. Canal water was used to leach these heavy montmorillonitic clayey soils through various leaching trials and amendments (gypsum, organic manures, green manures and their combinations). These leaching trials continued for a period from 1965 to 1974. The experiment was quite useful in determining the leaching requirements of these soils, and the results can be applied elsewhere.
- **Water management**. Water is wasted at field level. Farmers tend to overirrigate their crops. The advantage of adopting proper water management practices and the evils of excessive use of irrigation water should be demonstrated.
- **Rationalisation of cropping pattern**. Surveys have indicated that the maximum capacity for bearing sugarcane in the Kakrapar command area varied between 15 and 25 per cent. But the actual area under sugarcane had increased to even more than 50 per cent in some areas. The concentration of sugarcane plantation should therefore be reduced. In some areas of the command, farmers grow rice. The rice crop requires about 100–120 cm of water, but the farmers apply much more. Under these conditions, farmers should be encouraged to grow other crops which require less water. In recent years groundnut has emerged as the most profitable crop to grow in summer. This crop requires only 56 cm of water.
- **Conjunctive use of canal and groundwater**. Good quality groundwater is available in sufficient quantity in many parts of the command area. Conjunctive use of canal water with the available groundwater will lower the watertable. Although this approach has been successful in other irrigated areas with light soils, its applicability in lowering the watertable in this area, which has heavy soils, should be investigated.

- **Remote sensing.** With the help of LANDSAT facilities, the areas affected by waterlogging and salinity could be identified and monitored.

Salinity in the Mahi-Kadana Irrigation Project

The Mahi Canal Project covers an area of 212 000 ha. Cultivatable area is covered by a canal network of about 2600 km (Patel, 1984). Irrigation was started in 1958–59 in the initial reaches of the command area, and was extended gradually into larger areas on completion of conveyance systems. Since 1980–81, with completion of the Kadana Reservoir, water has been made available for irrigation throughout the year. The climate of the district is subtropical with an average annual rainfall of 775 m. Partial failure of the monsoon once in every three or four years is a common feature of the area. The crops grown are mainly cotton, tobacco, wheat, rice and vegetables.

The soils in the area under Mahi-Kadana command were covered by a detailed soil survey during 1957–64 and classified under a number of Land Irrigability classes. Of 311 414 ha surveyed, 180 574 ha were class I, 43 201 ha class II, 26 936 ha class III and 60 703 ha class IV. Thus about 72 per cent of lands were under classes I and II, and designated suitable for irrigation. However, the continuous use of excessive canal water, coupled with seepage losses, has resulted in a rise in the watertable at a rate of about 0.3 m per year. This has caused secondary salinisation and resulted in hundreds of hectares going out of cultivation.

Table 6-12 indicates increases in the area of watertable at various depths as measured for 1981 and 1986, relative to pre-irrigation levels in 1958. It reveals that the areas where the watertable reached within 1.5 m and 3 m of the surface were 901 ha and 13 752 ha respectively, during 1986, compared to nil and 2643 ha in 1958. However, the study shows that a total area of nearly 60 000 ha in 300 villages is in the grip of high salinity and alkalinity (Patel, 1984).

Table 6-12: Watertable depth in the Mahi command area before irrigation (1958) and after irrigation in 1981 and 1986 (areas in ha).

Year	Depth to watertable (m)					Total
	0–1.5	1.5–3	3–6	6–9	>9	
1958	–	2 643	18 058	15 277	257 982	293 860
1981	475	9 360	83 338	33 550	167 207	293 860
1986	901	13 752	98 910	63 923	116 554	293 860

Source: Mistry and Purohit (1989).

According to Mistry and Purohit (1989), the possible causes of the problem of waterlogging and salinisation are: inadequate open ditches for disposal of water; insufficient capacity of drains to remove water quickly;

excessive depth of irrigation application; seepage from canals; conversion of lands into paddy fields; negligence in the maintenance of the ditches; and blocking of natural drainage by construction of roads and other development works.

Patel (1984) provides characteristics of soil profile samples (depth, texture, clay content, pH, EC_e, Ca^{++}, Mg^{++}, Na^+, K^+ and ESP) and chemical composition of the saturated soil extract (Ca^{++}, Mg^{++}, Na^+, K^+, CO_3^-, HCO_3^-, Cl^-, SO_4^{--} and SAR). These analyses show that chlorides and sulphates are dominant. Patel (1984) describes the following ways to combat waterlogging and salinisation:

- **Lining of canal and branches.** The main canal, branches and network conveyance systems, totalling about 2600 km, are more or less completed. The main canal and branches are lined. Other conveyance systems are unlined.
- **Drainage system.** Major surface drains are more or less completed. However, to dispose of rainwater efficiently, some of the drains have been modified and/or are under modification. Lateral drains are also taken up to take care of disposal of water from low lying and problematic areas.
- **Groundwater utilisation.** Groundwater should be used for irrigation directly or could be mixed with canal water if the quality is not good (Patel, 1984). Investigations have shown that the available annual groundwater potential in the Mahi Right Bank Canal is about 265×10^6 m^3 yr^{-1}. This is about 1.6 times the existing level of groundwater abstraction (Sondhi et al., 1989).
- **Rainwater utilisation.** Canals should not be run during monsoon periods, except during dry spells. Canal water should be utilised only in the instance of the unavailability of the two alternatives (groundwater and rainwater).
- **Leaching trials and ameliorative measures.** The soils of the region are waterlogged and saline-sodic in nature. A suitable technique should be evolved for controlling the rise of watertable, leaching of salt and reducing the alkalinity. To fulfil these goals, the following measures/experiments should be taken up: development of rapid and efficient leaching methods; use of amendments like gypsum, organic manures and paddy straw to improve the infiltration rates and to reduce alkalinity; different cropping systems should be tried for efficient water use and higher economic return; and adoption of salt-tolerant crops.

According to Mistry and Purohit (1989), although the remedial measures of improved water management, canal lining and surface drainage have reduced the problem considerably in the upper areas, it has still remained unsolved to a certain extent in the lower reaches with very flat topography and heavy soils. About 20 000 to 30 000 ha remain submerged in the rainy season and even two months thereafter due to local depressions. A master plan by the Government of Gujarat, costing Rs 190 million (US$6.3

million), to solve the problem of waterlogging and salinity is under way. A drainage network costing Rs 110 million (US$3.6 million) has been completed. In addition, 275 tube-wells, drilled in the command area to augment the irrigation on higher lands, have reduced the problem to a great extent.

Salinity in coastal areas of Gujarat

A major salinity problem in coastal areas is due to overpumping of groundwater and seawater intrusion into coastal aquifers. The coastal area of Gujarat is an example of this problem. Similar problems exist in coastal areas of other states such as Tamil Nadu, Andhra Pradesh, West Bengal and Kerala State.

The narrow coastal plain of Saurashtra along the south coast of Saurashtra Peninsula has long been known as a very fertile and productive tract, with orchards and vegetable gardens in addition to intensive field crops. Due to the low reliability of monsoon rainfall and the lack of surface water for irrigation systems, farmers in the area are heavily dependent on groundwater. The water is obtained from shallow wells, tapping a productive aquifer which extends inland to 5–7 km along the entire coastal plain (Mistry, 1983).

Prior to the early 1960s, a delicate balance between the discharge and recharge of groundwater was maintained. However, since the introduction and expansion of motorised pumping, the discharge rate has exceeded the recharge rate and the underlying saline water of the aquifer system has intruded into the freshwater body. By 1970 the problem became serious, demanding immediate remedial measures. The most heavily affected area lies along a coastal strip of about 160 km between Madhavpur and Una (Figure 6-11). The area affected by intrusion of saline water has increased from about 35 000 ha in 1971 to 100 000 ha in 1977. The area irrigated by groundwater supply has been reduced from approximately 24 000 ha to 16 900 ha, as many wells had to be abandoned in the same period.

In this coastal belt four main geological units have been identified. These are: alluvium and coastal sand; cavernous limestone between 30 and 40 m depth; Dawark and Gaj-beds consisting of clay between 40 and 100 m; and hard and compact basalt forming the bedrock. Sand-dunes are restricted to a narrow strip along the coast. The limestone is well developed and is found about 7 km inland. In this area limestone forms an ideal groundwater reservoir, with an average hydraulic conductivity of 200 m d^{-1}, and is the main source of groundwater supply to the wells for agriculture. Overpumping from this aquifer (with a deficit of 23×10^6 m^3 in 1971 and 103×10^9 m^3 in 1977) has lowered the watertable and reversed the hydraulic gradient, which was initially toward the sea. This in turn paved the way for the intrusion of saline water inland. The use of water from the affected wells, with high percentages of total dissolved solids, caused deterioration in the quality of agricultural soils. Groundwater salinity was about 500 mg L^{-1} TDS in 1970, and increased to 2500 to 7000 mg L^{-1} in 1977. The effect of saline water intrusion in 1971 was observed within 2.5 to 4.0 km from the seacoast. By 1977 it had penetrated a further 5 to 7 km inland from the sea. Due to this galloping increase in the affected salinity area, about 12 500 wells of 120 villages have become inoperative, rendering 100 000 ha of cultivable land unproductive. More than 280 000 people inhabiting these 120 affected villages are facing an acute situation (Mistry, 1983).

According to Mistry (1989), in the coastal regions of Saurashtra and Kachchh, salinity has been caused by different processes. There are vast areas along the coast between Bhavnagar and Una which have turned saline due to inherent salinity in that area. These areas seem to have remained under the sea for quite a long time and the high saline watertable has caused salinity in the agricultural land. In the reach between Madhavpur and Dwarka, it is observed that the winds from the sea blow landwards. The winds bring salt from the sea which affects coastal agricultural lands up to 15 to 20 km inland.

To find means of combating the saline intrusion problems in the coastal areas of Saurashtra and Kachchh, the Government of Gujarat appointed a High Level Committee for the reach from Madhavpur to Una in 1976. A second committee was appointed in 1978 to suggest measures for the coastal reaches of Bhavnagar to Una, Madhavpur to Dwarka, Dwarka to Malia and Malia to Lakhpat (Fig. 6-11). These two committees studied the problem and recommended the following measures: management techniques (change in the cropping pattern and regulation of groundwater extraction by legislation); recharge techniques (dams, recharge tanks, recharge wells and spreading channels); and salinity control techniques (tidal regulators, freshwater barrier, etc). Mistry (1989) describes the progress of the measures undertaken and concludes that reversal of the effects of saline intrusion has been a slow process. Moreover, the effectiveness of the undertaken measures was arrested due to failure of the monsoon for three consecutive years in the state. However, the post-monsoon monitoring of watertable and water quality in October 1988 indicated a rise in watertable and improvement in water quality (TDS < 2000 mg L^{-1}) over an area of approximately 25 000 ha.

Salinity problems in other states

Salinity in India is not limited to the four states (Haryana, Punjab, Rajasthan and Gujarat) described in the previous sections. Major salinity and alkalinity problems exist in other states including Uttar Pradesh, West Bengal, Maharashtra, Orissa, Karnataka, Madhya Pradesh, Andhra Pradesh and Tamil Nadu. For a description of salinity problems in these and other states, readers are referred to other publications, such as: Government of India, Ministry of Irrigation (1984); Government of India, Ministry of Water Resources (1987); and Gupta and Pahwa (1978), which provides a comprehensive bibliographic reference of salinity problems in India. Also, Bhargava (1989) provides physico-chemical characteristics of saline and alkaline soils of India, under different agroclimatic conditions, and describes the appropriate management options for these soils.

General aspects of management options

Management and reclamation methods for saline and alkaline soils vary according to the nature of the problem and availability of good quality water for leaching. Amelioration of alkaline soils is possible on an individual farmer's holding basis, while saline soil reclamation needs to be initiated on a catchment basis. Although the management of salt-affected soils in four states has been described in the previous sections, it would be useful to describe the general aspects of saline and alkaline soil management in India. These descriptions are based mainly on Yadav (1987). Similar information is also available in Bhargava (1989).

Management of alkaline soils

Limitations in alkaline soils arise from high pH, high exchangeable sodium percentage (ESP), high sodium adsorption ratio (SAR) and severe nutritional deficiencies. The following options are available to overcome these problems.

Amendments

Application of a suitable amendment is essential to replace excessive amounts of exchangeable sodium. Several inorganic and organic amendments like gypsum, phosphogypsum, sulphur, sulphuric acid, pyrites, farmyard manure, crop residues and green manures have been used for this purpose. The main approach adopted at the Central Soil Salinity Research Institute (CSSRI), at Karnal, has been to reclaim initially only the top 10 to 15 cm of soil for growing shallow-rooted crops like rice and wheat; this has helped in reducing the dose of amendment applications. A simple method has been developed at CSSRI to determine the dose of gypsum by knowing the pH and texture of the top 15 cm of the soil profile (Table 6-13).

Table 6-13: Requirements of gypsum (t ha^{-1}) according to pH and texture of the soil.

Soil pH	Soil texture		
	Sandy loam	Loam	Clay loam
9.2	1.7	2.5	3.4
9.4	3.4	5.0	6.8
9.6	5.0	7.5	10.0
9.8	6.8	10.0	14.6
10.0	8.5	12.5	15.0
10.2 and above	10.0	15.0	15.0

Source: Yadav (1987).

The application of low grade pyrite has given encouraging results, especially in the calcareous sodic soils of Bihar and eastern Uttar Pradesh. Phospho-gypsum has been found effective for reclamation. The demonstrations conducted on farmers' fields in the states of Uttar Pradesh, Haryana and Punjab have shown good yields of rice and wheat. Also, the treatment of sulphuric acid in alkaline soils produced good yields of rice and wheat, but according to Bhargava (1989), its cost and the hazard involved in safe handling are serious constraints to its use.

According to Palaniappan and Budhar (1992), there is a renewed interest in the use of organic manures. Farmyard manure, compost and green manures are the organic materials commonly used. In addition to supplying nutrients, green manures enhance the soil productivity by reducing the problem of alkalinity in soils. During decomposition of green manures, considerable amounts of organic acids are liberated, bringing down the pH of the soil to some extent, besides forming a number of salts with sodium in the exchange complex. Green manures such as *Sesbania aculeata*, *Crotalaria juncea* and *Cymopis tetragonoloba* can be used for the management of alkaline soils.

Alkali-tolerant crops

Crops differ considerably in their tolerance to soil alkalinity. Therefore proper selection of crops, varieties and cropping sequence is of prime importance. Yadav (1987) provides the relative tolerance of 25 crops to exchangeable sodium (Table 6-14).

Rice hastens the reclamation process of alkaline soils. Its high irrigation requirement completes the dissolution of gypsum, and makes replacement of sodium and salt leaching possible (Bhargava, 1989: 236). According to Yadav (1987), the reclamation should invariably be started using rice as the first crop in the monsoon season. Wheat, berseem (*Trifolium alexandrinum*) and barley have given good results in winter, though most farmers prefer wheat.

Agronomic and cultural practices

The field to be reclaimed should be properly levelled to ensure uniform spread of irrigation water, amendment and fertilisers. Strong bunds around the field help in conserving rainwater and in preventing entry of run-off from the surrounding area. Rice-wheat-dhaincha (*Sesbania aculeata*) rotation should be followed for at least three years, after which other suitable crops can be introduced. The area under reclamation should not be left fallow for long periods. Since alkaline soils are deficient in nitrogen, an application of about 25 per cent more nitrogen, compared to normal soils, is highly beneficial (Yadav, 1987). Addition of 20 to 30 t ha^{-1} rice husk as mulch on the surface of sodic soil improves infiltration rate and increases yield (Bhargava, 1989: 237–238).

Table 6-14: Relative tolerance of crops to exchangeable sodium.

Relative tolerance	Common name	Botanical name
Tolerant:	Karnal grass (a)	*Diplachne fusca*
	Rhodes grass	*Chloris gayana*
	Para grass	*Brachiaria mutica*
	Bermuda grass	*Cynodon dactylon*
	Rice	*Oryza sativa*
	Sugar beet	*Beta vulgaris*
Semi-tolerant:	Dhaincha	*Sesbania aculeata*
	Wheat	*Triticum aestivum*
	Barley	*Hordeum vulgare*
	Oats	*Avena sativa*
	Raya	*Brassica juncea*
	Senji	*Melilotus indica*
	Berseem	*Trifolium alexandrinum*
	Sugarcane	*Saccharum officinarum*
	Pearl millet	*Pennisetum typhoides*
	Cotton	*Gossypium hirsutum*
Sensitive:	Cowpea	*Vigna unguiculata*
	Gram	*Cicer arietinum*
	Groundnut	*Arachis hypogaea*
	Lentil	*Lens esculenta*
	Black gram	*Phaseolus mungo*
	Green gram	*Vigna radiata*
	Pea	*Pisum sativum*
	Maize	*Zea mays*
	Cotton (at germination)	

(a) Kallar grass

Source: Yadav (1987).

Hydro-technical methods

Leaching and drainage are two important components of hydro-technical methods in the reclamation of salt-affected soils. Subsurface drainage in the sodic soils of the Indo-Gangetic Plains has not been found feasible because of poor hydraulic conductivity. Application of a suitable amendment like gypsum combined with leaching has been found to be an efficient method of reclamation in such sodic soils. Encouraging results have been obtained by manipulation of excess rainwater by storing the maximum possible water in rice fields, conserving most of the remaining water in ponds in nearby low lying areas, and finally allowing only excess water to go to the drain.

Revegetation

Some of the alkaline-affected areas, which cannot be utilised for production of agricultural crops, are often suitable for alternative uses such as growing trees and fodder grasses. In a field experiment undertaken at Karnal on a highly sodic soil, it was concluded that species like *Eucalyptus* hybrid, *Prosopis juliflora* and *Acacia nilotica* can be grown successfully by treating the alkaline soil of the planting pit (90 cm × 90 cm) with gypsum (3.5 kg) plus farmyard manure (25 kg), and with a little fertiliser application (Yadav,

1987). B. Singh (1987) described the results of a number of experiments for the rehabilitation of alkaline wastelands of Uttar Pradesh by plantation of trees (*Prosopis juliflora, Acacia nilotica, Pongamia pinnata, Terminalia arjuna, Eucalyptus tereticornis* and *Leucaena leucocephala*) and shrubs (*Sesbania sesban* and *Tamarix dioica*). The objective of the experiments was to identify suitable species, soil amendments (gypsum and pyrite) and planting density and to observe the effects of plantation on soil ameiloration.

Certain grasses can grow on alkaline soils. Para grass (*Brachiaria mutica*), kallar grass (*Leptochloa fusca*, synonym: *Diplachne fusca*) known as Karnal grass in India, Bermuda grass (*Cynodon dactylon*) and Rhodes grass (*Chloris gayana*) have been found to show greater tolerance to sodic conditions than other grasses (Yadav, 1987). Research indicates that, on a highly sodic soil (pH 10.6), *Leptochloa fusca* gave the highest yield of 18.8 t ha^{-1} without gypsum application, but *Chloris gayana* produced a maximum 27.1 t ha^{-1} with application of 5.2 t ha^{-1} of gypsum. *Brachiaria mutica* and *Cynodon dactylon* showed good performance even without or with very little addition of gypsum.

Management of saline soils

Saline soils have an excessive concentration of natural salts and are associated with a high watertable and, in many cases, poor groundwater quality. A brief description of a few important options for management of the saline soils of India is now given.

Drainage

In saline soils provision of a suitable drainage system to maintain the watertable below the critical depth and to facilitate leaching of salts out of the root zone is of paramount importance for salinity control. In a saline sandy loam soil with high watertable in Delhi, introduction of 1 m deep open drains, with a spacing of 32 m, significantly lowered the watertable and soil salinity. Provision of open drains, at 1.5 m depth and a spacing of 58.5 m in a saline area in Haryana, brought about substantial reduction in salt content, lowering of the watertable and increasing the crop yield. In a field experiment in a coastal saline area of West Bengal, a maximum reduction in salt content was achieved in a plot having open drains at 15 m spacing and laid at 1.75 m depth. In general terms, however, considerable loss of cultivable land limits the scope of open drain systems.

Placement of tiles, 10 cm in diameter, at 1.2 m depth and 12–24 m spacing, was found effective in lowering the watertable, increasing crop yield and leaching the salts from black clayey saline soils in Karnataka State. Subsurface drainage using cement concrete tiles (10 cm diameter), placed at 1.75 m depth and at a spacing of 25 m, 50 m and 75 m, has been tried at Sampla in Haryana State. Results showed that subsurface drainage with a spacing of 50–75 m was adequate.

Salt-tolerant crops

Different crops show different tolerance to soil salinity. Different varieties of the same crop differ widely in their tolerance. Tolerance also varies with stage of growth, being generally much lower at germination and increasing with age. Yadav (1987) provides the relative tolerance of crops to soil salinity under three groups: tolerant (5000–10 000 μS cm^{-1}), medium tolerant (3000–5000 μS cm^{-1}) and sensitive (1500–3000 μS cm^{-1}). As examples: barley, cotton, wheat and date palm are salt-tolerant; sunflower, sugarcane, tomato and cabbage are medium tolerant; and pea, pigeon pea and bean are sensitive.

Agronomic and cultural practices

In order to minimise salt injury at germination, adoption of suitable methods of sowing/plantation is of great significance. Also, mulching with crop residues has shown favourable effects in lowering temperature in surface soil, and thereby in reducing salt concentration. Urea has been found to be a better source of nitrogen compared to ammonium sulphate and calcium-ammonium nitrate. The highest rice yield in coastal saline soils of West Bengal was obtained when 75 per cent and 25 per cent of total nitrogen (100 kg ha^{-1}) as urea was applied at tillering and flowering stage, respectively.

Reclaimed saline and alkaline soils

Considering the effectiveness, as well as economic feasibility, of reclaiming alkaline soils, the state governments of Haryana, Punjab and Uttar Pradesh have established Land Reclamation Corporations for undertaking reclamation of alkaline soils on an extensive scale. As a result, more than 340 000 ha of alkaline soils are reported to have been brought under reclamation during the period 1974–85. The extents of the reclaimed land in the three states of Punjab, Uttar Pradesh and Haryana are 232 300 ha, 66 000 ha and 42 600 ha, respectively (Yadav, 1987). Recent statistics show that about 500 000 ha of land has been reclaimed so far in the three states mentioned earlier (Rao, 1992).

According to Yadav (1987), although saline soils occupy large areas in India (both inland and in the coastal areas), and this problem is further spreading with the expansion of canal irrigation, not much progress has been made regarding reclamation and crop production on these saline soils. Research results indicate that an appreciable portion of these soils can be reclaimed by adoption of proper management options (leaching, drainage, water management, crop management, etc.). However, difficulties, including the requirements of many options to be implemented at catchment scale, high cost, skill required and lack of infrastructural facilities, have been serious handicaps in large-scale amelioration of these soils. As a consequence, despite the vast magnitude of the problem, management activities have been on a limited scale.

Stream salinity

Rawal (1978) investigated the chemical characteristics of water samples taken from the important rivers of India. His investigation covered 41 rivers with 86 observation sites. Selection of sampling sites was made on a broad basis to provide a general outlook on the chemical characteristics of the major rivers of the country. The period of observation ranged from three months for the new sites to 64 months for the established sites, with an average of more than 22 months for the 86 sites. Water samples were analysed for electrical conductivity, pH, total dissolved solids, cations (Ca^{++}, Mg^{++}, Na^+ and K^+) and anions (CO_3^{--}, HCO_3^-, SO_4^{--}, Cl^-). Figure 6-14 shows the salinity values at a number of sites on major Indian rivers.

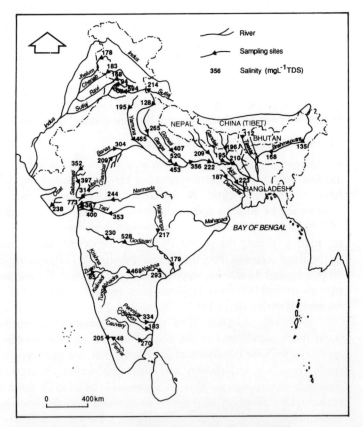

Figure 6-14: Salinity of major rivers of India (Data from: Rawal, 1978).

In general, the water quality of Indian rivers is suitable for irrigation. The northern and eastern rivers have a moderate salt content of 150–500 mg L^{-1}, and the salt concentration in the majority of cases does not increase appreciably during the dry period. Also, monthly as well as annual variations are low to moderate. Waters of the central India and the southern Peninsula rivers carry comparatively higher concentrations of dissolved salts than the northern and eastern rivers, and show an appreciable increase in salt concentration during the dry period. Salinity is low only during the monsoon period. Monthly variations are large and annual variations are appreciable in the majority of rivers. In the case of the Godavari River, soluble salts vary between moderate and high during different months of the year, and the lower reach of the Narmada River has a relatively high salinity.

References

Abrol, I.P. 1986. Fuel and forage production from salt affected wasteland in India. *Reclamation and Revegetation Research*. 5: 65–74.

Abrol, I.P. 1990. Problem soils in India. In. *Problem Soils of Asia and the Pacific*. Report of the expert consultation of the Asian network on problem soils, Bangkok, 29 August–1 September 1989. Bangkok: FAO Regional Office for Asia and the Pacific. 153–165.

Bapat, M.V. and Shah, R.C. 1984. Problems of waterlogging and salinity in the irrigated commands in Gujarat State with a case-study of Kakrapar Project. In. *Seminar on Problems of Waterlogging and Salinity in Irrigated Areas*. Krishnarajasagar, Karnataka State, 13–16 November 1984. New Delhi: Ministry of Irrigation, Government of India. 1–26.

Bhargava, G.P. 1989. *Salt-Affected Soils of India: A Source Book*. New Delhi: Oxford & IBH Publishing Co. 261 pp.

Bradley, E. and Phadtare, P.N. 1989. Paleohydrology affecting recharge to overexploited semi confined aquifers in the Mehsana area, Gujarat State, India. *Journal of Hydrology*. 108: 309–322.

Chatterji, P.C. 1980. Problems of ground water resources of Western Rajasthan and their possible management. In. Mann, H.S. Editor-in-Chief. *Arid Zone Research and Development*. Jodhpur: Scientific Publishers. 87–94.

Department of Technical Co-operation for Development. 1986. *Ground Water in Continental Asia (Central, Eastern, Southern, South-Eastern Asia)*. New York: United Nations. (Natural Resources/Water Series, no. 15). 391 pp.

Dhar, O.N. and Rakhecha, P.R. 1979. Incidence of heavy rainfall in the Indian desert region. In. *The Hydrology of Areas of Low Precipitation*. Proceedings of the Canberra Symposium, December 1979. (IAHS Publication no. 128). 33–42.

Food and Agriculture Organization of the United Nations (FAO). 1989. *Production Yearbook*. Rome: FAO. v.42. 350 pp.

Framji, K.K., Garg, B.C. and Luthra, S.D.L. 1982. *Irrigation and Drainage in the World: A Global Review*. New Delhi: International Commission on Irrigation and Drainage. v.II. 493–1159.

Gangwar, A.C. and van der Toorn, W.H. 1987. The economics of adverse groundwater conditions in Haryana State. *Indian Journal of Agricultural Economics*. 42(2): 160–172.

Ghosh, G. and Phadtare, P.N. 1990. Policy issues regarding groundwater management in India. In. *Proceedings of the International Conference on Groundwater Resources Management*. Asian Institute of Technology, Bangkok, 5–7 November 1990. Bangkok: Division of Water Resources Engineering, Asian Institute of Technology. 433–457.

Gore, S.R. and Bhagwat, K.A. 1991. Saline degradation of Indian agricultural lands: A case study in Khambhat Taluka, Gujarat State (India), using satellite remote sensing. *Geocarto International*. 6(3): 5–13.

Government of India, Ministry of Irrigation. 1984. *Seminar on Problems of Waterlogging and Salinity in Irrigated Areas*. Krishnarajasagar, Karnataka State, 13–16 November 1984. New Delhi: Ministry of Irrigation, Government of India. 180 pp.

Government of India, Ministry of Water Resources. 1987. *Special Course on Land Drainage and Salinity Control in Black Cotton Soils*. Water and Land Management

Institute (WALMI) Aurangabad, 16 February–6 March 1987. Aurangabad: WALMI. (Various pagings, about 350 pp.).

Gupta, I.C. and Pahwa, K.N. 1978. *A Century of Soil Salinity Research in India: An Annotated Bibliography 1863–1976*. New Delhi: Oxford & IBH Publishing Co. 400 pp.

Johnson, B.L.C. 1983. *India, Resources and Development*. Second edition. London: Heinemann Educational Books. 212 pp.

Kolvalli, S. and Chicoine, D.L. 1989. Groundwater markets in Gujarat, India. *Water Resources Development*. 5(1): 38–44.

Krishnan, A. 1971. Distribution of arid areas in India. In. Chatterjee, S.P., Roy, B.B. and Das Gupta, S.P., eds. *Proceedings of Symposium on Arid Zones*. Held at the Central Arid Zone Research Institute, Jodhpur, 22–29 November 1968. Calcutta: National Committee for Geography. 11–19.

Kulkarni, K.M., Rao, S.M., Singhal, B.B.S., Parkash, B., Navada, S.V. and Nair, A.R. 1989. Origin of saline ground water in Haryana State, India. In. *Regional Characterization of Water Quality*. (IAHS Publication no. 182). 125–132.

Mann, H.S. 1986. Arid land development in South Asia. In. *Arid Land Development and the Combat Against Desertification: An Integrated Approach*. United Nations Environment Programme (UNEP)/USSR Commission for UNEP. Moscow: Centre for International Projects GKNT. 84-87.

Mathur, C.M., Sharma, O.P. and Gaun, S.N. 1977. Saline waters of Rajasthan: Their chemical characteristics and utilization for agricultural purposes. In. Roonwal, M.L. ed. *The Natural Resources of Rajasthan*. Jodhpur: The University of Jodhpur. v.2. 653–662.

Mistry, J.F. 1983. Salinity ingress in the coastal area of Saurashtra-Gujarat State (India). *ICID Bulletin*. 32 (2): 52–60.

Mistry, J.F. 1989. Measures to overcome the salinity problems of the coastal areas in Gujarat. In. *Important Aspects of River Valley Projects*. Ahmadabad: Water Resources Department, Gujarat State. v.4. 41–50.

Mistry, J.F. and Purohit, M.U. 1989. Environmental effects of Mahi-Kadana and Dharoi Projects, Gujarat-India. *ICID Bulletin*. 38(1): 21–30.

Mukherjee, K.N. 1985. *River Basin Atlas of India: Based on Designated Best Use Classification of Streams*. New Delhi: Central Board for the Prevention & Control of Water Pollution. 36 pp.

Murty, K.S. 1990. India's large water projects and their impacts on environment. In. *The Impact of Large Water Projects on the Environment*. Proceedings of an International Symposium convened by UNESCO and UNEP, Paris, 21–23 October 1986. Paris: UNESCO. 127-136.

Muthiah, S. Editor-in-Chief. 1987. *A Social and Economic Atlas of India*. Delhi: Oxford University Press. 254 pp.

National Council of Applied Economic Research. 1963. *Techno-Economic Survey of Rajasthan*. New Delhi: National Council of Applied Economic Research. 316 pp.

Navalawala, B.N. 1992. Indian perspective in water resources planning. In. *Pre-Seminar Proceedings of Seminar on Irrigation Water Management*. Gandhinagar: Water Management Forum. v.1. 18–42.

Pal, I. 1977. Water resources for irrigation in Rajasthan. In. Roonwal, M.L. ed. *The Natural Resources of Rajasthan*. Jodhpur: The University of Jodhpur. v.2. 639–651.

Palaniappan, S.P. and Budhar, M.N. 1992. Role of green manure in management of salt affected soils. In. *Proceedings of the International Symposium on Strategies for Utilizing Salt Affected Lands*. Bangkok, 17–25 February 1992. Bangkok: Department of Land Development, Ministry of Agriculture and Cooperatives. 378–393.

Patel, C.L. 1984. Waterlogging and salinity problems in Mahi-Kadana Command area of Gujarat. In. *Seminar on Problems of Waterlogging and Salinity in Irrigated Areas*. Krishnarajasagar, Karnataka State, 13–16 November 1984. New Delhi: Ministry of Irrigation, Government of India. 161–174.

Patel, G. 1977. *Gujarat's Agriculture*. Ahmadabad: Overseas Book Traders. 363 pp.

Patel, V.B. and Shankara Iyer, D.R. 1987. Strategy for water planning and irrigation management during the scarcity year (1985-86) in the State of Gujarat, India. In. *Transactions, Thirteenth International Congress on Irrigation and Drainage*. Rabat, September 1987. v.1–B. 1027–1039.

Pathak, B.D. 1985. Groundwater resources development and management in India. In. Castany, G., Groba, E. and Romijn, E. eds. *Hydrogeological Mapping in Asia and the Pacific Region*. Proceedings of the ESCAP-RMRDC Workshop, Bandung 1983. (International Contribution to Hydrogeology. v.7). Hannover: Heise. 145–149.

Raman, S. 1984. Impact of canal irrigation on waterlogging and soil salinity in Ukai-

Kakrapar Command — A case study. In. *Seminar on Problems of Waterlogging and Salinity in Irrigated Areas*. Krishnarajasagar, Karnataka State, 3–16 November 1984. New Delhi: Ministry of Irrigation, Government of India. 131–145.

Rao, K.V.G.K. 1992. Central Soil Salinity Research Institute. Karnal, Haryana. (Personal communication).

Rao, K.V.G.K., Kumbhare, P.S., Kamra, S.K. and Oosterbaan, R.J. 1990. Reclamation of waterlogged saline alluvial soils in India by subsurface drainage. In. *Symposium on Land Drainage for Salinity Control in Arid and Semi-Arid Regions*. Cairo, 25 February–2 March 1990. Cairo: Drainage Research Institute. v.2. 17–25.

Rawal, N.C. 1978. Quality of river waters of India. In. *Proceeding of the 47th Research Session*. Central Board of Irrigation and Power. v.II. 139–162.

Reddy, S.T.S. 1989. Declining groundwater levels in India. *Water Resources Development*. 5(3): 183–190.

Roy, B.B. and Kolarkar, A.S. 1977. Saline and sodic soils of Rajasthan: A Review. In. Roonwal, M.L. ed. *The Natural Resources of Rajasthan*. Jodhpur: The University of Jodhpur. v.2. 1011–1021.

Sehgal, J.L. and Sharma, P.K. 1988. An inventory of degraded soils of Punjab (India) using remote sensing technique. *Soil Survey and Land Evaluation*. 8(3): 166–175.

Singh, B. 1989. Rehabilitation of alkaline wasteland on the Gangetic alluvial plains of Uttar Pradesh, India, through afforestation. *Land Degradation and Rehabilitation*. 1: 305–310.

Singh, K.B. 1979. Water management at the farm level in India. In. *Proceedings of the Workshop on Efficient Use and Maintenance of Irrigation Systems at the Farm Level in China*. New York: United Nations. (Water Resources Series no. 51). 72–75.

Singh, K.P. 1987. Behaviour of water levels in the canal irrigated tracts of Punjab State, India. In. *Groundwater and the Environment*. Proceedings of the International Groundwater Conference, 1987. Malaysia: Universiti Kebangsaan. B33–38.

Singh, K.P. 1989. *Technology Plan, Water Regime Management: Key Plan Area-I, Water Logging*. Chandigarh: Punjab State Council for Science and Technology. 39 pp.

Singh, K.P. 1990. Hydromorphological control of hydrogeological and hydrochemical regimes — a case study of Punjab State and adjoining areas, India. In. *Groundwater and the Environment*. Proceedings of the Second International Groundwater Conference, Malaysia, 25–29 June 1990. Bangi, Malaysia: Faculty of Engineering, Universiti Kebangsaan. A50–55.

Singh, S. and Ghose, B. 1980. Geomorphic significance in the environmental problems of the Rajasthan Desert, India. In. Mann, H.S. Editor-in-Chief. *Arid Zone Research and Development*. Jodhpur: Scientific Publishers. 103–108.

Sondhi, S.K., Rao, N.H. and Sarma, P.B.S. 1989. Assessment of groundwater potential for conjunctive water use in a large irrigation project in India. *Journal of Hydrology*. 107: 283–295.

Stone, I. 1984. *Canal Irrigation in British India: Perspectives on Technological Change in a Peasant Economy*. Cambridge: Cambridge University Press. 374 pp.

Szabolcs, I. 1989. *Salt-Affected Soils*. Boca Raton, Florida: CRC Press. 274 pp.

Tanwar, B.S. 1979. Effects of irrigation on the groundwater system in the semiarid zone of Haryana, India. In. *The Hydrology of Areas of Low Precipitation*. (IAHS Publication no. 128). 375–384.

Tanwar, B.S. and Kruseman, G.P. 1985. Saline groundwater management in Haryana State, India. In. *Hydrogeology in the Service of Man*. Memoirs of the 18th Congress of the International Association of Hydrogeologists. Cambridge, 8–13 September 1985. v.XVIII, Part 3 (Groundwater quality management). 24–30.

World Bank. 1992. *World Development Report 1992: Development and the Environment*. New York: Oxford University Press. 308 pp.

Yadav, J.S.P. 1987. Management of saline/alkaline soils in India. *Regional Expert Consultation on Management of Saline/Alkaline Soils*. Bangkok, 25–29 August 1987. Bangkok: FAO Regional Office for Asia and the Pacific. 42 pp.

Chapter 7: Iran

Introduction

Iran covers an area of 1 648 000 km² in south-western Asia. Some 16 per cent of the total land surface of the country is mountainous and rough with an elevation of more than 2000 m above mean sea-level. About 53 per cent of the country has an elevation of 1000 to 2000 m. High mountains extend along the western and northern margins of the Central Plateau, leaving the remaining part a broad upland with an average altitude of 1500 m. The main physiographic features of Iran are as follows (Figure 7-1):

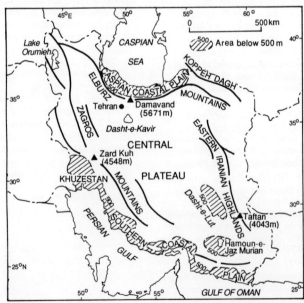

Figure 7-1: Main physiographic features of Iran (Beaumont et al., 1988: 20; Movahed-Danesh, 1972).

- **The Zagros Mountains** stretch from the north-west in a south-easterly direction to the region bordering the Persian Gulf and the Gulf of Oman. They form a continuous wall about 1000 km in length and often more than 200 km wide, lying between the plain of Mesopotamia and the Central Plateau of Iran. The highest peak of the Zagros Mountains is Zard Kuh with an elevation of 4548 m.
- **The Elburz (Alborz) Mountains** form a continuous wall along northern Iran from Ararat in Turkey to the north-eastern corner of Iran. Their central part is parallel to the southern shore of the Caspian Sea. Their highest peak is the snowcapped Mt Damavand, with an elevation of 5671 m. There is a sudden drop on the northern side to the flat plain bordering the Caspian Sea, which is about 27.2 m below mean sea-level.
- **The Central Plateau** is about 500 m to 2500 m high and is almost ringed by mountain chains of varying height and extent. Two great deserts extend over much of the Central Plateau. Dasht-e-Lut is covered largely with sand and rocks, while the Dasht-e-Kavir is covered mainly with salt. Both deserts are inhospitable and virtually uninhabited. In the winter and spring, small streams flow into the Dasht-e-Kavir creating lakes and swamps. At other times of the year both deserts are extremely arid.
- **The Caspian Coastal Plain** is a narrow plain with an average width of about 50 km, formed by the withdrawal of the sea. This plain has a high rainfall and fertile soils.
- **The Khuzestan and the Southern Coastal Plains** extend from south-western Iran to the coastal areas of the Persian Gulf and the Gulf of Oman.

Iran had an estimated population of 56 million in 1990, with a growth rate of 3.4 per cent over the period 1980–90 (World Bank, 1992: 269).

Rainfall and climate

Precipitation in Iran is predominantly in the form of rainfall except for snow in the high mountains and elevated areas. Rainfall occurs during early spring, late autumn and winter. The average annual precipitation ranges from less than 50 mm in the deserts to more than 1600 mm on the Caspian Coastal Plain (Figure 7-2). The average annual precipitation of the country is about 260 mm, and approximately 90 per cent of the country is arid or semiarid (Table 7-1). The portion of the precipitation that occurs as snow in the Elburz and Zagros Mountains has a major effect on regulating surface run-off. The water accumulated during the winter in snowfields at high elevations melts and is released during spring and early summer. The average annual potential evaporation of the country is very high, ranging from less than 700 mm along the Caspian Sea shore to over 4000 mm in the deserts and the south-western part of the Khuzestan Plain (Ministry of Energy, 1976b).

Figure 7-2: Average annual precipitation in Iran (Department of Technical Co-operation for Development, 1982: 5, and Ministry of Energy, 1976a).

Table 7-1: Distribution of precipitation in Iran.

Annual precipitation (mm)	Recipient area (km²)	(%)
<50	100 000	6
50–100	285 000	17
100–200	465 000	28
200–300	370 000	23
300–500	280 000	17
500–1000	130 000	8
>1000	18 000	1
Total	1 648 000	100

Source: Massoumi (1984).

The climate of Iran is one of great extremes due to its geographic location and varied topography. It ranges from extremely hot to Mediterranean and to extremely cold. The summer is hot to extremely hot with temperatures rising as high as 50°C in Khuzestan Province at the head of the Persian Gulf. In winter the temperatures are low. Temperatures of the order of –20°C are common in many places in north-western Iran.

Water resources

Surface water

Most of Iran's rivers flow mainly during winter and spring when precipitation is heaviest. The country's principal permanent rivers flow from the mountains on the slopes facing the Caspian Sea and the Persian Gulf. The Karun River, flowing from the Zagros Mountains to the Arvand-Rud (Shatt-al-Arab) at Khoramshahr, is the country's main navigable river with an annual flow of about 10×10^9 m^3. Based on hydrological conditions, Iran is divided into six major river basins (Figure 7-3). The main characteristics of these basins are summarised in Table 7-2.

Table 7-2: Characteristics of the major river basins in Iran during the water year 1988-89.

Basin	Area (km^2)	Annual rainfall (mm)	Volume of rainfall (10^9 m^3)	Annual run-off (10^9 m^3)
Caspian Sea	177 000	490	86.73	16.40
Persian Gulf and Gulf of Oman	430 000	369	158.67	40.50
Lake Orumieh	52 700	445	23.45	5.75
Central Plateau	831 000	166	137.95	5.55
Lake Hamoun	105 600	110	11.62	2.85
Kara Kum	43 900	243	10.67	0.45
Total	1 640 200	–	429.09	71.50

Source: Ministry of Energy (1990a).

In terms of water balance, of an average annual precipitation of 429×10^9 m^3, 305×10^9 m^3 (71 per cent) returns to the atmosphere as evapotranspiration. Of the remaining 124×10^9 m^3, some 38×10^9 m^3 recharge aquifers and 86×10^9 m^3 constitute run-off. The net inflow of surface water to the country is 4×10^9 m^3 (9×10^9 m^3 inflow and 5×10^9 m^3 outflow), adding to the potential surface water resources of the country. The total water resource potential of Iran is about 128×10^9 m^3, consisting of 90×10^9 m^3 of surface water and 38×10^9 m^3 of groundwater. Total water use in Iran is about 75×10^9 m^3, of which 72.5×10^9 m^3 is for agriculture, 2×10^9 m^3 for domestic and 0.5×10^9 m^3 for industrial purposes.

Because of the shortage of water and growing demand for water, the long-term objective of Iran's water resources development plan is based on control and regulation of water resources that can be economically utilised in agriculture, industry and urban development schemes. As a consequence of this policy 15 large dams have been constructed since 1957. These dams

Figure 7-3: Main river basins in Iran (Economic and Social Commission for Asia and Pacific, 1985: 231; Ministry of Water and Power, 1965).

control more than 22×10^9 m^3 of water (Table 7-3). There are also six large dams under construction which will control 1.8×10^9 m^3, and 43 dam projects are under study (Iranian National Committee on Large Dams, 1990; see also Economic and Social Commission for Asia and the Pacific, 1985: 234 for the location map of the existing, under-construction and under-investigation dams).

Aside from the Caspian Sea, Iran has a few large lakes. Most lakes shrink in size during the hot dry summer and have a high salt content, because they have no outlet to carry away the salt concentrated from summer evaporation. The largest water body entirely within Iran is Lake Orumieh in the north-west, at an elevation of about 1275 m. It covers an area of 5800 km^2, with a maximum depth of 16 m and volume of 45×10^9 m^3 (Korzun et al., 1978: 34). It is a saline lake with a very high salt content of more than 310 000 mg L^{-1} TDS (Hammer, 1986: 553).

Groundwater

The existence and occurrence of groundwater as a main source of water was known and understood by the Persians thousands of years ago. The traditional method of groundwater extraction is Qanat (Ghanat), which brings groundwater to the surface by gravity. This method is one of the oldest techniques of groundwater extraction and has its historical roots in the early kingdoms in the north-west of Iran (Beaumont et al., 1989; Bybordi, 1974).

Table 7-3: Characteristics of dams constructed since 1957 in Iran.

Dam	River	Period of construction	Effective capacity (10^6 m^3)	Regulated annual flow (10^6 m^3)	Irrigated area (ha)
Amir Kabir	Karaj	1958–61	195	450	21 000
Dez	Dez	1957–62	2 480	6 960	93 700
Droudzan	Kor	1966–72	860	640	41 000
Ekbatan	Abshineh	1959–63	5	17	200
Esteghlal	Minab	1974–82	271	240	14 000
Golpayegan	Golpayegan	1957–60	40	80	6 600
Kardeh	Kardeh	1978–88	35	32	3 700
Lar	Lar	1974–81	860	416	105 000
Latyan	Jaj-Rud	1963–67	85	220	30 000
Mahabad	Mahabad	1967–70	190	195	31 000
Sefid-Rud	Sefid-Rud	1957–61	1 650	2 000	250 000
Shahid Abas-Pour	Karun	1969–72	1 730	9 300	43 000
Torogh	Torogh	1978–87	35	13	13 000
Zarrineh-Rud	Zarrineh-Rud	1967–71	480	535	68 000
Zayandeh-Rud	Zayandeh-Rud	1967–70	1 090	1 208	90 000
Total			10 006	22 306	810 200

Source: Iranian National Committee on Large Dams (1990).

The groundwater resources of Iran occur mainly in alluvial and karstic limestone aquifers (Department of Technical Co-operation for Development, 1982: 42–56; Ministry of Energy, 1990b). The major alluvial aquifers are the main sources of groundwater supply and generally occur at the foothills of mountains in the form of alluvial fans and flood plains. They consist of boulder and gravel near the foothills, and the material becomes more finely textured toward the centre of the plains. The thickness of the alluvium is generally considerable and reaches more than 300 m in some plains such as Tehran, Ghazvin and Varamin. The depth of watertable normally ranges from more than 60 m to a few metres. Highly productive and extensive aquifers of this type occur at the southern foothills of the Elburz Mountains in Ghazvin, Karaj, Tehran, and the Varamin plains.

Groundwater quality in alluvial aquifers is usually good in recharge areas (TDS less than 500 mg L^{-1}) but increases along flow lines and in shallow aquifers due to evaporation (to 3000 mg L^{-1} or more). In some areas, particularly in the central and southern parts of the country, groundwater resources may become very saline due to the effects of saline geologic formations and salt domes.

The best karstified carbonate rocks of Iran belong to Jurassic, Cretaceous and Oligo-Miocene Periods, with Cretaceous and Oligo-Miocene formations displaying the largest and thickest of these formations. The extent of fractured carbonate rocks in Iran has been estimated to be about 150 000 km^2, or nearly 10 per cent of the total area of the country (Aghassi, 1990). Three distinct carbonate rock zones are observed: the western and south-western zone, covering an area of 85 000 km^2 in the Zagros Mountains; the northern zone in the Elburz Mountains, covering an area of 40 000 km^2; and the central zone, which consists of dispersed areas covering 25 000 km^2.

Karstic aquifers are well developed in the provinces of Lorestan, Bakhtaran, Fars, Hormozgan and in the northern part of Khorasan. In Fars Province, the best karstic limestone belongs to the Asmari Limestone from Oligo-Miocene. This is also the reservoir of the major Iranian oil fields (James and Wynd, 1965). In general, groundwater quality of the karstic formations is good and can be used for domestic, agricultural and industrial purposes.

Safe yield of groundwater resources has been estimated to be about 38×10^9 m^3 per annum. However, evidence shows that at present the annual extraction from the aquifers is more than their natural recharge (Table 7-4). This has caused significant decline in the groundwater level of many aquifers, land subsidence in heavily pumped areas, as well as intrusion of saline water from the sea or internal saline bodies. Some examples of decline and deterioration are the Kerman, Bardsir, Rafsanjan, Sirjan, Yazd, Mashad, Saveh, Qom, and Ardestan aquifers.

Table 7-4:   Groundwater extraction in Iran in 1991.

Means of extraction	Number	Volume of extraction (10^9 m^3)
Deep wells	65 448	22.07
Shallow wells	160 835	10.12
Qanats	27 210	8.25
Springs	25 082	8.53
Total	–	48.97

Source: Statistical Centre of Iran (1992: 26).

Land cover

In 1987, of a total land area of 163.6 Mha, about 14.83 Mha were under cultivation. Some 5.74 Mha were irrigated and the remaining 9.09 Mha non-irrigated (FAO, 1989). Permanent pasture covered 44.0 Mha, forest and woodlands 18.02 Mha and other lands 86.75 Mha. This latest figure includes 34 Mha of saline, sandy and rocky deserts of the Central Plateau (Ghobadian, 1990: 78). Agricultural land availability is not a major constraint in the development of Iranian agriculture where there are 32 Mha of suitable quality land. The major constraint is availability of water for development of these lands (Bybordi, 1989).

Irrigation

The problem of water supply has been a constant preoccupation since the beginning of the country's history, thousands of years ago. Its inhabitants learnt to design and implement efficient techniques for harnessing limited water resources and irrigating the land. Apart from Qanat, which has been a major source of irrigation and domestic water supply for centuries, Iranians have in the past 2500 years built dams of various types and weirs in Khuzestan, Fars and other provinces. Reza et al. (1976) have provided details of these ancient hydraulic structures. Some of them, built as long ago as 1000 years, are still in good condition.

Over the past few decades many important irrigation projects have been implemented throughout Iran. Framji et al. (1982: 610–629) have provided a brief historic background and some details of these projects. Table 7-3 provides the size of irrigated area downstream of existing dams. In these irrigation projects, drainage networks have been constructed. According to ICID (1977), for an irrigated area of 1.03 Mha developed or under development between 1966 and 1980, the lengths of the main and secondary irrigation canals were 1352 km and 4287 km, respectively, while the total length of drains was about 6520 km.

On-farm water application rates in the country are high and irrigation practice has a low efficiency of about 30 per cent (Ghobadian, 1990: 113–115). For example, the average annual application rates for major crops are as follows: barley, 4000 m^3 ha^{-1}; wheat, 5000 m^3 ha^{-1}; cotton, 13 000 m^3 ha^{-1}; sugar beet, 14 000 m^3 ha^{-1}; rice, 15 000 m^3 ha^{-1}; and vegetables, 17 000 m^3 ha^{-1} (Ghobadian, 1990: p. 47).

Apart from losses via unlined irrigation canals, the major part of the losses are at farm level via evaporation, due to inefficient irrigation practices (surface irrigation methods), and percolation to the shallow aquifers which causes watertable rise and salinisation. Bybordi (1989) discusses the inefficiency of irrigation projects in Iran and makes some recommendations to improve it. Major causes of inefficiency include: improper design of irrigation facilities; poor maintenance; careless operation; negligible water prices; fragmentation of responsibilities among different government agencies; and inadequate training of farmers.

Irrigation plays a vital role in the agricultural production of the country. Annual production of major crops includes: 6.01 Mt of wheat; 2.84 Mt of barley; 1.85 Mt of rice; 3.54 Mt of sugar beet; 2.03 Mt of potato and 2.43 Mt of vegetables (Directorate of Statistics of Agricultural Information, 1990: 27). Dryland farming is in practice in areas with relatively moderate rainfall and produces mainly wheat (1.88 Mt) and barley (0.77 Mt). These two amounts in brackets are included in the above mentioned statistics.

In the irrigated areas the average annual yield of major crops is: wheat, 2035 kg ha^{-1}; barley, 1980 kg ha^{-1}; and rice, 3570 kg ha^{-1}, while in the rain-fed areas annual yields are much lower and consist of 440 kg ha^{-1} of wheat and 480 kg ha^{-1} of barley (Directorate of Statistics and Agricultural Information, 1990: 27). These annual yields are much lower than the average yield of other countries, like Australia, where the average yields of wheat and barley are around 1200 kg ha^{-1} (Shaw, 1984: 8–9). This is mainly because of lower average annual rainfall in Iran and its high variability.

Salinity

The salinity of soils and water resources in Iran is one of the major problems inhibiting their effective utilisation in agriculture. These problems depend on a number of factors including widespread distribution of evaporites, low rainfall, high potential evapotranspiration, topographic conditions creating closed or semiclosed basins, irrigation with low quality water (surface and groundwater) and inadequate irrigation and drainage facilities. Evaporites occur as halite and gypsum interbedded with layers of marls. These are mainly Late Precambrian evaporites, widespread in the Persian Gulf region, forming over 150 salt domes, and the Miocene evaporites which are extensive in the central and southern parts of the country (National Iranian Oil Company, 1975–78).

The combined effect of the above mentioned factors has caused the formation of huge territories of saline and alkaline soils. The largest concentration of these is in the central deserts. An estimate of the total salt-affected soils exceeds 15.5 Mha or nearly 10 per cent of the surface area of Iran (Dewan and Famouri, 1964: 122; Szabolcs, 1989: 50). According to the Soil and Water Research Institute (1987: 46), the salt-affected soils of the country amount to about 18 Mha, including 7 Mha of salt marsh in Dasht-e-Kavir and Dasht-e-Lut.

In spite of the extent of natural and human-induced salinity in the country, there are very few publications describing the extent and characteristics of these soils. Here we refer to some of these publications. Further information is also available in Abtahi (1977) and Abtahi et al. (1979 and 1980).

Mahjoory (1979) investigated the characteristics of three representative sodic soils of the arid and the semiarid regions of Iran. Two were from the alluvial plains of Karaj and Ghazvin, while the third soil sample was from southern Iran about 25 km north of Shiraz. Laboratory analyses indicated that the soils contained large quantities of soluble salts, mainly carbonates and chlorides of calcium, magnesium and sodium. Soil samples had pH and EC_e values ranging from 8 to 10.2 and 9000 to 39 000 µS cm^{-1}.

Matsumoto and Cho (1985) describe the results of a soil profile investigation in the Shavour area, which is located about 70 km north of Ahwaz, in Khuzestan Province. In this area salts in the soil are supplied from sedimentary rocks and irrigation water which has an EC value of 1350 µS cm^{-1}. They selected four points according to the land conditions as follows: uncultivated land not influenced by leakage from irrigation canals; uncultivated land with a shallow watertable due to leakage from irrigation canals; irrigated land with tile drainage; and irrigated land without drainage. In this area tile drains had a spacing of 68 m and drainage water, which discharged to the Karkheh River, had an EC value of 5200 µS cm^{-1}. Matsumoto and Cho (1985) provide the chemical properties of the four soil profiles and demonstrate the effect of irrigation, shallow watertables and drainage on the distribution of salt in the soil profiles.

Hajrasuliha et al. (1991) analysed the results of Cl$^-$ and EC$_e$ measurements on 640 soil samples in the irrigated and non-irrigated areas at 13 sites. These sites were selected in an approximately 200 km wide by 2400 km long strip, running from south to north-east Iran along the Dasht-e-Kavir and Dasht-e-Lut. They developed linear and polynomial regression models to quantify the relationship between measured Cl$^-$ concentration and EC$_e$ values for a broad range of soils and soil salinity conditions in Iran. They concluded that these types of models can be used in soil salinity surveys to classify lands with respect to their Cl$^-$ concentration.

Irrigated land salinity

Secondary salinisation in the arid and semiarid parts of the country is a major problem, and considerable areas under the command of new dams have gone out of cultivation due to overirrigation and ensuing problems of waterlogging and salinisation (Bybordi, 1989; Framji et al., 1982: 618–619). The need for adequate drainage has been recognised in all irrigated lands but this does not mean that requirements have been adequately addressed. As mentioned earlier (see Irrigation, p. 345), in recently established projects drainage networks have been planned along with irrigation networks. However, the rise in watertables and development of waterlogging and secondary salinisation show the inadequacies arising from inappropriate design, construction and operation of the irrigation and drainage facilities and inefficient water management practices on irrigated fields. Development of secondary salinisation in the Khuzestan, Sistan, Moghan, Zarrineh-Rud, Doroudzan, Saveh and Zayandeh-Rud irrigation projects are good examples.

No comprehensive study has been undertaken regarding the extent of secondary salinisation in irrigated areas. However, considerable information in the form of reports by government agencies and consulting engineers has been accumulated on the subject for more than three decades, covering the periods of pre-development and post-development. According to an estimate for 1974, of the 4 Mha of land equipped with irrigation facilities, 2.5 Mha were on good quality soils with no particular constraints, while 1.5 Mha or 38 per cent of these lands had soils with considerable salinity and drainage problems (ICID, 1977). Extrapolating the 38 per cent to the current 5.74 Mha of the irrigated land in Iran indicates that about 2.2 Mha are salt-affected. In the following sections secondary salinisation in a number of irrigation projects will be briefly discussed.

The Moghan Irrigation Project

The Moghan Irrigation Project is located on the border of Iran with the Republic of Azerbaijan (CIS). The Moghan Plain covers 300 000 ha and has an arid climate with an average annual rainfall of about 300 mm per annum at Parsabad (Figure 7-4). Rainfall occurs in all seasons with about 64 per cent in winter and spring. The annual rainfall is variable and ranges from about 100 to 460 mm per year. Generally rainfall increases towards the west, east and the south. Annual potential evaporation is about 1485 mm per year with a maximum monthly value in July (224 mm) and a minimum value in January (34 mm). January and February are the coldest months with absolute values of –16°C and the warmest months are July and August with temperatures about 41°C. Originally, soils of the region had some salinity and alkalinity problems because of the geological characteristics of the region.

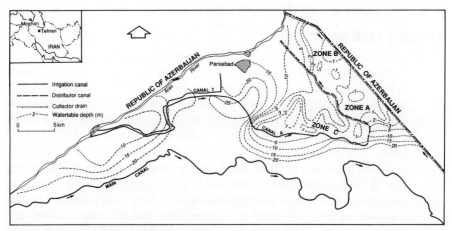

Figure 7-4: Watertable depth of the Moghan aquifer for May 1983 (Courtesy of Yekom Consulting Engineers, Tehran).

In spite of the availability of water from the Aras (Araks) River which flows from the south-west to the north-east at the western border of the plain, irrigation was not practised before the 1950s. Farmers produced some rainfed wheat and barley but the major part of the land was used for sheep and cattle grazing. The first irrigation canal (Canal T), with a capacity of 3 m^3 s^{-1}, was made available in about 1949. However, this canal was silted after construction and was abandoned. In 1959 a larger irrigation canal (Canal A), with a capacity of 17 m^3 s^{-1}, designed to irrigate 18 000 ha of land, and its irrigation network, was commissioned. This canal also had a siltation problem and its capacity reduced to 12–13 m^3 s^{-1}.

Following the 1963 agreement between Iran and the former USSR concerning the apportionment of water from the Aras River between the two countries, Aras Dam was built near Nakhichevan and a diversion dam was constructed about 250 km downstream of the Aras Dam (for the location of the dams, see Economic and Social Commission for Asia and the Pacific, 1985: 234). Also, plans were prepared for irrigation of another 72 000 ha of land, with the objective of increasing the extent of irrigation in the area to 90 000 ha. Aras Dam was commissioned in 1970 and part of the irrigation project was completed in 1974. The main irrigation canal is 116 km long, with a capacity of 80 m^3 s^{-1}. Gradually an intensive network of secondary canals (360 km) and open drains (417 km) has been developed in the region (ICID, 1977). Drainage facilities collect drainage waters with salinity values ranging from 600 to 4100 mg L^{-1} TDS and dispose them to the Aras River near the northern end of the area.

Aras water has an EC value of about 840–900 µS cm^{-1} and pH of 7.45–7.60, which means that the water is mildly alkaline. Groundwater extraction for irrigation water supply has not been attractive because of the availability of surface water and its higher quality. The watertable is more than 20 m deep in the southern and western parts of the region, but it is particularly shallow in the central and the north-eastern parts of the region (Figure 7-4, zones A, B and C). Groundwater salinity ranges from 3000 to 10 000 µS cm^{-1} in the areas with deep watertable, while in the areas with shallow watertable groundwater salinity ranges from 10 000 to 40 000 µS cm^{-1} (Figure 7-5).

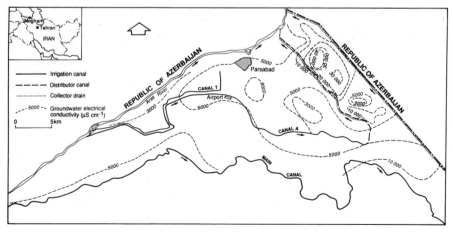

Figure 7-5: Groundwater electrical conductivity of the Moghan aquifer for May 1983 (Courtesy of Yekom Consulting Engineers, Tehran).

Water requirements of the crops within the 58 000 ha of land are estimated to be about 682×10^6 m^3 per year. The major crops are wheat (11 500 ha), lucerne (9500 ha), sugar beet (9500 ha), barley (8000 ha), corn (5000 ha), orchards (3400 ha) and soybean (3000 ha). The annual water application rates in the region are high and for the above mentioned crops are: 7600, 18 400, 16 900, 5300, 8900, 12 900 and 7600 m^3 ha^{-1}, respectively.

With respect to land salinity and alkalinity, a recent investigation, comparing the extent of different classes of soil salinity and alkalinity within the command areas of Canal A and the Main Canal before and after the development of irrigation, shows that generally irrigation had a beneficial effect on the reduction of soil salinity and alkalinity in the region, particularly in the areas with deep watertable (Tables 7-5 and 7-6). For example, in the command area of Canal A there is no more land with high salinity (class S3) and the extent of land with a salinity class of S2 has reduced from 2845 ha to 330 ha (Table 7-5). Also, areas affected by high alkalinity (class A3) were eliminated in this command area (Table 7-6).

Table 7-5: Changes in the extent of land salinity within the command areas of Canal A and the Main Canal before (1958) and after (1983) the development of irrigation in Moghan.

Salinity class	EC$_e$ (μS cm^{-1})	Canal A 1958 (ha)	Canal A 1983 (ha)	Main Canal 1968 (ha)	Main Canal 1983 (ha)
S0	< 4 000	5 100	13 125	16 750	45 763
S1	4 000–8 000	6 753	4 815	6 730	14 980
S2	8 000–16 000	2 845	330	31 170	2 874
S3	16 000–32 000	3 480	–	9 780	738
S4	>32 000	–	–	–	75
Total		18 178	18 270	64 430	64 430

Source: Ghandchi (1992).

Table 7-6: Changes in the extent of land alkalinity within the command areas of Canal A and the Main Canal before and after the development of irrigation in Moghan.

Alkalinity class	SAR	Canal A 1958 (ha)	Canal A 1983 (ha)	Main Canal 1968 (ha)	Main Canal 1983 (ha)
A0	<10	443	1 600	40 950	54 369
A1	10–15	14 027	6 490	19 180	4 736
A2	15–30	228	10 180	3 400	5 250
A3	30–50	3 480	–	900	75
Total		18 178	18 270	64 430	64 430

Source: Ghandchi (1992).

In the areas under the command of the Main Canal, salinity problems were more serious than alkalinity problems before the development of irrigation. In 1968 some 9780 ha had a high salinity class of S3, while only 900 ha of the area had a high alkalinity problem of class A3. Overall it could be said that the development of irrigation in the region had an ameliorative effect on the salinity and alkalinity of the soils by leaching the salts from the soil profile. These effects are particularly notable in areas with a deep watertable.

In some areas extensive irrigation, and subsequent watertable rise, has caused severe waterlogging problems. This problem surfaced towards the end of the 1970s. In 1980 a few thousand hectares became waterlogged, a number of villages were destroyed and some others were seriously damaged. In 1983 the watertable was just below the land surface in about 13 000 ha. To overcome the problem, subsurface drainage plans were approved in 1986 to drain 8400 ha of the seriously affected lands in the region, mainly within the shallow watertable zones A, B and C (Figure 7-4).

The Zarrineh-Rud Irrigation Project

The Zarrineh-Rud alluvial plain south-east of Lake Orumieh (Figure 7-6) covers an area of 1025 km^2, with a general topographic gradient from the south and south-east towards Lake Orumieh. It has an average annual rainfall of over 350 mm, decreasing gradually to 250 mm towards the lake. Two rivers flow in the region. These are the Zarrineh-Rud (rud means river) and the Simineh-Rud with respective annual flows of 1530×10^6 m^3 and 325×10^6 m^3 during the water year of 1988–89. The Zarrineh-Rud has a very good water quality, which is about 225 mg L^{-1} TDS at Nowrouzlu. This is also the case with the Simineh-Rud, where its salinity is about 290 mg L^{-1} TDS.

During 1967–71 a reservoir was constructed on the Zarrineh-Rud with an effective capacity of 480×10^6 m^3. Also a diversion dam was constructed at Nowrouzlu and an irrigation and drainage network was gradually developed. According to ICID (1977), the irrigated area of the plain has been 85 000 ha with 24 km of main irrigation canals, 216 km of secondary canals and 124 km of drains.

The Zarrineh-Rud plain is underlain by a multilayered aquifer system, less than 160 m thick, consisting of gravel, sand and clay with gradual decrease in particle size towards the lake. The aquifer system is mainly recharged by the rivers and discharges in the marshlands of Lake Orumieh. Water quality of the aquifers is good, particularly on the east side of the Miandoab to Mahabad road, where the salinity is less than 550 mg L^{-1} TDS, but the quality deteriorates towards the north-west and reaches high values of about 10 000 mg L^{-1} TDS. Although the watertable was initially shallow in the early 1960s and signs of salinisation existed in the region, excessive use of surface water via the irrigation network has caused a further rise of the watertable, causing widespread waterlogging and land salinisation.

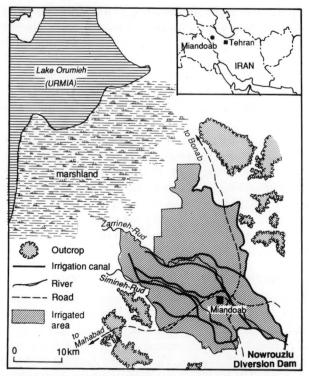

Figure 7-6: Zarrineh-Rud Irrigation Project (Mahab-Ghodss, 1984).

Management of waterlogging and salinity problems in the area embraces the following measures:

- The two rivers of the region (the Zarrineh-Rud and Simineh-Rud) flood their lower regions during periods of high rainfall. Flooding increases recharge to the shallow aquifers and aggravates waterlogging and salinity problems. Therefore flood control measures are required to overcome the problem.
- Loss in the main canals contributing to the recharge of shallow aquifers, rise of watertables and development of salinisation is estimated to be about 10–15 per cent of their capacity. Lining of these canals will reduce seepage and would alleviate salinity problems.
- Water application rates are high and irrigation practice has a low efficiency. Consequently improved agricultural practice and water use management are highly important.
- Because of the low topographic gradient and siltation problem current drainage facilities are not playing their role efficiently. These facilities need to be redesigned and maintained adequately.

- Groundwater resources of the region are partly adequate for agricultural production. Conjunctive use of surface and groundwater resources should be undertaken.

The Haft Tappeh Irrigation Project

The Haft Tappeh Project is located in Khuzestan Province. This province in the south-west of the country enjoys abundant surface water resources. Major rivers of the province are the Karun, Dez, Karkheh and Marun, which had annual run-offs of 9.7, 7.6, 5.0 and 1.2×10^9 m^3, respectively, during the 1988–89 water year (Ministry of Energy, 1990). In the flat area of this province, known as the Khuzestan Plain, the average annual rainfall decreases from 250 mm in the north and north-east to less than 150 mm towards the south-west. The potential evaporation also ranges from about 3000 mm to over 4200 mm in the same direction.

In Khuzestan, apart from the flood plains of the major rivers, considerable areas are severely saline (Dewan 1959; Dewan and Bordbar, 1961). Moreover, inefficient use of water resources for agricultural production and inadequate drainage facilities on the heavy soils have caused serious secondary salinity problems in most irrigation districts of the province.

For many centuries, Khuzestan was the land of sugarcane. Then the cane disappeared completely. According to Hajrasuliha (1970), this happened because of one or a combination of the following reasons: lands became salty due to lack of knowledge in soil and water management; complex water systems were disrupted by earthquake and/or invasions (14 since the time of Alexander); irrigation canals were filled by siltation; elaborate institutions needed to operate and maintain the complex water system were destroyed or broke down. However, after an absence of nearly 700 years, sugarcane has been cultivated in the region since the early 1960s. The first plantation was in Haft Tappeh, 45 km south of Dezful. Since then a number of plantations have been developed or are under development in the north-east and south-west of Ahwaz. Haft Tappeh area has an elevation of 43 to 82 m and receives an average annual precipitation of 258 mm, while the annual evaporation exceeds 2900 mm. The plantation takes water from the Dez River, which has an average electrical conductivity of 650 µS cm^{-1}. The area covers 24 000 ha, and annually some 8000–10 000 ha are planted with an average water application of 30 000 m^3 per ha. The method of irrigation is furrow, and irrigation furrows are 1.5 m apart and 40 cm deep. The irrigation system consists of 13 km of primary canals, 31 km of secondary canals and 39 km of tertiary canals. There exists an irrigation canal about 2500 years old, constructed at the time of Emperor Darius. About 800 m of this canal has been used as primary canal. Despite the accumulation of silt in some areas, Darius Canal is still 20 m deep and 100 m wide at the top. This canal is about 3 km in length today, but at one time it formed part of an 80 km canal carrying water from the Karkheh River southward (Hajrasuliha, 1970).

Soils of the plantation range from clay loam to loam with very low to moderate permeability. These soils are from 2 m to 6 m deep and are characterised by having low organic matter content. Although there are some non-saline, non-alkaline soils in the plantation, soils are saline in general. Hajrasuliha et al. (1980) provide an analysis of soil salinity measurements at three sites considered extremely saline, highly saline and low saline. These sites cover 150 ha, 232 ha and 440 ha and are 5 km, 9 km and 13 km apart with 232, 682 and 455 observation points, respectively.

A 1958 survey indicated existence of the watertable at a depth of 1.5 m. Also water analysis showed that the groundwater was saline and undesirable for domestic or irrigation supply. The high salt concentration in the soil and groundwater seems to have the following origins (Hajrasuliha, 1970): nature of the parent materials; transport of soluble salts from higher locations; irrigation of poorly drained lands in the past; and transport of some salt by dust storms (windborne salt) and deposition on the area. It seems that among these factors irrigation in the past had a fairly good share in salinisation of the poorly drained lands of the area. To control salinisation the following drainage facilities have been provided:

- Open drains were installed in 1962 with a spacing of 150 m, 200 m and 300 m and at a depth of 1.75 m. These drains were connected to a main drain 2.5 m deep. Measurements showed that the watertable was 0.75 m to 1.0 m at the midpoint between drains.
- Tile drains were installed in 1963 at a depth of about 2 m. The drains had a spacing of 80 m to 100 m in order to keep the watertable at least at one metre from the surface during summer. In areas where complete growth failure had occurred and which were subsequently tiled, leached and replanted, sugarcane grew again beautifully.

According to Hajrasuliha (1970), over 4000 ha of the plantation have been tiled and 470 km of drains and in excess of 134 km of open drains have been installed. Since the project started in 1961, tremendous improvements have been achieved. The yield of sugarcane has increased from 77.4 to 120.4 tonnes per ha. Much of this success has been attributed to the improvement of drainage facilities.

The Khalafabad Irrigation Project

The Jarrahi River in Khuzestan Province is formed from the junction of the Marun and Allah Rivers. It flows through Behbahan, Jayazan, Ramhormoz, Ramshir and Shadgan before being discharged to the Persian Gulf (Figure 7-7). In this catchment, salinity, alkalinity and waterlogging are major issues, particularly in the Khalafabad and Shadgan Irrigation Districts. The following brief description of salinity and alkalinity problems in Khalafabad is summarised from Mahab-Ghodss Consulting Engineers (1989). Problems are worse in Shadgan, located near the lower end of the catchment.

Khalafabad covers an area of 26 100 ha. It has a populatin of 26 800, with about 15 200 in Ramshir and the remaining 11 600 in the rural areas. The elevation of the area varies from about 25 m in the east to less than 10 m on the west. The area has long, hot summers and short, mild winters. The absolute minimum and maximum temperatures at Mahshahr, which is the closest climatological station, are about −3.5°C and 52°C, respectively. Average rainfall of the area is about 254 mm while its potential evaporation exceeds 2100 mm. The Jarrahi River crosses the area and its average annual flow, about 10 km upstream of Ramshir, is about 2×10^9 m^3, with an average flow rate of 67 m^3 s^{-1}. At this point, river water quality ranges from 2300 mg L^{-1} TDS in low flow to 800 mg L^{-1} TDS in high flow, with an average value of 1280 mg L^{-1} TDS at the average flow rate of 67 m^3 s^{-1}. These data indicate that the river carries about 2.7×10^6 tonnes of salt per annum. However, the quality of this water is considered acceptable for the agricultural requirements of the area.

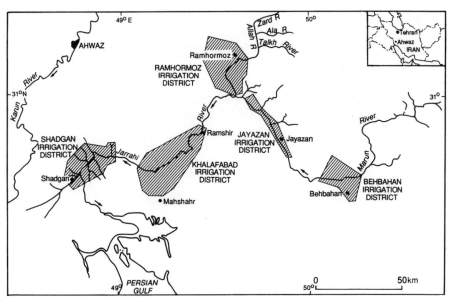

Figure 7-7: Jarrahi catchment irrigation districts (Mahab-Ghadss Consulting Engineers, 1989).

In this area 6340 ha are irrigated (5680 ha net irrigated and 660 ha as fallow), 15 725 ha non-irrigated (10 815 ha net cultivated and 4910 ha as fallow) and the remaining 4035 ha are occupied by the city of Ramshir, villages, roads, etc. The main crops consist of wheat (10 770 ha), barley (4080 ha) and vegetables (1365 ha). The annual water use for irrigation is about 85.4×10^6 m^3, with an efficiency of 34 per cent. Water requirements are maximum in March–April, with a monthly value of 16×10^6 m^3.

Groundwater levels of the area have annual fluctuations with high levels in March–April due to rainfall and irrigation, and low levels in October–November. As indicated in Table 7-7, in April 1985 the watertable was shallower than 3 m over some 20 558 ha, representing about 78.8 per cent of the area. Infiltrations due to rainfall and leakages from the irrigation canals have a rapid effect on the groundwater levels, while the effects of recharges from the boundaries of the aquifer are significantly slower. Groundwater flow direction on the left side of the river is towards the lower part of the river, while on the right side it is towards the swamps located in the western part of the area.

Table 7-7: Watertable depth in Khalafabad Irrigation District in April 1985.

	Depth of watertable (m)					
	0–1.5	1.5–2	2–2.5	2.5–3	>3	Total
Area (ha)	411	2525	8132	9490	5542	26 100
Area (%)	1.6	9.7	31.1	36.4	21.2	100

Source: Mahab-Ghodss Consulting Engineers (1989).

Groundwater resources of the area are saline with EC values ranging from 820 to 16 400 µS cm^{-1}. Table 7-8 represents the extent of groundwater salinity in the area.

Table 7-8: Extent of groundwater salinity in Khalafabad Irrigation District.

	Groundwater salinity (1000 µS cm^{-1})				
	<20	20–30	30–50	>50	Total
Area (ha)	4241	7003	11 802	3054	26 100
Area (%)	16.3	26.8	45.2	11.7	100

Source: Mahab-Ghodss Consulting Engineers (1989).

Soil salinity, alkalinity and waterlogging are widespread in the area (Figure 7-8) due to the combined effect of a number of factors including: climatic conditions, heavy soils, shallow watertables, high salinity of the groundwater, irrigation with brackish waters from the Jarrahi River, insufficient cross-section of the river causing flooding and waterlogging, inadequate drainage facilities, excessive irrigation, windborne salt and destruction of the native vegetation cover. Soil surveys have demonstrated that, of a total area of 26 100 ha, some 25 377 ha or 97 per cent of the area have low to high salinity and alkalinity problems.

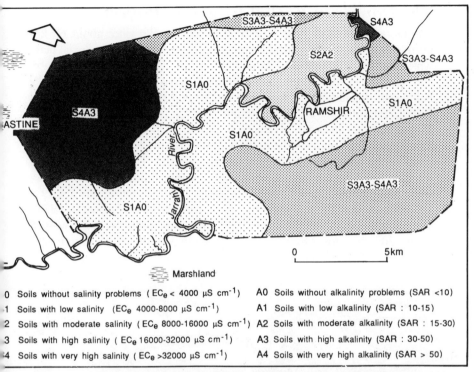

Figure 7-8: Soil salinity and alkalinity map of Khalafabad Irrigation District (Mahab-Ghodss Consulting Engineers, 1989).

Management plans have been prepared in order to overcome the above mentioned problems. These plans include the following measures: control of flooding via the construction of a diversion dam and diversion canal in order to divert the flood waters to the lower end of the area; and development of a modern irrigation and drainage (surface and subsurface) network.

In the designed project, the surface drainage system is to be developed at a depth of 2.1 m to 2.5 m, in order to keep the watertable 1.4 m below the ground surface. The distance between drains will vary from 40 m to 125 m depending on the soil hydraulic conductivities. Drainage tubes will be from high density polyethylene corrugated pipes with a diameter of 12.5 cm. The total length of pipes has been estimated to be about 2675.6 km, and the cost of the subsurface drainage project will be about 8603.5 million Rials (US$130 million at an exchange rate of 70 Rials for one US$), or 391 000 Rials (US$5600) per hectare. On the left bank of the river, drained waters will be discharged to the Jarrahi River, while on the right bank the final destination of the drained waters will be the river and the marshlands close to the village of Bastine, located in the extreme west of the area (Figure 7-8).

The Dorudzan-Korbal Irrigation Project

The Dorudzan-Korbal Plain is located in the Fars Province about 65 km north-east of Shiraz, the capital city of the province. It is a large alluvial plain which consists of a number of smaller plains, such as Marvdasht at the centre and Korbal in the south-east, extending from Band-e-Amir to Lake Bakhtegan (Figure 7-9). It is mainly surrounded by the mountains formed from Cretaceous and Eocene limestone (NIOC, 1975–78). The Dorudzan-Korbal Plain has an elevation ranging from 1620 m to 1560 m, with a general topographic gradient from the north-west towards the south-east. The major rivers of the plain are the Kor, Maeen and Seivand. The Kor River, which discharges to Lake Bakhtegan, has an average annual flow of 757×10^6 m^3, while the average annual flows of the Maeen and Seivand Rivers are 59×10^6 m^3 and 54×10^6 m^3, respectively (Abkav Consulting Engineers, 1975). The average annual rainfall of the area ranges from 600 mm in the west to about 200 mm towards the east and south-east. Potential evaporation increases from 1700 mm to more than 3200 mm in the same direction.

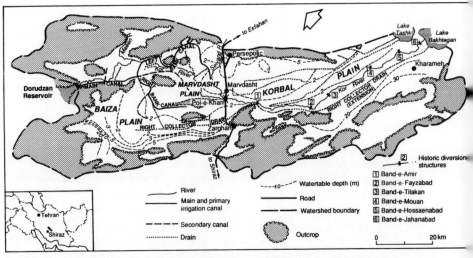

Figure 7-9: Main features of the Dorudzan-Korbal Irrigation Project and the watertable depth of the aquifer for 1986 (Fars Regional Water Authority, 1988).

The diversion structures on the Kor River, such as Band-e-Amir, Band-e-Fayzabad, Band-e-Tilakan, and Band-e-Jahanabad (Figure 7-9), which are about 1000 years old (Reza et al., 1976: 229–230), are indicative of the long history of irrigation practice in the region. In order to develop a modern irrigation system in the area and to secure a source of water supply and electricity for the city of Shiraz and the developing industry in the region, a multipurpose dam was designed for the Kor River at Dorudzan (Justin &

Courtny and Taleghani-Daftary, 1966). Construction works commenced in 1966 and the dam, with an effective capacity of 860×10^6 m^3, was commissioned in 1972. The irrigation project was designed initially to cover 41 000 ha and consisted of a 22.5 km main canal with a capacity of 40 m^3 s^{-1}, and 95 km of primary canals, 46 km of secondary canals and 400 km of tertiary canals. Later the irrigated area was expanded to 53 000 ha and the capacity of the main canal increased to 56 m^3 s^{-1}.

Kor River waters are mainly diverted via the Right and Left Bank canals for irrigation. Water quality of the river is good for irrigation all year round. A few kilometres down stream of the Dorudzan Reservoir, electrical conductivity ranges from 200–750 µS cm^{-1} with an average value of 500 µS cm^{-1} and SAR of 6.2 to 8.6. At Pol-e-Khan, EC values range from 400 to 1000 µS cm^{-1}. After Pol-e-Khan, because of the discharge of drainage water (particularly from the main collector drain at Band-e-Amir which drains about 28 000 ha of land) and industrial and municipal effluents to the river (at a rate of about 6.8 m^3 s^{-1} or 208×10^6 m^3 yr^{-1}), its quality deteriorates (Taleghani-Daftary Consulting Engineers, 1977; Khonssary, 1984).

Estimates show that water requirements for irrigation of 49 000 ha of land downstream of Band-e-Amir, and 54 000 ha upstream of this point, are 920×10^6 m^3 per year, of which about 60×10^6 m^3 can be provided by reuse of drainage water with suitable quality. Major crops in the region are wheat, barley, sugar beet, rice, cotton, potato, oil seeds and vegetables. Agriculture in the region is practised via traditional methods and for this reason production efficiency in terms of production per hectare is low. Both surface water and groundwater resources of the region are used for agricultural production.

Limestone formations surrounding the plain recharge the alluvial aquifer. Potentiometric head decreases from more than 1600 m in the north-east to less than 1550 m in the south-east (Abkav Consulting Engineers, 1975). The Kor and Seivand Rivers make a minor contribution of 8 to 10 per cent to the recharge of the aquifer, while vertical recharge contributes 80–90 per cent. Estimates show that, depending on the season, about 10 to 40 per cent of the irrigation water recharges the aquifer, and the lateral recharge of the aquifer from the limestone aquifers is about 43×10^6 m^3 per year. Basically the Kor River recharges the aquifer from downstream of the Dorudzan Reservoir to Band-e-Amir, and from about this point to Lake Bakhtegan the river acts as a drain.

Groundwater quality is relatively good at the margins of the plain (EC < 1000 µS cm^{-1}), but in general its salinity increases from the north-west towards the south-east (EC > 5000 µS cm^{-1}). Recent statistics show that annual extraction from 1840 shallow wells, 66 deep wells and 74 qanats in the plain were 256×10^6 m^3, 13×10^6 m^3 and 55×10^6 m^3, respectively (Fars Regional Water Authority, 1988). In addition, the annual discharge from 17 springs was about 122.5×10^6 m^3. Of an annual groundwater extraction of

446.5×10^6 m^3, about 437×10^6 m^3 (98 per cent) were used for irrigation. The range of EC values in µS cm^{-1} were as follows: wells (500 to 8140); qanats (380 to 12 000); and springs (460 to 1470).

Water losses from irrigation canals are high, with an average value of 40 per cent. These losses, estimated at 12 m^3 s^{-1} (or 422×10^6 m^3 yr^{-1}), plus losses in the irrigated fields, are the major contributors to watertable rise, which is shallower than 2 m in a large part of the plain (Figure 7-9). In some areas the watertable is even shallower than 1 m and, because of capillary rise from shallow aquifers, soils are highly salt-affected. Although a number of drains were designed to evacuate excess rainwater and irrigation water, control the watertable and prevent land salinisation (Justin & Courtny, Taleghani-Daftary, 1966), an investigation by Taleghani-Daftary Consulting Engineers (1979) indicated that 24 000 ha of land in the Baiza Plain and part of the Marvdasht Plain with watertable less than 2 m from the land surface required a network of horizontal subsurface drains. Their proposed design characteristics for these areas were as follows: diameter of the corrugated polyethylene pipes, 125 mm; depth of pipes about 1.8 m; and spacing 80 m to 150 m. Currently about 150 km of surface drains are available in the area, and these drainage facilities are not enough to control the rising watertable and expansion of land salinity in the area.

According to a 1972 soil survey undertaken over 46 000 ha of land in the region, approximately from the dam site to Band-e-Amir, about 16 000 ha of land had some salinity and alkalinity problems. These included 8500 ha of saline soils, 6500 ha of saline-alkaline soils and 1000 ha of saline soils with high alkalinity. Since then the situation has changed. Construction of the surface drainage facilities in the region and subsurface horizontal drains in some limited areas, accompanied by the leaching effects of applied irrigation water and rainfall, has reduced the soil salinity. In contrast, in some other areas without adequate drainage facilities, the rising watertable has caused land salinisation. However, no data is available about the current extent of different classes of salt-affected and alkaline soils in the area. Development of an efficient drainage system, accompanied by the lining of the unlined canals and efficient use of water at farm level, would alleviate the problem.

The Korbal Plain is also suitable for the development of modern irrigation systems. According to a project under construction, traditional diversion structures will be repaired and their heights will rise. About 68 km of irrigation canals with concrete lining, serving 32 000 ha of land, will be built and surface drainage facilites will be provided (Parab Fars Consulting Eng. Co., 1992). Also, in order to prevent the discharge of the drainage water to the Kor River at Band-e-Amir, the main collector drain on the right bank of the river will be extended to Lake Bakhtegan. It is expected that the project will be commissioned by September 1994.

The Zayandeh-Rud Irrigation Project

This project covers part of the Esfahan Plain, which is formed by the alluvial deposits of the Zayandeh-Rud (Figure 7-10). The average annual rainfall in the area decreases from about 300 mm in the west to less than 100 mm in the east. For example, Esfahan has an average annual rainfall of 120 mm while its potential evaporation is 1500 mm yr^{-1}. The Zayandeh-Rud is the main river of the area. Its average annual flow at Pol-e-Khajou was about 755×10^6 m^3 yr^{-1} over the period 1978–81. During the 1967–70 period, a reservoir was constructed on the river with an effective capacity of 1.1×10^9 m^3 and the irrigation project was completed in 1978. The designed project consists of two diversion dams at Neku-Abad and Abshar, 90 000 ha of irrigated land, 164 km of main canals, 129 km of secondary canals and 235 km of surface drains. Currently the project consists of 95 000 ha of modern irrigation system, 30 000 ha of traditional irrigation system, four main canals (two from each diversion dam), 280 km of irrigation canals and 256 km of drains. The main crops include wheat, barley, rice, sugar beet, vegetables and orchards.

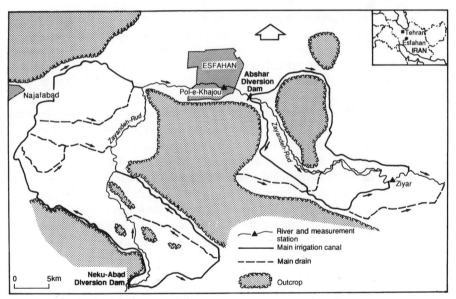

Figure 7-10: Main features of the Zayandeh-Rud Irrigation Project (Courtesy of Esfahan Regional Water Authority, Esfahan).

The watertable is shallow and the general direction of groundwater flow is towards the river, which acts as the main drain of the aquifer. In this project difficulties arose with the water distribution system. During the period of irrigation in spring and summer, the watertable becomes shallower and in some areas reaches 70 cm to 60 cm below the land surface. A modern drainage system collects drainage waters and discharges them to the river to be used by downstream farmers. However, part of the drainage water is reused locally and no part of the drainage water is disposed to evaporation basins.

Esfahan Regional Water Authority (1990) provides a detailed analysis of the chemical quality (Cl$^-$, SO$_4^{--}$, HCO$_3^-$, Na$^+$, K$^+$, Ca^{++}, Mg^{++}, pH, EC, SAR and TDS) of the Zayandeh-Rud water for the period 1976–1989. According to their analysis, water quality of the river is high in its headwaters down to Neku-Abad Diversion Dam, where the EC values ranged from 323 to 537 µS cm^{-1}. Downstream of this dam, water quality deteriorates because of the discharge of low quality groundwater and drainage water to the river. For example, EC values at Pol-e-Khajou and Ziyar stations ranged from 456 to 766 µS cm^{-1} and 526 to 1650 µS cm^{-1}, respectively, during the period 1976–89. Below Ziyar station, the increase in EC values has been faster and ranged from 1790 to 26 3000 µS cm^{-1} at Varzaneh, located about 30 km downstream of the Ziar, for the reasons mentioned before.

The Zabol Irrigation Project

The Zabol Irrigation Project is located in Sistan Province, in the south-east of Iran. It has an average low annual rainfall of 55 mm while its average annual potential evaporation is about 4800 mm. One of the climatic characteristics of the region is its annual windy period, which starts in early May and continues for about 120 days, with an average velocity of 26 km h^{-1} and maximum velocity of 110 km h^{-1} in July.

The Sistan Plain is formed by the alluvial deposits of the Hirmand (Helmand) River, which originates in the high mountains of Afghanistan. These deposits are relatively coarse towards the east of the plain but gradually become finer towards the west. The Sistan River, one of the tributaries of the Hirmand River, is the main source of irrigation water supply in the Zabol Irrigation Project. It crosses the project from east to west and discharges to the Hamun-e-Hirmand. The Sistan Diversion Dam was constructed on the river and water is distributed via two main canals and a number of secondary canals (Figure 7-11). Sistan River water, with a TDS value of 475 mg L^{-1}, is suitable for irrigation, while, because of low quality, groundwater is not a source of water supply for irrigation.

The irrigation project covers an area of about 100 000 ha, of which some 75 000 ha are cultivated annually. Wheat and barley are the major crops of the region and cover 90 and 5 per cent of the annual cultivated lands. Water and land use management, as well as agricultural practices, are based on traditional methods which have a low efficiency and are damaging the cultivated lands, developing salinity, alkalinity and waterlogging. The extent of land salinity and alkalinity within the 73 000 ha of land surveyed in 1987 is represented in Table 7-9. It indicates that salinity and alkalinity are widespread in the area, with only 15 190 ha (21 per cent) not affected by salinity and 3440 ha (5 per cent) not affected by alkalinity. Figure 7-11 shows the distribution of lands affected by very high salinity (class S4) and high alkalinity (class A4).

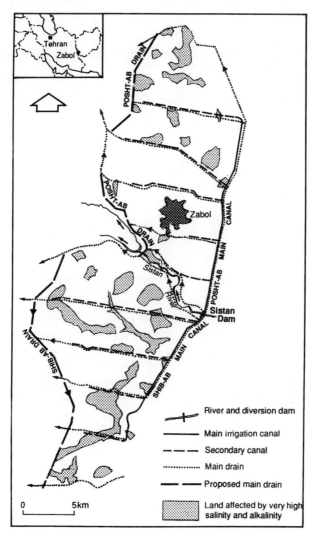

Figure 7-11: Main features of the Zabol Irrigation Project and the distribution of lands with very high salinity and alkalinity (Courtesy of Yekom Consulting Engineers, Tehran).

Analysis of soil profiles indicates that salts are mainly concentrated towards the top of the profiles and their concentrations decrease with depth. This is mainly because of the effects of capillary rise and high evaporation. In soil samples with high salinity and alkalinity, dominant anions and cations are Cl⁻, SO_4^{--}, Na⁺ and Mg⁺⁺, while in soils not affected or having a low salinity or alkalinity, dominant anions are SO_4^{--} and Cl⁻. There is not a significant difference between concentrations of the cations Na⁺, Mg⁺⁺ and Ca⁺⁺. A number of factors contribute to the development of secondary salinity and alkalinity in this project. These factors will be described now.

Table 7-9: Extent of land salinity and alkalinity in Zabol Irrigation Project.

	Salinity				Alkalinity		
Class	EC_e	Extent		Class	SAR	Extent	
	($\mu S\ cm^{-1}$)	(ha)	(%)			(ha)	(%)
S0	< 4 000	15 190	21	A0	< 8	3 440	5
S1	4 000–8 000	18 490	25	A1	8–13	12 580	17
S2	8 000–16 000	15 910	22	A2	13–30	28 600	39
S3	16 000–32 000	10 910	15	A3	30–70	14 880	21
S4	> 32 000	10 930	15	A4	> 70	11 930	16
Hills	–	620	1	Hills	–	620	1
Urban	–	950	1	Urban	–	950	1
Total		73 000	100	Total		73 000	100

Source: Yekom Consulting Engineers (1992).

The shallow watertable of the aquifer, which has a low quality (EC values ranging from 2000 to 90 000 µS cm⁻¹), plays a major role in the development of soil salinity and alkalinity. According to 1987 statistics, the percentages of the area within different ranges of watertable depth were: 0–1 m, 11.5 per cent; 1–2 m, 56 per cent; 2–3 m, 22 per cent; 3–4 m, 8 per cent; and the remaining 2.5 per cent of the area had a watertable deeper than 4 m. In September, a few months after the period of irrigation and significant evaporation from the aquifer, the watertable was shallower than 3 m over about 67 per cent of the area.

High potential evaporation (4800 mm) in the region, which is 87 times the average annual rainfall (55 mm), is another important factor. This factor, which generates capillary movement of shallow saline groundwater, is so strong that during the summer months it prevents the leaching of salt from the soil profile (particularly from the root zone) by irrigation water. Subsequently, salt accumulates in the upper part of the soil profile.

Strong winds, blowing from the north-north-east towards the south-south-east for about 120 days in summer, contribute to the expansion of soil salinity in the region by distributing the accumulated salts at the soil surface to a wider area. Moreover, farmers in the salt-affected areas traditionally remove the top 10–20 cm of soil and accumulate them at the edges of their farms. These soils, containing a high percentage of salts, are easily transported by winds. Observations show that, in some areas with relatively deep groundwater not affected by capillary rise, EC_e of soil samples within the top 10 cm exceeds 300 000 µS cm^{-1}, while at a depth of 10 to 120 cm, EC_e gradually decreases to 10 000 µS cm^{-1}. This can be explained only by windborne salt accumulation.

Management of water and land resources and agricultural production systems is based on traditional methods. Although the primary irrigation canals are lined, secondary canals and distributors are unlined and managed by the farmers. Seepage from these canals is significant and contributes to watertable rise. Flood irrigation is the dominant irrigation method in the region. It is expected to leach salts from the top soils and the root zone, but because of the insufficient water application and relatively long interval between irrigations, salts moved downwards come back to the surface.

Stream salinity

The process of salinisation of the surface water resources of Iran is mainly due to natural conditions, and to a lesser extent to the discharge of drainage waters into the river systems. The following paragraphs provide two examples of stream salinity in Iran.

Valles et al. (1990) describe the salinity of the Jaj-Rud (Djajerud) River Basin. The basin is located on the southern slopes of the Elburz Mountains, to the east of Tehran, and covers an area of 2900 km^2. The river is fed by the snowmelt and rainfall. The diversity of the water sources and the geological heterogeneity of the basin results in marked changes in the chemical composition of the river water in various parts of the basin. Valles et al. (1990) provide the chemical composition of the river water at 9 stations within the basin. Their data indicate that the concentration of dissolved compounds increases and the water quality deteriorates from upstream to downstream. For example, the EC values increase from 275 to 13 640 µS cm^{-1} (almost fiftyfold) during the summer (low flow period) and 255 to 1980 µS cm^{-1} (almost eightfold) during the spring (high flow period). The water quality deterioration has a major effect on the salinisation and alkalisation of the irrigated land in the basin, particularly the Varamin Plain, located in the lower end of the basin.

Shiati (1989 and 1991) has described the origin and management of salinity in the Shapur-Dalaki River Basin in the southern part of Iran. This river has a catchment area of 10 000 km^2, an average annual flow of 1×10^9 m^3 and discharges to the Persian Gulf. According to his analysis, water quality in the river is initially good, with an EC value of 350–500 µS cm^{-1}. However, along the route, over a distance of more than 200 km, water quality gradually deteriorates due to the confluence of saline springs and passage of the river through evaporites and salt domes. The extent of this deterioration is such that at the lower end of the catchment the EC value varies between 4000 and 8000 µS cm^{-1} and the salt load carried by the river is about 2.3×10^6 tonnes per year. According to Shiati (1989 and 1991), 6.7×10^9 m^3 of brackish water flows annually through 12 major rivers of the country. Any quality improvement of these water resources will have a beneficial influence on the agricultural and regional deveopment of Iran.

The discharge of drainage water to the river systems is a common practice in the country and in many cases, like the examples provided in the previous sections or the case of irrigation systems in Khuzestan, it can be repeated a number of times along the river. This practice seriously reduces the quality of irrigation water for farmers at the lower end of the river system. Use of such low quality water for irrigation contributes to the expansion of salinity and alkalinity as well as to a decline in agricultural production.

References

Abkav Consulting Engineers. 1975. *Groundwater Resources Development of the Kor River Basin.* Tehran: Abkav Consulting Engineers. 356 pp. (In Persian).

Abtahi, A. 1977. Effect of a saline and alkaline groundwater on soil genesis in semiarid southern Iran. *Soil Science Society of America Journal.* 41(3): 583–588.

Abtahi, A., Eswaran, H., Stoops, G. and Sys, C. 1980. Mineralogy of a soil sequence formed under the influence of saline and alkaline conditions in the Sarvestan Basin (Iran). *Pédologie.* 30(2): 283–304.

Abtahi, A., Sys, C., Stoops, G. and Eswaran, H. 1979. Soil forming processes under the influence of saline and alkaline groundwater in the Sarvestan Basin (Iran). *Pédologie.* 29(3): 325–357.

Aghassi, A.V. 1990. Groundwater evaluation in fractured zones with special emphasis on solubility of carbonated rocks. In: Parriaux, A. ed. *Water Resources in Mountainous Regions.* Memoires of the 22nd Congress of IAH, Lausanne, 27 August–1 September 1990. XXII(1): 480–485.

Beaumont, P., Blake, G.H. and Wagstaff, J.M. 1988. *The Middle East: A Geographical Study.* Second edition. London: David Fulton Publishers. 623 pp.

Beaumont, P., Bonine, M. and McLachlan, K. eds. 1989. *Qanat, Kariz and Khattara: Traditional Water Systems in the Middle East and North Africa.* London: The Middle East Centre, School of Oriental and African Studies, University of London. 305 pp.

Bybordi, M. 1974. Ghanats of Iran: Drainage of sloping aquifer. *Journal of the Irrigation and Drainage Division.* 100 (3): 245–253.

Bybordi, M. 1989. Problems in planning of irrigation projects in Iran. In: Rydzewski, J.R. and Ward, C.F. eds. *Irrigation: Theory and Practice.* Proceedings of the International Conference held at the University of Southampton, 12–15 September, 1989. London: Pentech Press. 115–123.

Department of Technical Co-operation for Development. 1982. *Ground Water in the Eastern Mediterranean and Western Asia.* New York: United Nations. (Natural Resources/Water Series no. 9). 230 pp.

Dewan, M.L. 1959. Soils of Iran, Part I: Soils of the Khuzistan plains. *Journal of the Indian Society of Soil Science.* 7: 127–135.

Dewan, M.L. and Bordbar, M. 1961. A summary of the report on leaching experiments in the Karkheh area, Khuzistan, Iran. In: *Salinity Problems in the Arid Zones, Proceedings of the Teheran Symposium, 1958.* Paris: UNESCO. 273–279.

Dewan, M.L. and Famouri, J. 1964. *The Soils of Iran.* Rome: FAO. 319 pp.

Directorate of Statistics and Agricultural Information. 1990. *Agricultural Statistics, 1989.* Tehran: Directorate of Statistics and Agricultural Information, Ministry of Agriculture. (Publication no. 7). 239 pp. (In Persian).

Economic and Social Commission for Asia and the Pacific. 1985. *Proceedings of the Tenth Session of the Committee on Natural Resources.* New York: United Nations. (Water Resources Series no. 59). 316 pp.

Esfahan Regional Water Authority. 1990. *Zayandeh-Rud Water Quality from Ghaleh-Shah-Rokh to Varzaneh*. Esfahan: Esfahan Regional Water Authority, Directorate of Investigations and Data Analysis. 24 pp. (Plus 42 Figures and 11 Tables). (In Persian).

Fars Regional Water Authority. 1988. *Groundwater Resources of the Dorudzan-Korbal Plain, Kor River Basin*. Shiraz: Fars Regional Water Authority, Water Resources Directorate. 223 pp. (In Persian).

Food and Agriculture Organization of the United Nations (FAO). 1989. *Production Yearbook*. Rome: FAO. v.42. 350 pp.

Framji, K.K., Garg, B.C. and Luthra, S.D.L. 1982. *Irrigation and Drainage in the World: A Global Review*. New Delhi. International Commission on Irrigation and Drainage. v.II. 493–1159.

Ghandchi, I. 1992. Yekom Consulting Engineers, Tehran. (Personal communication).

Ghobadian, A. 1990. *Iranian Natural Resources in Relation to Agricultural Utilization, Reconstruction and Reclamation*. Kerman: Kerman University. 480 pp. (In Persian).

Hajrasuliha, S. 1970. Irrigation and drainage practices in Haft Tappeh cane sugar project. In. *8th Near East, South Asia Regional Irrigation Practices Seminar, Kabul, Afganistan*. Sponsored by the United States Agency for International Development. Utah: Utah State University Printing Service. 117–143.

Hajrasuliha, S., Baniabbassi, N., Metthey, J. and Nielsen, D.R. 1980. Spacial variability of soil sampling for salinity studies in south west Iran. *Irrigation Science*. 1: 197–208.

Hajrasuliha, S., Cassel, D.K. and Rezainejad, Y. 1991. *Geoderma*. 49: 117–127.

Hammer, U.T. 1986. *Saline Lake Ecosystems of the World*. Dordrecht: Dr W. Junk Publishers. 616 pp.

International Commission on Irrigation and Drainage (ICID). 1977. *Iranian National Committee Report*. ICID Special Session, Tehran, May 1977. New Delhi: ICID. 13 pp.

Iranian National Committee on Large Dams. 1990. *An Overlook of Iran's Major Dams*. 58th International Commission on Large Dams (ICOLD) Executive Meeting, Sydney, 21–26 May 1990. (Leaflet).

James, G.A. and Wynd, J.G. 1965. Stratigraphic Nomenclature of Iranian Oil Consortium Agreement Area. *Bulletin of the American Association of Petroleum Geologists*. 46(12): 2182–2245.

Justin & Courtny, and Taleghani-Daftary. 1966. *Dorudzan Multipurpose Project (Final Report)*. Philadelphia: Justin & Courtny Consulting Engineers. 382 pp.

Khonssary, J. 1984. Pollution of the Kor River. *Scientific and Technical Bulletin of the Agricultural Problems*. Tehran: Yekom Consulting Engineers. (2): 81–85. (In Persian).

Korzun, V.I., Sokolov, A.A., Budyko, M.I., Voskresensky, K.P., Kalinin, G.P., Konoplyantsev, A.A., Korotkevich, E.S., Kuzin, P.S. and Lvovich, M.I. eds. 1978. *World Water Balance and Water Resources of the Earth*. Paris: UNESCO. 633 pp.

Mahab-Ghodss Consulting Engineers. 1984. *Zarrineh-Rud Irrigation and Drainage Project*. Tehran: Mahab-Ghodss Consulting Engineers. (In Persian).

Mahab-Ghodss Consulting Engineers. 1989. *Irrigation, Drainage and Diversion Dams of Jayazan, Khalafabad and Shadgan*. Tehran: Mahab-Ghodss Consulting Engineers. (In Persian).

Mahjoory, R.A. 1979. The nature and genesis of some salt-affected soils in Iran. *Soil Science Society of America Journal*. 43(5): 1019–1024.

Massoumi, A. 1984. Potential and limitations of the country's water resources and the need for saving water. In. *Papers of the Saving in Agricultural, Domestic and Industrial Water Use Conference*. Tehran, 18–20 December 1984 (27–29 Azar 1363). Tehran: Department of Water Affairs, Ministry of Energy. 50–65. (In Persian).

Matsumoto, S. and Cho, T. 1985. Field investigation on the agricultural development of arid region in Iran. III. Soil profile investigations and salt accumulation related to the depth of groundwater level. *Journal of the Faculty of Agriculture, Tottori University* (Japan). 20: 86–97.

Ministry of Energy. 1976a. *Rainfall Map of Iran (1: 2 500 000)*. Tehran: Directorate of Surface Water Resources, Ministry of Energy.

Ministry of Energy. 1976b. *Potential Evapotranspiration Map of Iran (1: 2 500 000)*. Tehran: Directorate of Surface Water Resources, Ministry of Energy.

Ministry of Energy. 1990a. *1988-89 Water Resources Bulletin*. Tehran: Bureau of Water Resources Investigation and Planning, Ministry of Energy. (In Persian).

Ministry of Energy. 1990b. *Water Resources Atlas of Iran*. Tehran: Bureau of Water Resources Investigation and Planning,

Ministry of Energy. Atlas (Hydrology, rainfall, hydrogeology, groundwater exploitation and water quality, 30 sheets, scale 1:1 000 000), v.1. Hydrology (63 pp. plus 240 pp. Appendices, in Persian), v.2. Hydrogeology (197 pp., in Persian).

Ministry of Water and Power. 1965. *Iran Main River Basins (1:2 500 000 Scale map)*. Tehran: Hydrological Service, Ministry of Water and Power.

Movahed-Danesh, A.A. 1972. Le Problème de l'eau en Iran. *L'Eau*. 67(1): 27-32.

National Iranian Oil Company (NIOC). 1975-78. *Geological Map of Iran* (1:1 000 000, six sheets published from 1975 to 1978). Tehran: NIOC, Exploration and Production Affairs.

Parab Fars Consulting Eng. Co. 1992. *Korbal Water Supply Project*. Shiraz: Parab Fars Consulting Eng. Co. (In Persian).

Reza, E., Kouross, G., Emam-Shoushtari, M.A. and Entezami, A.A. 1976. *Water and Techniques of Irrigation in Ancient Iran*. Tehran: Ministry of Water and Power. 299 pp. (In Persian).

Shaw, J. ed. 1984. *Collins Australian Encyclopedia*. Sydney: Collins. 848 pp.

Shiati, K. 1989. The origin and source of salinity in river basins: A regional case study in Southern Iran. In. *Regional Characterization of Water Quality*. (IAHS Publication no. 182). 201-210.

Shiati, K. 1991. A regional approach to salinity management in river basins: A case study in southern Iran. *Agricultural Water Management*. 19(1): 27–41.

Soil and Water Research Institute. 1987. *National Soil Policy and its Technical and Administrative Organisation in Iran*. Tehran: Soil and Water Research Institute. (Publication no. 725). 55 pp. (In Persian).

Statistical Centre of Iran. 1992. *Iran Statistical Yearbook 1369 (March 1990-March 1991)*. Tehran: Statistical Centre of Iran. 619 pp. (In Persian).

Szabolcs, I. 1989. *Salt-Affected Soils*. Boca Raton, Florida: CRC Press. 274 pp.

Taleghani-Daftary Consulting Engineers. 1977. *Korbal Water Supply Alternatives*. Tehran: Taleghani-Daftary Consulting Engineers. 169 pp. (In Persian).

Taleghani-Daftary Consulting Engineers. 1979. *Drainage Potential, Design of the Farm Drains and Solution for the Waterlogging Problem of the Dasht-e-Baiza*. Tehran: Taleghani-Daftary Consulting Engineers. 337 pp. (In Persian).

Valles, V., Gholami, M. and Lambert, R. 1990. Soil alkalinization and salinization in the Djajerud Basin, Iran. *Land Degradation & Rehabilitation*. 2(1): 43-55.

World Bank. 1992. *World Development Report 1992: Development and the Environment*. New York: Oxford University Press. 308 pp.

Yekom Consulting Engineers. 1992. Tehran. (Personal communication).

Chapter 8: Pakistan

Introduction

Pakistan covers an area of 796 095 km², stretches over 1600 km north to south and spans about 885 km from east to west (Figure 8-1). It comprises four provinces, Baluchistan, North-West Frontier (NWFP), the Punjab and Sind. Baluchistan is the largest province (347 188 km²), followed by Punjab (206 251 km²), Sind (140 913 km²), North-West Frontier (74 522 km²) and Federally Administered Tribal Areas (FATA) (27 221 km²). About 475 884 km² in the north-west and west of Pakistan form a highly differentiated mountainous terrain, while the remaining 320 211 km² present a flat and gradational surface. The whole land, excluding most of Baluchistan, is drained by the Indus River system. Pakistan comprises six major physiographic units:

- **The northern mountains** consist of the western ranges of the Himalaya with high peaks of over 8000 m. These mountains cover the northern part of the country and the major rivers of Pakistan originate in these mountains.
- **The Hindu Kush and the western mountains** form the boundary between Pakistan and Afghanistan and their highest peak is 7690 m above mean sea-level.
- **The Potwar Plateau and the Salt Range.** Potwar Plateau has an elevation varying from 305 to 610 m. South of the Plateau is the Salt Range with an average elevation of about 671 m and highest peak of 1525 m.
- **The Baluchistan Plateau** has an average elevation of about 600 m and consists of dry valleys, saline lakes and vast areas of desert with dry hills generally running from north-east to south-west.
- **The Indus Plain** covers an area of about 21 Mha and is the most prosperous agricultural region of Pakistan. It extends 1050 to 1130 km from the rim of the Potwar Plateau southward to the Arabian Sea. Its northern zone comprises the Province of the Punjab, while the southern zone is mainly the

Province of Sind. The Indus Plain has an average gradient of 19 cm per km towards the sea and is divided into the upper and lower plains. The upper Indus Plain is divided into a number of doabs, meaning the land lying between the two rivers.

- **The Thar Desert** lies in the south-east of the lower Indus Plain, is part of the larger Great Indian Desert which extends into Pakistan from India, and has salt lakes in its depressions.

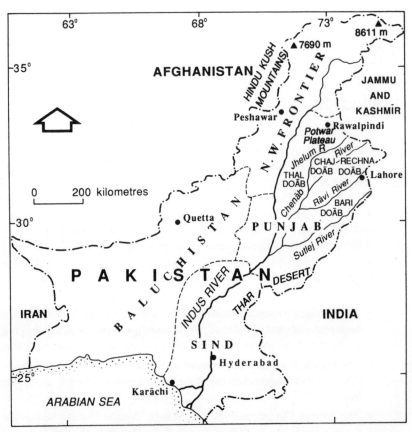

Figure 8-1:   Indus River and its tributaries in Pakistan.

At the time of independence in 1947, the population of Pakistan was about 36 million. According to Akhtar (1986: 9), Pakistan had an estimated population of 97.67 million in mid-1986. Punjab claimed 56 per cent of the total national population, Sind 23 per cent, NWFP 13 per cent, Baluchistan 5 per cent and FATA 3 per cent. The population of the country was estimated at about 112 million in 1990 with an average annual growth rate of 2.1 per cent over the period 1980–90 (World Bank, 1992: 268).

Some 72 per cent of the total population lives in rural areas and agriculture is the largest and most important single sector of Pakistan's economy. Agriculture accounts for about 31 per cent of the gross domestic product. Agricultural exports generate about 70 per cent of the total foreign exchange earnings of the country (Badruddin, 1987).

The 1990–91 production of the major crops of Pakistan included: wheat, 15.10 Mt; rice, 3.26 Mt; sugarcane, 36 Mt; and cotton, 1.63 Mt (Akhtar, 1991: 409–411 and 421). Crop yields per ha in Pakistan are low. For example, in 1990–91 the average productions of wheat, rice, sugarcane and cotton were 1920 kg, 1545 kg, 4100 kg and 607 kg per ha, respectively. Arshad (1991) identifies eight reasons, which include: waterlogging and salinity in the irrigated areas; low use of mechanical farm implements; and lack of farmers' knowledge about modern agricultural technology.

Rainfall and climate

Pakistan suffers from a general deficiency of rainfall. Its average annual precipitation varies largely, from more than 1200 mm in the Himalayan submountain area to less than 100 mm in the south. The average annual rainfall is about 200 mm at Karachi, 175 mm at Hyderabad and 90 mm at Jacobabad (see Figure 8-3). Around 75 per cent of the mean annual rainfall occurs during the months of July, August and September and is of monsoonal origin. It is estimated that 68.7 per cent of the whole area of Pakistan receives rainfall below 254 mm (10 inches), 22.2 per cent between 254–508 mm (10–20 inches), 5.4 per cent between 508–762 mm (20–30 inches) and only 3.7 per cent more than 762 mm (30 inches) annually. The estimated annual average of the country is 348 mm (13.7 inches).

There are two cropping periods in Pakistan. One, for summer crops, is called Kharif season and extends from April to September. The winter crop season is called Rabi. It extends from mid-October to March. The distribution and intensity of rainfall in Pakistan is too insufficient and irregular to allow the raising of satisfactory crops. Agricultural production is dependent on surface water and groundwater. The highest rainfall occurs during the monsoon circulation in the months of July and August. The intensity of rainfall and the volume of downpour is much more than can be utilised, therefore a large volume of water is wasted. The monsoon current originates in the Bay of Bengal and moves north-westward across the Indo-Gangetic Plains. Similarly, storms originating from the Arabian Sea cause heavy rains on the Western Ghats of India, but by the time these monsoon currents reach the Indus Plain their intensity is much reduced.

The total annual pan evaporation in the Indus Plain ranges from 1300 mm in northern Punjab to 2800 mm in Sind. Evaporation is generally lowest in January and highest in May, with 65 to 70 per cent occurring from

April to September (Department of Technical Co-operation for Development, 1986). In Pakistan, April, May and June are extremely hot and dry. July, August and September are hot and humid, while October and November are cool and dry. December, January and February are the coldest months of the year.

Pakistan possesses a great range of climatic diversity, from some of the hottest areas in the world in its south-western part to the snowy and cold region in the north. In the mountain regions of the north and west, temperatures fall below freezing during winter. In the Indus Plain area, temperatures range from about 32° to 49°C in summer and the average in winter is about 13°C. Along the coastal belt of the Arabian Sea, the climate is under the influence of sea breezes. In general, a major part of Pakistan has an arid to semiarid climate.

Water resources

Surface water

A major part of Pakistan is drained by the Indus River and its tributaries, with catchment areas extending beyond the borders. The Indus River originates in the Tibetan Plateau at an altitude of about 5486 m above mean sea-level. The Indus River is about 2800 km long and 62 per cent of its catchment lies in Pakistan (Shafique and Skogerboe, 1984). Its major tributaries are the Beas, Sutlej, Ravi, Jhelum and Chenab (Figure 8-1). The Beas, Sutlej and Ravi are called the eastern rivers, while the Indus, Jhelum and Chenab are called the western rivers. A number of comparatively smaller rivers join the Indus. These include the Kabul, Kurram, Tank and Gomal. In Baluchistan only two small streams, the Gaji and Baran, which are active only during monsoon, join the Indus River. Several rivers of Baluchistan disappear in the land surface and a few drain into the Arabian Sea (see Ahmad and Chaudhry, 1988: Figures 3.1 and 3.2 for a detailed map of the Indus River tributaries and the rivers of Baluchistan). The Indus River system is fed by glaciers, snowmelt and rainfall. It is the principal source of water supply in the country. According to the 1960 Indus Waters Treaty between India and Pakistan, India is entitled to divert the entire flow of the eastern rivers, namely the Ravi, Beas and Sutlej (for the text of the Treaty see Michel, 1967: 557–572). Pakistan is left with the Indus River and its other tributaries, namely the Jhelum and Chenab.

Discharge of the rivers is recorded at suitable sites, generally called rim stations, where the rivers debouch into plains. Apart from rim stations, discharges and gauges are observed at all barrages and at other suitable sites. The average annual flow recorded at rim stations is given in Table 8.1. The average annual flow of the Indus at Tarbela is 82.6×10^9 m^3. The Kabul River contributes 32.1×10^9 m^3, which makes up the total flow above Attock, equal to 114.7×10^9 m^3.

Table 8-1: Mean annual flow of the Indus and its tributaries.

River	Rim station	Catchment area (km^2)	Average annual flow (10^9 m^3)
Indus	Attock	168 820	114.7
Jhelum	Mangla	33 410	28.4
Chenab	Marala	29 530	32.1
Ravi	Madhopur	8 030	8.6
Beas	Mandi Plain	16 840	16.0
Sutlej	Rupar	27 200	17.3

Source: Ahmad and Chaudhry (1988: Table 3.2).

The average annual river flow from the Indus River and its tributaries is about 175.2×10^9 m^3, comprising 114.7×10^9 m^3 for the Indus, 28.4×10^9 m^3 for the Jhelum and 32.1×10^9 m^3 for the Chenab. The average annual flow of 175.2×10^9 m^3 is equivalent to 220 mm or 63 per cent of the average annual rainfall of the country (348 mm). There is a marked seasonal variation in the river flows, with 84 per cent flow during the summer cropping season and only 16 per cent during the winter cropping season. Although there is a significant variation in flow from year to year, the Indus River system has a high degree of natural regulation, generally attributed to a large component of snow and glacier melt (Badruddin, 1987). Statistics for the natural regulation, measured as the ratio of the minimum annual flow to average annual flow, are as follows (Ahmad, n.d.): Indus, 63 per cent; Jhelum, 68 per cent; and Chenab, 70 per cent. Because of the high natural regulation of the Indus River system and the absence of large reservoir sites, there is little practical likelihood of use of surface reservoirs to carry over water from high flow to low flow years. However, reservoir storage for transfer of water from the summer high flow season to the winter low flow season will continue to be an essential element of water management. Table 8-2 provides some of the features of the major dams in Pakistan.

Table 8-2: Features of some important dams in Pakistan.

Name	Location	River	Year of completion	Capacity (10^9 m^3)
Mangla	Jhelum	Jhelum	1967	7.15
Chashma	Mianwali	Indus	1971	1.00
Tarbela	Haripur	Indus	1976	13.69
Hub	Karachi	Hub	1979	1.14

Source: Framji et al. (1982: 1055).

According to Ahmad (n.d.), out of an average annual flow of 175×10^9 m^3, some 125×10^9 m^3 is diverted to canal irrigation systems and

50×10^9 m³ flows to the Arabian Sea. Also, in order to satisfy irrigation requirements, 44×10^9 m³ per year is extracted from public and private tube-wells.

The apportionment of the Indus water was a source of conflict for more than 70 years. Finally, on 21 March 1991, an accord on the distribution of Indus waters among the four provinces was unanimously approved by the Council of Common Interests. The sharing of the Indus waters would be as follows: Baluchistan, 4.77×10^9 m³; NWFP, 7.13×10^9 m³; Sind, 60.12×10^9 m³; and Punjab, 68.97×10^9 m³ (Akhtar, 1991: 413 and 509).

The chemical quality of water samples of the Indus, Jhelum and Chenab Rivers, collected at the rim stations, shows that the salt content commonly falls in the range of 150 to 250 mg L⁻¹ total dissolved solids (TDS), reflecting high flow conditions in summer to low flow conditions in winter. Near the outfall of the combined waters of the Indus and its tributaries into the sea, the average salt content is about 300–420 mg L⁻¹ TDS (Department of Technical Co-operation for Development 1986; Ashraf and Abdul Qayyum, 1978). In other words, the chemical quality of water of the Indus River system is excellent for irrigation, drinking or industrial purposes. However, the surface water resources are mainly used for irrigation and the total diversion of these resources in the irrigated system is about 125×10^9 m³ per year or about 70 per cent of the average annual flows.

Groundwater

Almost all of the Indus Plain is underlain by a thick alluvial complex of Holocene and late Pleistocene age, deposited in a depression. Groundwater is contained in interbedded deposits of alluvial sand and silt which are present almost everywhere. These deposits were laid down by the Indus River and its tributaries in a vast plain that extends from the foothills of the Himalaya to the Arabian Sea. Although the alluvium may locally contain a large proportion of silt and silty clay, the occurrence of these deposits is generally in the form of thin lenticular beds (Swarzenski, 1968). Hydrogeologically the whole of the Indus alluvial complex can be treated as a huge, single unconfined aquifer, with high hydraulic conductivity in the range 30–60 m d⁻¹ and an average storage coefficient of 12 per cent. These values indicate a high yielding aquifer with substantial storage capacity (Rathur, 1987). As reported by Gazdar (1987), the aquifer is believed to extend 300 m in depth over most of the area and the water stored in the aquifer is 50 to 100 times the annual surface flow of the Indus River system. The aquifer is recharged by rain and rivers and seepage from surface storage reservoirs, canals, watercourses and fields. Seepage from these sources contributes 40 to 60×10^9 m³ of water annually to the aquifer. Discharge from the aquifer is by evaporation, base flow to the streams and rivers, basin discharge downstream and withdrawals by wells and horizontal drains (Gazdar, 1987).

Groundwater quality through the Indus Basin aquifer varies depending on the climatic parameters, nature of the surface flow, topography, extent of seepage and irrigation practice. Generally, the quality deteriorates as one traverses the plain from upstream to downstream towards the Arabian Sea. Fresh groundwater with a salinity of less than 1000 mg L^{-1} occurs in the upper north-eastern part of the interfluvial region, where the precipitation is relatively high. Fresh groundwater is also found in belts parallel to the major rivers of the region, where infiltration from rivers crossing the present flood plains provides recharge (Figure 8-2). Zones of saline groundwater are found in the central and lower parts of the interfluvial regions. In the lower Indus Plain and particularly in Sind Province groundwater quality is poor and in large areas TDS values are greater than 3000 mg L^{-1} (see Ahmad and Chaudhry, 1990 for a coloured groundwater quality map of the Indus Plain).

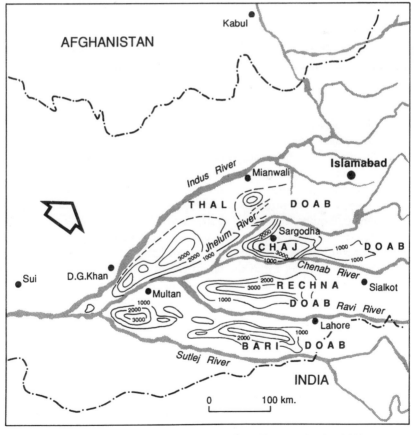

Figure 8-2: Average groundwater salinity (in mg L^{-1} TDS) to the depth of 110 m (Rathur, 1987).

In general, saline groundwater in the Indus Plain occurs in the following forms: in the form of lenses near or at some depth below the surface surrounded by fresh groundwater; underneath of a large freshwater zone; and, finally, saline groundwater may start near the surface with a gradual increase in salinity with depth. Table 8-3 shows areas within different ranges of average TDS of groundwater to a depth of about 110 m for the irrigated areas in Pakistan.

Table 8-3: Extent of groundwater salinity in the irrigated areas of Pakistan (areas within different ranges of average TDS of groundwater to a depth of about 110 m, in Mha).

Province	Gross area	Area with TDS range in mg L^{-1}		
		< 1000	1000–3000	> 3000
Punjab	9.84	5.48	2.69	1.67
NWFP	0.36	0.30	0.06	–
Sind and Baluchistan	6.32	0.62	0.90	4.80
Total	16.52	6.40	3.65	6.47

Source: Water and Power Development Authority (1988).

Because the Indus Plain sediments were deposited by the ancestral Indus River and its tributaries in a subsiding tectonic depression covered by sea which existed in the area, it is believed that the existing saline water may be a remnant of seawater. Recently Sajjad et al. (1993) investigated the origin of saline groundwater in three areas in the upper Indus Basin. Their investigation provided evidence that the salinity in the studied areas is not of connate marine origin. It is mainly due to the dissolution of sediment salts in the infiltrating freshwater, which carried away these salts towards the deeper groundwater regions. Apart from salinisation, groundwater quality deterioration is also due to the use of fertilisers. The nitrate content, which was up to 3 mg L^{-1} in the pre-irrigation period, has increased considerably and is now hundreds of mg L^{-1} at many places (Sajjad et al., 1993).

Estimates show that the annual potential of potable groundwater of the country is around 56.9×10^9 m^3, consisting of 46.9×10^9 m^3 in Punjab, 7.41×10^9 m^3 in Sind, 1.85×10^9 m^3 in NWFP and 0.74×10^9 m^3 in Baluchistan (Khan, 1992). Pumped groundwater serves a number of objectives, including: lowering the watertable; reclaiming salt-affected areas; supplementing canal irrigation supplies during critical summer and winter periods; irrigation water supply in areas not served by canals; and domestic and industrial water supply.

According to Ahmad and Chaudhry (1988: 6.33), irrigation water supply at farm gates in the canal commanded areas of Pakistan was about 120.55×10^9 m^3 in 1980–81, of which 80.4×10^9 m^3 was supplied by canal water and 40.15×10^9 m^3 by groundwater (consisting of 9.15×10^9 m^3 by public tube-wells and 31×10^9 m^3 by private tube-wells). In other words, groundwater supply was about 50 per cent of the canal supplies and about 33 per cent of the total supplies. Recently, 183 000 tube-wells in the private sector and 16 000 tube-wells in the public sector, with an estimated annual rate of 43.2×10^9 m^3, were reported to be in operation (Hussain, 1992).

Further information concerning the groundwater resources of Pakistan is available in Greenman et al. (1967), Department of Technical Co-operation for Development (1986), Directorate General of Hydrogeology (1989), Khan (1992) and particularly in Ahmad and Chaudhry (1988), which describes this issue in two lengthy chapters.

Land cover

According to FAO (1989), of a total land area of 77.09 Mha (total area of the country excluding areas under inland water bodies), about 20.76 Mha or 27 per cent of the land area were under cultivation in 1987. Of these lands, 16.08 Mha (77.5 per cent) were irrigated and 4.68 Mha (22.5 per cent) non-irrigated or rainfed. Permanent pasture covered 5 Mha, forest and woodlands 3.14 Mha and other lands 48.19 Mha. Estimates show that 44.6 Mha of land are unsuitable for agricultural production and forestry.

Other statistics provided by Akhtar (1986: 515) give the cultivated area in Pakistan as 20.43 Mha, while another 10.91 Mha are suitable for agricultural production. In the cropped area, food grains (wheat, rice, maize, barley, etc.) claim 61 per cent of the area, cash crops (sugarcane, sugar beet, cotton, tobacco, etc.) occupy 16 per cent, while the remaining 23 per cent is shared by other crops like oilseeds, vegetables, fruit and fodder.

Irrigation

Although irrigation has a 3000 year history in the Indus River Valley, the first canals were constructed some five or six centuries ago. In the nineteenth century the British initiated a massive irrigation and development program which gradually became the world's largest irrigation system. Ahmad and Chaudhry (1988: 4.1–4.83) describe in detail the ancient irrigation system, as well as the irrigation systems, canals and dams developed by the British Army engineers and those constructed since independence. Information on the development of irrigation in Pakistan is also available in Framji et al. (1982: 1040–1069).

The Indus irrigation system is gigantic by any measure. Its irrigated land comprises the largest single irrigation system in the world and has some of the largest infrastructural works found anywhere (Fairchild, 1984). Today the irrigation system includes three major storage reservoirs (Tarbela, Mangla and Chashma) on the western rivers, 20 diversion structures, 12 link canals, 48 main canals and some 89 000 watercourses, each serving 150 to 200 ha of land on average (Arshad, 1991; Rathur, 1987). The length of the irrigation network of canals is about 62 730 km, which is more than 1.5 times the length of the equator. It consists of 36 950 km in Punjab, 20 700 km in Sind, 2420 km in NWFP and 2660 km in Baluchistan (Arshad, 1991). Surface drains are about 15 171 km long, consisting of 2768 km of main drains and 12 403 km of branch drains (Framji et al., 1982: 1049). The system also contains 186 000 tube-wells, including 155 000 in Punjab (Gazdar, 1987), with a discharge capacity of 30 to 150 L s^{-1} each (Shafique and Skogerboe, 1984).

Tarbela Reservoir on the Indus River and Mangla Reservoir on the Jhelum River have very large capacities (Table 8-2). Because of the variation in supplies between different rivers, and also as a result of the Indus Waters Treaty with India, the system of link canals functions as a water transfer mechanism from the western rivers to the eastern rivers (Indus to Jhelum, Jhelum to Chenab, Chenab to Ravi, etc.) in order to provide an equitable distribution of water supply to the irrigated areas (see Framji et al., 1982: 1054 or Badruddin 1987, for the schematic diagram of the Indus River irrigation systems. Scholz 1984, provides a schematic diagram and coloured map of the system).

Agricultural development in Pakistan owes a great deal to the development of the irrigation network. A major part of the total irrigated area is under the canal command area, while 2.8 Mha are irrigated by tube-wells (Akhtar, 1986). Figure 8-3 shows the canal irrigated areas in Pakistan. The perennial canals supply water for both the summer and winter crops. One-third of the total irrigation system, termed non-perennial, receives irrigation supplies only in the summer. Cropping in the winter season therefore is limited by partial supplies, residual moisture or groundwater supplement.

Delivery losses between watercourse heads and the field are high and vary considerably, averaging about 40 per cent (Framji et al., 1982: 1047). Ahmad (n.d.) estimated the total water losses to be about 105×10^9 m^3, consisting of 31×10^9 m^3 conveyance losses, 54×10^9 m^3 watercourse delivery losses and 20×10^9 m^3 field irrigation losses. These low efficiencies are mainly due to unlined earthen canals, earthen watercourses and unlevelled fields. Reduction of these losses and the improvement of water management towards meeting crop needs present a great opportunity and challenge.

The crop water requirements for net crop needs have been computed for Pakistan based on the existing cropping intensity of around 105 per cent in the irrigated areas of the country. The net water requirement comes to around 85.1×10^9 m^3 whereas the net availability is around 77.5×10^9 m^3, consisting of 61×10^9 m^3 irrigation water and 16.5×10^9 m^3 effective rainfall. This leaves a shortage of 7.6×10^9 m^3 (Ahmad, n.d.). Experts in irrigation

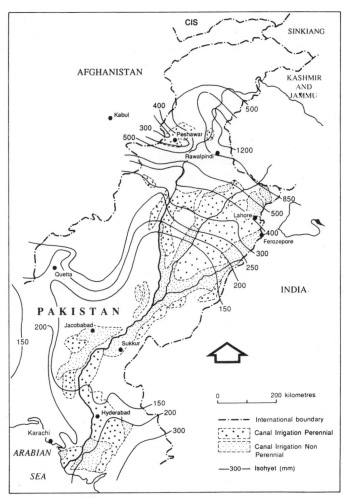

Figure 8-3: Canal irrigated areas and rainfall distribution in Pakistan (Badruddin, 1987).

management have identified a number of areas for improvement, including the following (Arshad, 1991): minimising losses in various canal systems; remodelling various canal systems to enable them to carry additional supplies; and precision land levelling.

Another problem in the Indus River Basin irrigation system is related to inequity in water distribution. According to Asghar et al. (1992), findings from the International Irrigation Management Institute (IIMI) in Pakistan, on the operation and performance of distributary (secondary) canals in the Lower Chenab Canal System over several years, show that the system performance objective of equity in water distribution is rarely achieved. When distributaries are operating at or near the full supply levels, outlets in the tail reaches seldom obtain more than a fraction of their authorised discharge at the watercourse head, in contrast to outlets in the upper reaches which commonly receive more than their designed discharge.

Salinity

Salinisation in Pakistan occurs both naturally and also as a consequence of poor irrigation practice (Sandhu and Qureshi, 1986). Part of the Indian subcontinent including the Indus Plain was formed from sediments transported by rivers into a shallow sea. The receding sea has left behind residues of salt, both in the soil profile and in the aquifer. In addition, minerals in parent rocks release significant quantities of salt into the soil solution during weathering. Under the prevailing arid and semiarid climatic conditions, salts released through weathering are not leached out of the soil profile. The natural salinisation occurs around the margins of natural depressions. Secondary salinisation, which is related to the modern irrigation system, is the result of either accelerated redistribution of salts in the soil profile due to high watertable or the use of insufficient water to leach salts out of the soil. Although the salt content of the Indus River water lies between 150 to 250 mg L^{-1}, irrigation of land for years using insufficient water without proper drainage has caused accumulation of salts in the soil profile. For example, application of 600 mm of water with a salinity of 200 mg L^{-1} will add 1.2 t of salt per ha. According to Asghar et al. (1992), canal water allocated to the farmers in perennial and non-perennial command areas is limited and is only 0.2 L s^{-1} ha^{-1}, compared to about 1 L s^{-1} ha^{-1} in the United States. Therefore farmers can neither meet the optimum requirements of crops nor increase their cropping intensity. To meet these deficiencies, groundwater is being used and sodic soils have been formed where groundwater with high SAR is applied for crop production.

Hydrogeologic investigations in the Indus Plain show that under pre-irrigation conditions, the watertable was about 30 m deep (Figure 8-4) and, in interfluvial areas, groundwater flow was from rivers towards the interfluvial areas. Even at the beginning of this century, the watertable over most of the Indus Plain was below 15 m (Fairchild, 1984). However, the watertable has risen due to irrigation, causing waterlogging and salinity problems. The average rate of the watertable rise has been estimated to be between 15 to 70 cm per year (Awan and Latif, 1982: 4).

According to Shafique and Skogerboe (1984), waterlogging has been recognised as a problem since 1841. However, no practical solution was applied at that time. After substantial irrigation began in 1856, the intensified problem of waterlogging became a major concern. The following is a chronological summary of the major mitigation activities:

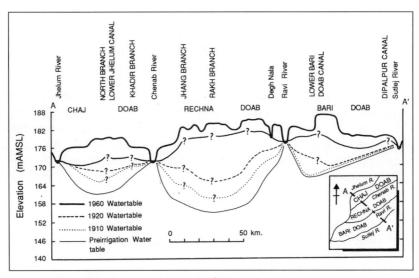

Figure 8-4:   Watertable profiles along line A-A¹ (Greenman et al., 1967).

- 1870, systematic observation of groundwater levels was initiated.
- 1917–26, comprehensive schemes were prepared to combat the waterlogging damage, but financial conditions after World War I did not permit significant improvement works.
- 1926–47, Upper Chenab Canal, Rechna Doab drains and many surface drains were excavated and tube-wells installed.
- 1952–58, an aerial photography survey of the Indus Plains showed that, of 29 Mha surveyed, 4.9 Mha were severely saline and 4.6 Mha were covered with saline patches.
- 1954, a detailed investigation was initiated by the Punjab Department of Irrigation and later by the Water and Power Development Authority (WAPDA) with the help of the International Cooperation Administration (now USAID).
- 1959, WAPDA engaged two consultant firms (Hunting Technical Services Ltd and Tipton and Kalmbach Inc.) to develop a detailed program for alleviating waterlogging and salinity throughout the Indus Basin.
- 1961, a master plan for the control of waterlogging and salinity was formulated by WAPDA, which consisted of 31 000 tube-wells, 12 000 km of main drainage channels and 40 000 km of secondary drains. However, this ambitious program suffered a serious setback because of later political developments.

In order to investigate the extent of soil salinity, many surveys have been undertaken, including the 1953–54 reconnaissance soil and salinity survey carried out under the Colombo Plan in the Indus Plain. The survey was based on air photo interpretation supported by field traverses. Most of the photographs were taken during the winter months when salinity is extensive and can be effectively recorded. This survey showed that 2.03 Mha (7 per cent) of land were severely saline, and in 4.58 Mha (16 per cent) saline patches were common. Also 4.86 Mha (17 per cent) of the surveyed area were waterlogged or poorly drained (Ahmad and Chaudhry, 1988: 2.30). According to this survey Rechna Doab, Upper Sind Plain, Lower Sind Plain and the Indus Delta had about 458 000 ha, 438 000 ha, 400 000 ha and 336 200 ha of severely saline soils respectively, and in these four regions the percentages of each region affected by severe salinity were 15, 33.5, 50 and 35 per cent, respectively.

A survey carried out during the 1977–79 period showed that of 16.72 Mha of irrigated areas in Pakistan, 12.07 Mha (72 per cent) were salt free, 1.89 Mha (11 per cent) were slightly saline, 1.0 Mha (6 per cent) moderately saline and 1.33 Mha (8 per cent) strongly saline (Table 8-4). In other words, 4.22 Mha (or about 25 per cent) of the irrigated area in the country are in the grip of salinity to varying degrees. Among the four provinces of Pakistan, Sind with 2.62 Mha has the largest area of irrigated land affected by salinity. It is followed by Punjab (1.43 Mha), Baluchistan (0.09 Mha) and NWFP (0.08 Mha). The ratio of salt-affected land to irrigated land is the highest in Sind (47 per cent), followed by Baluchistan (26 per cent), Punjab (14 per cent) and NWFP (13 per cent).

Table 8-4: Irrigated soil salinity in different provinces of Pakistan (1977–79).

Province	Total irrigated area (Mha)	Salt-free[a] (Mha)	Slightly saline[b] (Mha)	Moderately saline[c] (Mha)	Strongly saline[d] (Mha)	Misc.[e] (Mha)
Punjab	10.17	8.54	0.71	0.41	0.31	0.20
NWFP	0.62	0.48	0.06	0.01	0.01	0.06
Sind	5.58	2.79	1.06	0.56	1.00	0.17
Baluchistan	0.35	0.26	0.06	0.02	0.01	–
Total	16.72	12.07	1.89	1.00	1.33	0.43

(a) $EC_e < 4000$ μS cm^{-1}; (b) $4000 < EC_e < 8000$;
(c) $8000 < EC_e < 15\,000$; (d) $EC_e > 15\,000$ μS cm^{-1};
(e) Areas under canals, roads, settlements, etc.

Source: Ahmad and Chaudhry (1988: Table 2.26).

Results of a study of the chemical status of soil profiles with respect to salinity and alkalinity are represented in Table 8-5. It shows that of 63 866 soil profiles, 10.7 per cent (6814) were saline, 23.6 per cent (15 100) were saline-alkaline and 3.5 per cent (2226) were alkaline.

Table 8-5: Extent of salinity and alkalinity in soil profiles in different provinces of Pakistan (1977–79).

Province	Non-saline, Non-alkaline	Saline	Saline-alkaline	Alkaline	Miscellaneous	Total
Punjab	29 334	2 803	5 757	1 813	256	39 963
NWFP	1 553	216	138	28	23	1 958
Sind	7 918	3 430	8 677	373	145	20 543
Baluchistan	497	365	528	12	–	1 402
Total	39 302	6 814	15 100	2 226	424	63 866
Per cent	61.5	10.7	23.6	3.5	0.7	100

Source: Ahmad and Chaudhry (1988: Table 2.28).

The extent of waterlogging and land salinisation depends on the watertable depths, which are subject to seasonal variations. They are lower in the pre-monsoon period from April to June and are higher in the post-monsoon period in October. Tables 8-6 and 8-7 show this seasonal variation, with groundwater depths ranging from less than 1.5 m to more than 6 m. As shown in these tables, the area with groundwater depth less than 1.5 m varied

Table 8-6: Area (in 1000 ha) under various watertable depths in April/June 1987 (pre-monsoon period).

Province	Area under various depths to watertable ranges					Total
	0.0–1.5 m	1.5–3.0 m	3.0–4.5 m	4.5–6.0 m	> 6.0 m	
Punjab	961.80	2973.50	2259.80	1481.00	2287.90	9964.00
NWFP	41.80	135.40	78.60	64.90	220.30	541.00
Sind	1079.37	3608.63	633.88	258.23	155.79	5735.90
Baluchistan	50.98	106.01	112.70	56.47	72.94	399.10
Total	2133.95	6823.54	3084.98	1860.60	2736.93	16640.00

Source: Water and Power Development Authority (1988).

Table 8-7: Area (in 1000 ha) under various watertable depths in October 1987 (post-monsoon period).

Province	Area under various depths to watertable ranges					Total
	0.0–1.5 m	1.5–3.0 m	3.0–4.5 m	4.5–6.0 m	> 6.0 m	
Punjab	1059.40	2815.30	2380.70	1540.40	2168.20	9964.00
NWFP	52.80	122.20	77.40	62.20	226.40	541.00
Sind	3109.63	1459.62	608.55	366.29	191.81	5735.90
Baluchistan	118.70	59.70	63.60	71.40	85.70	399.10
Total	4340.53	4456.82	3130.25	2040.29	2672.11	16640.00

Source: Water and Power Development Authority (1988).

from 2.1 Mha in April/June to 4.3 Mha in October, while the area with watertable depth less than 3 m did not change significantly and remained close to 9 Mha.

Tables 8-6 and 8-7 also show that the Sind Province, located at the lower end of the Indus River system, has a large area of very shallow watertables (less than 1.5 m), both in pre-monsoon (1.08 Mha) and post-monsoon periods (3.1 Mha). For this reason, soil salinity is very severe in this province, affecting about 2.62 Mha (see Table 8-4). Special measures are required to alleviate the problem.

Management options

Management of saline soils

In the early days of the occurrence of irrigated land salinity in Pakistan, when knowledge of the processes of salinisation was limited, suggestions were put forward to scrape the salt-affected soil surface to combat against salinity. A trial at one or two sites confirmed that this method was ineffective. Also an attempt was made to decrease the toxic concentration of salts by mixing good soil with low salinity silt and even sand. The measure gave a temporary relief but was impracticable on a large scale (Ahmad and Chaudhry, 1988: 7.14–7.15). The following sections describe some of the measures undertaken for the management and reclamation of salt-affected soils.

SCARP Projects

Management options to control waterlogging and salinisation are drainage through tube-wells, surface drains, subsurface tile drains and the conjunctive use of surface and groundwater for irrigation and on-farm water management. These options have been implemented in a series of Salinity Control and Reclamation Projects (SCARP).

The first SCARP, extending over an area of 486 826 ha in the middle area of the Rechna Doab in Punjab, was started in 1956 and completed in 1963. It was considered as a pilot project and became the forerunner of similar large-scale projects in which drainage was to be achieved by pumping out groundwater. Pumped water was to be utilised for supplementary canal supplies for a more intensive irrigated agriculture. Awan and Latif (1982) describe the SCARP I project and analyse its technical, social and economic aspects. The following description is based on their account.

The objective of SCARP I was to preserve/restore/improve and develop the land and agricultural water resources for national welfare through:

- Tube-wells to lower the watertable and to provide more water for extensive and/or intensive agriculture.
- Reducing salinity hazard by leaching, adding amendments, increasing cropping intensity, but avoiding land deterioration by hazardous tube-well waters.

- Helping the farmers by supplying recent technical information to them, removing their financial hurdles through loans, subsidies and so on, and ensuring correct return for their agricultural production sent to markets.

The project represented the culmination of many years of study and investigation of the means to alleviate problems associated with waterlogging and salinity, and embodied construction of some 2068 tube-wells ranging in capacity from 84 to 140 L s^{-1}.

Reclamation of the area affected by waterlogging and salinity and increasing agricultural productivity were the major objectives of the project. Agricultural productivity of the project area was about 53.2 per cent in 1959 and it was estimated that, without the SCARP Project, productivity would drop to 10 per cent from 1959 to 1980, while with the project it would increase to 100 per cent over the same period. Table 8-8 shows the effect of the project on the extent of different classes of soil salinity. It indicates a major expansion of the areas with non-saline soils and a significant decrease in the areas affected by moderate to strongly salt-affected soils. With respect to waterlogging, about 73 per cent of the gross area or some 360 000 ha had a watertable shallower than 3 m in 1959, while in 1977–78 this area was reduced to 37.6 per cent or about 161 000 ha. Estimates show that productivity increased to 80.4 per cent in 1977–78.

Table 8-8: Land salinity status of the SCARP I Project with a gross area of 493 700 ha between 1959 and 1977–78 as per cent of the gross area.

Year	Non-saline[a] (%)	Moderately saline[b] (%)	Strongly saline[c] (%)	Highly saline[d] (%)	Total
1959	65	18	14	3	100
1977–78	86	4	5	5	100

(a) Soil with 0–0.2 per cent of salts;
(b) Soils with 0.21–0.5 per cent of salts;
(c) Soils with 0.51–1.0 per cent of salts;
(d) Soils with more than 1 per cent of salts.

Source: Awan and Latif (1982: 34-35).

At this stage it is important to refer to some of the problems related to tube-wells in the project. The acute shortage of canal supplies, because of their limited conveyance capacity and the urgency of control of waterlogging and salinity, gave tremendous impetus to large-scale groundwater exploitation by large capacity turbine-type tube-wells. However, as pumping equipment of different makes was imported, enormous management problems were created later, during the operation of the project. The problem was aggravated further because of the large gap between the tube-well life estimated by the consultants (30–40 years) and that actually achieved during the project (less than 15 years). Other technical problems consisted of corrosion and encrustation of strainers and improper design of gravel packs. Subsequently tube-well discharges reduced to 50 per cent of their initial discharge and, even worse, a significant number of them had to be closed.

Location of high capacity tube-wells also created major social problems for the utilisation of water for irrigation. For example, the location of a high capacity tube-well at the head of a watercourse necessitated enlarging the watercourse, which was previously designed only for canal irrigation supplying less than one-fourth the combined tube-well and canal discharge. Being the responsibility of the farmers, remodelling of the watercourse took a long time, with the result that wastage of water had increased. In situations of one tube-well serving more than one watercourse, the tube-well water from one side of the canal is conveyed to the other side through a pipe siphon. This siphon gets choked by silt and debris floating in the watercourse. As a result, the farmers on the side of the distributary opposite to the tube-well receive less discharge. This led to disputes between farmers on the two sides of the distributary as well as among those on one side, especially when a watercourse served two to three villages. These social problems could have been eliminated had the tube-well been of small capacity and dispersed over the whole length of the watercourse. This would have reduced the large investment made by the farmers in enlarging their watercourses and, at the same time, ensured equitable distribution of groundwater from head to tail of a watercourse.

After completion of SCARP I, similar projects were developed by consultants for other areas. These projects included: SCARP II for Chaj Doab; SCARP III for Lower Thal Doab; SCARP IV for the northern area of Rechna Doab; SCARP V for the southern area of Rechna Doab; SCARP VI for Panjnad Abbasia; SCARP Khairpur and SCARP Shikarpur (see van der Leeden, 1975: 245 for a location map of SCARP projects). According to Ahmad and Chaudhry (1988: 7.45), WAPDA set up its own project planning organisation from 1970 onward and prepared projects for several areas in various provinces. By the end of June 1978 there were 25 completed SCARP projects consisting of 11 projects in Punjab, 9 projects in Sind and 5 in NWFP. At the same time there were 17 projects under construction consisting of 9 projects in Sind, 5 in Punjab and 3 in NWFP (see Ahmad and Chaudhry, 1988: 7.46–7.53 for the features of these projects).

According to recent statistics (Water and Power Development Authority, 1988), 32 projects were successfully completed during the period 1960–88, covering an area of 3.8 Mha. Another 13 projects were under construction in 1988 and these will serve an additional area of 3.3 Mha. Some 13 000 tube-wells, 5000 km of surface drains and 2000 km of tile drains are in operation in completed projects, while 5000 tube-wells and 8000 km of surface and 8000 km of tile drains will be put into operation with the completion of ongoing projects.

The cost of maintenance and operation of SCARP tube-wells has continued to increase and has become exorbitant. During the 13-year period

1972–1985, the cost of annual operation and maintenance of Punjab's canals increased six times, while similar costs for tube-wells increased 14 times (Ahmad and Chaudhry, 1990: 10.8; Ahmad and Chaudhry, 1988: 7.67–7.63). For this reason, the government decided to transfer freshwater pumping tube-wells to the private sector by offering 90 per cent of the cost as loan and by providing electrical transmission lines. The government also helps farmers to install their own tube-wells.

Pumping saline groundwater

In the Indus Plain, water-bearing medium possessing groundwater with a TDS value of 3000 mg L^{-1} or more, within a depth of 110 m, is classified as a zone of saline groundwater. The extent of saline groundwater is about 1.67 Mha in Punjab and 4.80 Mha in Sind and Baluchistan Provinces (Table 8-3). These areas lie scattered at various sites in the irrigated parts of the northern Indus Plain and particularly in the middle of Doabs (Figure 8.2). Saline groundwater covers a large area in the southern Indus Plain.

WAPDA has undertaken responsibility for drainage of areas of saline groundwater, to lower the level of saline groundwater and to protect the useable freshwater zones against the intrusion of saline groundwater. According to Ahmad and Chaudhry (1990: 10.7), 365 tube-wells were installed in the Khairpur area of Sind and about 1070 in Punjab, and about 577 were under construction. These tube-wells are of high capacity, pumping 55 to 85 L s^{-1}, and are installed in deep saline groundwater. The quality of pumped water is about 10 000 to 20 000 mg L^{-1} TDS. The highly saline waters pumped out by public tube-wells are disposed to main canals or a river. In the case of the SCARP VI project, saline groundwater is to be discharged into evaporation basins spreading over an area of 23 000 ha (Ahmad and Chaudhry, 1988: 7.65). In the case of the Sind Province, the highly saline water will be discharged to the Arabian Sea via the Left Bank Outfall Drain.

Left Bank Outfall Drain

Disposal of saline drainage effluent into canals or into drains discharging into a river is possible only on a limited scale. Disposal into a river merely transports the salts further downstream where it may end up on irrigated land at the tail of the system. These lands tend to be less well-watered and thus less able to cope with the additional salinity. As well as being considered environmentally undesirable, the disposal by evaporation from ponds on the edge of irrigated areas is able to deal with only limited quantities of water.

In some cases, a satisfactory solution is disposal to the sea. This is the primary purpose of the Left Bank Outfall Drain (LBOD) located in the Lower Indus region (Figure 8-5). This region includes 5.66 Mha of cultivable area, of which 85 per cent is underlain by saline to ultra-saline groundwater. In one respect at least, this region is fortunate in having access to coastal zones where saline water can be discharged to the open sea. According to Lee et al. (1987), the potential catchment of LBOD initially comprised about 2 Mha of perennially irrigated land underlain by saline groundwater (SGW) on the left bank of the Indus River below Sukkur. Stage 1 of the LBOD project, prepared with UNDP assistance, is to alleviate waterlogging and land salinisation on 577 000 ha. This area covers a substantial part of the irrigated area in Nawabshah, Sanghar and Tharparker districts.

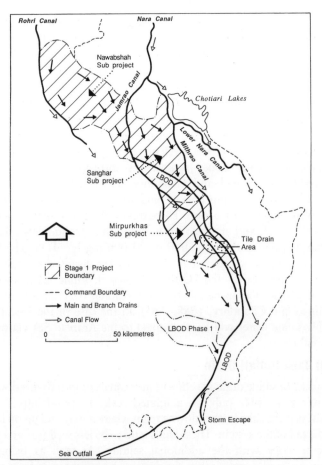

Figure 8-5: Left Bank Outfall Drain project (Lee et al., 1987).

These three districts together produce about 20 per cent of the cotton, 10 per cent of the wheat and 5 per cent of the sugarcane produced in Pakistan each year. The project will control present waterlogging by installing 2000 wells on 395 000 ha, an area which would otherwise have a watertable within 1.5 m of the surface. A further 22 000 ha, also enduring high watertables, will be provided with tile drains. Both wells and drains will be pumped into a disposal system providing surface drainage for the whole project area. The disposal system will discharge into a spiral drain started previously under the LBOD Core Programme. The total length of this drain will be extended to 290 km to provide a sea outfall for saline effluent. It is expected that Stage 1 of this project will take at least 6–7 years to implement, starting in 1987 (Lee et al., 1987).

Improved water management

Because of the low overall efficiency of irrigation water application in Pakistan, available water resources are less than the requirements. Improvement in water management, both in canal systems as well as on-farm water management (OFWM), will save water for increasing agricultural production and would alleviate salinity problems. Arshad (1991) describes the objectives as well as the current activities in this area in Pakistan. A brief description of the OFWM Projects is as follows:

- The pilot OFWM Program was launched by the Government of Pakistan in 1977, under the financial and technical assistance of USAID. Improvement of 1319 watercourses, precision levelling of 30 294 ha of land and training of 6000 people were completed by June 1981.
- OFWM-I project was launched under the assistance of the World Bank for a period of 3 years (1982 to 1984) at a total cost of US$80 million. This project aimed to renovate 2065 watercourses, improve 16 500 watercourses under the accelerated watercourse improvement program and precisely level 4875 ha of land. Establishment of 700 demonstration plots and 17 trickle irrigation plots and construction of 37 storage tanks was also included in the project. The project targets were achieved in all the provinces.
- The Asian Development Bank is financing two programs in Punjab and the North-West Frontier Province. The World Bank-assisted OFWM-II Program is being implemented in all four provinces.
- In addition to the above main water management programs, 3124 watercourses are being renovated and 6250 ha of land are being levelled in different regions with financial assistance from the World Bank and other agencies.

The basic aim of the Command Water Management Project (CWMP) is the integration of specialised agencies in water and agriculture, at the command level, with the expressed objective of increasing agricultural production by combining institutional, administrative and technical expertise and control over surface and groundwater supplies (Arshad, 1991). Improvements undertaken through the project include: rehabilitation of distributaries; canal lining, channel protection and water control structures; remodelling of outlets; watercourse improvements; surface drainage; tube-well drainage; and conjunctive use of groundwater (from shallow and deep tube-wells) and canal water.

The targets of the CWMP in Punjab, Sind, NWFP and Baluchistan include: 251.36 km of canal rehabilitation, 190.25 km of canal lining, surface drainage of 24 240 ha and 1050 wells. The estimated cost of the project including its engineering and non-engineering components is about US$81.9 million.

With respect to on-farm water management to combat irrigated land salinisation in Pakistan, Asghar et al. (1992) describe the following options:

- Pre-sowing irrigation for leaching. Salts tend to accumulate near the soil surface during the fallow period, especially when watertables are high or if there is little winter rain. This happens frequently during the winter season when more than 30 per cent of the fields remain without crop. For these fields, both germination and seedling growth can be seriously reduced unless the soil is leached before planting. A pre-sowing irrigation, applied before cultivation, and seedbed preparations may increase the percentage of seed germination and impact positively on growth of seedlings.

- Frequent irrigations. Field studies show that salt concentration in the top soil layer are lowest following an irrigation and highest just before the next irrigation. Frequent irrigations bring down the salt concentration in the root zone and increase average soil water content. This is particularly relevant in Punjab where heavy irrigations are applied to most of the crops in order to reduce the overall number of irrigations and work involved.

- Sprinkler irrigation. With appropriate system design and management, sprinklers can apply water with good uniformity at application rates low enough to prevent surface run-off. The depth of water applied to supply the crop's water requirement plus leaching can be controlled accurately by adjustments in the duration and frequency of application.

- Land levelling. It has been observed frequently that fields are not levelled or graded properly to allow uniform water distribution by surface irrigation. High spots in the field reduce the opportunity for water intake, which may lead to accumulation of salts and, hence, poor yields. In some cases, cultivators apply excessive amounts to cover the high spots, which results in wastage of scarce water resources.

Farmers' participation

Asghar et al. (1992) also emphasise the fundamental importance of farmers' participation in managing salinity problems because their knowledge and constraints are key factors in the success of any management options. Therefore farmers need to be educated on optimal land and water use, and to be convinced of the economic benefits of adopting proper management options.

Revegetation

Apart from engineering solutions to combat waterlogging and salinisation, efforts have been made towards plantation of salt-tolerant plants. Malik et al. (1986) describe successful results with plantation of kallar (i.e. salt) grass (*Leptochloa fusca*, synonym: *Diplachne fusca*). It has been known to grow in salt marshes and salt-affected areas of Pakistan and is called 'Australian grass'. According to Qureshi et al. (1982), there is hearsay evidence that the species now prevalent in Pakistan was probably introduced from Australia 80 years ago, though when this occurred and who was responsible are not known.

Kallar grass is a deep-rooted perennial tall grass which grows to a height of 1.5 m; its native distribution area includes Egypt, India, Sri Lanka and Australia. It is a highly salt-tolerant grass which can survive at a very high salinity (EC_e) of 40 000 µS cm^{-1}, but remains economical up to a salinity of 20 000 µS cm^{-1}. It also tolerates sodic soils and waterlogged conditions. Kallar grass possesses an excellent salt excretory mechanism through its leaves, harbours nitrogen-fixing bacteria and obtains 60 to 80 per cent of its nitrogen requirements from the air. Therefore it grows very well without fertiliser. Experiments indicate that, in general, different fertilisers have little effect on the growth of kallar grass. It produces about 50 tonnes of biomass per hectare, even when irrigated with brackish water.

Kallar grass roots are profuse and deep and greatly help to open up the impermeable sodic soils. Along with leaf litter, they add a considerable amount of organic matter to the soil. The growth of the grass produces CO_2 at root level, which converts the insoluble calcium carbonate to bicarbonate. Calcium thus solubilised would exchange with sodium, producing sodium bicarbonate, which is soluble in water and can be easily leached down. Exchange of Na^+ with Ca^{++} would also restore the soil structure and thus its permeability to water, and leaching of salts would be enhanced.

The growth of the grass has improved soil conditions in Pakistan to the extent that some moderately salt-tolerant crops, and a few selected varieties of barley and some tree species, have also been grown on the ameliorated soils.

Kallar grass is most commonly used as a forage plant. It does not retain most of the salts taken up, hence it remains reasonably palatable for farm animals. Due to the usefulness of the grass as a forage species, farmers have been encouraged to plant their salt-affected soils with this plant. As a result its use as a fodder is generally increasing. For example, in the Jhang district, large tracts of kallar grass are being used as a sole source of fodder for livestock. So far no adverse effects of feeding this grass to animals have been reported. The economic returns to private farmers of raising buffaloes with this grass are particularly encouraging.

Although kallar grass can be propagated through seeds, the best field establishment is obtained by the sowing of stem cuttings or root stubbles.

Kallar grass can be sown at any time of the year in the plains of Pakistan. However, the best sowing time is in March (see Malik et al., 1986 and Qureshi et al., 1982 for further information concerning kallar grass and its plantation for amelioration of salt-affected soils).

According to Sandhu and Qureshi (1986), kallar grass has been shown to be the most promising species for production of summer fodder from waterlogged and saline sodic soils. However, the problems of fodder supply during winter and utilisation of dry salt-affected areas remain unsolved. Work is in progress on salt-tolerant winter fodders, like *Agropyron* spp., *Salicornia* spp. and *Atriplex* spp., as possible sources of fodder from dry salt-affected areas. Research indicates successful results growing salt-tolerant plants like *Prosopis juliflora*, *Eucalyptus camaldulensis* and *Tamarix aphylla* on sandy soils using highly saline groundwater for irrigation (EC values of 10 000–15 000 µS cm^{-1}).

Ahmad et al. (1990) reported the results of their field experiments at Faisalabad during 1988–90. In the experiment area, the soil is loam to sandy clay loam, and is saline-alkaline with some variations with respect to EC_e (7400 to 9000 µS cm^{-1}) and SAR (55.6 to 73.0) in the upper 30 cm layer. The objectives of their study were: to compare the effectiveness of sesbania (*Sesbania aculeata*), sordan (*Sorghum bicolor* × *Sorghum sudanense*) and kallar grass (*Leptochloa fusca*) with gypsum, to monitor the performance of the plant species for biomass production; to monitor the soil quality changes; and to observe the growth response of wheat cultivation following reclamation treatments. In this experiment canal water with an EC value of 270 µS cm^{-1} was used for irrigation and leaching. Agricultural grade gypsum powder (70 mesh with 90 per cent purity) was used at 100 per cent of the gypsum requirement in the top 15 cm of the soil (13 t ha^{-1}). Results of the study demonstrated that the forage plant growth was an effective reclamation tool. Sesbania and kallar grass emerged as potential biotic materials that can be included in a reclamation program. The crops not only produced substantial biomass but also improved soil quality, which in turn gave a good yield of wheat, comparable to that of the gypsum treatment.

Management of alkaline soils

Malik (1987) also provides a short review of some of the research undertaken in Pakistan for the management of alkaline soils using gypsum, sulphuric acid and hydrochloric acid and growing rice and green manure crops. Overall application of gypsum has been adopted because it is the most economical amendment.

Malik (1987) also describes the results of some field experiments on saline-alkaline soils. The soil texture of the land under experiment varied from loam to clay loam. Thus the results could be used for almost all types of soils in the country. Soil samples were taken every 15 cm from the soil surface, before and after the experiment, and analysed for pH, EC_e and SAR as well as cations and anions. Various rates of gypsum powder (100 mesh) were applied to different plots to meet 100, 50 and 25 per cent requirements of the top 30 cm of the soil. Gypsum was applied before the start of the leaching process and was mixed with the upper 7.5 cm of the soil. Based on the results of the experiment, Malik (1987) concluded that: reclamation of alkaline soils can be achieved easily and in a short period of time with the application of powder gypsum; the application of gypsum to meet 100 and 50 per cent of the requirements are equally useful; and the reclaimed lands can be kept in fit condition by providing crop cover to them through the year.

Reclaimed saline and alkaline soils

As a result of SCARP projects, expansion of groundwater extraction from public and private tube-wells and other measures undertaken, watertables have been lowered to a reasonable extent and salinity and alkalinity have been reduced. A comparison of the data provided in Table 8-4 for the period of 1977–79 with those available for earlier periods indicates that the salt-free soil category was the highest in Punjab during 1977–79 (84 per cent), and was 70 per cent in 1953–65, followed by NWFP where salt-free soils were 77 and 75 per cent in the two periods. In Sind this type of land was 50 per cent during 1977–79 and 26 per cent during 1953–54 (Ahmad and Chaudhry, 1988: 2–61). These data show that the salt-free area increased during 1977–79 compared to the early surveys.

Overall during the 1953–65 survey, 32 and 45 per cent of the Indus Basin was affected by surface and profile salinity, while during the 1977–79 survey, surface and profile salinity were 26 and 39 per cent, respectively (Ahmad, n.d.). These data indicate that about 6 per cent of the Indus Basin affected by surface and profile salinity was reclaimed. Akhtar (1986) reported that on average about 81 000 ha of affected land is being reclaimed annually in the country, while according to Asghar (1992), about 0.5 Mha has been reclaimed in Punjab since 1945.

Economic damage

An indication of the economic costs of salinisation can be obtained from a study conducted by Water Resources Planning, Planning Division of the Water and Power Development Authority of Pakistan. They considered the entire irrigated area of the upper Indus Plain, comprising the two Provinces of Punjab and NWFP. It was estimated that the economy of the country suffers a loss of approximately 4.3 billion rupees (about US$300 million) annually in this area on account of decreases in farm production on soils slightly to moderately affected by salinity. The loss on a country-wide basis would be much higher (Water and Power Development Authority, 1988).

Apart from economic damage due to production loss, Pakistan has spent and allocated a considerable amount of its financial resources to alleviate salinity problems. According to Fairchild (1984), Pakistan had invested over US$550 million in SCARP projects. UNDP assisted Pakistan by providing the funds and the World Bank had an executive role in these projects.

With respect to the Left Bank Outfall Drain Project, the cost of Stage 1 will be US$640 million, jointly financed by the World Bank, Asian Development Bank, Islamic Development Bank, OPEC, Saudi Arabia, Britain, Canada and Switzerland (Lee et al., 1987).

References

Ahmad, S. n.d. *Viability of Agricultural Resource Base: A Critical Appraisal.* Islamabad: Water Resources Research Institute, National Agricultural Research Centre (NARC). 30 pp. (The estimated date of publication is 1985 or post-1985).

Ahmad, N. and Chaudhry, G.R. 1988. *Irrigated Agriculture of Pakistan.* Lahore: Shahzad Nazir, 61–B/2, Gulberg-3. (Various pagings, about 810 pp).

Ahmad, N. and Chaudhry, G.R. 1990. *Summaries of Irrigated Agriculture of Pakistan.* Lahore: Shahzad Naziv, 61–B/2, Gulberg-3. (Various pagings, about 130 pp).

Ahmad, N., Qureshi, R.H. and Qadir, M. 1990. Amelioration of a calcareous saline-sodic soil by gypsum and forage plants. *Land Degradation & Rehabilitation.* 2: 277–284.

Akhtar, R. 1986. *Pakistan Year Book, 1986–87.* Fourteenth edition. Karachi: East and West Publishing Company. 688 pp.

Akhtar, R. 1991. *Pakistan Year Book, 1991–92.* Nineteenth edition. Karachi: East and West Publishing Company. 551 pp.

Arshad, M.S. 1991. Pakistan. In. *Farm-Level Irrigation Water Management.* Tokyo: Asian Productivity Organization. 171–184.

Asghar, Ch.M., Bhatti, M.A. and Kijne, J.W. 1992. Management of salt affected lands for crop production in Punjab of Pakistan. In. *Proceedings of the International Symposium on Strategies for Utilizing Salt Affected Lands.* Bangkok, 17–25 February 1992. Bangkok: Department of Land Development, Ministry of Agriculture and Cooperatives. 241–258.

Ashraf, M. and Abdul Qayyum, M.A. 1978. Criteria for irrigation water quality in Pakistan. In. Framji, K.K. ed. *State-of-the-Art Irrigation, Drainage and Flood Control.* New Delhi: International Commission on Irrigation and Drainage. v.3. 185–201.

Awan, N.M. and Latif, M. 1982. *Technical, Social and Economic Aspects of Water Management in Salinity Control and Reclamation Project No. 1 in Pakistan.* Lahore: Centre of Excellence in Water Resources Engineering, University of Engineering and Technology. 142 pp.

Badruddin, M. 1987. Policy framework and institutions for water resources planning, development and management in Pakistan. In. Ali, M., Radosevich, G. and Khan, A.A. eds. *Water Resources Policy for Asia.* Rotterdam: A. A. Balkema. 105–117.

Department of Technical Co-operation for Development. 1986. *Ground Water in Continental Asia (Central, Eastern, Southern, South-Eastern Asia).* New York: United Nations. (Natural Resources/Water Series, no. 15). 391 pp.

Directorate General of Hydrogeology. 1989. *Hydrogeological Map of Pakistan (Scale 1:2 000 000).* Lahore: Water and Power Development Authority.

Fairchild, W.D. 1984. Drainage and salinity control programs in Pakistan. In. French, R.H. ed. *Salinity in Watercourses and Reservoirs: Proceedings of the 1983 International Symposium on State-of-the-Art Control of Salinity.* Salt Lake City, Utah, 13–15 July 1983. Boston: Butterworth. 43–52.

Food and Agriculture Organization of the United Nations (FAO). 1989. *Production Yearbook.* Rome: FAO. v.42. 350 pp.

Framji, K.K., Garg, B.C. and Luthra, S.D.L. 1982. *Irrigation and Drainage in the World: A Global Review.* Third edition. New Delhi: International Commission on Irrigation and Drainage. v.II. 1161–1667.

Gazdar, M.N. 1987. Groundwater and the environment: Pakistan scenario. In. *Groundwater and the Environment: Proceedings of the International Groundwater Conference 1987.* Malaysia: Universiti Kebangsaan. H12–21.

Greenman, D.W., Swarzenski, W.V. and Bennett, G.D. 1967. *Ground Water Hydrology of the Punjab, West Pakistan, with Emphasis on Problems Caused by Canal Irrigation.* (US Geological Survey Water-Supply Paper 1608–H). Washington DC: US Government Printing Office. 66 pp.

Hussain, N. 1992. Water and Power Development Authority, 94 Ferozepur Road, Lahore, Pakistan. (Personal communication).

Khan, L.A. 1992. Groundwater investigation in Pakistan: An emphasis on desert areas. *Water Resources Development.* 8(4): 257–269.

Lee, P.S., Shaikh, A.R. and Youssef, A.N. 1987. Left bank outfall drain: Integrated irrigation and drainage in Pakistan. In. *Transactions: Thirteenth International Congress on Irrigation and Drainage.* Rabat, Morocco, September 1987. New Delhi: International Commission on Irrigation and Drainage. V.I–D. 33-63.

Malik, N.A. 1987. Management of saline/alkaline soils in Pakistan. Paper presented at the Regional Expert Consultation on Management of Saline/Alkaline Soils. FAO Regional Office for Asia and the Pacific (RAPA), 25-29 August 1987, Bangkok. Bangkok: FAO-RAPA. 22 pp.

Malik, K.A., Aslam, Z. and Naqvi, M. 1986. *Kallar Grass: A Plant for Saline Land.* Faizalabad, Pakistan: Nuclear Institute for Agriculture and Biology. 93 pp.

Michel, A.A. 1967. *The Indus Rivers: A Study of the Effects of Partition.* New Haven: Yale University Press. 595 pp.

Qureshi, R.H., Salim, M., Abdullah, M. and Pitman, M.G. 1982. Diplachne fusca: An Australian salt-tolerant grass used in Pakistan agriculture. *The Journal of the Australian Institute of Agricultural Science.* 48(4): 195–199.

Rathur, A.Q. 1987. Groundwater management to eradicate waterlogging and salinity in the Upper Indus Basin, Punjab, Pakistan. In. *Groundwater and the Environment: Proceedings of the International Groundwater Conference 1987.* Malaysia: Universiti Kebangsaan. G96–107.

Sajjad, M.I., Hussain, S.D., Tasneem, M.A., Ahmad, M., Akram, W. and Khan, I.H. 1993. Isotopic evidence of the source of groundwater salinity in the Indus Basin, Pakistan. *The Science of the Total Environment.* 129(3): 311–318.

Sandhu, G.R. and Qureshi, R.H. 1986. Salt affected soils of Pakistan and their utilization. *Reclamation and Revegetation Research.* 5: 105–113.

Scholz, F. 1984. Bewässerung in Pakistan: Zusammenstellung und Kommentierung neuester Daten. *Erdkunde.* 38(3): 216–226.

Shafique, M.S. and Skogerboe, G.V. 1984. Planning and implementation framework for salinity control in the Indus River Basin. In. French, R.H. ed. *Salinity in Watercourses and Reservoirs: Proceedings of the 1983 International Symposium on State-of-the-Art Control of Salinity.* Salt Lake City, Utah, 13–15 July 1983. Boston: Butterworth. 93–102.

Swarzenski, W.V. 1968. *Fresh and Saline Ground Water Zones in the Punjab Region, West Pakistan.* (US Geological Survey Water-Supply Paper, 1608–I). Washington DC: US Government Printing Office. 24 pp.

van der Leeden, F. ed. 1975. *Water Resources of the World: Selected Statistics.* Port Washington, New York: Water Information Center. 568 pp.

Water and Power Development Authority (WAPDA). 1988. Lahore: WAPDA, 94 Ferozepur Road. (Personal communication).

World Bank. 1992. *World Development Report 1992: Development and the Environment.* New York: Oxford University Press. 309 pp.

Chapter 9: South Africa

Introduction

South Africa covers an area of 1 221 037 km² and according to van Pletsen (1989) has the following physiographic features (Figure 9-1):

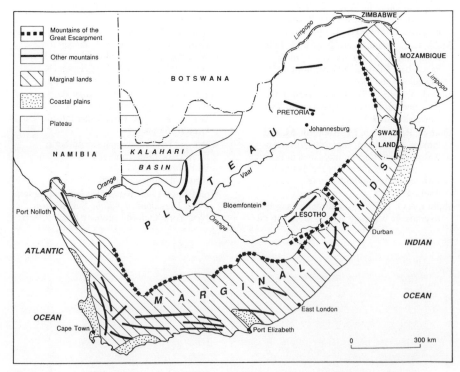

Figure 9-1:   Main physiographic features of South Africa (van Pletsen, 1989: 5).

- **The Great Plateau** dominates the topography of South Africa and occupies about two-thirds of the country. It has a mean altitude of 1200 m above sea-level. The Plateau is divided into two subregions, the Kalahari Basin and the Peripheral Highlands. The Kalahari Basin is for the most part between 650 m and 1200 m above sea-level and only the southern part extends into South Africa. The Peripheral Highlands, in contrast with the Kalahari, generally exceed 1200 m in elevation with a maximum elevation of 2500 m and are traversed by shallow river valleys.
- **The Great Escarpment** forms the boundary between the Plateau and the Marginal Lands area. This is the outer edge of the Plateau and the most conspicuous topographical feature. The highest points in South Africa, with elevations above 3300 m, are located on the Escarpment. Towards the south, the Escarpment declines in height.
- **The Marginal Lands** lie between the Great Escarpment and the coast. This zone varies in width from 80 to 240 km in the east and south to a mere 60 to 80 km in the west.

South Africa had a population of 36 million in 1990 and a growth rate of 2.4 per cent over the period 1980–90 (World Bank, 1992: 269). It consists of four provinces: The Cape Province, The Orange Free State (OFS), Natal and Transvaal, covering, respectively, 57.1, 11.2, 8.1 and 23.6 per cent of the country's surface area. Further information concerning the geography of South Africa is available in Mountjoy and Embleton (1965: 461–537).

Rainfall and climate

The wide expanses of ocean on the east, south and west sides of South Africa have a certain moderating influence on its climate. The east coast air masses are relatively warm, tend to be less stable and are more likely to give rise to abundant rain. Thus Durban records 1018 mm of rain annually. In general, the eastern coast has a humid type of climate. Over the west coast, climatic conditions discourage rain-forming processes. Hence Port Nolloth, for example, has a mean annual rainfall of 63 mm. The interior experiences dry sunny conditions, typical of the Plateau area in winter. Rain over this interior occurs almost exclusively in summer (Figure 9-2). Exactly opposite conditions prevail in the south-western Cape and adjacent areas. Here the rain occurs in winter and drought characterises the summer. The winter rainfall region is a relatively small area along the Cape west and south-west coasts. The summer rainfall region covers most of the rest of the country. Between the winter and summer dominant rainfall regions lies a transitional area where rain falls in all seasons.

South Africa has an average annual precipitation of 464 mm (Keyter, 1992: 4). Twenty-one per cent of the country receives less than 200 mm

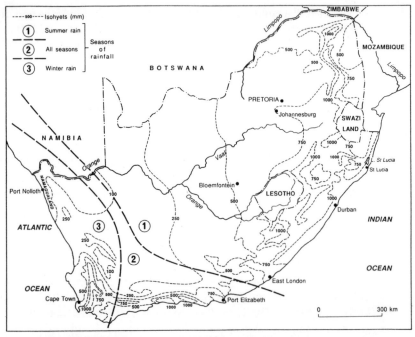

Figure 9-2: Generalised mean annual precipitation of South Africa (Keyter, 1990: 9) and seasonal rainfall zones (Le Roux, 1990).

annually, 48 per cent between 200 mm and 600 mm, while only 31 per cent receives more than 600 mm. Generally some 65 per cent of the country has a precipitation of less than 500 mm per year.

The distribution of annual precipitation in South Africa has two main features. Firstly there is a fairly regular decrease over the Plateau from east to west. The moist Indian Ocean air masses which are the main sources of rain over most of the country gradually lose their moisture as they move towards the western interior. The very lowest rainfall occurs on the west coast in Namaqualand (Figure 9-2). The second main feature of the distribution pattern is the strong orographic influence. The highest rainfall occurs on the windward slopes of the Cape ranges, and the eastern Transvaal Escarpment. In the mountains of the south-west Cape, figures as high as 3200 mm have been recorded, while the eastern Transvaal Escarpment has a mean annual rainfall of 2088 mm.

Throughout South Africa the rainfall is unreliable and unpredictable. Large fluctuations around the mean annual figure are the rule rather than the exception. South Africa is periodically affected by severe and prolonged droughts. Droughts often end in severe floods. The highest rainfall recorded in the country in a 24-hour period is 597 mm and was measured at Lake St Lucia on 31 January 1984 (Keyter, 1990: 14). This was associated with the tropical cyclone Domoina which also caused widespread flood devastation in north-eastern Natal. A detailed description of the climate and rainfall in South Africa is available in Tyson (1986).

Water resources

Surface water

The Great Escarpment divides the river systems of South Africa into two groups: the Plateau rivers and those of the Plateau slopes and Marginal Lands. The slopes of the eastern Plateau cover 13 per cent of the country but account for 43 per cent of the total run-off. This is distributed over a large number of river basins (Figures 9-3 and 9-4). Table 9-1 provides basic data concerning the drainage area, length and annual flow of the ten major rivers in South Africa.

The Orange River originates in Lesotho, flows in a north-westerly direction and discharges into the Atlantic Ocean (Figure 9-4). The Vaal River originates in the north-east of the country, flows in a south-westerly direction and joins the Orange River. The Limpopo River originates in the north-east, enters Mozambique and discharges into the Indian Ocean. The Orange and Vaal Rivers have a combined catchment area of 0.6×10^6 km^2 with an average annual flow of 12.06×10^9 m^3. The Limpopo River has a catchment area of 0.11×10^6 km^2 and an average annual flow of 2.29×10^9 m^3. The Orange and Vaal Rivers drain almost the entire central plateau, or 48 per cent of the total area of the country, but account for only 22.5 per cent of the total

Table 9-1: The ten largest rivers in South Africa in order of run-off volume.

River	Drainage area (km^2)	Length (km)	Annual flow (10^6 m^3)	Percentage of total run-off (%)
Orange and Vaal	606 700	2 340	12 057	22.5
Tugela	29 000	520	4 589	8.6
Olifants (Letaba included)	68 300	760	3 103	5.8
Umzimvubu	19 800	470	2 968	5.5
Komati (Crocodile included)	21 700	340	2 681	5.0
Limpopo main basin	109 600	960	2 290	4.3
Greater Usutu	16 700	290	2 005	3.1
Breede	12 600	310	1 785	3.3
Umzimkulu	6 700	380	1 472	2.8
Pongola	11 800	440	1 253	2.3

Source: Keyter (1992: 6).

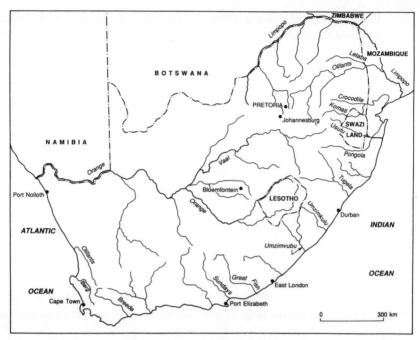

Figure 9-3: Major rivers of South Africa (Framji et al., 1983: 1223).

Figure 9-4: Main river basins of South Africa (van Pletsen, 1989: 14).

annual run-off. This is mainly due to relatively lower precipitation and higher evaporation over their catchments.

The combined annual run-off of all South African rivers amounts to 53.5×10^9 m^3 or about 9 per cent of the total precipitation. About 33×10^9 m^3 of this annual run-off can be utilised economically (Keyter, 1992: 83). Currently about 55 per cent or 18×10^9 m^3 of this water is under control. South Africa has 132 major reservoirs with capacities ranging from less than 100×10^6 m^3 to more than 2000×10^6 m^3.

Table 9-2 summarises the pattern of water use in South Africa from 1965 to 2000 in the agricultural sector and the domestic, municipal, industrial and mining sectors. It shows almost a threefold increase in total water use over a period of 35 years and a relative decline in the share of agricultural water use compared to other sectors.

Table 9-2: Pattern of water use in South Africa from 1965 to 2000.

Water use sector	1965[a]		1989[b]		2000[a]	
	(10^9 m^3)	(%)	(10^9 m^3)	(%)	(10^9 m^3)	(%)
Agriculture (irrigation, stock watering and rural supply)	6.9	83	8.5	71	11.0	49
Domestic, municipal, industrial and mining	1.4	17	3.5	29	11.4	51
Total	8.3	100	12.0	100	22.4	100

Source of data: (a) van Pletsen (1989: 336); (b) Department of Water Affairs (1989: 4).

A large number of government water-related projects either have been completed or are under construction (Keyter, 1990: 365–369). The Orange River Project is the largest water development project that has been undertaken in South Africa. It consists of a number of dams, canals, pipelines, pumping stations and tunnels. The Orange-Fish Tunnel is the largest irrigation water tunnel in the world. It transfers water from the Orange River to the Great Fish River Valley. The tunnel is 82.5 km long and has a diameter of 5.35 m. The main objectives of the project are twofold. First, existing irrigation is to be stabilised and water provided for more irrigation in the lower Orange River. Second, water is to be transferred from the Orange to the Great Fish and Sundays River Basins to revitalise irrigation development in the valleys. The plan originally envisaged eventual irrigation of some 174 000 ha in the Orange River Valley and 76 000 ha in the Great Fish and Sundays River Valleys. However, due to other demands for water and factors such as high development cost for irrigation projects, these figures will probably be considerably reduced.

Groundwater

Before 1880, South Africa was virtually dependent on springs and surface water only. A century later, in 1980, about 1.8×10^9 m^3 or nearly 15 per cent of South Africa's total annual consumption was obtained from groundwater resources (Vegter, 1987). Nearly 80 per cent of this volume was used for irrigation (Table 9-3). South Africa has about one million water boreholes and some 20 000 to 30 000 boreholes are constructed every year (Department of Technical Co-operation for Development, 1989: 255).

Table 9-3: Use of groundwater in South Africa in 1980.

Water use section	Annual use	
	(10^6 m^3)	(%)
Irrigation	1400	78.2
Rural domestic use	120	6.7
Stock watering	100	5.6
Mining and quarries[a]	100	5.6
Urban use[b]	70	3.9
Total	1790	100.0

(a) Total pumped water from mines, quarries and boreholes was about 405×10^6 m^3.
(b) Including industrial and mining use from public sources.

Source: Department of Water Affairs (1986: Table 3.4.3).

Although the contribution of groundwater to total consumption is small, owing to a lack of perennial streams, the semiarid to arid western two-thirds of the country is largely dependent on groundwater for farming and urban supplies. From 1964 to 1976 the total area irrigated with groundwater grew at an average rate of 4.9 per cent per year. This growth rate may be maintained for at least another decade or two. The area irrigated from groundwater abstraction in 1980 is estimated to have been 215 000 ha (Department of Water Affairs, 1986: 3–19).

Groundwater occurs in hard-rock formations over more than 80 per cent of the area of South Africa. These are weathered and fractured rocks which lie directly beneath the surface to depths of less than 50 m. At greater depths unweathered rocks occur which contain very little groundwater because of their dense nature. The quantity of water stored in the saturated portion of these weathered and fractured rocks is generally very limited and the permeability of these aquifers is usually low, highly variable and discontinuous. The possibility of exploitation offered by these aquifers is limited and they are used mainly to provide water on farms for domestic use, stock watering, irrigation on a small scale and to provide water for small communities. By contrast, appreciable quantities of groundwater can be abstracted at high rates from karstic Precambrian, dolomitic strata, particularly in the southern and western Transvaal (Department of Water Affairs, 1986: 3.16; Vegter, 1987). Aquifers comprising porous deposits and granular materials also contain substantial quantities of exploitable water, but the size and distribution of this type of aquifer are limited.

The yield of aquifers depends on various factors including the recharge rate. Estimates of the average annual volume of recharge varies from 16×10^9 m^3 to 37×10^9 m^3 (Department of Water Affairs, 1986: 3–16). However, a small portion of the infiltrated water can be recovered by means of boreholes because of high evapotranspiration and seepage losses which do not contribute to stream flow.

In spite of considerable research, it is still difficult to estimate the quantities that could be abstracted on an economic basis from aquifers. In the present circumstances and as a rough estimate only, it is not likely that groundwater abstraction could exceed 5.4×10^9 m^3 per year. It is therefore likely that groundwater will eventually not constitute much more than 15 per cent of the supply from conventional sources (Department of Water Affairs, 1986). Further information concerning the groundwater resources of South Africa is available in Department of Technical Co-operation for Development (1989: 251–258) and Geological Society of South Africa (1993).

Land cover

According to FAO (1989), of a total land area of 122.10 Mha, about 13.17 Mha were under cultivation in 1987. Of these cultivated lands 1.13 Mha were irrigated. Permanent pasture covered 81.38 Mha, forest and woodlands 4.51 Mha and other lands 23.04 Mha.

Considering South Africa's soil quality, rainfall and climatic and other conditions, the total surface area suitable for dryland crop production is about 14.2 Mha (Table 9-4).

Table 9-4: Land area suitable for dryland crop production in South Africa.

Region	Area	
	(ha)	(Per cent of country)
Summer rainfall	12 650 000	10.34
All year rainfall	281 400	0.23
Winter rainfall	1 248 000	1.02
Total	14 179 400	11.59

Source: van Pletsen (1989: 336).

Irrigation

Irrigation by individual effort, by means of shallow wells and small diversions, is fairly old in South Africa, but the construction of community-size irrigation works began in the later part of the nineteenth century (Framji et al., 1983: 1221). Since then several projects have been built. During the first half of this century development of water resources was primarily for agricultural purposes, but in the past 30 to 40 years the increase in urban and industrial demand has been four times that of the agricultural sector. The current total rate of increase is of the order of 1.6 per cent per year. Developments have therefore been oriented towards satisfying these demands. Major projects include the Orange River Project, Welbedacht-Bloemfontein Project, Tugela-Vaal Project, Lesotho Highlands Water Project and many others (Keyter, 1990: 365–369). Because of the unreliability and variability of rainfall, irrigation is a vital component of the agricultural industry in South Africa. Irrigated agriculture is the largest single user of surface water and will continue to be so for some time. In view of the growing demand for water by urban, industrial and mining sectors, irrigation will have to move towards unprecedented levels of efficiency.

The area under irrigation was about 1 Mha in 1970 (Framji et al., 1983: 1222), which slowly increased to about 1.13 Mha in 1980–81 (Ninham Shand Consulting Engineers, 1985). A similar figure of 1.13 Mha is provided by FAO (1989) for 1987. Table 9-5 shows the irrigated area and the corresponding water consumption for 1980–81 in the government, irrigation boards and private irrigation schemes. Ninham Shand Consulting Engineers (1985) noted that the volume of irrigation water consumed in 1980–81 (7700×10^6 m^3) was about 75 per cent of total water use in the country.

Table 9-5: Irrigated area and irrigation water consumption in South Africa (1980–81).

Scheme	Estimated area (ha)	Total water consumed (10^6 m^3)	Water consumed per ha (m^3)
Government water schemes	230 013	1564	6800
Irrigation boards	295 697	1895	6410
Private irrigation schemes	608 264	4232	6960
Total	1 133 974	7691	6780

Source: Ninham Shand Consulting Engineers (1985).

Most of the irrigated lands are scattered in the subhumid and semiarid parts of the country along the major rivers. According to du Plessis (1992), surface irrigation systems account for 48 per cent, followed by conventional and mechanised sprinkler (20 per cent each) and microjet and drip (12 per cent). Surface irrigation methods (especially flooding) and losses in distribution systems are mainly responsible for progressive deterioration of some soil types on a number of irrigation projects.

According to Dent et al. (1988), much of South Africa's dryland farming is carried out on lands which, in terms of typical soil water content, can be described as marginal. Therefore the practice of supplementary irrigation is increasing rapidly, and crop yields are boosted by supplementary irrigation.

Salinity

Irrigated land salinity

Waterlogging and land salinisation occur on some irrigation projects, but from a national perspective the situation seems to be largely under control. In old irrigation schemes, salinisation of irrigated land is mainly the result of overirrigation of soils with impervious subsurface layers, poor drainage and excessive salt content in the subsoils. In dry parts of the country, high evaporation causes salt accumulation at or near the surface.

No program exists to determine the extent of waterlogged and salt-affected soils of South African irrigation schemes on a regular basis. However, a recent survey of five major irrigation projects indicated that, on average, 28 per cent of irrigated land shows signs of either waterlogging or harmful high salt contents or both, which ultimately could render the soil unproductive (van Pletsen, 1989: 337). The Soils and Irrigation Research Institute estimated that the extent of saline land in the country was approximately 111 000 ha in 1975 and about 100 000 ha in 1985 (Department of Agriculture and Water Supply, 1988). However, these figures may increase during years with above average rainfall and decrease during periods of low rainfall.

With respect to waterlogging, Ninham Shand Consulting Engineers (1985) estimated that there were some 148 000 ha of waterlogged irrigated land in the country in 1975 and some 37 000 ha in 1980. The large waterlogged area in 1975 is attributed to abnormally high rainfall during 1974–76. Table 9-6 provides a summary of an investigation, carried out by the Department of Agriculture in 1982 and quoted by Ninham Shand Consulting Engineers (1985), concerning the occurrence of waterlogging in a number of irrigation schemes. It shows that of 135 868 ha of irrigated land in the surveyed schemes, some 11 946 ha or 9 per cent of irrigated land were waterlogged.

In general terms the relative extent of waterlogged and salt-affected lands in South Africa is much less than many other countries. According to du Plessis (1994), possible explanations for this favourable state of affairs are: the emphasis on potential for waterlogging and salinisation and their prevention in selection criteria for irrigated soils; government assistance with the financing of drainage works; and the fact that the South African irrigated areas are generally smaller in size, having a greater slope and good surface drainage. However, there is ample room for improvement. Estimates show that water losses are significant. For example, losses at the farm border are approximately 30 per cent of gross releases from dams, on-farm distribution losses amount to about 20 per cent of the volume delivered to the farm, while a further 10–15 per cent are lost by over-irrigation (du Plessis, 1994). Improved water management at both farm and irrigation scheme level would thus undoubtedly result in a further improvement in the situation.

Table 9-6: Occurrence of waterlogging problems in a number of South African irrigation schemes as determined by the Department of Agriculture in 1982.

Irrigation scheme	Irrigated area (ha)	Waterlogged area (ha)	Year
Oudtshoorn	1 900	50	1982
Klaasvoogds	545	405	1982
Hex River	5 000	40	1982
Elgin	15 000	0	1982
Olifants River (Western Cape)	12 000	200	1982
Golden Valley	2 523	1 800	1980
Douglas	1 200	512	1978
Vaalharts (North Canal)	25 000	1 500	1977
Riet River (B-Farms)	2 600	2 000	1974
Haartebeespoort	22 000	154	1982
Marico	4 600	135	1980
Loskop	25 200	1 250	1980
Pongola	9 200	2 300	1978
Sterk River	2 500	600	1980
Blyde River	6 600	1 000	1979
Total	135 868	11 946	–

Source: Summarised from Ninham Shand Consulting Engineers (1985: Table 3.2).

Stream and reservoir salinity

The quality of water resources supplying the main industrial and economic centres of South Africa has gradually deteriorated over the past few decades. The main degradation is due to increasing salinity, but to a lesser extent, there has been eutrophication and pollution by trace elements and micro-pollutants (Department of Water Affairs, 1986: 4.3).

In recognition of the fact that the quality of the country's water resources was deteriorating despite the enforcement of water pollution control measures, the Department of Water Affairs and Forestry recently changed its approach to the management of water quality to the Receiving Water Quality Objectives (RWQO) approach for non-hazardous substances and the Pollution Prevention approach for hazardous substances (Department of Water Affairs and Forestry, 1991).

The RWQO approach to water pollution control involves the specification of water quality requirements in receiving waters and to control point and non-point sources of pollution to such an extent that these water quality requirements are met. It is based on the acceptance of the principle that receiving waters have a capacity to assimilate pollution without detriment to accepted use for the water concerned (van der Merwe and Grobler, 1990).

Some pollutants are regarded as hazardous or dangerous because they represent a major threat to the water environment as a result of their toxicity persistence and capacity for bio-accumulation. The RWQO approach to managing these pollutants is considered to be inappropriate. It is often difficult to set safe receiving water quality objectives because too little is known about the properties of these pollutants and the potential long-term risk they pose to humans and the environment. Therefore, an approach based on limiting or, preferably, preventing their input into the water environment was developed to control the presence of hazardous or dangerous substances in the water environment.

The monitoring of water quality in South Africa started during the 1970s, when the Department of Water Affairs initiated a regular sampling and analytical program for surface waters. As part of this monitoring program, water samples from 180 dams and 800 river sampling points are collected on a regular basis at various frequencies ranging from daily to once a month (du Plessis and van Veelen, 1991). From analysis of the collected data, it has been concluded that, with the exception of the dry interior parts of the country, South Africa is generally endowed with low salinity waters for 90 per cent of the time (Figure 9-5), and that the observed quality differences could probably be explained by natural conditions such as rainfall, aridity, geology and proximity to the sea. However, relatively high salinities occur in a number of river systems and in the industrial areas and population centres.

In the industrial complex area of Witwatersrand and Pretoria, the coalfields of the eastern Transvaal and part of northern Natal, saline discharges from both point and non-point sources associated with industrial mining operations have been largely responsible for the increase in salinity (Hall and Görgens, 1978; Cass, 1984). While the mining sector uses 100×10^6 m^3 of groundwater per year, the total groundwater pumped from mines, quarries and boreholes is about 405×10^6 m^3 per year (Department of Water Affairs, 1986: Table 3.4.3). Discharge of this volume of water, usually with a

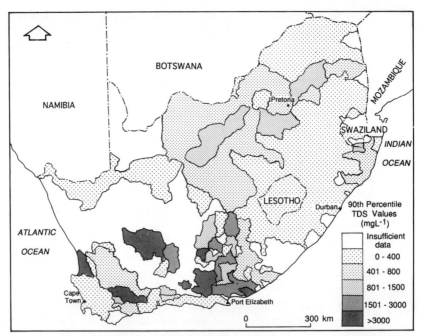

Figure 9-5: Surface water salinity distribution for South Africa for the period 1979 to 1988 (du Plessis and van Veelen, 1991).

high salt content, is one of the sources of river salinity. According to Williams (1987), the salinity of effluents from heavy industry is about 1000 mg L^{-1} and mining effluents may have salinities of 10 000 mg L^{-1}. However, salinisation from natural and human-induced non-point sources has also given cause for concern in rivers draining certain semiarid areas. The Great Fish, Sundays, Berg and Breede Rivers are among the rivers affected by salinity. In these and other river systems where the salt concentration due to natural processes is high, impounding the river flow and utilising the water for irrigation has resulted in progressively higher salt concentration downstream. Sometimes this occurs to the extent that the water is rendered useless both for downstream irrigation and municipal supplies (Hall and Görgens, 1978).

According to Görgens and Forster (1989), many of the serious salinity problems experienced in South Africa to date have occurred during, or as a result of, extreme climatic events. For example: the rapid rise in the salinity of water in the dams of the middle and lower Vaal catchment as well as certain dams in the Eastern Cape was directly related to the 1979–84 drought; and the rise in the salinity of Lake Mentz during 1975–78 was a direct result of successively heavy seasonal rains in the Sundays River catchment. It should be noted that the salinisation impact of extreme events is often more pronounced in systems where water utilisation and human activity are the greatest. Hence, the total responsibility for a salinity problem cannot always be ascribed to extreme events only.

In essence, increase in salinity during drought conditions is primarily a result of the shortage of dilution water which under normal circumstances is generally available to maintain salinities at non-problem levels. If dilution water shortages occur while the discharge of saline effluent remains relatively constant, then salinity levels will increase. It should be noted that during drought conditions, the mechanisms which mobilise and transport salts from diffuse sources are less developed. Consequently, the contribution to salinity problems is reduced.

Salinity problems resulting from extreme rainfall conditions are generally associated with increased groundwater levels in aquifers which often contain brackish water. Subsequent to heavy rainfall or flooding events these aquifers may discharge saline water to the surface drainage system for several months. Such discharges may occur during baseflow conditions, thereby resulting in high river salinity levels.

Table 9-7 shows the salinity status of a number of reservoirs in South Africa. Lake Mentz on the Sundays River has a salinity level exceeding 955 mg L^{-1} for 50 per cent of the time.

Table 9-7: Salinity status of a number of reservoirs in South Africa (see Figure 9-6 for locations).

Name of reservoir	Capacity	TDS concentration (mg L^{-1}) equalled or exceeded for various percentages of time			Sampling period
	(10^6 m^3)	10%	50%	90%	
1. Vaal	2190.6	177	137	99	4/1968–11/1984
2. Vaal Barrage	61.5	710	495	210	1/1973–9/1985
3. Bloemhof	1273.0	511	305	240	11/1972–6/1984
4. Hartebeespoort	193.7	446	417	382	6/1962–8/1985
5. Loskop	180.0	182	139	106	5/1968–9/1985
6. Krugersdrift	76.7	1007	849	195	3/1975–7/1983
7. Kalkfontein	321.8	710	594	330	4/1968–8/1983
8. Stompdrif	55.3	1312	591	381	4/1968–1/1983
9. Miertjieskraal	1.6	1163	797	402	8/1977–11/1983
10. Floriskraal	52.2	599	409	297	4/1968–9/1983
11. Lake Mentz	192.5	1470	955	738	1/1969–10/1983
12. Laing	21.7	811	526	335	10/1968–11/1983
13. Bethulie	2.0	2220	709	463	2/1978–10/1983

Source: Department of Water Affairs (1986: Table 4.1.1).

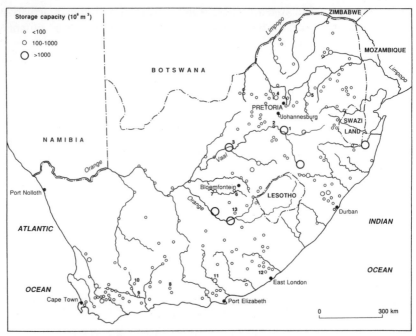

Figure 9-6:   Major reservoirs in South Africa (van Pletsen, 1989: 356). Numbers indicate location of the reservoirs in Table 9-7.

As South Africa is a water deficient country, the Water Act 1956 (Act 54 of 1956) requires that all effluent be purified and returned to the stream of origin (Department of Water Affairs, 1986: 4.8). An effect of this stipulation, however, is that the quality of water in many areas is deteriorating as a result of the increased return flows of saline effluent from growing urban, industrial and agricultural areas. In time, quality may become a more important factor than quantity in some areas, particularly in the interior. Increasing salinity limits the number of cycles of industrial reuse. Because of salt build-up, water in recycling circuits has to be blended more often and replaced by freshwater in order to maintain tolerable salt concentration, and the intake of freshwater is increased as a result. Freshwater supplies for irrigation also have to be increased to meet the greater requirement for leaching associated with higher salinity. There is also an ever increasing need to release part of the water stored in dams to control downstream salinity. Apart from salts added by effluent, the effect of the evaporation of water in storages can account for considerable increase in concentration of dissolved salts.

Görgens and Forster (1989) summarised the sources of salt and causes of salinity problems as follows: natural conditions, such as geology and soils containing quantities of soluble salts, together with associated saline aquifers; reduction in river flow due to the upstream diversion of dilution water;

evaporative losses from surface waters such as rivers, large dams, canals and farm dams; irrigation return flow, including seepage losses from irrigation water storage and distribution systems; run-off from dryland agriculture; saline industrial effluents, including the discharge of water from surface and deep mines; urban development including both point source and diffuse source contributions; and atmospheric deposition, from both natural and anthropogenic origins.

In view of the known trends in water use it is reasonable to assume that in real terms salinities are increasing and that such increases are an inevitable consequence of increased abstractions, consumptive use and return flows. The effect of flood and drought sequences, dam construction and operation, water supply quality management strategies and pollution control measures tends to conceal this overlying trend.

In brief, the deterioration in water quality, mainly as a result of salinisation, is caused by different factors depending on the type of area in which it occurs. The following sections describe salinity problems in a number of river systems and urban and industrial areas.

Salinity problem in the Breede River

The salinity problem in the Breede River catchment is an example of salinisation caused by irrigation. The following description of the situation in the catchment is summarised mainly from Flügel (1989) and Flügel and Kienzle (1989).

The Breede River is the main river system and an important agricultural production area in the Western Cape. It drains a catchment of about 12 600 km^2 (Figure 9-7). Three main relief units can be differentiated: high mountain ranges between 10 and 20 km north and south, parallel to the river, formed by resistant quartzites and sandstones, reaching heights of up to 1800 m in the north-west of the catchment; south of the Breede River, in front of the high mountain chain, sandstone and shales border the hill zone, reaching heights of between 300 and 800 m; and the Breede River Valley, which has a gentle hilly relief.

The rainfall season is in the winter, between May and October. The spatial rainfall distribution is very closely associated with the three relief units. The north-west of the catchment receives between 800 and 2500 mm per year. However, the main irrigation area in the centre of the valley has a semiarid character and receives only between 200 and 400 mm of rain, with evaporation exceeding rainfall considerably. As the Breede River catchment lies in the winter rainfall area, most of the run-off occurs in the winter months, while water is mainly needed during the irrigation season between October and April. Winter run-off is therefore stored in dams. The major storage dam is the Brandvlei Dam with a full supply level capacity of 474.7 × 10^6 m^3. Water from the dam is released into the Le Chasseur canal and into the Breede River from where irrigation water is pumped by a number of stations or diverted into a number of canals. Other sources of irrigation water are dams supplied by non-saline tributaries of the Breede River, a few high yielding springs and a number of boreholes.

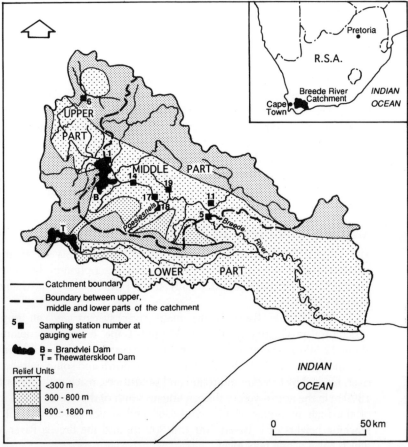

Figure 9-7: Breede River Catchment (Flügel and Kienzle, 1989: courtesy of the authors).

Irrigation began in the eighteenth century and about 45 000 ha of vines, orchards and alfalfa are presently under irrigation. These are concentrated in the middle part of the catchment. Since the irrigation areas are being expanded further away from the Breede River, higher lying soils with a high salt yielding potential are increasingly coming under irrigation.

In the flood plain of the river, the groundwater flow direction is always towards the river. In the non-irrigated land, groundwater is deeper than 15 m below the surface and shows negligible change. Groundwater salinity varies, but is higher than 5000 µS cm^{-1}, reaching a maximum of 20 000 µS cm^{-1}. The groundwater level under irrigated alluvial land is seldom deeper than 3 m from the surface and the quality varies between 1000 and 2000 µS cm^{-1}. Moreover, the shallow watertable shows periodic variations corresponding with irrigation cycles.

The Breede River functions as both supply channel and receiver of irrigation return flows. Due to the inflow of saline irrigation return flow from various irrigation districts, the water quality of the irrigation supply in the Breede River deteriorates in a downstream direction, creating problems for downstream irrigation users. Between the two gauging weirs 17 and 5 (see Figure 9-7) four saline tributaries join the Breede River. These tributaries collect saline return flows from the drainage system as well as seepage of saline groundwater from the bedrock and the alluvial aquifers. This causes a stepwise salinity increase from less than 250 µS cm^{-1} (≈ 160 mg L^{-1} TDS) to about 1000 µS cm^{-1} (≈ 640 mg L^{-1} TDS) within a distance of just 40 km (Figure 9-8).

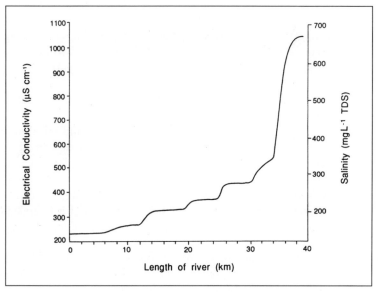

Figure 9-8: Stepwise salinisation of the irrigation supply water in the Breede River between the measuring weirs 17 and 5 (Flügel, 1989).

In order to control river salinity, freshwater is released from Brandvlei Dam to maintain salinity within the water quality goal of less than 390 mg L^{-1} for at least 50 per cent of the time and 680 mg L^{-1} for no more than 20 per cent of the time. With expansion of the irrigated area, saline return flows to the river are increasing, requiring more freshwater from Brandvlei Dam to be released. Approximately 10 million m^3 of water is reserved annually in the Brandvlei Dam for freshening purposes (dilution flow).

The significance of the contribution of return flows to the salinisation process in the Breede River and its tributaries has motivated investigations showing that a transition from flood to drip irrigation will improve irrigation water use efficiency. This would result in an increase of soil salt concentration. However, due to lower return flows, less salt would enter the receiving rivers.

Kirchner and Walraven (1993) used the strontium isotopes ratio together with the strontium concentration to assess the groundwater contribution to the salinisation of the Breede River. They observed a noticeable difference between the strontium ratios of groundwater and surface water and concluded that groundwater cannot be a significant source of the salt load of the Breede River.

In order to simulate the impact of irrigation on Breede River flow and salinity, in 1987 the Department of Water Affairs initiated the development of a physically-based computer model, which is supported by a comprehensive database. The model, named DISA (Daily Irrigation and Salinity Planning), is described by Beuster et al. (1990). The model is a decision making tool designed to aid the future planning and management of the Breede River system. The catchment has been subdivided into a number of homogeneous units. Each of the units comprises a part of the catchment with relatively homogeneous agricultural and hydrological features. The 33 units, covering an area of approximately 2900 km^2, were then subdivided into 66 return flow cells on the basis of soil type.

The physical processes governing water and salt movement through the system include: irrigated soil profile and aquifer dynamics, which simulates water and salt distribution inside an irrigated model element as well as the root zone; and the river channel processes, which serve as a link between neighbouring model elements. The surface run-off generation processes are not included in DISA, as the model is intended for irrigation season modelling in the Western Cape, where rainfall during the irrigation season is minimal.

Data requirements of the model include: rainfall and evaporation; irrigation (irrigated and irrigable areas); crop distribution and irrigation techniques; soil information; aquifer characteristics and groundwater use in the alluvium in the river valleys; saline groundwater seepage from the fractured bedrock aquifer; river channel geometry; and configuration of the water supply distribution network.

The DISA model was used to simulate a number of management options (Forster et al., 1991). The options investigated included: the abandonment of the Breede River as a water supply conduit downstream of the Brandvlei Dam; the phased abandonment of certain lower reaches of the river on the basis of its salinity; the interception and diversion of some of the more significant saline return flows by a common drain; and the use of freshening releases (dilution flow) from the Brandvlei Dam to assist the control of river salinity.

According to Forster et al. (1991), the DISA model has proved ideal for assessing physical modification to the system, such as the installation of drains to divert selected return flows, and operational changes, such as the release of freshening water. However, the most important outcome of the model was the indication of the level of development at which water quality problems can be expected. They conclude that the present level and rate of irrigation development in the Breede River do not justify decisions on major capital-intensive works for salinity control. Initial indications are that the use of freshening releases is a promising short-and medium-term water quality control option, provided that the necessary volume of water is available. When used in conjunction with a return flow interceptor drain, the lifespan of the freshening release strategy might be extendable.

Moolman et al. (1983) describe the salinity problem in the Poesjesnels River, a right bank tributary of the Breede River (Figure 9-7). The area has a mean annual rainfall of 273 mm, most of which precipitates in winter during the four months from May to August. The annual rainfall during 1979 was 209 mm and the total class-A pan evaporation 1979 mm. These figures are representative of a so-called 'dry' year. January and February are the two hottest months of the year, with the mean maximum temperature exceeding 30°C, while July is the coldest month, with a mean minimum temperature of 5.2°C.

The Poesjesnels River originates on mountain slopes. It can therefore be expected that the run-off from the mountains during a rain event will be of a good quality. In the valley, however, the river flows over shale to join the Breede River and this would increase the river salinity. The soils of the flood plain are alluvial. Most of these soils are deeper than 1.5 m and have a fluctuating watertable which normally subsides during the first half of the year. Analysis of the TDS content of this river for the dry period 1978–1980 revealed that the mean TDS content of the first six months of each year was lower than that of the last six months (Figure 9-9). The trend was for the TDS values to decrease from January to June, followed by a substantial increase from July to September. With the exception of 1980, the maximum values were recorded in either August or September.

A conceptual computer model was used to simulate the movement of water and solutes in 769 ha of land in an area where vineyards are irrigated intensively. Important input requirements of the model are the physical and chemical properties of the irrigated soils, consumptive use data and root distribution patterns. The model simulates chemical and physical processes associated with agricultural lands drained by subsurface tile drainage systems. The simulation starts with field applications of water, salts and nutrients (nitrogen) and ends with predictions of flow and water quality from the drains.

The model used in this study found that the present system of vineyard irrigation, which involves a large amount of water being applied as a pre-bud-burst irrigation in late August or early September, followed by more or less

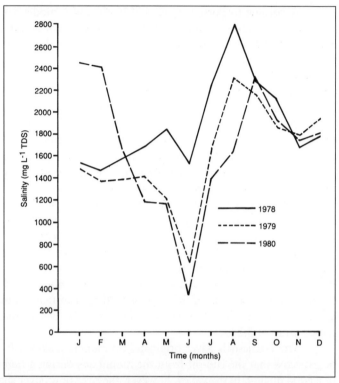

Figure 9-9: Observed mean monthly TDS content of the Poesjesnels River for the period 1978 to 1980 (Moolman et al., 1983).

fixed amounts of water being applied at set frequencies during the rest of the irrigation season, results in deep percolation losses and accompanying salt loads which are much bigger during the last six months than during the first six months of the year. The results further indicate that during the period January to June, capillary rise will exceed deep percolation. The study concluded that irrigation return flow plays an important role in the salinisation of the Poesjesnels River. Further information concerning the salinity problem in the Poesjesnels River is available in Greeff (1994).

Salinity problem in the Berg River

The Berg River is one of the major river systems in the western part of the Cape Province. It originates in the north-east of Cape Town and flows towards the Atlantic Ocean (Figure 9-3). Agriculture in the upper part of the catchment is characterised by irrigation of vines and orchards, and in the middle part by dryland wheat cropping with summer fallowing. The salinity of the Berg River increases as it passes through the wheat region. A research project in the Sandspruit River, a representative tributary of the Berg River, began in 1984. Flügel (1991a and b) describes the dryland salinity problems in this catchment. The following summary is based on his accounts.

The climate of the Sandspruit catchment is semiarid and the mean annual rainfall, which occurs only in winter, is about 400 mm; consequently the river flows periodically only from May to October. The underlying strata in the catchment is Precambrian shale which is deeply weathered. The catchment has an area of 152 km². It has a gentle hilly relief with 0–4° inclination occurring in 61 per cent of the catchment and 4–7° occurring in 27 per cent. The valleys have a shallow watertable during the winter rain season. Only halophytic vegetation grows on the valley floor and it is used to feed sheep. Saline seeps occur in depressions spread all over the catchment and these seeps have no vegetation cover but are characterised by crystallised salt patches during the hot summer from November to March. The groundwater in these seepage areas is shallow and very saline.

Seven field stations were installed at which rainfall, soil water and groundwater were sampled every three days. During winter and spring, when the soil contained sufficient water, soil water was collected from 15, 30 and 45 cm depths. Analysis of the collected data shows that salt concentrations increase with depth. Groundwater was sampled from boreholes drilled for stock watering. Groundwater levels were measured periodically during the winter season and these measurements indicate that considerable recharge occurs to the fractured shale aquifer. A summary of the salt content of soil water and groundwater is as follows: the TDS of soil water and groundwater were highly variable, which is consistent with evaporation and the variable infiltration of precipitation; stations at the valley floor had a higher salt content in the soil water than those on the slopes; soils with higher clay content had higher salt concentration than the sandy soils; in general, stations with higher salt concentrations in the soil were underlain by more saline groundwater.

Apart from groundwater and soil water sampling, 67 samples of atmospheric precipitation were collected. The salt concentration in precipitation ranged from 14 to 125 mg L^{-1} and averaged 37 mg L^{-1}. The average concentration was used to compute atmospheric contribution to the salt balance.

The results of the soil and groundwater investigations reflect the variability of salinity within the catchment. They also indicated that the salt distribution is associated with the topography. Soil salinity on both slopes of the valley was similar and varied from 2000 to 10 000 µS cm^{-1}. A considerable salinity increase was found in the sandy alluvial deposits in the valley floor of the Sandspruit River.

The Sandspruit flows only periodically from May to October and salt transport mainly occurs from July to September. The interactive hydrological and salinity dynamics of the river based on the August 1986 data are as follows:

- Two separate rainfall events occurred during the first half of August and each generated a flood. Each flood was characterised by a steep linear discharge increase from about 20 to 7000 L s^{-1} followed by a more gentle exponential decrease. The floods were generated by surface run-off and interflow from the slopes because the shallow soils were saturated quickly during the rain.
- Corresponding to the rapid increase in flow, the EC of the discharging water decreased from 10 000 to 2000 µS cm^{-1} as surface run-off from the slopes diluted the saline base flow between the rainfall events.
- After surface run-off from the slopes had ceased, the relative contribution of interflow and groundwater increased, causing the EC of the discharging water to increase, approaching the 10 000 µS cm^{-1} pre-event base-flow concentration in about 14 days after the second flood.

The water and salt balance was computed for the 1986 water year from October 1985 through September 1986. Rainfall in this period was 325 mm, which was 19 per cent below the long-term annual mean but within the annual variability. Therefore, this period was considered as representative for a hydrological year in this region. The results of the salt balance are as follows: atmospheric salt input was 1829 tonnes concentrated in the three months from June to August 1986, but it never exceeded 400 tonnes per month; river flow started in May and peaked in August. Run-off never exceeded 12 per cent of the monthly rainfall input and only 11 mm or 3.5 per cent of the total rainfall discharged from the catchment. The remainder of 314 mm evaporated or recharged the groundwater; the major salt transport from the catchment occurred during floods in July (1371 tonnes) and August (5627 tonnes); the total salt output was 8052 tonnes of which atmospheric input accounted for only 1829 tonnes or 23 per cent of the total salt transport.

Based on the results of the investigation, Flügel (1991a and b) concluded that:

- Due to the seasonal rainfall distribution there is a distinct difference between summer and winter. The rains of May and June fill up the shallow soils on the slopes. Rain events in July and August then saturate the soils and create surface run-off which generates rapid increases in discharge. Infiltrating rainfall at the slopes also recharges the fractured shale aquifer and the watertable rises in the valley floor. The groundwater provides flow after rainfall has ceased.
- Salts from the recharge areas on the slopes are leached out of the soils and the weathered shale into the underlying groundwater. They are transported to the valley floor where they seep into the river.
- In the summer season, the salt in the soil and groundwater is concentrated by evapotranspiration. In the valley floor, salts are transported by capillary rise from the saline groundwater and are concentrated in the soil water or

crystallise on the soil surface. The groundwater level beneath the river bed decreases and flow in the river eventually ceases, as occurred in October 1986.
• The soil salinity distribution within the catchment is a result of the hydro-saline dynamics. Because the salts are leached during winter, soil on the slopes is less saline and consequently is only moderate to medium saline. In the valley floor, where salt is transported both upwards due to capillary rise and downwards due to infiltration, salts are concentrated and characteristic saline soils develop there.

Salinity problem in the Great Fish River

Irrigation of land along the Great Fish River, which commenced during the nineteenth century, has a history of water shortage and poor water quality. During the 1920s two dams (Grassridge and Lake Arthur) were constructed to stabilise the water supply to irrigation farms. Continued short supplies of water led to the construction of the Kommandodrift Dam in the 1950s and finally the purchase by government and withdrawal from irrigation of about a third of the land listed as irrigable, thereby alleviating the water shortage for the remaining irrigators (du Plessis, 1990). The Orange River Project (ORP) has resulted in an abundant supply of low salinity water for irrigation. Figure 9-10 shows the diluting effect of the imported water on the salinity of the river for both the upper and lower regions. The increase in salinity of water as a result of salt pick-up as it flows down the river is also clear (higher TDS in the lower than the upper Fish River).

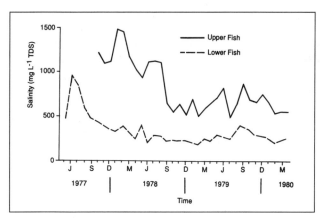

Figure 9-10: Change in the salt content (TDS) of the Upper and Lower Great Fish River from May 1977 to April 1980 (du Plessis, 1990).

Since the implementation of the first phase of the ORP in late 1977, the Great Fish River has served both to transport water and to accept natural and artificial drainage water from surrounding land. Concern about the level of river water salinisation led to the establishment of a multidisciplinary research group to study the phenomenon and quantify the various contributing processes. This group adopted a mathematical modelling approach as the only means of combining the contributions of many interrelated processes in describing the behaviour of the system as a whole. The model was used successfully to simulate river flow and salinity at a number of points along the river over the period from May 1977 to April 1980. This covered the period just prior to, and after the commencement of, importation of Orange River water.

Du Plessis (1990) provides some references and background information about the model used to predict the volume and composition of water draining from irrigated land, and describes the predicted changes in ionic composition of drainage water as obtained with the irrigation return flow submodel.

A 30-year simulation of the drainage water composition from irrigated land, due to soil chemical processes, predicted that salts which accumulated during previous periods of water shortage and poor quality water will be leached from the root zone during the first three years of availability of larger quantities of better quality waters. The three-year period of rapid decline in overall salinity was followed by a long period of stable salinity conditions under a new leaching regime and water quality. The model predicted that during the period of relatively stable salinity conditions, the ionic composition of drainage water would continue to change slowly (du Plessis, 1990).

Salinity problem in the Sundays River

The salinity problems in the Sundays River are another example of salinisation caused by irrigation. According to Hall and Görgens (1979), this river drains a relatively arid basin in the Cape Province. Lake Mentz is a major impoundment on the river and is the supply source for the irrigation of extensive citrus orchards in the Lower Sundays River Valley. The drainage basin above Lake Mentz and below Van Rynevelds Pan Dam (a smaller impoundment in the upper reaches of the Sundays River) is about 12 000 km^2 in area and experiences a mean annual precipitation of about 280 mm. Although run-off in the Sundays River has always been known for its relatively high salinity, chloride concentrations in Lake Mentz have been quite satisfactory for citrus irrigation. However, January 1974–May 1977 constituted a period of abnormally high rainfall and during this period steady saline base flow led to a severe deterioration in the quality of water in Lake Mentz. Chloride concentration rose to 700 mg L^{-1} by the end of September 1978. This phenomenon had harmful effects on the irrigated citrus trees. Fortunately, ameliorative measures were possible from August 1978, in that the first stage of a canal link had been completed, enabling water of a low salinity to be transferred from the Orange River via the Great Fish River to Lake Mentz.

Hall and Görgens (1979) also describe the simulation of the monthly chloride concentrations in Lake Mentz for a 22-year period. The mathematical model represents the principal hydrological and salinisation processes in a drainage basin, including surface run-off, infiltration, soil water storage, seepage to the river, removal of chloride from the basin surface by rain water (surface washing), soil leaching and groundwater compartment spillage. Application of the model has shown that occurrences of highly saline base flow in the Sundays River can be attributed to the activation during wet years of saline groundwater compartments which are relatively inactive during normal or dry years.

Salinity problems in other river systems

Stream salinity caused by irrigation return flow and dryland salinity is not confined to the previously described river systems. Similar problems exist in other rivers of South Africa (Hall and Görgens, 1978; Cass, 1984; Ninham Shand Consulting Engineers, 1985).

Ninham Shand Consulting Engineers (1985) analysed available salinity records at river monitoring points and reservoirs in 28 selected irrigations schemes for the period 1970–83. They concluded that:

- Significant improvements in irrigation water quality were evident in the three schemes or subschemes: Buchuberg (since 1977); Lower Great Fish River (since 1977); and Lower Sundays River (since 1974). The improved water quality in the Great Fish and Lower Sundays schemes is a direct consequence of the high quality of water imported from Orange River since 1978.
- Significant deterioration was evident in irrigation water quality in three schemes: Hartz River (since 1975); Loskop (since 1975); and Lower Vaal River (Bloemhof Dam since 1977).
- Significant deterioration was evident in the water quality of outflows from five schemes: Blyde River (since 1976); Hartebeespoort (since 1975); Hluhluwe (since 1977); Lower Vaal River (since 1977); and Pongola River (since 1977). It is possible that the increased irrigation was leading to increased return flow salt load in the receiving rivers. The drought experienced in most of the summer rainfall area, resulting in low river flows and consequently higher salinity, may also be a factor in the above deterioration.
- Over the period of record, 10 schemes (Breede River, Great Fish River, Groot River, Hluhluwe River, Lower Vaal River, Modder River, Pongola River, Riet River, Tarka River and Vaalharts) showed high increases in mean river water salinity from upstream to downstream over the length of the scheme.

Ninham Shand Consulting Engineers (1985) also estimated that of a total 7691×10^6 m^3 of water used for irrigation in 1980–81, some

4407×10^6 m^3 or 57 per cent constituted the minimum return flow from all sources. Table 9–8 shows the components of the total return flow.

Table 9-8: Estimated components of total irrigation return flow in South Africa.

Return flow component	Per cent
Network losses	39.8
On-farm losses	37.0
Overirrigation	18.1
Tail waters and rejects	5.1
Total	100.0

Source of data: Ninham Shand Consulting Engineers (1985: I).

Ninham Shand Consulting Engineers (1985) reported some of the estimates of the salt leaching rates in a number of irrigation schemes, ranging from 1 to 24 kg ha^{-1} d^{-1} (0.36 to 8.76 t ha^{-1} yr^{-1}) in the Great Fish River, up to 85 kg ha^{-1} d^{-1} (31 t ha^{-1} yr^{-1}) in the Olifants River, but they could not provide any estimates of salt load associated with their estimate of 4407×10^6 m^3 return flow.

Salinity in urban and industrial areas

In urban and industrial areas, including those subject to mining activities, the main causes of increased salinisation are underground water pumped to the surface by mines, run-off and seepage from industrial (including mining) areas and direct municipal and industrial effluents (Stander, 1987).

A recent investigation of 185 mine dumps in the catchment of the Vaal Barrage was carried out by Jones et al. (1989) and reported by Pieterse (1989). It showed that these dumps discharged approximately 50 000 t of salt in 1985 into the near surface environment, of which an unknown proportion reached the Vaal Barrage itself. The direct run-off from the mine dumps was not considered to be the major source of the high salt concentration in the stream flow. The probable source of the increased salt load in the base flow of streams is seepage from the deposits either directly into the streams or indirectly via groundwater which ultimately feeds the streams. Although the total salt load generated in the Vaal Barrage catchment during the period 1984–86 is not available, studies show that for 1977–78 the total salt load generated was approximately 698 000 t, with the diffuse source contribution being 398 000 t. The latest available figures for the 1982–83 hydrological year show that the total salt load generated in the Vaal Barrage catchment was about 198 000 t, with a contribution from the diffuse sources of 94 000 t (Pieterse, 1989).

To illustrate the nature of regional deterioration under heavily urbanised and industrialised conditions, water quality problems in the Rand

Water Board (RWB) supply area and Buffalo River Catchment will be summarised from Department of Water Affairs (1986: 4.8–4.15), Stander (1987) and Herold et al. (1991).

The RWB supplies a region which accounts for more than half of the gross domestic product of the Republic of South Africa and is the largest user of water for urban, industrial and mining purposes. The quality of the water supplied to this region is thus of the utmost importance. Since local water sources are inadequate to meet growing demand, water is increasingly being imported from further afield at a high cost. The supply area generates large volumes of domestic, industrial and mining effluent, which is discharged into local streams. Most of this effluent reaches the Vaal Barrage and is abstracted, together with the natural run-off, and returned to the water supply system after treatment. The increased recycling of effluent via the Vaal Barrage, combined with the wash-off of large quantities of dissolved solids and nutrients from the catchment, have resulted in increased salinity (Figure 9-11) and eutrophication.

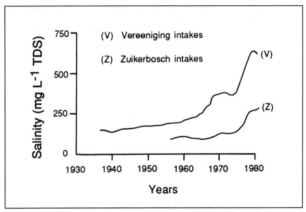

Figure 9-11: Five-year moving average of salinity concentrations in the Vaal Barrage (Department of Water Affairs, 1986: Figure 4.2.2).

Concern about this situation led to the establishment of an advisory committee to examine various management options for controlling the salinity of water supplied for the Vaal River system. The best solution to the problem was found to be to blend appropriate quantities of Vaal Dam water, having a low TDS concentration, with Vaal Barrage water with a high TDS concentration. In this way, acceptable TDS levels in the range of 250–300 mg L^{-1} could be maintained. Moreover, stricter control over industrial discharges of highly saline effluents into municipal sewers will be gradually implemented and, as soon as desalination techniques are considered economically viable, industries will be required to implement these techniques to remove salt from effluents.

Another example of water quality deterioration occurs in the Buffalo River Catchment. This catchment covers a small area of 1250 km² and supports a large and increasing urban and rural population of about 500 000. The two major impoundments on the river are the Laing and Bridle Drift Dams (Figure 9-12), both situated downstream of the urban and industrial area of King William's Town and Zwelitsha. The Upper Buffalo River Catchment includes the Maden and Rooikrantz Dams, which supply water to King William's Town and neighbouring industries. The water quality in these dams is good, having typical TDS concentrations in the range of 40 to 60 mg L^{-1}. Below the Rooikrantz Dam, however, there is a marked increase in TDS concentrations. The increase is attributed to natural leaching of salts from soils and weathered geological formations, which is made worse by irrigation return flow. Immediately upstream of King William's Town the average TDS concentration increased from 128 mg L^{-1} in 1964 to 306 mg L^{-1} in 1983. In the middle reach of the Buffalo River, the water quality downstream of King William's Town and Zwelitsha has been consistently poor. The average TDS concentration ranged from 782 mg L^{-1} in 1970 to 994 mg L^{-1} in 1983. High nitrogen and phosphate levels also existed. The salinity exceeds recommended levels for potable use. The cause of this high salinity was primarily return flows from industries, in particular industries which irrigated pasture with saline waste water.

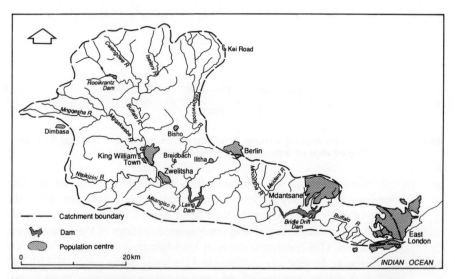

Figure 9-12:   Map of the Buffalo River Catchment (Courtesy of H.M. du Plessis).

In the Laing Dam, the TDS concentrations have increased from an average of 384 mg L^{-1} in 1980 to an average of 804 mg L^{-1} in 1983. Low flow conditions have led to an almost closed circuit of water use. Water from the dam is pumped upstream to King William's Town and to Bisho and Zwelitsha, while the treated sewage effluent and seepage from irrigation schemes and industrial effluent return to the dam. Run-off from the irrigated areas and seepage and overflows from evaporation ponds constitute the major threat to the water quality of Laing Dam. One of the strategies to control salinity is the construction of a pipeline to convey effluent to the coast for disposal at sea. Although such a scheme could be implemented in a very short time, the justification for and impact of disposal at sea would have to be studied very carefully.

The Institute of Water Research (Rhodes University) and the Division of Water Technology (CSIR) recently completed a, as yet unpublished, situation assessment of the water quality in the Buffalo River on behalf of the Water Research Commission (du Plessis, 1994). They found that the fears, expressed particularly during the early 1980s, that salinity in Laing Dam would continue to increase over time, did not come true. While the salinity levels in Laing Dam did increase from very low levels immediately after the dam was built, there is no discernible long-term increase in TDS in either the river or Laing Dam. Temporary increases do occur during droughts, and the river is then flushed out by floods.

Management options

According to Müller (1989), the South African public in general is unaware of the effect of salinity, although this could very easily change in future. The politicians, on the other hand, usually only act on something forced upon them by public opinion. Based on present media coverage, salinity is of no major concern to them, compared with other forms of pollution. The public and politicians are sensitive to cost increases brought about by more strict pollution control measures. It would be extremely difficult to persuade the agricultural sector to assume responsibility for diffuse source pollution, despite being very much aware of the salinity in some catchments. The scientific and engineering community is fully aware of this problem and solving it will pose a tremendous challenge.

Stander (1987) summarised government strategies for dealing with salinity at point sources. These strategies include: the intention to apply government policy more strictly and make more use of the special provision of the Water Act in order to control the discharge of saline effluents more efficiently; the government will continue to undertake and fund studies for the development of river basin management systems; the government will encourage and support impact and feasibility studies as well as research and development in the field of desalination; the government will give special attention and support to the search for and establishment of safe sites for the disposal of brines; the government will increasingly require polluters to undertake and fund monitoring programs and ecological impact studies in order to assess the environmental effects of saline effluent discharge.

As far as the reduction of land salinisation is concerned, the adopted strategies consist of providing surface drainage, and where necessary subsurface drainage, on irrigation schemes, as well as improving farm management practices. In government-planned irrigation schemes, off-farm surface drainage is always provided. Disposal of subsurface drainage water is also planned for in case this becomes necessary. Government subsidies are available for the installation of on-farm drains and the upgrading of irrigation equipment (Department of Agriculture and Water Supply, 1988).

Other strategies to manage salinity problems consist of:

- Developing more effective operating strategies for water supply systems. There is a great need for operating strategies for dams, canals, pipelines and rivers to reduce salinity levels to acceptable standards. Improvements can be achieved by mixing, blending and importing freshwater to certain dams or supply systems. An operating strategy may advocate using canals or pipelines for good quality water and rivers for poorer quality. At present there is substantial experience and knowledge of the blending of water, as performed by the RWB in the Vaal River, which has resulted in the lowering of the maximum TDS level in the RWB supply to below 300 mg L^{-1} since its implementation in January 1988 (Müller, 1989).
- Developing and evaluating management systems to enable use of water which is more saline than that used by traditional water users (Moolman, 1989), such as: selecting crops according to their relative salt tolerance and the salt content of the available water, and switching between water of different salt contents according to the crop salt sensitivity at various growth stages.

Mathematical models for salinity management

Apart from the models described in previous sections (e.g. Breede River and Great Fish River), mathematical models of salinisation have been applied with reasonable success to mimic past salinity trends, identify important salt sources, quantify contributions and integrate their effects. They have also yielded predictions on which salinity management decisions can be based (du Plessis and van Veelen, 1991). The success achieved in South Africa with hydro-salinity models applies to areas that are both predominantly industralised and predominantly rural.

One of the models that has been developed simulates the daily fluctuations of discharge and salt concentration at any point in the tributaries of the Vaal River that drain the southern Pretoria-Witwatersrand-Vereeniging region as well as at the Vaal Barrage and in the Vaal River between the Barrage and Bloemhof Dam (Herold, 1981). It was designed to enable the effects of a wide variety of management and planning options to be fairly easily compared with one another. This model and its further extensions have been used extensively for such purposes. A successful model has also been developed to simulate the salinisation of water in the Great Fish and Sundays River system resulting from natural processes, irrigation and other development in the area. The calibrated model has been used to test various planning options and operating rules aimed at minimising the quantity of imported Orange River water, while at the same time meeting the demand and maintaining salinity at acceptable levels (Hall and du Plessis, 1984).

Economic damage

No estimate of the damage to the national economy as a whole due to salinisation is available. However, Heynike (1981) has studied the economic effects of the salt content present in the Vaal Barrage on the community of the Pretoria, Witwatersrand, Vereeniging and Sasolburg Complex. The objective of this study was to estimate the cost to the community as a result of the rise in the salt content of the water in the Vaal Barrage basin. It was assumed that the initial salt content of the water was 300 mg L^{-1} TDS and the effects of an increased salt content to 500 and 800 mg L^{-1} TDS were investigated. Table 9-9 summarises the results of this study.

Table 9-9: Annual direct costs for increases in water salinity from 300 to 500 and 800 mg L^{-1} TDS on the Pretoria, Witwatersrand, Vereeniging and Sasolburg Complex community.

Sector	Water salinity			
	500 mg L^{-1} TDS		800 mg L^{-1} TDS	
	10^6 Rand	10^6 US$[a]	10^6 Rand	10^6 US$[a]
Industry and power stations	36.30	14.00	57.10	22.00
Households	33.00	12.70	62.45	24.00
Rand Water Board: Additional water	5.37	2.00	15.15	5.80
Corrosion in pipe networks and additional reservoir	1.56	0.60	4.56	1.80
Total	76.23	29.30	139.26	53.60

(a) Costs in US$ have been converted from estimates in Rand using a factor of 2.6.

Source: Heynike (1981: v).

Table 9-9 shows that the annual direct cost related to these increases in salt content will be about 29 and 54 million US dollars per year. Industry and households will suffer the most from additional costs, followed by the RWB, which has to supply additional industrial water when TDS increases and accelerates corrosion in pipe networks. Local authorities supplying industries with water containing higher TDS would have to increase their storage capacity because the industry would not be able to recycle the water to the same extent as it could prior to the increase in salt content.

References

Beuster, J., Görgens, A.H.M. and Greyling, A.J. 1990. *Breede River System: Development of a Daily Hydro Salinity Model (DISA)*. Cape Town: Ninham Shand Consulting Engineers. (Report no. 1674/4747). 49 pp. (plus Appendix).

Cass, A. 1984. *Irrigation Return Flow Water Quantity and Quality Modelling in South Africa*. Pietermaritzburg: Department of Soil Science and Agrometerology, University of Natal. 170 pp.

Dent, M.C., Schulze, R.E. and Angus, G.R. 1988. *Crop Water Requirement, Deficits and Water Yield for Irrigation Planning in Southern Africa*. Report to the Water Research Commission. (WRC Report no. 118/1/88). Natal: Department of Agricultural Engineering, University of Natal. 184 pp.

Department of Agriculture and Water Supply. 1988. Pretoria. (Personal communication).

Department of Technical Co-operation for Development. 1989. *Ground Water in Eastern, Central and Southern Africa*. New York: United Nations. (Natural Resources/Water Series no. 19). 320 pp.

Department of Water Affairs. 1986. *Management of the Water Resources of the Republic of South Africa*. Pretoria: Department of Water Affairs. (Various pagings, about 300 pp. plus Appendices).

Department of Water Affairs. 1989. *Department of Water Affairs, Annual Report 1988/1989*. Pretoria: Department of Water Affairs. 126 pp.

Department of Water Affairs and Forestry. 1991. *Water Quality Management Policies and Strategies in the RSA*. Pretoria: Department of Water Affairs and Forestry. 32 pp.

du Plessis, H.M. 1990. Prediction of changes in the ionic composition of irrigation return flow using a mathematical model. In. *Transactions, 14th International Congress of Soil Sciences*. Kyoto, Japan, 12-18 August 1990. VI. 125-130.

du Plessis, H.M. 1992 and 1994. Water Research Commission, Pretoria. (Personal communication).

du Plessis, H.M. 1994. Researching and applying measures to conserve natural irrigation resources. In. Green, G.C. ed. *Proceedings of the Southern African Irrigation Symposium*. Durban, 3-6 June 1991. Pretoria: Water Research Commission. (In press).

du Plessis, H.M. and van Veelen, M. 1991. Water quality: Salinization and eutrophication time series and trends in South Africa. *South African Journal of Science*. 87(1–2): 11–16.

Flügel, W.-A. 1989. Groundwater dynamics influenced by irrigation and associated problems of river salinisation; Breede River, Western Cape Province, Republic of South Africa. In. Abriola, L.M. ed. *Groundwater Contamination*. (IAHS Publication no. 185). 137–145.

Flügel, W.-A. 1991a. River salinisation due to dryland agriculture in the Western Cape Province, Republic of South Africa. In. Peters, N.E. and Walling, D.E. eds. *Sediment and Stream Water Quality in a Changing Environment: Trends and Explanation*. (IAHS Publication no. 203). 191–200.

Flügel, W.-A. 1991b. Wasserwirtschaft und probleme der dryland salinity in der Republik Südafrika. *Geographische Rundschau*. 43(6): 374–383.

Flügel, W.-A. and Keinzle, S. 1989. Hydrology and salinity dynamics of the Breede River, Western Cape Province, Republic of South Africa. In. Ragone, S. ed. *Regional Characterisation of Water Quality*. (IAHS Publication no. 182). 221–228.

Food and Agriculture Organization of the United Nations (FAO). 1989. *Production Yearbook*. Rome: FAO. v.42. 350 pp.

Forster, S.F., Beuster, H., Greyling, A. and Görgens, A.H.M. 1991. Application of the DISA hydrosalinity model to irrigation water supply planning in the Breede River. Pretoria: In. *Proceedings of the Fifth South African National Hydrological Symposium*. Stellenbosch, November 1991. Pretoria: South African National Committee for the International Association of Hydrological Sciences (SANCIAHS) and the Division of Water Engineering of the South African Institute of Civil Engineers (SAICE). 5A-1-1 to 5A-1-9.

Framji, K.K., Garg, B.C. and Luthra, S.D.L. 1983. *Irrigation and Drainage in the World: A Global Review*. Third edition. New Delhi: International Commission on Irrigation and Drainage. v.3. 1161–1667.

Geological Society of South Africa. 1993. *Africa Needs Ground Water*. An international ground water convention, Johannesburg, University of the Witwatersrand, 6-8 September 1993. Johannesburg: Geological Society of South Africa, Ground Water Division. Convention papers: v.I. (Various pagings, about 300 pp.); v.II (Various pagings, about 290 pp.).

Görgens, A.H.M. and Forster, S.F. 1989. *Establishment of a Master Plan for Salinity Research: Future Salinisation Trends and Impacts.* Pretoria: Co-ordinating Committee for Salinity Research, Water Research Commission. 24 pp. (plus Tables).

Greeff, G.J. 1994. Ground-water contribution to stream salinity in a shale catchment, R.S.A. *Ground Water.* 32(1): 63–70.

Hall, G.C. and du Plessis, H.M. 1984. *Studies of Mineralisation in the Great Fish and Sundays Rivers: v.II, Modelling River Flow and Salinity.* Pretoria: Council for Scientific and Industrial Research (CSIR), Division of Water Technology. 377 pp. (Special Report WAT 63).

Hall, G.C. and Görgens, A.H.M. eds. 1978. *Studies of Mineralisation in South African Rivers.* South African National Scientific Programmes, Report no. 26. Pretoria: Cooperative Scientific Programmes, Council for Scientific and Industrial Research. 24 pp.

Hall, G.C. and Görgens, A.H.M. 1979. Modelling run-off and salinity in the Sundays River, Republic of South Africa. In. *The Hydrology of Areas of Low Precipitation: Symposium.* Proceedings of the Canberra Symposium, December 1979. (IAHS Publication no. 128). 323–330.

Herold, C.E. 1981. *A Model to Simulate the Daily Water and Salt Balance in the Vaal River Water Supply.* Johannesburg: University of Witwatersrand, Hydrological Research Unit. Report no. 5/81. 189 pp.

Herold, C.E., Görgens, A. and van Vliet, H.R. 1991. Vaal Dam Salinity Trend: A cause for Concern? In. *Proceedings of the Fifth South African National Hydrological Symposium.* Stellenbosch, November 1991. Pretoria: South African National Committee for the International Association of Hydrological Sciences (SANCIAHS) and the Division of Water Engineering of the South African Institute of Civil Engineers (SAICE). 3A–2–1 to 3A–2–10.

Heynike, J.J.C. 1981. *The Economic Effects of the Mineral Content Present in the Vaal River Barrage on the Community of the PWVS Complex (A Desk Study).* Pretoria: Water Research Commission. 131 pp.

Jones, G.A., Brierley, S.E., Geldenhuis, S.J.J. and Howard, J.R. 1989. *Research on the Contribution of Mine Dumps to the Mineral Pollution Load in the Vaal Barrage.* Pretoria: Water Research Commission, Report no. 136/1/89. 114 pp. (plus Appendices).

Keyter, E. ed. 1990. *South Africa 1989-90: Official Yearbook of the Republic of South Africa.* Fifteenth edition. Pretoria: Bureau of Information. 842 pp.

Keyter, E. ed. 1992. *Official Yearbook: South Africa 1992.* Eighteenth edition. Pretoria: South African Communication Service. 267 pp.

Kirchner, J. and Walraven, F. 1993. Use of strontium isotopes to establish the influence of ground water on the salinization of the Breede River. In. *Africa Needs Ground Water.* An international groundwater convention, Johannesburg, University of the Witwatersrand, 6-8 September 1993. Johannesburg: Geological Society of South Africa, Groundwater Division. Convention papers, v.2. 11 pp.

Le Roux, J.S. 1990. Spatial variations in the rate of fluvial erosion (sediment production) over South Africa. *Water SA.* 16(3): 185–194.

Moolman, J.H. 1989. *An Assessment of the Situation with Regard to the Goal to the Alternative Uses of Saline Water.* Pretoria: Co-ordinating Committee for Salinity Research, Water Research Commission. 17 pp.

Moolman, J.H., van Rooyen, P.C. and Weber, H.W. 1983. The effect of irrigation practices in the Breë River Valley on the salt content of a small river. *Irrigation Science.* 4: 103–116.

Mountjoy, A.B. and Embleton, C. 1965. *Africa: A Geographical Study.* London: Hutchinson Educational. 688 pp.

Müller, H. 1989. *An Assessment of the Situation with Regard to the Goal to Define, Develop and Apply Management Strategies for Salinity Control.* Pretoria: Co-ordinating Committee for Salinity Research, Water Research Commission. 6 pp. (plus Appendix).

Ninham Shand Consulting Engineers. 1985. *Report on a Situation Study of Irrigation Return Flow Quantity and Quality in River Basins with Extensive Irrigation Development in South Africa.* Cape Town: Ninham Shand Consulting Engineers. Report no. 943/4061. (Various pagings, about 80 pp).

Pieterse, T. 1989. Salt pollution of mine dumps monitored. *SA Water Bulletin.* 15(2): 4–7.

Stander, J.v.R. 1987. Fighting SA's salinity problems. *SA Water Bulletin.* 13(5): 10–13.

Tyson, P.D. 1986. *Climatic Change and Variability in Southern Africa.* Cape Town: Oxford University Press. 220 pp.

van der Merwe, W. and Grobler, D.C. 1990. Water quality management in the RSA: Preparing for the future. *Water SA.* 16(1): 49–53.

van Pletsen, D. ed. 1989. *South Africa 1988-89: Official Yearbook of the Republic of South Africa.* Fourteenth edition. Pretoria: Bureau of Information. 816 pp.

Vegter, J.R. 1987. The history and status of groundwater development in South Africa. *Environmental Geology and Water Sciences.* 10(1): 1–5.

Williams, W.D. 1987. Salinisation of rivers and streams: An important environmental hazard. *Ambio.* XVI(4): 180–185.

World Bank. 1992. *World Development Report 1992: Development and the Environment.* New York: Oxford University Press. 308 pp.

Chapter 10: Thailand

Introduction

Thailand covers a total area of 513 115 km². The country's extreme dimensions are about 1700 km from north to south and about 805 km from east to west. The coastline is 2100 km long and borders the Gulf of Thailand in the east and the Andaman Sea in the west. Thailand has the following principal physiographic regions (Figure 10-1):

- **The Central Plain** occupies the lower central part of Thailand. It is bordered by hilly areas to the east, north and west and by the Gulf of Thailand to the south. The south-central part of the region is occupied by the delta of the Chao Phraya River system and is known as the Bangkok Plain. The Central Plain is the richest agricultural and most densely populated part of the country.
- **The Southeast Coast** is bordered to the north by hills, to the south and west by the Gulf of Thailand and to the east by the Thai-Cambodia border.
- **The Northeast (Khorat) Plateau** is composed mainly of the broad river terraces of the Mekong River and its tributaries.
- **The Central Highlands** region has a complex physiography including hills, plateaus, peneplains and a number of valleys.
- **The North and West Continental Highlands** region is usually divided into two subregions: the western mountains and the northern hills and valleys. The Doi Inthanon Peak, with an elevation of 2595 m, is located in this region and is the highest point in Thailand.
- **Peninsular Thailand** is characterised by a number of distinct ranges of hills and mountains. The landscape between the main ranges is mainly made up of lower hills and undulating terraces of fluvial origin.

Thailand had a population of 56 million in 1990 and a growth rate of 1.8 per cent for the period 1980–90 (World Bank, 1992: 268). The population is unevenly distributed with the greatest concentration in the central region. The population density was 109 persons per square kilometre in 1990. Thailand's population in the years 2000 and 2025 is projected to be 64 and 84 million, respectively.

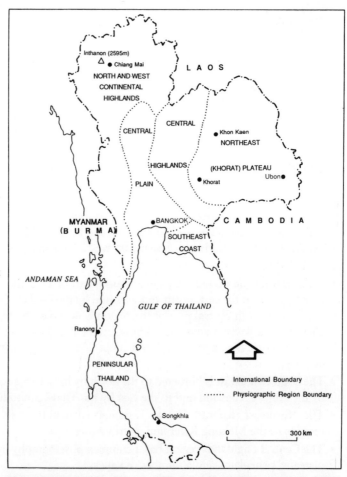

Figure 10-1: Main physiographic regions of Thailand (Arbhabhirama et al., 1987: Figure 2-1)

Rainfall and climate

The climate of Thailand is tropical and determined primarily by its location in tropical latitudes, and generally by a monsoonal or seasonal reversal of the prevailing winds, blowing south-westerly from November to January and north-easterly from May to October (Sternstein, 1976: 24–31; Donner, 1978:

25–32). Thailand has the following climates: savanna for the areas above Peninsular Thailand, excluding the eastern section of the Southeast Coast; tropical rainforest for the lower east coast of the Peninsula; tropical monsoon climate for the rest of the Peninsula and also the eastern section of the Southeast Coast. In addition, the humid subtropical climate is applicable to the area in the far north-eastern part of the Continental Highlands.

In most parts of Thailand three major seasons are generally recognised: the cool season from November through February; the hot season from March to May; and the rainy season from May through October.

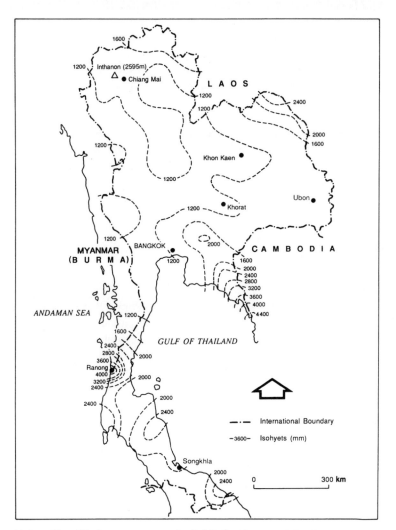

Figure 10-2: Simplified average annual rainfall distribution of Thailand for 1951–75 (Arbhabhirama et al., 1987: Figure 3-1).

Temperature in Thailand shows slight variations with the seasons. In the lower elevations, the minimum temperature occurs during the cool season, and mean monthly temperatures during January range from 26°C to 28°C for most of the country. In the hot season, during April, the temperature ranges from 28°C to 32°C. Southern Thailand shows even less variation, with mean monthly temperatures remaining between 26°C and 30°C throughout the year. Temperatures in the highlands of the north and north-east are lower due to the altitude.

The average annual rainfall in Thailand is about 1485 mm (Economic and Social Commission for Asia and the Pacific, 1991: 8–12); however, its temporal and spatial distribution is highly uneven. In temporal terms some 85 per cent of the rainfall comes in the rainy season from May to October. In the south of the country, over 50 per cent of the annual rainfall may come between October and January. In spatial terms the level of rainfall tends to increase from northern to southern regions, with the average annual rainfall around 1250 mm in the north, 1417 mm in the north-east, 1239 mm in the central region, 2047 mm in the south-east coast and 1742 mm to 2719 mm in Peninsular Thailand (from the east to the west). The west coast town of Ranong has the country's highest rainfall average of 4700 mm per year (Figure 10-2). In brief, Thailand's relatively abundant rainfall is concentrated both in time and in space.

Water resources

Surface water

Thailand has two main watersheds: the Chao Phraya River and its tributaries, which drain into the Gulf of Thailand, and part of the watershed of the Mekong, which drains into the South China Sea (Figure 10-3). In the north-western corner of the country a strip of Thailand drains into the Salween River and thus into the Andaman Sea. Rivers on the west side of Peninsular Thailand also drain into the Andaman Sea. Many small and medium-sized rivers on the eastern side of Peninsular Thailand and in the Southeast Coast region shed their water into the Gulf of Thailand.

The average volume of rainfall in Thailand is about 761.7×10^9 m^3. However, after substracting losses such as evaporation and infiltration, the average annual run-off is approximately 189.8×10^9 m^3, which constitutes 25 per cent of the annual precipitation (Economic and Social Commission for Asia and the Pacific, 1991: 12).

The most important and largest river basin in Thailand is the Chao Phraya Basin. It originates in the mountain ranges in the north and covers nearly all areas in the northern and central regions. The major tributaries are the Ping, Wang, Yom and Nan Rivers. Their confluence occurs at Nakhon Sawan, about 200 km north of Bangkok, forming the Chao Phraya River. The Pa Sak joins the Chao Phraya River about 55 km north of Bangkok. The Chao Phraya Basin has a drainage area of 178 000 km^2. The average annual run-off at the river mouth is 30.3×10^9 m^3 (Arbhabhirama et al., 1987: 49). The majority of Thailand's agricultural products are produced in this basin.

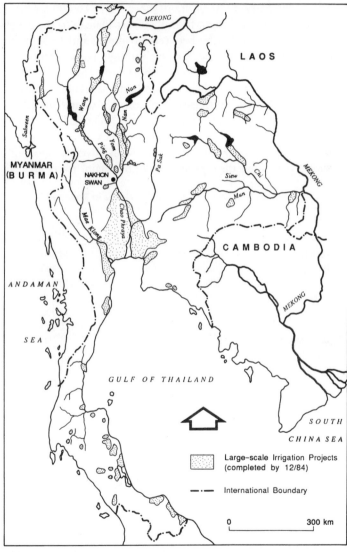

Figure 10-3: Simplified hydrographic network of Thailand and large-scale irrigation projects completed by December 1984 (Arbhabhirama et al., 1987: Figure 3-2).

The second most important river basin is the Mae Klong Basin. Its tributaries originate in the mountain range in the west near the Myanmar border and join together to form the Mae Klong River. The drainage area of the Mae Klong Basin is 33 000 km² and the average annual run-off is 13.4×10^9 m³. This basin has the highest potential in terms of water resources development, since it has much more water than would be needed by the land resources in the basin.

The Mekong River drains more than 810 000 km² of land stretching along its 4350 km length from the Tibet Plateau to the South China Sea. It forms part of the border between Myanmar and Laos, as well as between Laos and Thailand, also flowing through Laos, Cambodia and Vietnam before draining into the South China Sea. The mean annual flow of the Mekong River at Krâchéh in Cambodia is about 14 200 m³ s⁻¹ (448×10^9 m³ yr⁻¹). In the north-eastern region of Thailand, the Chi and Mun Rivers are the most important tributaries of the Mekong River from Thailand. The Chi River joins the Mun River and flows eastward to join the Mekong River. The average annual run-off of the Mun River is about 28.5×10^9 m³. Table 10-1 shows the typical flows of the major rivers in Thailand.

Table 10-1: Mean annual flow of some major rivers in Thailand.

River	Gauging station[a]	Drainage area (km²)	Annual flow (10^9 m³)
Mae Kok	Ban Tha Kok, Chiang Rai	10 300	5.50
Mae Ing	Therng, Chiang Rai	5 400	2.20
Chao Phraya	Nakhon Sawan	110 570	24.20
Sakaerung	Uthai Thani	3 250	0.55
Pa Sak	Saraburi	14 500	2.40
Mun	Phibulmungsahan, Ubon	115 690	28.50
Songkhram	Seka, Nong Khai	4 650	0.50
Mae Klong	Kanchanaburi	26 450	12.30
Phetchaburi	Phetchaburi	4 190	1.05
Bang Pakong	Kabinburi, Prachinburi	7 500	3.90
Rayong	Ban Khai, Rayong	1 255	0.47
Khlong Yai	Trad	920	1.56
Lang Suan	Chumpon	1 240	1.78
Ta Pi	Surat Thani	6 200	11.00
Trung	Trung	1 800	1.33
Pattani	Yala	3 300	2.45

(a) Gauging station does not necessarily cover the whole basin area.

Source: Arbhabhirama et al. (1987: Table 3.3).

In Thailand 648 reservoirs have been constructed on the rivers for different purposes, such as irrigation, flood control and electricity generation. Their storage capacity is about 80.8×10^9 m³. The regional distribution of these reservoirs is shown in Table 10-2.

Table 10-2: Regional distribution of the reservoirs in Thailand.

Region	Number	Volume (10^9 m^3)
North-east	552	6.23
North	44	48.72
East	27	18.78
Central	19	0.33
South	6	6.71
Total	648	80.77

Source: Arbhabhirama et al. (1987: Table 3.2).

In terms of water use, in 1990, 30.05×10^9 m^3 were withdrawn for irrigation, 1.53×10^9 m^3 for domestic use and 1.55×10^9 m^3 for industry, making a total of 33.13×10^9 m^3. Given the current trends, by the year 2000 the corresponding figures will be 38.48×10^9 m^3, 3.90×10^9 m^3 and 2.34×10^9 m^3 per annum, respectively, making a total of 44.72×10^9 m^3 (Economic and Social Commission for Asia and the Pacific, 1991: Table 35). The demand for water for the Metropolitan Water Works Authority (MWWA) is projected to grow from 0.94×10^9 m^3 in 1990 to 1.37×10^9 m^3 by the year 2000. To meet the projected long-term excess demand, the MWWA is planning to transfer 0.95×10^9 m^3 of water from the Mae Klong River Basin through a 100 km long canal (Phantumvanit and Panayotou, 1990: 53).

Groundwater

Groundwater resources exist throughout Thailand. However, the quantity and quality of groundwater vary according to local hydrogeological conditions. According to the Department of Technical Co-operation for Development (1986: 277–308), groundwater occurs in all rock types, but the water-bearing characteristics of each type are quite different. The alluvial, fractured rocks and the carbonate rocks are the important aquifers of Thailand. Within the group of alluvial aquifers, with the exception of those in the lower part of the Central Plain and in the coastal plains of Peninsular Thailand, the alluvial aquifer occurs on narrow to broad flood plains or meander belts or as channel fillings. The thickness rarely exceeds 50 m in the intermontane basins or river valleys. In the lower part of the Lower Central Plain and the adjacent coastal regions, the alluvial and deltaic sediments with minor estuarine deposits occur at a depth of about 120 m. Successive layers of sand and gravel form multiple aquifers separated by compact and/or leaky clay beds. Groundwater also exists in cracks, joints and bedding planes of shale, siltstone and sandstone as well as in the fractured metamorphic and volcanic rocks.

Limestones of different age are extensively represented and exposed in sizeable areas of the country. The Permian limestone is cavernous and numerous solution cavities have been encountered in the course of drilling, principally in south and central Thailand. Wells that have penetrated the cavernous zones have yields of 7 to 55 L s^{-1}. Usually large and high-yielding aquifers occur in alluvium and terrace deposits. To a lesser extent, groundwater also exists within fractured or cavernous limestone and sandstone (Piancharoen, 1982; Wongsawat, 1985). Groundwater is not widely used for agriculture, but it is more widely used for domestic water supply.

The Bangkok area is underlain by a thick sequence of consolidated or semiconsolidated sediments with more or less alternating layers of sand, gravel, silt and clay. Eight relatively well-defined aquifers have been found in the upper 550 m (Das Gupta and Yapa, 1982). All the aquifers are productive and yields of 80 to 110 L s^{-1} are quite common. Groundwater was used in Bangkok as early as 1914. Early on, there were only private shallow wells. In 1954 the MWWA began pumping groundwater for its water supply at a rate of 8360 m^3 per day. Since then groundwater extraction has increased rapidly.

The increasing spread of industry into the outlying areas of the Bangkok Metropolitan Region (BMR) which are less well served by the MWWA has resulted in increased use of groundwater. However, even in areas served by the MWWA, 9 out of 10 industries prefer groundwater to pipe water because of its lower cost. It is estimated that 2.87×10^6 m^3 of groundwater were pumped per day in 1989 by the industrial sector in the BMR (Phantumvanit and Panayotou, 1990: 55).

The estimated safe yield of the Bangkok aquifer system is about 0.8×10^6 m^3 per day. However, with the excessive rate of pumping, groundwater levels in all aquifers have fallen considerably from the original free flowing artesian condition to 45 m below ground surface in 1979 (Das Gupta and Yapa, 1982) and 50 m in 1982 (Das Gupta et al., 1987). This fall caused land subsidence which in turn aggravated the problem of flooding and contamination of the groundwater resources of the area by seawater intrusion (Das Gupta and Yapa, 1982; Das Gupta et al., 1987; Milliman et al., 1989).

The rate of subsidence has been estimated at 5–10 cm per year in the inner city and some areas subsided 1.4 m between 1940 and 1980. Studies using mathematical models suggest that unless the use of groundwater is greatly reduced in the Bangkok area, the city will subside by at least a further two metres by the year 2000 (Arbhabhirama et al., 1987: 62–64).

With respect to the problem of seawater intrusion into the aquifers of the Bangkok area, the rate of intrusion is as great as 500 m per year in areas of heavy groundwater extraction. As a result of this problem, hundreds of public supply wells are known to have been abandoned (Department of Technical Co-operation for Development, 1986: 303).

The Quaternary deposits of Thailand cover approximately one-third of the whole country (Dheeradilok, 1987). In addition to the Bangkok area, in other parts of the country, such as the northern Central Plain and Chiang Mai Basin, these sediments are a major source of good quality groundwater (Stewart et al., 1987).

Land cover

Of a total land area of 51.1 Mha, 20.0 Mha were under cultivation in 1987 (FAO, 1989). Of this land 4.0 Mha were irrigated and 16.0 Mha non-irrigated. Permanent pasture covered 0.8 Mha, forest and woodlands 14.4 Mha and other lands 15.9 Mha.

With respect to land suitability for agriculture, 10.82 Mha are suitable for upland crops such as fruit and vegetables, 13.51 Mha are suitable for rice paddies, 2.62 Mha are suitable for perennial crops, 7.98 Mha are generally unsuitable for economic crops but potentially suitable for cultivating special crops if appropriate measures are taken, 15.98 Mha are unsuitable for agriculture and the remaining 0.39 Mha are occupied by water bodies (Asian Institute of Technology, 1983).

In the coastal regions of Thailand, land is used by mangrove forests, shrimp farms, rice fields, coconut plantations and recreational facilities. The coastal lands are continuously being put to other uses. Mangrove forests, in particular, as well as rice fields are increasingly being converted to shrimp farms, which yield a high profit. The total mangrove area of Thailand reduced from 367 900 ha in 1961 to 180 559 ha in 1989, while shrimp farming increased threefold from 26 036 ha in 1980 to 81 552 ha in 1989 (Arunin and Im-Erb, 1991).

Agricultural growth in Thailand was accomplished largely through land expansion into forest reserves. From 1961 to 1988 forest cover declined from 28.67 Mha to 14.38 Mha (8.02 Mha in the north, 2.36 Mha in the north-east, 1.72 Mha in the central region and the remainder in the south and the east). During the same period cultivated land increased by about 9.7 Mha. However, estimates show that the demand for agricultural land will level off during the early 1990s and will decline through the 1990s and the early part of the twenty-first century. By the year 2010 the demand for cultivation will stand at 14.08 Mha (Phantumvanit and Panayotou, 1990: 8–11 and 80). The reasons for such a decline include the structural transformation of the economy from agriculture to industry and diversification away from land-extensive crops such as cassava, maize and rainfed rice towards land-saving crops such as vegetables and fruits.

Irrigation

Deficiency and irregular distribution of rainfall are inhibiting factors for successful agricultural production in Thailand. Although irrigation on a small scale had been practised for centuries in the Northeast Plateau and in the Central Plain, it was only in 1888 that the first serious large-scale irrigation was attempted in the Central Plain. In 1959 the irrigated area was about 1.5 Mha, which increased to 1.9 Mha in 1966, served 6100 km of irrigation canals (Framji et al., 1983: 1349). According to the Royal Irrigation Department, reported by Arbhabhirama et al. (1987: 52), in 1984 there were 3.46 Mha of completed irrigation projects (large, medium and small-scale) including 1.95 Mha in Chao Phraya and 0.56 Mha in the Mekong River Basin (Table 10-3). During the same year, 0.58 Mha of irrigation projects were under construction, including 0.13 Mha in Chao Phraya and 0.16 Mha in the Mekong River Basin.

Table 10-3: Irrigation projects in Thailand completed and under construction in 1984 by river basin.

River basin	Completed and under construction (ha)	Completed (ha)
Mekong	718 801	560 344
Chao Phraya	2 082 039	1 951 791
Bang Pakong	230 030	221 790
Mae Klong	451 008	277 344
Phetchaburi	57 942	57 030
Pranburi	43 902	41 902
Southern Region	406 226	318 871
Eastern Coast	40 316	31 360
Khmer Lake	9 331	4 547
Total	4 039 595	3 464 979

Source: Arbhabhirama et al. (1987: Table 3.5).

Rice is the major crop and due to the expansion of irrigated and rainfed culture and increased yield from 2.36 t ha^{-1} to 3.07 t ha^{-1} in the irrigated areas, Thailand has become a leading country in rice export. The quantity of rice exported in 1981 was 3.03 Mt, accounting for 22.6 per cent of the world total export of rice (Kanoksing, 1987). Other crops include maize, cassava, sugarcane, peanuts, coconuts, soybeans, rubber, cotton and kenaf. According to Phantumvanit and Panayotou (1990: 3), only 20 per cent of agricultural land in Thailand is irrigable. Virtually all suitable sites for construction of irrigation dams have been used. Additional capacity can only come at a higher price in terms of construction and environmental costs. Further information regarding the development of irrigation in Thailand is available in Framji et al. (1983: 1342–1356), while Sternstein (1976: 143–155) and Donner (1978: 70–132) describe the development of agriculture in the country.

Salinity

In Thailand 3.61 Mha of lands are salt-affected (Arunin, 1992a). There are two major types of saline soils. One is inland in the north-east of Thailand. The other type occurs along the coastal areas where the salt originates from seawater. Saline soils in the north-east cover an area of approximately 2.85 Mha, while the coastal saline soils cover a total area of 0.58 Mha. Also 0.18 Mha of salt-affected lands are found in the Central Plain. The salt-affected soils in the north-east are generally sandy, low in fertility and high in sodium and chloride content. About 75 per cent of the salt-affected lands in this region is under rainfed rice cultivation and 1.5 per cent is regarded as wasteland. The Northeast Plateau has approximately 7 per cent of irrigated area. About 10 per cent of this area has been salt-affected (Arunin, 1992a). The coastal saline soils may be characterised as heavy clays and they are subject to tidal influences and saline seawater intrusion.

Acid sulphate soils are generally characterised by low pH and pyrite content. Oxidation of pyrite forms sulphuric acid which produces high acidity and low pH. In Thailand acid sulphate soils occur in coastal plains and a high proportion of them are located in the Central Plain. Estimates of acid sulphate soils in Thailand vary from 0.6 Mha (Attanandana and Vacharotayan, 1984), to 0.8 Mha (Uwaniyom and Charoenchamratcheep, 1984) and even 1.5 Mha (Krishnamra, 1990). Further information regarding the characteristics of acid sulphate soils in Thailand and their amelioration is available in Uwaniyom and Charoenchamratcheep (1984) and Attanandana and Vacharotayan (1984).

As human-induced salinisation is widespread in north-east Thailand, some background information about this region will be provided before describing the causes of salinisation and its management.

North-east Thailand

The north-east region of Thailand (Figure 10-4), also known as the Khorat Plateau, comprises an area of about 170 000 km^2 (one-third of the country), supports more than 18 million people and is the most arid region. Agriculture is the dominant occupation in north-east Thailand and the majority of farmers own their land, averaging around three hectares per household. Due to low and unstable agricultural production resulting from erratic rainfall and generally poor soils, per capita income in this region is the lowest in the country. In 1985, average per capita annual income of the agricultural population in north-east Thailand was US$121 compared to US$320 in Central Thailand (Johnson III, 1993). During the wet season farmers normally plant rainfed rice, while during the dry season they plant cassava, kenaf, watermelon, tobacco, or leave the ground fallow. With irrigation water, farmers tend to grow rice, sweet corn, soybeans, peanuts and tomatoes.

Figure 10-4: Administrative map of north-east Thailand (Donner, 1978: Figure 102).

The Khorat Plateau is relatively flat with low elevation. Approximately 63 per cent of the Plateau has an elevation of 100–200 m, 28.4 per cent 200–500 m, 6.8 per cent 500–1000 m and 1.8 per cent has an elevation of above 1000 m (Donner, 1978: Table 114). The mean annual rainfall of the north-east during the period 1965–80 was 1447 mm and mean evaporation was 1729 mm (Arunin, 1984). The average annual rainfall represents about 246×10^9 m^3 of water over the Plateau. However, rainfall is unevenly distributed and occurs in the rainy season as torrential rains. More than 80 per cent of the rainfall occurs from May to September. During this period, precipitation exceeds evaporation, while the region suffers from soil moisture deficit during the other months and it is at this time that soil salinity is the most widespread.

Geology and hydrogeology

The geology of the Khorat Plateau is dominated by a sequence of conglomerated sandstone, siltstone, shales and evaporites of Mesozoic age, overlying Palaeozoic sediments, and is gently folded into two basins, the

Sakon Nakhon Basin in the north and the Khorat Basin in the south, separated by the Phu Phan Range (see Figure 10-7). The upper part of the Khorat Group, consisting of Khok Kruat and Salt Formations from the Cretaceous Period, underlie large areas of the Plateau. The Khok Kruat Formation (700 m thick) consists mainly of sandstone, siltstone and shale occurring mainly on the outer periphery of the basins. The Salt Formation (Maha Sarakham Formation), with a thickness of 600–1000 m, derives its name from the salt-bearing beds in the uppermost sediments of the Khorat Group. The Maha Sarakham Formation consists predominantly of shale and siltstone. This formation contains many beds of gypsum and anhydrite up to 15 m or more in thickness and beds of rock salt 2–200 m or more thick. One of the most distinctive features of the Khorat Plateau is the relatively thick and uniform sand cover that veneers ridge, hills and undulating plains. This sand cover may be between 1 and 3 m thick and overlies partly weathered shale and sandstones of the Maha Sarakhan Formation (see Löffler and Kubiniok, 1988, and Löffler et al., 1984, for a discussion of the origin of the sand cover). The north-east Plateau of Thailand belongs to the watershed of the Mekong River and drains its water into this river. Within the region the Mun and Chi Rivers serve as the main collectors of water in the Khorat Basin and represent the most important tributary of the Mekong from Thailand, while the Song Khram River drains the Sakon Nakhon Basin and discharges to the Mekong River also.

Hydrogeologic information on north-eastern Thailand (Phiancharoen, 1973; Piancharoen, 1982; Ramnarong, 1985) shows that groundwater resources of the region are in Quaternary alluvial deposits and in sandstone beds. The Khorat Aquifer consists of Upper Triassic to Cretaceous rocks and is subdivided into three units (Wongsawat, 1985), the upper (shale, siltstone and sandstone), the middle (massive sandstone and conglomerate) and the lower (shale and soft sandstone). The alluviums of the Chi and Mun Rivers are of limited extension and yield less than 7 L s^{-1}, while the ancient alluvium of the Mekong River can extend approximately 5–20 km from the river bank and normally yields up to 28 L s^{-1}. Almost 90 per cent of the groundwater in the Khorat Plateau is found in the continental sedimentary rocks of the Khorat Group. However, the groundwater from this type of aquifer is variable in both quality and quantity. Besides, more than 90 per cent of wells drilled in the flood plains and relatively flat lands yield salty water (Chulalongkorn University Social Research Institute, 1981: v.1. 2–61).

An early investigation of the geology and hydrogeology of north-east Thailand by La Moreaux et al. (1958) showed that the chloride content of water samples from 145 wells was in the range of 10 to 1000 mg L^{-1}, with the exception of 10 samples having chloride contents of 1000 to 4000 mg L^{-1}. A recent investigation by Williamson et al. (1989) in an area near Khorat indicated groundwater salinity is highly variable and ranges between 200 and 60 000 mg L^{-1} TDS. Water with salinity less than 800 mg L^{-1} occurred in recharge areas, while the more saline waters are in the discharge zones. Also an increasing salinity with depth was observed as a general trend. Buaphan et al. (1990) investigated the groundwater quality in a fractured shale aquifer in an area located about 17 km west of Khon Kaen. According to their results, groundwater electrical conductivity ranged from 1200 to more than 10 000 µS cm^{-1} and TDS ranged from about 670 to 7900 mg L^{-1}.

Salinity in north-east Thailand

Soil salinisation in north-east Thailand is not a new problem and is not entirely induced by human activities. Soil salinisation in this region is at least partly a natural phenomenon. There is also archaeological evidence that localised occurrences of surface salt have been utilised for salt making for 2000 years in very much the same way as it is practised today (Löffler and Kubiniok, 1988).

Because of the widespread salinity problem in the Khorat Plateau, much research has been undertaken to explain the causes of the problem, estimate its extent and address its management. The following is a brief description of some of these research efforts.

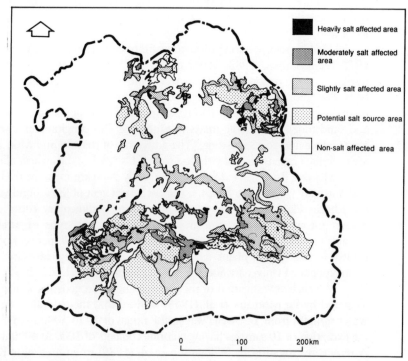

Figure 10-5: Soil salinity distribution in the north-east of Thailand (Rimwanich and Suebsiri, 1984).

Sinanuwong et al. (1980) describe the application of LANDSAT imagery of July 1975 and 1976 coupled with ground check sampling in about 2000 km² of the Nakhon Ratchasima Province (Figure 10-4). The soil salinity classes were arbitrarily divided into areas of non-, slightly, moderately and strongly salt-affected, based on the interpretation of the imagery. Then in July 1977 a field survey was conducted to verify the results of the interpretation corresponding to the time of the year when the image used was taken. Soil samples were taken at the surface and subsurface. Sinanuwong et al. (1980) presented the results of their investigation for 14 soil profiles in tabulated form and in form of the soil salinity class for the study area.

Rimwanich and Suebsiri (1984) describe the nature and management of problem soils in Thailand, which include saline soils, and provide a soil salinity distribution map of north-east Thailand (Figure 10-5).

According to Limpinuntana and Arunin (1986), estimates based on satellite imagery show that 2.85 Mha of salt-affected soils in north-east Thailand, which represents approximately 17 per cent of the total area, can be subdivided into three classes as shown in Table 10-4. They also show another 10.8 Mha being potentially saline. Severely salt-affected areas in the north-east are the low-lying alluvial plains, while the potentially saline soils are elevated areas containing either salt-bearing shale, siltstone or sandstone in the subsoil. However, it should be noted that Limpinuntana and Arunin (1986) have not separated the salt-affected soils into primary and secondary classes. This is also the case in Figure 10-5.

Table 10-4: Salt-affected soils in north-east Thailand.

Class	Electrical conductivity (EC_e) (µS cm^{-1})	Area (Mha)
Slightly affected	4000–8000	2.02
Moderately affected	8000–16 000	0.59
Severely affected	> 16 000	0.24
Total		2.85

Source of data: Limpinuntana and Arunin (1986).

Causes of salinisation

Various human activities are responsible for the spread of salinisation in north-east Thailand. These activities include deforestation, the construction of reservoirs, salt making and irrigation (Arunin, 1987). However, it is important to note that the largest part of the area affected by salinity has been caused by forest clearing. The following sections describe the effects of the above mentioned causes.

Deforestation

As mentioned previously, the total forest area in Thailand was approximately 28.67 Mha in 1961 or 56 per cent of the country. By 1988 this area had decreased to 14.38 Mha or 28 per cent. In the north-east, forest cover had declined from 42 per cent (7.4 Mha) to a mere 13 per cent (2.36 Mha) during the period 1961–85. Deforested areas have been used for upland cash cropping. According to Rigg (1987), since the end of the Second World War, upland cash cropping has spread in Thailand. These crops are grown on higher ground unsuited for cultivation of rice. In north-east Thailand there has been an accelerating expansion in the land planted to kenaf, cassava (*Manihot utilissima*) and other crops. Kenaf was far more suited than maize to environmental conditions over much of the north-east and with the rise in the kenaf price between 1959 and 1961 much upland crop farming shifted from maize to kenaf. Like kenaf, cassava is agronomically suited to the north-east, but, in addition, it is an extremely flexible crop that can be cultivated with a minimum of labour and cash input. Cassava can be grown on the poorest of soils without the need for fertiliser. Moreover cassava plantation is more profitable than kenaf. Therefore large areas recently cleared and never previously planted to kenaf were used for cassava cultivation.

Subsequent changes in the hydrologic conditions following land clearing and replacement of deep rooted trees with shallow rooted crops (cassava, kenaf, etc.) increased the natural recharge of aquifers and resulted in saline seepage on lower slopes and valley floors. The deeper saline groundwater is well below the range of capillary movement, but as it is commonly artesian, it is possible that saline groundwater moves upwards under pressure. In localised areas, shallow saline groundwater may be formed by the groundwater flowing through a shallow salt dome and moving up from depths below 10–15 m (Arunin, 1984).

According to Löffler and Kubiniok (1988), although there is little disagreement amongst researchers on the relationship between deforestation and salinity, the main controversy lies in the explanation of how salt reaches the surface. In this respect there are two lines of thought. The first concept envisages that the salt originates directly from weathering of the Maha Sarakham Formation and is transported from nearby uplands through shallow interflow to low-lying areas. Areas with bedrock close to the surface also contribute to the movement of salt to the surface by shallow interflow. The second concept claims that the main cause of salt reaching the surface is a general rise of watertables to the capillary fringe and consequently the rise of salt to the surface. However, there is no doubt that both mechanisms operate in the region, but under different geomorphological and hydrological conditions. Generally speaking, salinisation by shallow interflow is responsible for a great number of saline areas, while watertable rise is responsible for the salinisation of large alluvial plains such as Tung Kula Ronghai. The following paragraphs describe the results of salinity investigation in two areas, one near

Khorat and the other in the Tung Kula Ronghai region. In the first case salinisation is caused mainly by shallow interflow, while in the second case watertable rise is the main cause of salinisation.

Williamson et al. (1989) made an intensive study of the groundwater hydrology and salinity of a 5000 ha area near Khorat, from June 1985 to September 1988. Their objective was to determine the source and flow path for water and salt of the upper 20 m of the regolith and to provide a basis for establishing appropriate management strategies. The land use in the study area was divided about equally into an agricultural woodland in the uplands cropped to cassava and kenaf and salt-affected rice paddy fields in the valleys with a traditional salt making operating at one valley location. They showed that groundwater flow from the uplands converges in the central valley, where the major salinisation occurs. In a larger context, groundwater flows from this area to the north-east where terraces and flood plains form part of the upper reaches of the Mun River.

Three recharge areas contributing to the movement of salt into the salinised valley were identified and the salt outflow estimted to be about 32.3 t d^{-1} (Williamson et al., 1989). Their conclusions can be summarised as follows: deforestation has substantially decreased transpiration and interception and provided a source of water for the salinity problems; halite in the Maha Sarakham Formation or secondary deposits of salt in more recent overlying unconsolidated formations were identified as the source of salt; the mechanism for salt redistribution is the water flow down a vertical displacement of 15–30 m, from recharge area to valley; the saline areas have been found where the flow path is upward towards the soil surface or where the spatial variations in hydraulic conductivity produce a convergence of flow; and that the most appropriate approach to the management of the problem would be the use of trees as biological drains with appropriate species, placement and density.

The Tung Kula Ronghai (TKR) is a vast alluvial plain covering approximately 337 000 ha of land in the southern part of the Khorat Basin; because of its poverty it bears the name which means '*Plain of the weeping Kula people*'. The TKR covers parts of five provinces (Roiet, Surin, Sisaket, Maha Sarakham and Yasothon) and is bordered from the south by the Mun River. Most of the area is flat with a gentle slope downward from the west to the east. The difference in ground elevation within the area is about 12 m. The area had a population of 370 900 in 1980. The 1981 land use data indicates that 83.6 per cent of the area (281 750 ha) was used for agriculture. Of this, about 80 per cent was used for paddy rice cultivation and the remainder of the cultivated area was planted with kenaf, legumes, vegetables, fruits and others. Although the area has been extensively used for agricultural production, it is considered to be marginal for future agricultural development due to a number of problems including lack of rainfall, flooding, salinity and infertility of the soils (see Chulalongkorn University Social

Research Institute, 1981, for further information about the TKR area and its development plan).

In 1977 the Australian Development Assistance Bureau, in response to the Thai Government, funded a number of joint projects for development of the area. These projects included a feasibility study of soil salinity (Australian Development Assistance Bureau, 1978). This study was followed by another investigation (McCowan International, 1983). The following brief description of salinity in the TKR area is based on the McCowan International (1983) and Löffler and Kubiniok (1988).

Salinity in TKR has been studied using a multidisciplinary approach. The relationship of excess salinity to stratigraphy, hydrogeology, land use and soil has been assessed at specific study sites and at a more general level for all of TKR. TKR is extensively deforested with less than 2 per cent under forest. It is estimated that only 6 per cent of TKR is totally free of salinity. The most severely salt-affected lands are located within the alluvial plain, where rice cultivation is dominant.

The main cause of salinity is watertable rise (between 3 and 8 m in various sites) and the presence of shallow saline watertables from which salts rise to the root zone. The shallow saline watertable has developed because of increased recharge to the aquifer resulting from forest clearing, mainly in upland areas both within and to the north-west of TKR. Within TKR about 20 per cent of the area provides recharge to shallow, local, low salinity flow systems. Recharge on uplands supply deeper, long distance, more saline flow systems. Shallow flow systems are divided into unconfined and semiconfined flows. Deeper longer flow systems flow through shales and sandstones to emerge in the deeper parts of the Quaternary sequence away from uplands in TKR. These long, deep flow systems dissolve salts from a variety of sources of solid salts in the Cretaceous and Tertiary Maha Sarakhan Formation. Upward movement of these saline waters contributes to the salinisation of the area.

Management options applicable to the salinity problems of TKR are likely to be those which have low capital, operation and maintenance costs. The options recommended are mainly land use changes which aim at reducing groundwater recharge and improving water control and soil physical conditions within the salinised rice fields. Specific recommendations include: reforestation of recharge areas; development of effective mulching/amendment and rice field levelling strategies; design of a surface water control scheme within saline lands; modification of rice field layout within water control schemes to conform to natural contours and drainage patterns; and species evaluation for trees, crops, pasture and rice varieties.

Construction of reservoirs

Construction of reservoirs in areas with shallow saline groundwater also caused salinisation and widespread environmental problems. Salinisation has occurred in the vicinity of these reservoirs and has spread out over an even wider area from the reservoirs. Arunin et al. (1988) investigated the impact of 132 reservoirs on salinisation in north-east Thailand during the three-year period 1984–86. Reservoirs have been located on soils with six salinity levels as defined by the soil salinity map produced by the Land Development Department. As Table 10-5 shows, reservoirs have been classified into five groups according to their capacity, ranging from 50 000 to more than 1 000 000 m^3.

Arunin et al. (1988) showed that: the capacity of the reservoirs had no effect on water and soil salinity; electrical conductivity of reservoir water has been increasing every year due to the inflow of saline water into reservoirs; and salinities increased during dry periods and decreased during wet periods. Table 10-5 shows that seven reservoirs (5.3 per cent) have an EC value of greater than 3000 µS cm^{-1} and 27 reservoirs (20.5 per cent) have a lower EC of 700 to 3000 µS cm^{-1}, while in 98 reservoirs (74.2 per cent) the EC has been lower than 700 µS cm^{-1}. Arunin et al. (1988) demonstrated that a high percentage of reservoirs constructed on saline areas are salt-affected. For example, all reservoirs built on the severely saline areas have moderate to high salinity levels.

Investigations show that water in the reservoirs tends to become saline a few years after construction. Figure 10-6 shows fluctuations in the electrical conductivity of water in six reservoirs in the Maha Sarakham Province (Figure 10-4) from 1968 to 1977. In the case of the Nong Bo and Ekasatsuntorn

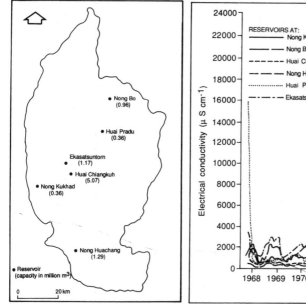
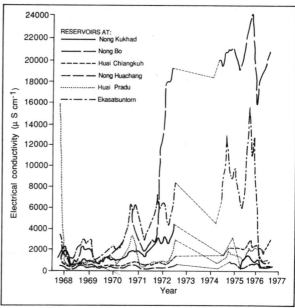

Figure 10-6: Location, capacity and fluctuations in the electrical conductivity of water in six reservoirs, Maha Sarakham Province, Thailand (Arunin, 1984).

Table 10-5: Capacity and average electrical conductivity of water of 132 reservoirs in north-east Thailand.

Salinity level[a]	Capacity (m³)															
	50 000			50 001–100 000			100 001–500 000			500 001–1 000 000			> 1 000 000			Total
	<.7	.7-3	>3	<.7	.7-3	>3	<.7	.7-3	>3	<.7	.7-3	>3	<.7	.7-3	>3	
	Electrical conductivity of water (1000 μS cm⁻¹)															
1	–	2	–	–	2	–	–	3	–	–	–	1	–	–	–	8
2	3	1	–	3	2	2	1	1	–	2	–	–	–	1	–	15
3	3	–	–	6	2	–	4	1	–	1	2	–	5	1	–	25
4	14	2	–	4	1	1	8	–	–	2	–	1	12	2	–	47
5	8	–	–	–	–	–	8	2	2	2	–	–	4	3	–	29
6	–	–	–	1	–	–	4	–	–	2	–	–	1	–	–	8
Sub-total	28	5	–	14	7	3	25	7	2	9	2	2	22	6	–	132
Total	33			24			34			13			28			132

(a) 1. Severely saline, 2. Moderately saline, 3. Mildly saline, 4. Low land areas having potential for salinisation, 5. Plateau areas with salt layer beneath the surface, 6. Non-saline.

Source: Arunin et al. (1988).

Reservoirs salinity reached high values of 24 000 and 16 000 µS cm^{-1}, respectively. It should be noted that in the case of the Nong Bo Reservoir, discharge of the effluents from the salt making operation had a major effect on the salinity of the reservoir's water.

Salt making

Salt making is another cause of human-induced salinisation. Krairapanond et al. (1992) describe salt production in north-eastern Thailand and its impact on land and water resources. The following summary is based on their account.

North-east Thailand contains more than 4700 Mt of rock salt deposits in five provinces as follows: 2000 Mt in Nong Khai, 1500 Mt in Chaiyaphum, 700 Mt in Maha Sarakham, 300 Mt in Ubon Ratchathani and 200 Mt in Nakhon Ratchasima (Figure 10-4).

Salt production in north-east Thailand began in a primitive manner for household consumption. Soils with patchy crusts of salt on the surface were dissolved and then boiled to crystallise soluble salt. Salt produced by this method was practical to every local livelihood in this region. In 1969, saline groundwater with a TDS of 50 000 to 60 000 mg L^{-1} was found during a water well drilling to a depth of 27 m in Borabu District, Maha Sarakham Province. This discovery changed salt production in the region from a primitive livelihood to a commercial one. Rock salt mining operations eventually dominated rice farming in some areas. It is estimated that a hectare of land used for salt production produces about 62.5 t yr^{-1}, which is worth US$10 000. The same size of land would give a rice farmer a return of about US$1000 per year.

Three different types of rock salt mining have been practised in northeast Thailand. Two traditional operations are boiling and sunning methods for local investors and one newly introduced advanced technology is called solution mining. These salt making processes are as follows.
- The boiling method is a primitive approach to producing salt. In this case the soil is used, the white crust of salt is collected from the soil surface, dissolved in water and filtered. Then the dissolved salt is crystallised by boiling. When high saline groundwater is used, the operators just boil it to produce salt.
- The sunning or on-farm method has two types of operations. In the first type, natural brine or saline water is pumped out from 30–190 m below the surface and is processed into salt on a farm by sunlight-induced evaporation. In the second type, freshwater is pumped down to dissolve the underground rock salt and then the water containing dissolved salts is crystallised by sunlight.
- Solution mining consists of pumping water down to about 200 m below the land surface to dissolve the rock salt, which is then pumped up in the form of salt water to the surface. The salt water is later treated at a high temperature to produce salt. This method seems to effectively protect the environment from the adverse effects of the operation.

In general, rock salt mining can be found scattered mainly in the areas of Sakon Nakhon, Nakhon Ratchasima, Udon Thani, Maha Sarakham, Nong Khai and Roiet Provinces. In these areas shallow deposits of rock salt are observed at depths of less than 100 m (Figure 10-7).

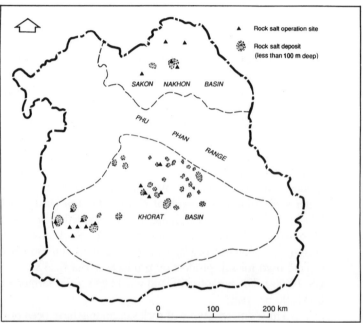

Figure 10-7: Rock salt deposits and rock salt mining operations in the north-east of Thailand (Krairapanond et al., 1992).

When salt production escalated after 1971, many forest areas were largely deforested because of the boiling method used to crystallise salt water. This boiling method was soon limited by a shortage of fuel wood.

When controls were imposed on the burning wood method, salt farmers turned to sunning the salt water. This commercial application was widely practised in most of the rock salt producing areas and later generated serious environmental problems such as degradation of agricultural lands and water resources, deforestation and land subsidences. Therefore in 1980 the Thai Government prohibited salt mining practices in Maha Sarakham, Roiet and Sisaket Provinces. However, the environmental problems as well as soil and water resources degradation were continuously aggravated because of an ineffective enforcement. As a result, in 1989 the government issued another order banning rock salt mining practices, especially in the Borabu District, Maha Sarakham Province. This time the order was effective. However, rock salt mining operations had spread to some other areas of Udon Thani and Sakon

Nakhon Provinces, where the banning was not in force, and the problem continued to grow. At present the Thai Government has ordered a halt to rock salt mining operations in every salt farming area. This ban excludes the large firms using solution mining, with which the environment can be protected from the adverse effects of operations. However, it seems that the illegal salt making has continued because of the substantial profits involved.

The practice of salt making pollutes downstream rivers and paddy fields, destroying rice crops and killing fishes. For example, the Nong Bo Reservoir in Maha Sarkham Province which is a middle-sized reservoir (see Figure 10-6), was damaged by the high salinity of waste waters from rock salt farming. The salinity of water in the reservoir reached nearly 100 000 mg L^{-1} in 1990, affecting farming families along the Siew River (a tributary of the Mun River, Figure 10-3) by the release of the water from the reservoir for agricultural purposes. Tasker (1990) reported that an unofficial estimate has put the number of farmers affected by the salinised Siew River at 300 000 in the three provinces of Maha Sarakham, Roiet and Sisaket, covering 500 villages and 50 000 ha of rice field. Affected farmers from 500 villages formed a committee to assess the impact of mining on rice farms and to seek ways to desalinate the Siew River and restore rice fields. At the Siew's headwaters in Nong Bo Reservoir the water is so saline that it resembles a mini Dead Sea. The water was initially fresh and drinkable and fish were abundant. Now only one type of small inedible fish survives in the lake.

Although salt making has been blamed for the expansion of salinity of the Siew River, Williamson (1992) expects that the groundwater inflow along the river with salt sources at depth in the Maha Sarakham Formation could be as great if not much greater than the salt making input. However, the effect of different processes contributing to the salinity of the river needs to be quantified.

Irrigation

Salinisation of land and water resources of north-east Thailand caused by irrigation is not well documented. However, secondary salinisation in irrigated areas has been reported from the Nong Wai irrigation area in the Khon Kaen Province, Pumpawapi south of Udorn, the Lam Pao irrigation scheme at Kalasin Province and Un in Sakon Nakhon Province (Arunin, 1984) but no detailed information is available. According to Arunin (1992b), in the Lam Pao and Nong Wai, with respective irrigated areas of 41 000 ha and 11 000 ha, about 10 per cent of each area is affected by secondary salinisation.

Management options

Management of salinity problems requires a reasonable knowledge of the extent and severity of the problem. In this regard between 1982 and 1986 the Land Development Department (LDD) of the Ministry of Agriculture and Cooperatives produced soil salinity maps of north-east Thailand at scales of 1:250 000 and 1:100 000. These maps identified areas of high, medium and low salinity in the region. The LDD is now developing a new series of detailed soil salinity maps at a scale of 1:50 000. These maps will be used to identify priority areas in the north-east where the Royal Forestry Department will implement reforestation programs to alleviate salinity problems (see Hunt, 1992, for two poster-sized salinity maps of the Nakhon Ratchasima and Chaiyaphum Provinces).

Effective management options cannot be implemented without considering the socioeconomic conditions of the north-east which suffers from poverty. According to Arunin (1987), human activities such as cutting trees, building reservoirs and salt making, leading to land and water resources salinisation, have not been done with malicious intention but rather for survival. These activities created immediate benefits for the participants. Cutting trees for wood or charcoal production provides an income. Selling salt collected from saline soil or saline water is profitable. Building reservoirs usually provides a year-round supply of water for fish culturing and rice paddy irrigation. To change or reclaim and manage saline land will necessarily require that alternative income be provided for the people plus education to dissuade current and traditional behaviour.

Arunin and Im-Erb (1991) provide a comprehensive list of the screened imported and native salt-tolerant trees for the management of salinity problems in Thailand. The recommended species include *Azadirachta indica*, *Acacia auriculiformis*, *Casuarina glauca*, *Cassia siamea* and *Eucalyptus camaldulensis*.

Reforestation of the recharge areas with trees to lower the watertable in lowlands and reduce the pressure of artesian aquifers (as in the case of the Tung Kula Ronghai area) is the major thrust of the salinity control program of the LDD. Revegetation of the discharge lowlands is also included in the salinity control program. According to Arunin (1992a), experiments on changes of groundwater levels by eucalypt tree plantations showed that the trees at the age of 3 to 6 years at Phra Yun, Khon Kean Province, have been successfully lowering the watertable from 1.6 m to 3.5 m during three years of experiments from 1986 to 1988. Another experiment at Kularonghai, Roiet Province, during 1986–1988 showed that eucalypt trees of 5 and 6 years old could lower the groundwater levels by 1 and 3 m, respectively, while the groundwater level under native dipterocarp forest dropped by 2 m.

According to Löffler and Kubiniok (1988), the social acceptance of reforestation is a major obstacle even on public land, since these lands are generally used as grazing land by the villagers. Pure reforestation will therefore only be possible where large areas of vacant public land are available. Therefore agroforestry with a combination of tree planting and pasture will be most suitable in order to offset the loss of grazing land.

One of the first large-scale attempts to use reforestation as an appropriate measure to alleviate soil salinity was undertaken within the upland reforestation component of the Thai–Australian Tung Kula Ronghai Project. The aim of this component was to reforest 360 ha of upland annually over a period of five years. In contrast to many other reforestation projects, which are purely for the purpose of reforestation, this one is an integrated one and aims at producing timber, creating adequate pasture and lowering the watertable. Initially the main tree used for reforestation was *Eucalyptus camaldulensis*, which does grow well on poor soils. It is hoped to gradually replace the exotic eucalypts by native trees (Löffler and Kubiniok, 1988).

For revegetation of the heavily salt-affected lands considered as wasteland, halophytes are introduced as forage crops. Presently, plants from the United States and Australia are being screened with some native species (see Limpinuntana and Arunin, 1986, for the list of native salt-tolerant grass, pasture, tree and shrub species in some areas of north-east Thailand). Selected species should be salt-tolerant as well as tolerant to waterlogging conditions, which often occur during the wet season. The promising species are *Sporobolus virginicus* and *Distichlis spicata* (Arunin, 1992). *Sporobolus virginicus*, imported from Delaware, United States, showed the greatest tolerance among others while all native grasses died in the dry season (Yuvaniyama and Arunin, 1992a and b). Preliminary testing of *Atriplex* spp. and other salt-tolerant species from Australia were conducted without success by the LDD and by Khon Kaen University. Their establishment faced major problems, presumably because of several factors including poor germination, excess soil moisture and infestation by insect pest and native weeds (Limpinuntana and Arunin, 1986).

In slightly or moderately salt-affected areas, land is used for rice production. Tolerant varieties of rice have been screened and recommended to farmers. Salt-tolerant species of sesbania, used as green manure for rice, were even imported from Senegal in the mid-1980s and distributed free to farmers (Hunt, 1992). In the areas described as slightly or moderately salt-affected, other measures such as leaching, land levelling and use of organic amendments (rice hull, animal manure and green manure) will alleviate the salinity problem and increase rice production. Arunin (1984) reports that rice hull, at the rate of 3125 kg ha^{-1}, mixed with manure, at a rate of 6560 kg ha^{-1} was found to give the highest rice yield. According to Arunin (1992a), salt-tolerant varieties of others crops such as tomatoes, sweet potatoes and corn have been selected and recommended to farmers.

According to Löffler and Kubiniok (1988), the practice of trying to grow rice in banded fields in upland areas should be discontinued, since this practice increases the recharge to aquifers. Replacement of rice with other crops, preferably tree crops, should be encouraged. Although over 95 per cent of rice fields in the north-east are dependent on rain and flood water, the distribution of water is not efficient, so that some fields receive too much and others too little. Therefore water distribution and water management needs to be improved. Finally, in some areas, especially those that are severely salinised and where there is evidence that salinity has been a long-standing occurrence, it may not be possible or worthwhile to combat salinity at all.

References

Arbhabhirama, A., Phantumvanit, D., Elkington, J. and Ingkasuwan, P. eds. 1987. *Thailand Natural Resources Profile: Is the Resource Base for Thailand Development Sustainable?* Bangkok: Thailand Development Research Institute. 310 pp.

Arunin, S. 1984. Characteristics and management of salt-affected soils in the northeast of Thailand. In. *Ecology and Management of Problem Soils in Asia.* Taiwan: Food and Fertilizer Technology Center for the Asian and Pacific Region. (FFTC Book Series no. 27). 336–351.

Arunin, S. 1987. Management of saline and alkaline soils in Thailand. Paper presented at the Regional Expert Consultation on the Management of Saline/Alkaline Soils. Bangkok: FAO Regional Office for Asia and the Pacific, 25–29 August 1987. 15 pp.

Arunin, S. 1992a. Strategies for utilizing salt affected lands in Thailand. In. Proceedings of the International Symposium on Strategies for Utilizing Salt Affected Lands. Bangkok, 17-25 February 1992. Bangkok: Department of Land Development, Ministry of Agriculture and Cooperatives. 26–37.

Arunin, S. 1992b. Soil Salinity Research Section, Land Development Department, Bangkok. (Personal communication).

Arunin, S. and Im-Erb, R. 1991. Agricultural use of coastal reclaimed lands including salt-affected areas. Paper presented at the International Training Course on Agriculture Use of Reclaimed Land, Kwangju, Republic of Korea, 18–24 October 1991. Kwangju: Chonnam National University. 12 pp.

Arunin, S., Rungsangchan, P., Dissataporn, C. and Yuvaniyama, A. 1988. Impact of reservoirs on salinization in Northeast Thailand (in Thai). *Journal of Agricultural Sciences of Thailand.* 21(5): 331–345.

Asian Institute of Technology. 1983. *Policy Study on Agricultural Development and Related Activities.* v.III. Policy on Land, Water and Forest Resources. Bangkok: Asian Institute of Technology. 56–163.

Attanandana, T. and Vacharotayan, S. 1984. Rock phosphate utilization on acid sulphate soils of Thailand. In. *Ecology and Management of Problem Soils in Asia.* Taiwan: Food and Fertilizer Technology Center for the Asian and Pacific Region. (FFTC Book Series no. 27). 280–292.

Australian Development Assistance Bureau (ADAB). 1978. *Feasibility Study of Soil Salinity in North East Thailand.* Canberra: ADAB. 112 pp.

Buaphan, C., Boonsener, M., Archwichai, L., Wannakao, P. and Youngme, W. 1990. Groundwater quality variation and fluctuation in fractured shale aquifer of northeastern Thailand. In. *Proceedings of the International Conference on Groundwater Resources Management.* Bangkok, 5–7 November 1990. Bangkok: Division of Water Resource Engineering, Asian Institute of Technology. 291–298.

Chulalongkorn University Social Research Institute (CUSRI). 1981. *Kula Ronghai Project: Review for Implementation.* Bangkok: CUSRI. v.1. (Various pagings, about 270 pp), v.2. (Various pagings, about 280 pp).

Das Gupta, A., Paudyal, G.N. and Seneviratne, A.P.R. 1987. Optimum groundwater pumping pattern from a confined aquifer. In. Awadalla, S. and Ismail Mohd. Noor. eds. *Groundwater and the Environment: Proceedings of the International Groundwater Conference 1987.* Malaysia: Kebangsaan Universiti. G60–G68.

Das Gupta, A. and Yapa, P.N.D.D. 1982. Saltwater encroachment in an aquifer: A case study. *Water Resources Research.* 18(3): 546–556.

Department of Technical Co-operation for Development. 1986. *Ground Water in Continental Asia (Central, Eastern, Southern, South-Eastern Asia).* New York: United Nations. (Natural Resources, Water Series, no. 15). 391 pp.

Dheeradilok, P. 1987. Review of Quaternary geological mapping and research in Thailand. In. Wezel, F.W. and Rau, J.L. eds. *Progress in Quaternary Geology of East and Southeast Asia.* Bangkok: Co-ordinating Committee for Offshore Prospecting (CCOP) Technical Secretariat. (CCOP Technical Publication, no. 18). 141–167.

Donner, W. 1978. *The Five Faces of Thailand: An Economic Geography.* London: C. Hurst & Company (For the Institute of Asian Affairs, Hamburg). 930 pp.

Economic and Social Commission for Asia and the Pacific. 1991. *Assessment of Water Resources and Water Demand by User Sectors in Thailand.* New York: United Nations. 99 pp.

Food and Agriculture Organization of the United Nations (FAO). 1989. *Production Yearbook.* Rome: FAO. v.42. 350 pp.

Framji, K.K., Garg, B.C. and Luthra, S.D.L. 1983. *Irrigation and Drainage in the World: A Global Review*. Third edition. New Delhi: International Commission on Irrigation and Drainage. v.III. 1161–1667.

Hunt, P. 1992. Salt of the earth. *Manager: Thailand Business Monthly*. 43 (July): 40–43.

Johnson III, S.H. 1993. Determining irrigation service area under monsoonal rainfall conditions: An example from northeast Thailand. *Water International*. 18(4): 200–206.

Kanoksing, P. 1987. Water resources utilization for agriculture in Thailand. In. Ali, M., Radosevich, G.E. and Khan, A.A. eds. *Water Resources Policy for Asia*. Proceedings of the Regional Symposium on Water Resources Policy in Agro-Socio-Economic Development, Dhaka, Bangladesh, 4–8 August 1985. Rotterdam: A.A. Balkema. 127–135.

Krairapanond, N., Krairapanond, A., Sinthuwanich, D. and Junpet, T. 1992. Environmental impact of rock salt mining operations on land and water resources of northeast Thailand. In. *Proceedings of the International Symposium on Strategies for Utilizing Salt Affected Lands*. Bangkok, 17–25 February 1992. Bangkok: Department of Land Development, Ministry of Agriculture and Cooperatives. 309–322.

Krishnamra, J. 1990. Problem soils of Thailand. In. *Problem Soils of Asia and the Pacific: Report of the Expert Consultation of the Asian Network on Problem Soils*. Bangkok, Thailand, 29 August to 1 September 1989. Bangkok: Regional Office for Asia and the Pacific (RAPA), Food and Agriculture Organization of the United Nations. 261–274.

La Moreaux, P.E., Charaljavanaphet, J., Jalichan, N., NaChiengmai, P.P., Bunnag, D., Thavisri, A. and Rakprathum, C. 1958. *Reconnaissance of the Geology and Ground Water of the Khorat Plateau, Thailand*. (US Geological Survey Water-Supply Paper 1429). Washington DC: US Government Printing Office. 62 pp.

Limpinuntana, V. and Arunin, S. 1986. Salt affected land in Thailand and its agricultural productivity. *Reclamation and Revegetation Research*. 5: 143–149.

Löffler, E. and Kubiniok, J. 1988. Soil salinization in north-east Thailand. *Erdkunde*. 42(2): 89–100.

Löffler, E., Thompson, W.P. and Liengsakul, M. 1984. Quaternary geomorphological development of the lower Mun River Basin, northeast Thailand. *Catena*. 11(4): 321–330.

McCowan International. 1983. *Thai-Australian Tung Kula Ronghai Project: Tung Kula Ronghai Salinity Study*. Canberra: Australian Development Assistance Bureau. 145 pp.

Milliman, J.D., Broadus, J.M. and Gable, F. 1989. Environmental and economic implications of rising sea level and subsiding deltas: The Nile and Bengal examples. *Ambio*. XVIII(6): 340–345.

Phantumvanit, D. and Panayotou, T. 1990. *Natural Resources for a Sustainable Future: Spreading the Benefits*. Bangkok: Thailand Development Research Institute. 95 pp.

Phiancharoen, C. 1973. *Hydrogeological Map of Northeastern Thailand (Scale 1:500 000)*. Bangkok: Department of Mineral Resources, Ministry of Industry.

Piancharoen, C. 1982. *Hydrogeology and Groundwater Resources of Thailand*. Bangkok: Department of Mineral Resources, Ministry of Industry.

Ramnarong, V. 1985. Review on groundwater development in the northeastern Thailand. In. Thanvarachorn, P., Hokjaroen, S. and Youngme, W. eds. *Proceedings of the Conference on Geology and Mineral Resources Development of the Northeast, Thailand*. Khon Kaen, Thailand, 26–29 November 1985. Khon Kaen: Department of Geotechnology, Khon Kaen University. 113–125.

Rigg, J. 1987. Forces and influences behind the development of upland cash cropping in north-east Thailand. *The Geographical Journal*. 153(3): 370–382.

Rimwanich, S. and Suebsiri, B. 1984. Nature and management of problem soils in Thailand. In. *Ecology and Management of Problem Soils in Asia*. Taiwan: Food and Fertilizer Technology Center for the Asian and Pacific Region. (FFTC Book Series no. 27). 13–26.

Sinanuwong, S., Wichaidisdha, P., Pramojanee, P. and Trakuldist, P. 1980. The use of LANDSAT imagery for soil salinity study in the northeast of Thailand. *Thai Journal of Agricultural Science*. 13 (July): 227–237.

Sternstein, L. 1976. *Thailand: The Environment of Modernisation*. Sydney: McGraw-Hill. 200 pp.

Stewart, M.R., Tamblyn, W.L. and Charukalas, B. 1987. Evaluation of the shallow Quaternary aquifer for irrigation in northern Thailand. In. Thiramongkol, N. ed. *Proceedings of the Workshop on Economic Geology, Tectonics, Sedimentary Processes and Environment of the Quaternary in Southeast Asia*. Haad Yai, Thailand, 3–7

February 1986. Bangkok: Department of Geology, Chulalongkorn University. 233–245.

Tasker, R. 1990. Salt in the wound. *Far Eastern Economic Review*. (7 June): 28–29.

Uwaniyom, J. and Charoenchamratcheep, C. 1984. Liming of acid sulfate paddy soils in the Bangkok Plain. In. *Ecology and Management of Problem Soils in Asia*. Taiwan: Food and Fertilizer Technology Center for the Asian and Pacific Region. (FFTC Book Series no. 27). 265–279.

Williamson, D.R. 1992. CSIRO Division of Water Resources, Perth, Western Australia. (Personal communication).

Williamson, D.R., Peck, A.J., Turner, J.V. and Arunin, S. 1989. Groundwater hydrology and salinity in a valley in northeast Thailand. In. *Groundwater Contamination*. (IAHS Publication no. 185). 147–154.

Wongsawat, S. 1985. Status of hydrogeological mapping in Thailand. In: Castany, G., Groba, E. and Romijn, E. eds. *Hydrogeological Mapping in Asia and the Pacific Region*. Proceedings of the ESCAP-RMRDC Workshop, Bandung, 1983. (International Contributions to Hydrogeology; vol. 7). Hanover: Heise. 231–246.

World Bank. 1992. *World Development Report 1992: Development and the Environment*. New York: Oxford University Press. 308 pp.

Yuvaniyama, A. and Arunin, S. 1992a. Physiological changes of *Sporobolus virginicus* (L.) Kunth under various salinity levels. In. *Proceedings of the International Symposium on Strategies for Utilizing Salt Affected Lands*. Bangkok, 17–25 February 1992. Bangkok: Department of Land Development, Ministry of Agriculture and Cooperatives. 426–432.

Yuvaniyama, A. and Arunin, S. 1992b. Adaptability of some exotic halophytes to strongly salt affected soil in northeast Thailand. In. *Proceedings of the International Symposium on Strategies for Utilizing Salt Affected Lands*. Bangkok, 17–25 February 1992. Bangkok: Department of Land Development, Ministry of Agriculture and Cooperatives. 550–555.

Chapter 11: United States of America

Introduction

The United States of America covers a total area of 9 372 570 km². To this, Alaska contributes 1 530 690 km² and Hawaii 16 760 km². Areas outside the 48 contiguous or conterminous states include Puerto Rico, American Samoa, Guam and the Virgin Islands. The east-west and north-south dimensions of the country are approximately 4517 and 2572 km. The United States has diverse physiographic features with broad plains, plateaus and mountain areas (Figure 11-1). The country's surface elevation ranges from 86 m below sea-level in Death Valley in California to 6198 m at Mount McKinley in Alaska, with an approximate mean elevation of 763 m (US Bureau of Census, 1993: Table 360). Its main physiographic regions are listed below (Murphy, 1978):

- **The Coastal Plain** is located along the Atlantic Ocean and the Gulf of Mexico. Midway along the Coastal Plain, the Florida Peninsula separates the waters of the Atlantic from those of the Gulf of Mexico. On the west of the peninsula, the Coastal Plain is wide, but towards the north-east it gradually narrows and in the neighbourhood of New York the plain comes to an end.
- **The Applachian Highlands** make up the dominant relief feature of the eastern United States, but they have relatively low elevation. Mount Michell in North Carolina, with an elevation of 2038 m, is the highest peak in eastern North America. The Applachian Highlands form an effective watershed between the streams that flow to the Atlantic Ocean and those that flow westward to the Mississippi River.
- **The Central Lowlands** resemble a vast saucer, rising gradually to higher lands on all sides. Southward the land climbs to the Ozark Plateau, Quachita Mountains and the Interior Low Plateau. On the eastern side it climbs towards the Appalachian Plateau and on the western side towards the Great Plains. The Central Lowlands constitute the nation's greatest agricultural resource and the richest single agricultural region in the world.

Figure 11-1: The main physiographic regions of the contiguous United States (Murphy, 1978: 500).

- **The Great Plains** stretch along the eastern edge of the Rocky Mountains. The eastern edge of the Great Plains in most places is marked by a steep escarpment. Its climate is generally semiarid and it is noted for its dryland wheat farms and cattle ranches.
- **The Rocky Mountains** are the backbone of the North American Continent. They form what is known as the Continental Divide, or the Great Divide. West of the Divide, streams find their way to the Pacific or into lakes and sinks. East of the Divide, they flow to the Gulf of Mexico. The Rockies are higher than the Appalachians. Mount Elliert, with an elevation of 4399 m, is the highest peak. The Rockies have three main divisions, the Northern, the Middle and the Southern.
- **The Intermontane Plateau** is located between the Rocky Mountains and the Pacific Mountains area. It includes the Great Basin, the Columbia Plateau and the Colorado Plateau. It contains the driest part of the United States. The Mojave Desert and Death Valley are here as well as the Sonoran Desert of south Arizona. The Great Basin occupies more than half of this area.
- **The Pacific Mountain Area** is a complicated assembly of mountains and lowlands to the west of the Intermontane Plateau. Along the Pacific coast extend the Pacific Coast Ranges. About 160 km towards the east and parallel to this range lie the Cascade and Sierra Nevada Ranges. Between the Sierra Nevada and Coast Ranges lies the Central Valley of California,

while the Williamette and Puget Sound Lowland is located further north, between the Cascade Mountains and the Pacific Coast Ranges.

The United States had an estimated population of 250 million in 1990 with a growth rate of 0.9 per cent for the period 1980–90 (World Bank, 1992: 269). More recent statistics indicate that the country had a population of 255.4 million in 1992 with an estimated increase to 273.6 million by the year 2000 and to 339 million by the year 2025 (US Bureau of Census, 1993: Tables 2 and 4).

Rainfall and climate

The average annual precipitation of the conterminous United States is about 762 mm and ranges from less than 100 mm in the desert areas of southern Arizona and California to over 5000 mm in some parts of the coastal area of the Pacific Northwest (US Department of Agriculture, 1989: 182). In general, the average annual precipitation is high in the eastern states (1345 mm in Florida and Mississippi, 1270 mm in Georgia, Tennessee and North Carolina) while it is low in the central and western states (230 mm in Nevada, 330 mm in Utah and 355 mm in Arizona) (Geraghty et al., 1973: Plate 3). Figure 11-2 shows the average annual precipitation of the country.

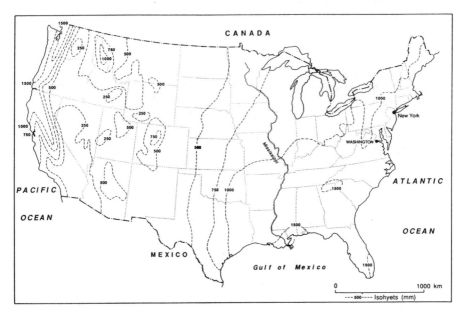

Figure 11-2:   Average annual precipitation of the contiguous United States (El-Ashry and Gibbons, 1988a: Figure 1.1).

Snowfalls represent a very important source of surface water throughout much of the country. Extremely heavy snowfalls occur in the Rocky Mountains and also on the eastern shores of the Great Lakes.

The United States has a wide variety of climatic types, ranging from subarctic regions on the higher mountain peaks to wet, subtropical regions of the Gulf Coast and the southern tip of Florida, to the dry west. The arid and semiarid west covers a large area of the country from the Great Plains to the Intermontane Plateau. The Gulf Stream warms the east coast slightly. The California Current cools the southern part of the western coast while the North Pacific Drift warms the northern part of the west coast. In general the country has a north-south gradient of increasing temperature.

Water resources

Surface water

The United States has a large number of rivers. The largest one is the Mississippi River, which is 5970 km long and drains an area of about 3.2×10^6 km^2 or 34 per cent of the country between the Appalachian and Rocky Mountains (Figure 11-3). The country is divided into 21 Water Resource Regions, 18 in the contiguous United States and one each in Alaska, Hawaii and the Caribbean.

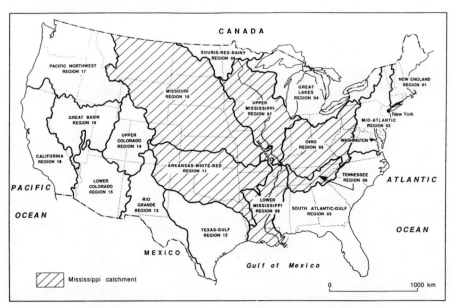

Figure 11-3: Water Resources Regions of the contiguous United States (Carr et al., 1990: 552).

In an average year a volume of about 5800×10^9 m^3 falls as precipitation on the conterminous United States (US Department of Agriculture, 1989: 181). About two-thirds of this volume, or 3870×10^9 m^3, returns to the atmosphere through evaporation and transpiration. The remaining 1930×10^9 m^3 replenishes surface and groundwater resources. The average annual run-off of the 18 Water Resources Regions is about 1840×10^9 m^3 (Table 11-1) and the groundwater recharge is 90×10^9 m^3. The average annual flow of water through streams and rivers into the oceans is about 1725×10^9 m^3.

Table 11-1: Natural annual run-off in the 18 Water Resources Regions of the contiguous United States.

Water Resources Region	Natural annual run-off	
	Average year (10^9 m^3)	Dry year (10^9 m^3)
1. New England	108.55	87.23
2. Mid-Atlantic	111.11	86.33
3. South Atlantic-Gulf	320.90	232.76
4. Great Lakes	103.85	82.56
5. Ohio	191.82	141.11
6. Tennessee	56.73	49.97
7. Upper Mississippi	104.91	85.09
8. Lower Mississippi	103.53	21.27
9. Souris-Red-Rainy	8.47	4.82
10. Missouri	84.91	64.45
11. Arkansas-White-Red	93.41	58.64
12. Texas-Gulf	49.17	27.08
13. Rio Grande	7.33	6.00
14. Upper Colorado	19.26	15.15
15. Lower Colorado	3.02	3.02
16. Great Basin	8.25	6.98
17. Pacific Northwest	370.56	312.69
18. California	93.91	69.61
Total	1839.69	1354.76

Source: US Department of Agriculture (1989: 256).

In the United States, surface water resources are highly regulated and transfer of water between catchments is a common practice (see Geraghty et al., 1973: Plates 15, 16 and 18).

In the United States, water withdrawal was in continuous increase up to 1980, when it reached a peak value of 614×10^9 m^3. It then started to decline, reaching 564×10^9 m^3 in 1990. Table 11-2 shows the estimated annual withdrawal, source of supply and the consumptive use for 1980 and 1990. In 1990 the major part (55 per cent) of the water withdrawn was for industry (about 85 per cent for thermoelectric power industry and 15 per cent for other industries), followed by irrigation (34 per cent). Surface and groundwater had a contribution of 452×10^9 m^3 (80 per cent) and 111×10^9 m^3 (20 per cent), respectively.

Table 11-2: The estimated water withdrawal, source of supply and the consumptive use in the United States for 1980 and 1990 (in 10^9 m^3 yr^{-1}).

		1980	1990
a.	**Withdrawal:**		
	Public supply	47	53
	Rural domestic and livestock	8	11
	Irrigation	207	189
	Industry	352	311
	Total	614	564
b.	**Source of water supply:**		
	Surface water	498	452
	Groundwater	115	111
	Reclaimed waste water	1	1
	Total	614	564
c.	**Consumptive use**	138	13

Source: Solley et al. (1993: Table 31).

Groundwater

Groundwater is available nearly everywhere in the United States, but the quantity available and the conditions controlling its occurrence and development differ from one part of the country to another. Detailed information concerning the hydrogeology of the country is available in Back et al. (1988) and Heath (1984).

Goundwater is a major source of water supply in the United States (Table 11-2). The amount of groundwater withdrawal in 1980 was about 115×10^9 m^3. In 1990, out of a total withdrawal of 564×10^9 m^3, about 111×10^9 m^3 were from groundwater. These statistics show that the annual withdrawal from groundwater resources is declining and this is mainly due to a significant drop in groundwater levels. More than 80 per cent of all the water pumped from wells comes from unconsolidated sediments (sand and gravel). Consolidated rock aquifers may yield only enough water for modest scale irrigation or for industrial uses.

Many states are relying on groundwater to meet most of their projected increase in water use, but groundwater availability is an issue in 47 states (US Department of Agriculture, 1989: 71). The most common problem of groundwater availability is water-level decline caused by intensive pumping (35 states) or increasing competition for available groundwater supplies (26 states). Major areas of groundwater decline are concentrated in the irrigation-dependent south-west. According to Postel (1989), more than 4 Mha or roughly a fifth of the nation's irrigated area is watered by pumping in excess of recharge. By the early 1980s the depletion was already particularly severe in Texas, Kansas and Arizona (Table 11-3).

Table 11-3: Estimated area irrigated with overpumped groundwater, in the United States, 1982[a].

State	Area irrigated by over-pumping groundwater (1000 ha)	Total irrigated area in state (1000 ha)	Share of total irrigated by overpumping (%)
Texas	1 543	2 144	72
California	644	3 221	20
Kansas	604	1 060	57
Nebraska	412	2 423	17
Colorado	196	1 088	18
Arizona	175	426	41
Arkansas	121	809	15
New Mexico	101	281	36
Florida	100	528	19
Idaho	87	1 237	7
Oklahoma	87	189	46
Other	(b)	6 448	–
Total	4 070	19 854	21

(a) Areas where watertable was falling at an average rate of at least 15 centimetres per year. Includes only areas irrigated every year.
(b) Local groundwater depletion scattered throughout other states; no figure available.

Source: Postel (1989: Table 3).

Estimates show that pumping exceeded aquifer recharge in 1975 by more than 27×10^9 m^3, equivalent to 40 per cent of the annual groundwater withdrawal for irrigation (Frederick, 1988). In the Texas and Oklahoma High Plains alone, overdraft from the Ogallala Aquifer was about 17.3×10^9 m^3 yr^{-1}, which is close to the average annual flow of the Colorado River (Frederick, 1988).

Groundwater extraction at a higher rate than the long-term natural recharge resulted in lowered watertables, increased pumping costs and contributed to land subsidence, diminished stream flow and saline water intrusion into freshwater aquifers.

To alleviate these problems to some extent, artificial recharge is in practice in many parts of the country. In 1988, about 558 out of the 719 injection wells surveyed in 14 states were recharging aquifers (Bouwer et al., 1990). The others were being used for saltwater intrusion barriers, drainage and subsidence control. In 1990 the fastest growing type of recharge was the aquifer storage recovery. Unlike other wells, these are dual purpose: they both store and recover water from the same wells according to supply and demand.

Land cover

Of a total land area of 916.7 Mha, some 189.91 Mha were cropland in the United States in 1987. Permanent pasture covered 241.5 Mha, forest and woodlands 265.2 Mha and other lands 220.1 Mha (FAO, 1989).

According to the US Department of Agriculture (1993: Tables 1 and 2), in 1992 some 139 Mha were used for crop production and 22 Mha were idled by federal programs including the Commodity Program and the Conservation Reserve Program, which makes ideal use of highly erodible and other sensitive croplands. In the same year (1992), land areas planted to major crops were as follows: corn, 3201 Mha; wheat, 29.3 Mha; hay, 24.1 Mha; soy beans, 24.0 Mha; sorghum, 5.4 Mha; cotton, 5.4 Mha; oats, 3.2 Mha; barley, 3.2 Mha; and rice, 1.3 Mha.

Irrigation

Irrigation in the United States had its earliest beginning among the Indian settlements of the south-west. Ruins of early Indian civilisation in Arizona and New Mexico show that Indians practised irrigation long before the first European settlement (Framji et al., 1983: 1484). Heavy immigration of Europeans from the east marked the period from 1870 to 1900. Estimates indicate that 111 400 ha were under irrigation in 1870. This figure increased to approximately 405 000 ha in 1880 and 1.47 Mha in 1889 (see Framji et al., 1983: 1452–1542 for a description of the development of irrigation in the United States).

Irrigated area in the United States is about 18.7 Mha. Irrigation takes place mainly in the 17 western states, plus Arkansas, Florida and Louisiana (Bajwa et al., 1992: 283). These 20 states account for 91 per cent of all irrigated area in the United States and 82 per cent of all irrigated farms. The extent of irrigated land in the top five irrigated states is as follows: California, 3.1 Mha; Nebraska, 2.3 Mha; Texas, 1.7 Mha; Idaho, 1.3 Mha; and Colorado, 1.2 Mha. These five states contain over half of all irrigated areas of the United States.

Although irrigation is not essential for most crops in humid regions of the United States, supplementary irrigation increases yields and reduces risks associated with drought and rainfall variability (Frederick, 1988). According to Bajwa et al. (1992: 111), irrigation in humid regions has been increasing consistently. Between 1982 and 1987, the north-east has had a 20 per cent increase; the Lake States a 10.75 per cent; and the south-east a 5.2 per cent increase.

Irrigation is the largest water user in the west. The nine western Water Resources Regions (Figure 11-3) account for 90 per cent of the total water withdrawn for irrigation (Solley et al., 1993: Table 15). In the eastern Water Resources Regions most of the water withdrawn for irrigation is in the Lower Mississippi (10.2×10^9 m^3) and South Atlantic-Gulf (6.15×10^9 m^3). By state, California, Idaho and Colorado (Table 11-4) are the largest users of irrigation water and together account for 43 per cent of the national total. Florida withdraws the most water for irrigation in the east, although it ranks thirteenth nationwide (Table 11-4).

Table 11-4: Irrigation water withdrawal and source of supply in 16 states in 1990, in order of the volume of withdrawal.

No.	State	Withdrawal (10^9 m^3)	Source s.w.(a) (%)	Source g.w.(b) (%)	No.	State	Withdrawal (10^9 m^3)	Source s.w.(a) (%)	Source g.w.(b) (%)
1	California	38.54	62	38	9	Washington	8.33	88	12
2	Idaho	25.83	65	35	10	Arizona	7.32	61	39
3	Colorado	16.02	78	22	11	Arkansas	7.25	18	82
4	Montana	12.43	100	-	12	Kansas	5.79	5	95
5	Texas	11.67	34	66	13	Florida	5.15	48	52
6	Wyoming	9.89	97	3	14	Utah	4.96	86	14
7	Oregon	9.48	92	8	15	New Mexico	4.16	54	46
8	Nebraska	8.43	29	71	16	Nevada	3.89	69	31

(a) Surface water; (b) Groundwater   Source of data: Solley et al. (1993: Table 16).

As mentioned previously, surface water and groundwater contributed 118×10^9 m^3 (63 per cent) and 71×10^9 m^3 (37 per cent), respectively, to the source of water for irrigation in 1990. However, as shown in Table 11-4, in some states the share of groundwater as the source of irrigation water supply is quite significant (e.g. 95 per cent in Kansas, 82 per cent in Arkansas, 71 per cent in Nebraska, 66 per cent in Texas, 52 per cent in Florida, 46 per cent in New Mexico and 38 per cent in California). Of 189×10^9 m^3 of water withdrawn for irrigation in 1990, 20 per cent (38×10^9 m^3) was lost in conveyance, 56 per cent (106×10^9 m^3) was consumptive use and 24 per cent

(45×10^9 m^3) was returned to surface water or groundwater (Solley et al., 1993: 34).

In terms of irrigation method, in 1988 about 11.1 Mha was irrigated by gravity systems (gated pipe, ditch with siphon tube, and flooding), 7.4 Mha by sprinkler systems (centre pivot, mechanical move, hand move, solid set and permanent), 0.75 Mha by drip or trickle method and 0.24 Mha by subirrigation method (Bajwa et al., 1992: Table 1).

In 1977, 322 storage reservoirs, 345 diversion dams, 23 319 km of canals, 1492 km of pipelines, 56 297 km of laterals, 174 major pumping plants and 24 994 km of drains were in operation on Bureau of Reclamation Projects (Framji et al., 1983: 1492).

The nation's total irrigated area has declined in the past decade. In the mid-1980s irrigated area was almost 10 per cent below the late 1970s peak of just over 20.24 Mha. According to Frederick (1988), a number of factors contributed to this decline: western water supplies have become increasingly scarce and costly (the rapid growth of irrigation was based on easy access to inexpensive water, a condition that no longer exists); groundwater overdraft or pumping in excess of aquifer recharge has decreased groundwater levels and increased energy costs; problems such as rising salinity levels in several western river basins pose the principal environmental threat to irrigation.

Based on the 1982 extent of irrigated lands and irrigation water use, projections suggest that in the future under the intermediate condition fewer areas might be irrigated and therefore less irrigation water will be used (US Department of Agriculture, 1989: 10–11). According to these projections, in 2030 about 12 Mha will be irrigated (compared to 19.8 Mha in 1982) and 60×10^9 m^3 of irrigation water (compared to 102×10^9 m^3 in 1982) will be used.

Like irrigation, development of drainage made a major contribution to the expansion of agricultural land in the United States. As described by van Schilfgaarde (1987), without drainage it is hard to imagine the US Midwest as it is in the twentieth century. Much of Ohio, Indiana, Illinois and Iowa originally was swamp or at least too wet to farm. Also, without drainage, irrigation development in the western United States would have failed through waterlogging and salinisation.

In 1985, an estimated 45 Mha benefited from artificial drainage in the United States, mainly in the eastern states (Pavelis, 1987a). At least 70 per cent of the drained land is in croplands, 12 per cent in pasture, 16 per cent in woodland and 2 per cent in miscellaneous uses. Drained areas, include 4 Mha in Illinois, 3.2 Mha in Indiana, 3.2 Mha in Iowa, 3 Mha in Ohio, 2.6 Mha in Florida, 2.4 Mha in Texas and 1.2 Mha in California. Surface and subsurface drainage account for 66 and 34 per cent of the drained areas, respectively (for further information concerning farm drainage in the United States, see Pavelis, 1987b).

Salinity

Saline land in the United States is largely found in the western arid and semiarid areas of the country. It is also scattered through the Northern Great Plains of the USA, extending into Canada (McKell et al., 1986; Szabolcs, 1989: 58–64). Saline soils are both natural and human-induced, due to irrigation activities and dryland farming. Climate is a major factor in the creation and distribution of saline soils. In the east of the United States soil salinity is generally not a problem because the annual rainfall is high and salts do not accumulate in the soil profile. However, in the western semiarid and arid parts of the country, annual rainfall is low and salt accumulates in the soil profile.

In the 20 Water Resources Regions about 19.6 Mha of cropland and pastureland are either sodic or affected by various degrees of salinity or sodicity (Table 11-5). These conditions are lowering the productivity of about 9 per cent of US cropland and pastureland. The extent of the affected lands is particularly high in the following Water Resources Regions: Missouri (8.6 Mha), Souris-Red-Rainy (3.1 Mha) and California (1.6 Mha). The percentage of the affected land is 72.9 per cent in the Rio Grande, followed by the Lower Colorado (64.4 per cent) and Great Basin (49.7 per cent). According to the US Department of Agriculture (1989: Appendix, Table 30), the top 14 states with the largest extent of saline/sodic land are: North Dakota (3.81 Mha), South Dakota (3.67 Mha), Montana (2.8 Mha), Texas (1.82 Mha), California (1.60 Mha), Minnesota (1.12 Mha), Colorado (0.74 Mha), Idaho (0.67 Mha), Kansas (0.47 Mha), Utah (0.46 Mha), Wyoming (0.45 Mha), Nevada (0.35 Mha), Arizona (0.34 Mha) and New Mexico (0.27 Mha).

Table 11-6 shows that 23 per cent or about 5.6 Mha of irrigated cropland and pastureland in the continental United States are affected by salinity or sodicity. The proportion of affected irrigated lands is significantly high in a number of Water Resources Regions such as the Rio Grande (76 per cent), Lower Colorado (66 per cent), Great Basin (58 per cent), Upper Colorado (41 per cent) and California (35 per cent). Table 11-6 also shows the share of affected irrigated lands to the total cropland and pastureland. For example, this proportion is about 99 per cent in the Lower Colorado, 91 per cent in California, 87 per cent in Upper Colorado and 82 per cent in the Pacific Northwest Water Resources Region. Inefficient irrigation practice accelerates the rate of salinisation. Among locations where salinity is increasing because of irrigation practices are the San Joaquin Valley in Central California, areas in the Upper and Lower Colorado River Basin, parts of the Rio Grande Basin and Western Texas.

In the Upper Mississippi, Souris-Red-Rainy and Missouri Water Resources Regions (Figure 11-3), higher percentages of non-irrigated land than irrigated land are affected. Salinity in non-irrigated land increases as a result of saline-seeps.

Table 11-5: Cropland and pastureland soils affected by salinity and sodicity in 20 Water Resources Regions in 1982.

Water Resources Region[a]	Saline (1000 ha)	Slightly saline (1000 ha)	Sodic (1000 ha)	Total affected (1000 ha)	Total crop/ pastureland (1000 ha)	Percentage affected
New England	0.0	0.0	0.0	0.0	1 261.4	0.0
Mid-Atlantic	0.0	0.0	0.0	0.0	7 629.5	0.0
South-Atlantic-Gulf	8.6	4.9	1.6	15.1	17 292.3	0.0
Great Lakes	0.0	0.0	0.0	0.0	11 405.0	0.0
Ohio	0.0	0.0	13.2	13.2	19 730.1	0.0
Tennessee	0.0	0.0	0.0	0.0	3 355.6	0.0
Upper Mississippi	6.2	476.6	96.4	579.2	32 406.6	1.8
Lower Mississippi	16.5	21.4	256.1	294.0	11 520.8	2.6
Souris-Red-Rainy	251.5	2649.7	213.9	3 115.1	9 118.9	34.2
Missouri	1456.8	6101.1	1123.0	8 680.9	50 760.9	17.1
Arkansas-White-Red	67.6	475.4	228.9	771.9	24 500.4	3.2
Texas-Gulf	444.0	903.2	64.4	1 411.6	15 864.1	8.9
Rio Grande	213.4	510.1	41.7	765.2	1 049.7	72.9
Upper Colorado	166.3	129.7	12.8	308.8	948.9	32.5
Lower Colorado	79.1	321.9	9.4	410.4	637.3	64.4
Great Basin	385.4	302.8	61.6	749.8	1 509.1	49.7
Pacific Northwest	343.6	514.5	51.8	909.9	9 482.8	9.6
California	786.9	448.5	348.0	1 583.4	4 938.7	32.1
Hawaii	0.0	0.0	0.0	0.0	529.0	0.0
Caribbean	6.4	0.0	0.0	6.4	551.7	1.2
Total	4232.3	12 859.8	2522.8	19 614.9	224 492.8	8.7

(a) See Figure 11-3 for location.   Source: US Department of Agriculture (1989: Table 11).

In the following sections we will define irrigated salinity problems in the Colorado River Basin and the Central Valley in California, and dryland salinity problems in the Northern Great Plains.

Colorado River Basin

The Colorado River is about 2250 km long and originates as snowmelt-fed streams in the Rocky Mountains of Colorado, Wyoming and New Mexico and discharges to the Gulf of California. It drains about 632 000 km^2 in seven states (Colorado, Wyoming, Utah, New Mexico, Arizona, Nevada and California) before entering Mexico (Figure 11-4). Its major tributaries are the Green, San Juan and Gila Rivers. The Colorado River Basin is divided into the upper and lower basin in the USA. Colorado, New Mexico, Utah and Wyoming are generally known as the Upper Basin States, while Arizona, California and Nevada are generally referred to as the Lower Basin States.

Table 11-6:   Salinity and irrigation in 12 selected Water Resources Regions in 1982.

Water Resources Region[a]	Irrigated cropland and pastureland		Affected portion of irrigated soils (%)	Irrigated portion of total affected cropland and pastureland (%)
	Total (1000 ha)	Saline, slightly saline or sodic (1000 ha)		
Upper Mississippi	373.1	1.5	0.4	0.3
Lower Mississippi	2 108.6	178.4	8.5	60.7
Souris-Red-Rainy	39.1	9.6	24.5	0.3
Missouri	5 637.0	837.0	14.8	9.6
Arkansas-White-Red	3 038.7	150.7	5.0	19.5
Texas-Gulf	2 816.2	421.9	15.0	29.9
Rio Grande	782.1	590.8	75.5	77.2
Upper Colorado	650.8	269.2	41.3	87.2
Lower Colorado	615.7	406.8	66.1	99.1
Great Basin	996.7	580.6	58.3	77.4
Pacific Northwest	3 410.5	745.7	21.9	81.9
California	4 048.0	1435.3	35.4	90.6
Total	24 516.5	5627.5	23.0	

(a) See Figure 11-3 for location.   *Source:* US Department of Agriculture (1989: Table 12).

The regional economy of the basin is based on irrigated agriculture, livestock grazing, mining, forestry, manufacturing, oil and gas production and tourism. Of the land within the basin, about 75 per cent is owned and administered by the Federal Government or held in trust for Indian tribes (Miller et al., 1986: 21). By far the greatest portion of the naturally occurring salt load originates on these federally owned and administered lands.

The headwaters of the Colorado River and its major tributaries, the Green and San Juan Rivers, lie in the high peaks of the Rocky Mountains where precipitation averages 1000 to 1500 mm per year. Most of its course, however, crosses the semiarid Colorado Plateau and the desert, where average annual precipitation may be as low as 60 mm (Mueller and Moody, 1984). The basin has a population of more than a million and through export projects its water provides either a full or supplemental supply to over 18 million people. Moreover, Mexico receives about 1.85×10^9 m^3 yr^{-1} for irrigation and to provide municipal water supply to nearly 1.7 million people (for a brief history of water apportionment in the basin and conflict resolution between the United States and Mexico, see p. 94).

Historically, the run-off of the river was very sporadic with the majority of the run-off occurring as floods in the spring and early summer and very little flow during the rest of the year. The annual flow has varied from 6.75×10^9 m^3 to 29.65×10^9 m^3 (Barton, 1981). This highly variable flow was a problem as most of the agricultural land and major cities in the area including Denver, Los Angeles, Phoenix and San Diego depend upon the Colorado River or its tributaries for their water supply. Ten major storage dams such as the Glen Canyon Dam in the Upper Basin and the Hoover Dam in the Lower Basin have been built on the river to stabilise its flow. They have a reported storage capacity of 75.94×10^9 m^3 (Barton, 1981). However, construction of these reservoirs had some environmental impacts which include: sediment trapping; increased evaporation and salt concentration; thermal stratification of water in the reservoirs; and changes in aquatic species (El-Ashry and Gibbons, 1990).

Figure 11-4:   Colorado River Basin (Jensen and Leach, 1984).

The Colorado River picks up large quantities of sediments. These sediments gave the river a muddy red colour and suggested the Spanish name 'Colorado' or 'coloured red'. Now much sediment is trapped behind the dams that have been constructed throughout the basin and salinity has replaced the sediment problem as the water quality characteristic of most concern in the Colorado's lower reaches (Miller et al., 1986: 1).

Irrigation

Irrigation development in the Upper Basin began gradually in the 1850s and 1860s. Farmers began to settle along the tributaries and parts of the main stems of the Green and Colorado Rivers in Wyoming, Utah and Colorado. By 1905 some 324 000 ha were irrigated. Between 1905 and 1920 the development of irrigated land continued at a rapid pace, and by 1920 nearly 567 000 ha were being irrigated. The development then levelled off because of physical and economical limitations (Trueman and Miller, 1984).

Major irrigation development in the Lower Basin began near the end of the nineteenth century. In this part of the basin farmers were faced with the siltation of their irrigation facilities. The effect of silt on waterworks was of greater concern in the early period than the effect of salt on crops. Silt control was one of the reasons for the construction of the Imperial Dam (Miller et al., 1986: 23). Although the post-Second World War developments did give some consideration to salinity, the problem was not thought to be severe.

In 1982, about 1.03 Mha were irrigated within the basin in the USA, 0.56 Mha in the Upper Basin and 0.47 Mha in the Lower Basin. Moreover, about 182 000 ha of crops are irrigated with Colorado River water in Mexico. Table 11-7 shows the irrigated area, irrigated water use efficiencies and volumes in the Colorado River Basin in 1982. It indicates that a large volume (12.95×10^9 m^3) of water was diverted (withdrawn) for the irrigation of 1.03 Mha of cropland and pastureland, or on average 12 570 m^3 ha^{-1} (11 300 m^3 ha^{-1} for the Upper Basin and 14 000 m^3 ha^{-1} for the Lower Basin). Within the basin, consumptive use is 58 per cent of the gross diversion (53 per cent for the Upper Basin and 63 per cent for the Lower Basin) and the return flows are 47 and 35 per cent for the Upper and the Lower Basin, respectively.

Table 11-7: Irrigated area, irrigation water use efficiencies and volumes in the Colorado River Basin in 1982.

River Sub-basin	Irrigated area[a] (Mha)	Off-farm conveyance efficiencies	Crop and pasture	Incidental loss	Return flows	Gross diversion	Consumptive use
			Irrigation water efficiencies[b] (Percentage of gross diversion)			Annual water use[b] (10^9 m^3)	
			Net depletion				
Upper Colorado	0.56	75	38	15	47	6.35	3.32
Lower Colorado	0.47	86	55	10	35	6.60	4.14
Total	1.03					12.95	7.46

Source: US Department of Agriculture (1989, (a): page 260, (b): page 261).

Irrigated land salinity

Salinity has long been recognised as one of the major problems in the irrigated areas of the Colorado River Basin. By 1929 irrigators in the Imperial Valley had initiated the installation of underground pipe drains to remove the excess water (Holburt, 1984). As shown in Table 11-5, in the Upper and Lower Colorado River Basin, 309 000 ha and 410 000 ha or about 33 and 64 per cent of the cropland and pasturelands are affected by salinity and sodicity. Irrigated areas of the basin (Figure 11-7) include the Grand Valley (Colorado), Lower Gunnison (Colorado), McElmo Creek (Colorado), Uinta Basin (Utah), Wellton-Mohawk (Arizona) and Imperial Valley and Coachella Valley (southern California). Highly saline return flows from the irrigated areas to the Colorado River are a major source of the Colorado River salinity. In the irrigated areas of southern California, irrigated with water diverted from Colorado River, drainage waters are collected and disposed of to the Salton Sea, which has a salinity concentration of about 40 000 mg L^{-1} TDS.

River salinity

Salinity has long been recognised as one of the major problems of the Colorado River. At its headwaters in the mountains of Colorado, the Colorado River has a salinity concentration of less than 30 mg L^{-1}. The salinity concentration progressively increases downstream as a result of two processes: salt loading, or the addition of soluble salt to the river, and salt concentration, caused by a reduction in the volume of river water as a result of evaporation, transpiration or withdrawal. At Lees Ferry (Arizona, see Figure 11-7) salinity concentration is more than 500 mg L^{-1} (Liebermann et al., 1989: 60). By the time the Colorado River reaches Hoover Dam (Figures 11-4 and 11-7) its salinity increases to about 800 mg L^{-1} TDS, carrying more than 8.2 Mt of salt annually.

The Bureau of Reclamation has developed a computer program and databases known as the Colorado River Simulation System (CRSS), as a tool for use by water resources managers dealing with water-related issues and problems in the Colorado River Basin. The CRSS simulates the flow of water and salt through the system and the operation of the major reservoirs and hydroelectric power plants (Colorado River Basin Salinity Control Forum, 1993: 2–9). Figure 11-5 shows the historical record of salinity at Imperial Dam from 1940 to 1982 and the CRSS projection over 1982 to the year 2010. It shows a dramatic change in the historical pattern of variation since 1972. Statistical analysis shows that TDS has been increasing at an average rate of 3.9 mg L^{-1} per year for the 1941–82 period. The decline in TDS from 1972 to 1980 is likely to be the result of several combined factors. The most significant factor is filling of the Colorado River storage system, which began discharging water which would have otherwise been stored (Trueman and Miller, 1984). In 1982 the salinity concentration averaged 825 mg L^{-1} at

Imperial Dam, the last major diversion point on the Colorado River in the United States. The CRSS projection to the year 2010 shows that without control measures the concentration could possibly reach a level of 1089 mg L^{-1} at Imperial Dam.

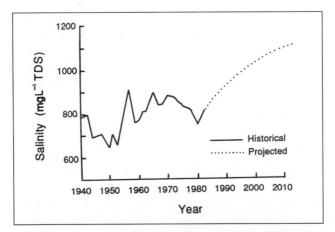

Figure 11-5: Salinity levels at Imperial Dam and Colorado River Simulation System projection to the year 2010 (US Bureau of Reclamation, 1983).

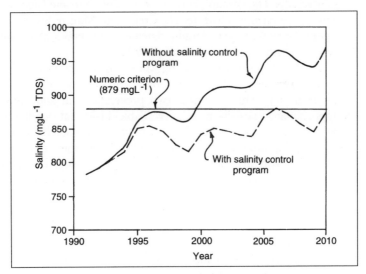

Figure 11-6: Salinity projections with and without further controls at Imperial Dam (Colorado River Salinity Program Coordinator, 1993: Figure 2).

According to Colorado River Basin Salinity Control Forum (1993: 2–4), salinity concentration increased in 1981–1982 and decreased significantly from 1983 to 1986. The period 1983 through 1986 was a period of above-normal rainfall. Each of the four years had an estimated natural flow at Lees Ferry in excess of 24.7×10^9 m^3 with the four-year average of 27.9×10^9 m^3. The record high flows during the period 1983–86 produced a reduction in salinity concentration of approximately 250 mg L^{-1} at Imperial Dam from 826 mg L^{-1} in 1982 to 577 mg L^{-1} in 1986 (see Colorado River Basin Salinity Control Forum, Table 2.1, for average annual salinity records at below Hoover Dam, below Parker Dam and at Imperial Dam from 1972 to 1992). However, the salinity concentration is projected to increase, following the overall rising trend shown in Figure 11-6, to a level of about 970 mg L^{-1} at Imperial Dam by the year 2010. This estimate is about 120 mg L^{-1} below the previous estimate (see Figure 11-5).

Sources of salinity

Salinity of the Colorado River is closely related to the geology of the basin (Miller et al., 1986: 3). In the Upper Basin, the Rocky Mountains are composed of granites, schists, gneisses, lava and sedimentary rocks including sandstone, limestone and shale. Much of this region is resistant to weathering and erosion, and contains few soluble salts. Farther downstream in the plateau area of south-western Wyoming, western Colorado, eastern Utah and northern Arizona, the geologic formations contain thick layers of sediments deposited in ancient seas. These formations include high concentrations of soluble salts. The Paradox Formation (in Colorado and Utah) is composed of halite, gypsum and anhydrite. The Cretaceous Mancos Shale, which underlies a good portion of the Colorado Plateau, also contains abundant amounts of gypsum (Azimi-Zonooz and Duffy, 1993).

As mentioned previously, salinity problems in the Colorado River Basin are the result of two basic processes: salt loading and salt concentrating. Through these two processes, man and nature contribute approximately equally to salinity (Table 11-8). Salt loading occurs as the river passes over soils or through rocks containing soluble salts. Discharge of saline springs and saline groundwater also contribute to the salt loading of the river. Soils in much of the basin have developed on gypsum-bearing shales. Irrigation water applied to these lands promotes weathering and dissolution of salts from the soil and underlying shales and returns to the river with a greater salt load than was diverted. Salt concentration occurs when the same salt load is carried in a reduced volume of water. Evaporation of water from the reservoirs, diversion of water for irrigation and export of water to other basins reduce the volume of water in the river and contribute to the salt concentration.

Table 11-8 shows the relative contribution of natural and human-induced sources (irrigation, reservoir evaporation, exports and municipal and industrial uses) to the salinity of the river. Natural sources contribute 47 per cent, followed by irrigation (37 per cent) and reservoir evaporation (12 per cent).

Table 11-8: Estimated sources of salinity in the Colorado River Basin.

Source	Contribution (%)
Natural sources	47
Irrigation	37
Reservoir evaporation	12
Exports	3
Municipal and industrial uses	1
Total	100

Source: Colorado River Salinity Program Coordinator (1993: Figure 1).

Natural sources

Significant natural salt loading occurs at various locations (Miller et al., 1986: 5; Liebermann et al., 1989: 61). For instance, Blue Springs in Arizona constitute the largest point source in the basin, yielding 500 000 t yr^{-1} of salt to the river. In the Glenwood Dotsero area of Colorado, 18 springs, known as Glenwood Springs, combine to discharge water containing 430 000 t yr^{-1} of salt into the river. Seepage of groundwater into the Dolores River in Colorado's Paradox Valley, which overlies a salt dome near the surface, adds about 200 000 t yr^{-1} of salt to the river. LaVerking Springs on the Virgin River in Utah adds 90 000 t of salt annually.

Irrigation

Salt loading occurs also due to the irrigation process. The groundwater system in many of the irrigated valleys in the Colorado River Basin is derived almost entirely from deep percolation of irrigation water and seepage from irrigation distribution systems (El-Ashry, 1980). Irrigation efficiencies in the basin are not very high and salt pick-up rates from irrigated soils and particularly underlying geological formations are generally significant. As reported by El-Ashry (1980), the dissolved solids yield from irrigated lands in the basin generally ranges from 0.2 t ha^{-1} yr^{-1} in the headwater areas to 20 t ha^{-1} yr^{-1} in some of the interior valleys. In most irrigated valleys, lands adjacent to the river or a tributary were the first to be irrigated. Later, water distribution systems were developed to irrigate lands at a higher elevation. As excessive water was applied to the higher elevation lands, percolated water from these lands caused high watertables and salt problems on the lower lands. In the Grand Valley this process has created waterlogging and salinity problems on irrigated lands adjacent to the river.

According to Liebermann et al. (1989: 61), where irrigation occurs on areas underlain by Mancos Shale large salt loads discharge to the Colorado River tributaries. The largest of these loads come from the Uncompahgre River (312 000 t yr^{-1}), the Price River (218 000 t yr^{-1}) and the San Rafael River

(174 000 t yr^{-1}). The Grand Valley area produces a salt load of 400 000 to 528 000 t yr^{-1}, almost entirely attributable to irrigation return flow. Irrigation of areas overlying the Uinta and Green River Formations contributes to a large salt load from the Duchesne River (319 000 t yr^{-1}) and the Big Sandy River (148 000 t yr^{-1}). Because of the importance of river salt pick-up in irrigated areas, two cases, in the Grand Valley and Wellton-Mohawk Irrigation and Drainage District, will be described briefly in the following paragraphs.

The Grand Valley in western Colorado near the Colorado-Utah border is a significant contributor to salt loading in the Colorado River. The soils of the valley are highly saline. The soils are derived from the Mancos Shale, a marine deposit with a very high percentage of soluble salt, mostly derived from the dissolution of gypsum. An analysis of the salt and water budgets in the valley by Inman et al. (1984), for a 29-year period from 1952 to 1980, shows that the mean annual salt pick-up is about 528 000 t yr^{-1} or 6 per cent of the 8.2 Mt annual average of salt loading to the Colorado River. Given that there are 26 300 ha of irrigated land in the valley, the average annual salt pick-up is therefore about 20 t ha^{-1}.

Irrigation water in the Grand Valley is applied in an unlined canal and earthen lateral system, and nearly all farmers in the area apply water using the furrow irrigation method (Walker et al., 1979). The groundwater system in the Grand Valley is derived almost entirely from deep percolation of irrigation water and seepage from irrigation distribution and tail-water collection systems (El-Ashry, 1980). Irrigation-related water entering the groundwater systems in the valley amounts to about 179–185 × 10^6 m^3 per year. The groundwater serves as a conduit for water flow to the Colorado River through the weathered Mancos Shale, the alluvium adjacent to the natural drainage and the Cobble Aquifer.

The Wellton-Mohawk Irrigation and Drainage District located east of Yuma comprises about 30 000 ha. Historically, the Wellton-Mohawk area was irrigated with Gila River water. Around 1915 the farmers began to irrigate with the local groundwater, but deep percolation increased the salinity of the groundwater to the point where it reached about 6000 mg L^{-1} in the 1940s. In 1952 the Gila Project delivered the first Colorado River water to district lands, with an average salt content of 820 mg L^{-1} (Sabol et al., 1987). Shortly after, the watertable rose and by 1957 crops were threatened by waterlogged soils. Some of the old irrigation wells were reactivated as drainage wells and others were constructed to remove the excess groundwater and lower the watertable. Drainage water was quite saline and averaged about 6000 mg L^{-1} TDS. The effect of discharge of this saline drainage into the Gila River, near its confluence with the Colorado River above the Mexican border, was that salinity of the water delivered to Mexico increased from an annual average of about 800 mg L^{-1} to nearly 1500 mg L^{-1}.

Salt load of the Gila River was about 454 000 t yr^{-1} (Hedlund, 1988a). In 1973 an Interagency Advisory Committee was organised to investigate the potential for reducing the return flow by improving irrigation efficiency. Since 1975 farmers have implemented the Interagency Committee Plan. The plan included installation of 420 km of concrete-ditch lining, 18 000 ha of laser land levelling, 1520 ha of soil improvement and retirement of 2500 ha of irrigated land. Implementation of the plan has reduced the irrigation return flow by half. Moreover, the Bureau of Reclamation has constructed the world's largest reverse osmosis desalting plant at Yuma to treat drainage water. It produces 100×10^6 m^3 yr^{-1} of water with a salinity of 295 mg L^{-1} TDS for delivery to Mexico (for further information about the Yuma Desalting Plant see the section on desalination on p. 68).

The successful application of a water management program in the Wellton-Mohawk Irrigation and Drainage District attracted countless water management specialists from all over the United States and several hundred visitors from more than 50 countries to see first-hand the excellent irrigation water management in this area (Hedlund, 1988a).

Evaporation

Another process of salt concentration is evaporation from water storage and conveyance systems. Almost 2.5×10^9 m^3 of annual evaporation occurs at the basin's reservoirs.

Export

Export or transfer of water from the basin reduces the amount of freshwater available in the basin and contributes to the process of salt concentration. Some 1×10^9 m^3 of water are exported to eastern Colorado, the Great Basin of Utah and the Rio Grande Valley of New Mexico (Miller, 1986: 7). Another 5.2×10^9 m^3 are exported from the Lower Basin, mainly to California. The Central Arizona Project began to divert additional water in 1985. This diversion will eventually grow to about 1.5×10^9 m^3. These diversions remove water that otherwise might dilute downstream salinity.

Municipal and industrial uses

Discharge of municipal and industrial uses contributes to the salinity of the Colorado River. This contribution is minimal and is estimated to be about one per cent.

Management options

It was a Lower Basin project that first made salinity a major issue. In 1961 saline concentration of about 6000 mg L^{-1} in the drainage water of the Wellton-Mohawk Irrigation and Drainage District in Arizona caused Colorado River water flowing to Mexico to reach 2700 mg L^{-1} (Miller et al., 1986: 24). Mexico expressed alarm and outrage at the resulting rise in salinity of Colorado River water it received. The United States negotiated a five-year agreement with Mexico in 1965 involving the bypassing of drainage water around Mexico's diversion point above Morelos Dam.

In 1971 the Environmental Protection Agency (EPA) advised the basin states to take the steps necessary to establish a numeric objective. The states supported the concept of maintaining salinity at 1972 levels, but they argued that the salinity should be regarded as a basin-wide problem. In 1973 the basin states organised themselves into a Colorado River Basin Salinity Control Forum. In 1976 the EPA and the states agreed on the numeric criteria listed in Table 11-9 for three Lower Basin locations.

On 24 June 1974 the United States Congress recognised the severity of the salinity problems and passed the Colorado River Basin Salinity Control Act (Public Law 93-320). The Act provides for the construction, operation and maintenance of certain works in the seven Basin States to control the salinity of water delivered to users in the United States and Mexico (Colorado River Salinity Program Coordinator, 1993: 4 and 5). Under the Act, the Secretary of the Interior is authorised and directed to proceed with a basin-wide program for enhancement and protection of the quality of Colorado River water. Title I of Public Law 93-320 enables the United States to comply with its obligations under the agreement with Mexico dated 30 August 1973. Title II of the Act, which is essentially the Colorado River Water Quality Improvement Program, provides for programs upstream of Imperial Dam necessary to stabilise the salinity of the Colorado River. Title II directed the Secretary of the Interior to expedite the completion of planning reports on 12 salinity control units and to proceed with the construction of the Paradox Valley, Grand Valley, Crystal Geyser and Las Vegas Wash Units.

Table 11-9: Salinity criteria and observed salinity at three locations in the Lower Colorado River Basin.

Location	Salinity criteria (mg L^{-1} TDS)	1972 salinity (mg L^{-1} TDS)
Below Hoover Dam	723	724
Below Parker Dam	747	734
Imperial Dam	879	861

Source: Miller et al. (1986: Table 2).

On 30 October 1984, Public Law 93-320 was amended to Public Law 98-563. This law amended the original salinity control program by authorising construction of additional units by the Bureau of Reclamation and de-authorising Crystal Geyser because of poor cost-effectiveness. The Secretary of Agriculture was directed to establish a major voluntary on-farm cooperative salinity control program. The authorising legislation provides for cost-sharing and technical assistance to participants for planning and installing needed salinity reduction practices. Participants pay at least 30 per cent of the costs to install salinity reduction facilities. The Public Law directed the Bureau of Land Management to develop a comprehensive program for minimising salt contributions from 19.5 Mha of basin lands which it administers.

The present program of activities for controlling salinity includes a variety of projects designed to reduce salt input from return irrigation flow, from localised point sources such as saline springs or abandoned oil wells, and from diffuse sources. This program involves state governments and federal government agencies.

At the state level, all seven states have joined efforts to adopt standards and to implement a plan to meet the criteria. To accomplish needed coordination, the Governors of the Basin States appointed representatives to the Colorado River Basin Salinity Control Forum and the Colorado River Basin Salinity Control Advisory Council.

At federal level salinity control requires coordination among various agencies including the US Geological Survey, the Bureau of Land Management, the Bureau of Reclamation, the Environmental Protection Agency and the US Department of Agriculture and its agencies (e.g.: Soil Conservation Service; Agricultural Stabilization and Conservation; Agricultural Research Service; Cooperative State Research Service and Extension Service). The federal agencies are coordinated through an Interagency Salinity Control Coordinating Committee (see Colorado River Salinity Program Coordinator, 1993, for further information on the Colorado River Basin Salinity Control Forum and Advisory Council).

The involved federal agencies, working in close cooperation with the Colorado River Basin Salinity Control Forum, have identified potentially cost-effective, viable salinity control units that will be implemented over the next two to three decades. The collective efforts of the US Bureau of Reclamation, the US Department of Agriculture and the Bureau of Land Management are summarised in Table 11-10. The location of the salinity control units is illustrated in Figure 11-7.

Table 11-10 shows that federal projects completed or under construction at January 1993 remove 237 500 t yr^{-1} of salt. The designed units will remove 1.25 Mt yr^{-1} of salt by the year 2015. Implementation of the Price-San Rafael Project, which by June 1993 had not received congressional authorisation, will remove 75 500 t yr^{-1} of salt by the year 2015 and 146 000 t yr^{-1} when the project is complete. Apart from the units in Table 11-10, several other units are under consideration or needing congressional authorisation. Also some units, such as the Dirty Devil and Palo Verde Irrigation Districts, have been investigated but currently are not being given further consideration.

Important components of the plan of implementation for salinity control are the Basin States' activities associated with the control of total dissolved solids through the National Pollutant Discharge Elimination System (NPDES) Permit Program and the Water Quality Management Plans.

Table 11-10:   Federal programs for salinity control in the Colorado River Basin.

Unit and Federal Agency	Begin implementation or complete	Project date of completion	Salt removal at Jan 93 (t yr^{-1})	Projected salt removal by 2015 (t yr^{-1})	Cost effectiveness ($ t^{-1} yr^{-1})
Meeker Dome (USBR)	Complete	1983	43 500	43 500	15
Grand Valley Stage One (USBR)	Complete	1984	19 900	19 900	133
Las Vegas Wash Pittman (USBR)	Complete	1985	3 500	3 500	27
Grand Valley (USDA)	1979	2010	51 400	148 000	30
Uinta Basin (USDA)	1980	2010	50 400	97 000	88
Nonpoint Sources (BLM)	1983	2015	2 500	37 200	33
Well Plugging (BLM)	1984	2015	7 300	12 700	22
Grand Valley Stage Two (USBR)	1985	1998	23 200	105 000	125
Paradox Valley (USBR)	1988	2001	0	163 300	54
Big Sandy River (USDA)	1988	2003	11 300	48 000	30
Lower Gunnison (USDA)	1989	2016	22 400	252 200	77
McElmo Creek (USDA)	1990	2009	2 100	34 500	92
Lower Gunnison Winter Water (USBR)	1991	1994	0	67 100	42
Dolores Project (USBR)	1991	1995	0	20 900	93
Moapa Valley (USDA)	1994	2006	0	17 000	42
Lower Gunnison-Laterals (USBR)	1994	2007	0	54 400	66
Uinta Basin 1[a] (USBR)	–	2004	0	23 100	97
San Juan-Hammond[a] (USBR)	–	2007	0	25 400	41
Price-San Rafael[a] (USBR/USDA)	–	2021	0	75 500	43
Total			237 500	1 248 200	–

USBR: US Bureau of Reclamation
USDA: US Department of Agriculture
BLM: Bureau of Land Management
(a):   Units that have been investigated and are in the Salinity Control Plan of Implementation, but require congressional authorisation.

Source: Colorado River Basin Salinity Control Forum (1993: Table 4.1).

Most of the salinity control projects consist of irrigation water management with the objective of reducing highly saline irrigation return flow. In these cases management options include: lining of irrigation canals; installation of pipe laterals; land levelling; and installation of sprinkler and drip irrigation systems. Control of saline springs and plugging of oil and gas wells are in practice in the Bureau of Land Management Units.

Paradox Valley Unit involves pumping saline groundwater and thereby lowering the watertable and reducing saline inflows to the Dolores River. The pumped brine is injected into a deep well in the Paradox Valley. The injection test well, the brine pipeline, the surface treatment building and the injection building were completed in June 1993. System testing and shakedown are scheduled to be completed in 1995 (see p. 65, the section on 'Deep-well disposal', for a brief description of the unit).

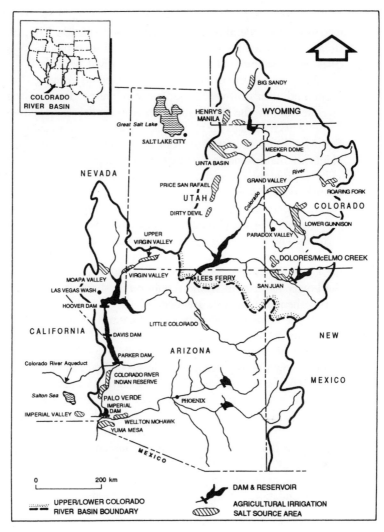

Figure 11-7: Agricultural irrigation salt source areas (Hedlund, 1984) and the Colorado River Basin salinity control units (Colorado River Basin Salinity Control Forum, 1993: Figure 4.1).

Further information about the federal and state programs is available in Colorado River Basin Salinity Control Forum (1993) and Soil Conservation Service (1977, 1981 and 1983) for Grand Valley, Lower Gunnison and McElmo Creek Units, respectively. A brief description of some of the projects is also available in Howe and Ahrens (1988).

Financing of the salinity control activities is an important issue. According to Colorado River Basin Salinity Control Forum (1993: 6-2) and

Miller et al. (1986: 44 and 45), in enacting Public Law 93-320, Congress recognised the federal responsibility for the Colorado River as an interstate stream. It adopted a cost-sharing formula which provides 75 per cent of the costs of the salinity control projects authorised by Title II of the Act. The remaining 25 per cent of the costs are to be repaid from the Upper and Lower Basin funds over a 50-year period without interest. The two basin funds draw revenue from the the Bureau of Reclamation hydroelectric power sales. Of the 25 per cent charged to the two funds, 15 per cent was to come from the Upper Basin fund and 85 per cent from the Lower Basin fund. Although power plants are owned and for the most part financed by the federal government, salinity program costs charged against these funds are regarded as state contributions because salinity control costs are built into rates paid by electricity consumers in the affected states who buy the power.

The 1984 amendment to Public Law 93-320 changed the cost-sharing formula. The federal government share was reduced to 70 per cent, with the remaining 30 per cent to come from Upper and Lower Basin funds. However, the Upper Basin fund would repay its share over 50 years with interest and the Lower Basin fund would reimburse its share of the annual expenditure during the year that costs are incurred.

The voluntary on-farm salinity control program of USDA requires a minimum 30 per cent cost-share from the local participants unless the Secretary of Agriculture finds that such cost-sharing requirements would result in a failure to proceed with the needed on-farm measures.

The success of federal/state cooperation via the Colorado River Basin Salinity Control Program is contingent upon sufficient funding to allow the plan of implementation to proceed as scheduled. In each fiscal year the Colorado River Basin Salinity Control Forum (Forum) urges Congress to provide the Bureau of Reclamation, the Bureau of Land Management and the Department of Agriculture with the adequate funds to implement the authorised salinity control program. Table 11-11 shows the Forum's funding recommendation and the federal appropriation for the 1993 year.

Table 11-11: Colorado River Salinity Control Program funding for the fiscal year 1993.

Agency	Forum recommendation ($ million)	Federal appropriation ($ million)
Bureau of Reclamation	36.9	33.8
Bureau of Land Management	7.3	0.9
Department of Agriculture	18.5	13.8
Total	62.7	48.5

Source: Colorado River Basin Salinity Control Forum (1993: Table 1.1).

Economic losses

Salinity causes millions of dollars of damage to water users in Colorado, Utah, Arizona, Nevada, California, Wyoming, New Mexico and the Republic of Mexico. The estimated 1986 economic losses in the Lower Colorado Basin are about US$311 million (Colorado River Salinity Program Coordinator, 1993: 3). About 50 per cent of the economic losses (US$156 million) are associated with household use and 3 per cent (US$9 million) with the industrial sector. The losses are primarily from increased water treatment costs, accelerated plumbing replacement and appliance wear, increased soap and detergent needs and decreased water palatability. Agricultural damages are estimated at US$113 million (36 per cent). For irrigation, the higher concentrations cause decreased crop yields, alter cropping patterns, necessitate higher leaching and drainage and raise management costs.

With the current salinity concentration of the Colorado River, a recent report indicates that economic losses for the water users in the Lower Basin are estimated to be over US$750 million per year, which is more than double the estimate for 1986 (Colorado River Basin Salinity Control Forum, 1993: 2–8). This damage could double by the year 2015, if the proposed plan for salinity control is not implemented.

San Joaquin Valley

The Central Valley of California is about 640 km long and 80 km wide. The valley contains the Sacramento Valley in the north and the San Joaquin Valley in the south. The Sacramento and San Joaquin Rivers are the largest and second largest rivers in California. The Sacramento Valley occupies the northern 40 per cent of the Central Valley. Rainfall and frequent flooding prevent the build-up of salt in the Sacramento Valley's soils except at localised sites.

Because of its climatic characteristics and fertile soils, the San Joaquin Valley is one of the most productive and diverse agricultural areas in the world. Over 200 different crops are grown commercially with 125 of these contributing significantly to the food supply and economy of the area, the state of California and the nation (San Joaquin Valley Drainage Program, 1987: 8). For many decades the San Joaquin Valley has been faced with the problem of rising watertables and salinisation. The discovery of migratory bird deaths and deformities linked to high selenium levels in agricultural drainage water in 1983 aggravated the situation. In the following sections some aspects of these problems are discussed.

Surface water

Annual rainfall in the Valley averages about 125 to 355 mm from the south to the north. Rainfall provides a minor portion of the Valley's total water supply. Additional water supplies average 14.8×10^9 m^3 annually. Of this, 5.18×10^9 m^3 are from groundwater, 5.18×10^9 m^3 from surface water and

4.44×10^9 m³ imported from north California through the Delta to the west side of the Valley. About 95 per cent of total water use is for agriculture (San Joaquin Valley Drainage Program, 1987: 12).

The water resources of the San Joaquin Valley have been subject to intensive development over the past 60 years. During this period all of the major rivers in the Valley, which have a combined natural annual run-off of more than 7×10^9 m³, have been regulated for power generation, water supply, flood control and irrigation (Orlob and Ghorbanzadeh, 1981). Consumptive use of water, principally by irrigated agriculture, has increased steadily with the most dramatic changes occurring with the advent of the Central Valley Project (CVP) in the 1940s. One of the main features of this project is the Delta-Mendota Canal which supplies irrigation water to the semiarid western side of the Valley (Figure 11-8).

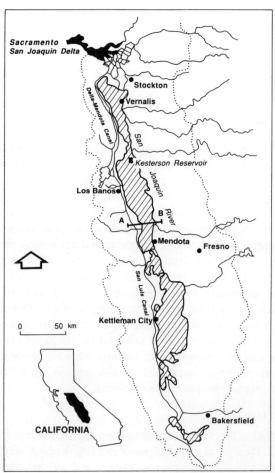

Figure 11-8: San Joaquin Valley drainage problem area (Coppock, 1988c: Figure 1).

As a result of exploitation of the Upper San Joaquin River and the subsequent development of saline lands along the Valley's western side, the main system downstream of Mendota has been deprived of the natural runoff which it received prior to the CVP. At the same time it has become the principal agricultural drainage course of the Valley. The progressive diminution of run-off and deterioration in its quality have seriously impacted agriculture in the northern portion of the Valley, particularly in the southern section of the Sacramento-San Joaquin Delta for which the San Joaquin River is the unique source of water supply.

Changes in water quality are attributed to many factors. The most prominent factor appears to be reduction in the dilution flow of natural runoff and the increase in salt accretion to natural drainage courses from irrigation. Analysis of mean monthly salinity data at Vernalis at the northern extremity of the Valley since the 1930s shows the increasing trend of salinity, from a maximum value of about 330 mg L^{-1} TDS in the 1930s to about 600 mg L^{-1} TDS in the 1970s (Orlob and Ghorbanzadeh, 1981: Figure 4).

Groundwater

Geologically, the Central Valley is a large asymmetric north-westward trending structure. It has been filled with about 10 km of sediments in the San Joaquin Valley. These sediments range in age from Mesozoic to Quaternary and include both marine and continental deposits (Page, 1986). Marine deposits of pre-Tertiary age crop out along most of the western flank of the Valley, and marine sediments of Tertiary age crop out along the western, south-western, southern and south-eastern flanks of the Valley. The post-Eocene continental deposits contain most of the fresh groundwater in the Central Valley. The continental deposits consist of gravel, sand, silt and clay.

There are important differences between the soils of the two valley floor slopes of the San Joaquin Valley. Soils on the west side of the Valley tend to contain large amounts of soluble salts and trace elements, because they are derived from marine sedimentary rocks. Salinity and contaminant problems are most pronounced in these fine-textured soils and in the groundwater on the west side of the Valley. In contrast, on the east side the soil tends to be more coarse in texture and almost free of salt and trace elements (San Joaquin Valley Drainage Program, 1987: 12). The Sierran Sand on the east side of the Valley contains groundwater of good quality and is essentially low in selenium.

Figure 11-9 shows a generalised hydrogeologic cross-section of the Valley. The Corcoran Clay layer of Pleistocene age lies approximately at a depth of about 120 to 180 m below the land surface. It is an important geological feature which divides the groundwater flow systems.

Three bodies of groundwater exist in the western San Joaquin Valley. First, there is an unconfined and semiconfined freshwater aquifer above the Corcoran Clay. Second, there is a confined freshwater aquifer beneath the

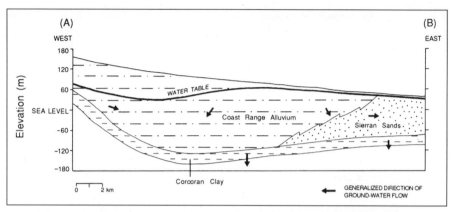

Figure 11-9: Generalised hydrogeologic cross-section of the San Joaquin Valley (Belitz, 1988: Figure 24). Location of the cross-section is shown on Figure 11-8.

Corcoran Clay. Third, a saline body of water underlies the confined aquifer. The groundwater flow system of the western San Joaquin Valley has undergone considerable change since the development of irrigated agriculture (Belitz, 1988: 10). The current flow system is in a transient state and is responding to both historic and present stresses imposed upon it. Activities that have affected the semiconfined aquifer include percolation of irrigation water, historic pumping from below Corcoran Clay, delivery of surface water and installation of subsurface tile drainage. Under natural conditions, recharge was primarily from infiltration of stream water from intermittent streams and perhaps from smaller ephemeral streams. Rainfall was an insignificant mechanism for recharge. Discharge was primarily by evapotranspiration and stream flow along the valley. Development of agricultural activities in the area increased significantly with groundwater pumping from the confined aquifer, which altered the groundwater system. Percolation of irrigation water became a major source of recharge. It has been estimated that post-development recharge during 1961–77 was more than 40 times greater than the estimated pre-development values (Belitz, 1988: 14).

Irrigation, drainage and salinity

After the discovery of gold in 1848 in Sierra Nevada, irrigation began to be practised in the San Joaquin Valley. As early as 1886 a widespread need for agricultural drainage in the Valley was evident. In the 1890s and early 1900s some cultivated lands were forced out of production because of salt and drainage problems (Hartshorn, 1986: 11). The invention of the turbine pump in the 1920s and the consequent pumping of groundwater expanded irrigated agriculture further into the west side of the San Joaquin Valley. Irrigated agriculture continued to flourish in spite of some setbacks with salinity and drainage problems in localised areas. In the mid-1950s the California Department of Water Resources noted the increasing degradation of water quality in the middle reach of the San Joaquin River caused by irrigation return flows and the absence of water release from dams (Tanji et al., 1986). Development of irrigation has caused the rise of watertables. Approximately 200 000 ha of west side agricultural lands are currently affected by groundwater levels that have risen to within 1.5 m or less of the land surface (San Joaquin Valley Drainage Program, 1987: 8).

Irrigation adds salts to the soil. In the San Joaquin Valley, according to one estimate, about 6.3 Mt salts are introduced into farmlands by irrigation water every year (Coppock, 1988a: 3). Nearly half of this salt load comes in water imported through the Central Valley Project. Another major source is the chemical weathering of soil minerals, particularly gypsum, which is common in the west side.

The Federal Service Area consists of 486 000 ha of irrigated lands (Tanji et al., 1986). This area is made up of about 243 000 ha located between the Sacramento-San Joaquin Delta (Figure 11-8) and a point south of Los Banos (Delta-Mendota Service Area), and another 243 000 ha between Mendota and Kettleman City (San Luis Service Area). The northern portion of the Federal Service Area drains irrigation return flows to the middle reaches of the San Joaquin River. Some of this drainage water is used in wetland habitats. The southern portion has no existing drainage outflow to the San Joaquin River. Except for the period between 1978 and June 1986, subsurface drainage water was discharged into the San Luis Drain with Kesterson Reservoir as the terminal evaporation pond (Kesterson Reservoir, initially planned as a flow-regulation reservoir, consists of 12 ponds with an average depth of 1.2 m and a total water surface area of 386 ha). Drainage of the irrigated lands is a major problem in the San Joaquin Valley. In 1985 about 34 000 ha were being drained while an additional 68 000 ha required drainage (Tanji et al., 1986). Figure 11-8 shows the areas of the Valley with drainage problems.

The concrete-lined San Luis Drain, built in 1971, conveyed surface run-off to the Kesterson Reservoir between 1971 and 1977. In 1978 discharge of subsurface drain water was added to some surface run-off. From 1981 the San Luis Drain conveyed water exclusively from tile drainage.

For many years San Joaquin Valley farmers hoped for a master drain. In the 1970s the Interagency Drainage Program was formed and investigated possibilities for valley drainage which included a master drain northward to Suisun Bay (for further information see the section on 'Disposal to seas and lakes' on pages 61–62). A feasibility study of the disposal method was commenced, but in 1983 the situation changed suddenly. Discovery of mosquito fish (*Gambusia*) with high selenium levels and selenium-deformed waterfowl at Kesterson revealed the double-barrelled threat of high watertables and salinity on the one hand and toxic trace elements on the other (Coppock, 1988a). Finally in March 1985 the Department of Interior called for the cessation of all drainage flows to the Kesterson Reservoir by 30 June 1986.

Selenium in the San Joaquin Valley environment

In July 1982 high selenium concentrations were discovered in mosquito fish in Kesterson Reservoir. The concentration was 100 times greater than that in similar fish at the nearby wildlife refuge which receives a mixture of surface run-off and water from the Central Valley Project. In samples drawn between August and December 1983 from Kesterson Reservoir and San Luis Drain, selenium concentration ranged from 140 to 1400 ppb (parts per billion). Further sampling in 1984 showed that the samples also contained significant concentrations of boron, chromium, iron, lead, lithium, molybdenum, vanadium and zinc. Another sampling of 130 shallow groundwater sites in the Delta-Mendota and San Luis Service Areas confirmed the presence of abnormally high concentrations of selenium.

The US Geological Survey tested 433 water wells. Of these, 16 exceeded the federal guideline of 20 ppb for selenium in irrigated water and 37 exceeded the 10 ppb standard for drinking water (Coppock, 1988b: 4). However, these wells are used primarily for irrigation and stock watering, not domestic water supply.

Table 11-12 summarises the concentration of selected trace elements from 40 water sampling sites on the west side of the Valley (see San Joaquin Valley Drainage Program, 1987: Figure 8 for the location of the areas with the highest concentrations of selenium). These data were collected by the US Bureau of Reclamation and the US Geological Survey during 1984–86 in the Federal Service Area and Kings County (see Deverel et al., 1984 and Deverel, 1989 for further information). The problem of selenium concentration is in fact part of a drainage problem that involves waterlogging and associated accumulation of excess salt in the irrigated crop root zone (Tanji et al., 1986).

Selenium coexists with sulphur in pyrite. Selenium is also commonly associated with sulphate in minerals like gypsum. In the environment of the western San Joaquin Valley, the most common chemical form of selenium is selenate (SeO_4^{2-}), which is highly mobile. However, another form of selenium called selenite (SeO_3^{2-}) is less mobile and adsorbed to soil more tightly.

The pathways through which selenium accumulates in shallow groundwater are not entirely clear. One hypothesis is that, in the Moreno Shale of the Pacific Coast Range Mountains, seleniferous pyrite seams were exposed on the land surface and subjected to oxidation, and that soluble forms of selenium were transported by natural run-off. Surface soils have a low selenium concentration because of leaching from natural rainfall and/or applied irrigation (Tanji et al., 1986). However, geostatistical analysis of groundwater and soil samples shows that the highest soil salinity and selenium concentrations occur in alluvial fans deposited by ephemeral streams and at the margins of the alluvial fans deposited by intermittent streams (Deverel, 1989; Deverel and Gallanthine, 1989).

Table 11-12: Concentrations of trace elements in drainage water samples.

Constituent	Concentration (parts per billion)		
	Minimum	Maximum	Median
Arsenic	<1	50	<1
Boron	40	84 000	6500
Cadmium	<1	4	<1
Chromium	<1	800	2
Copper	<1	510	6
Iron	<3	360 000	250
Lead	<1	73	<1
Manganese	<1	9 000	80
Mercury	<0.1	1.6	<0.1
Molybdenum	<1	1 500	13
Nickel	<1	900	16
Selenium	<1	4 700	25
Silver	<1	10	<1
Zinc	<3	1 000	11

Source: San Joaquin Valley Drainage Program (1987: Table 4).

Preliminary human epidemiological surveys in the Kesterson area have not shown selenium poisoning. The cumulative environmental impacts of other trace elements are not well known. Historically most evidence of both selenium toxicity and deficiency involved livestock and wildlife. Human effects are little known, probably because of a more varied diet involving widely separated food source materials. The average daily intake of selenium in US diets is between 60 and 176 micrograms per day. The US National Academy of Science suggested a range of 50 to 200 micrograms per day of selenium as acceptable (Coppock, 1988b: 6). A survey by the University of California in the San Joaquin Valley found that no food produced under commercial conditions contained selenium at concentrations considered hazardous to health (Coppock, 1988b: 8).

Management options

For decades low-lying San Joaquin Valley farmlands have been threatened by rising watertables and increasing salinity. Management of these problems has become more complex since the discovery of high levels of selenium and the subsequent closure of Kesterson Reservoir and plugging of the drain lines connected to it. In the San Luis Service Area, which was served previously by the San Luis Drain and Kesterson Reservoir, farmers are now either reusing saline water from farm drains or disposing it in evaporation ponds. Further north in the Delta-Mendota Canal Service Area, where agricultural drainage traditionally has flowed into the San Joaquin River, farmers are under increasing pressure to reduce or eliminate that discharge. In order to meet the objective of reducing the selenium content of public surface waters to less than 5 ppb, drainage water would need to be reduced significantly.

It is estimated that tile drains in the area average 2110 m^3 of outflow per hectare annually. To meet the river water quality objective, the output would have to drop to about 1360 m^3 ha^{-1} (Coppock, 1988c: 2). Such drastic changes are mandated only where selenium might reach the river. However, similar reductions in farm output of drainage would do much to ease the serious high watertable and salinity problems elsewhere in the Valley. In fact the most promising immediate approach to the Valley's overall salinity and selenium problem is to reduce the flow of drainage from individual farms by at least half. On-farm water management, improved irrigation systems, reuse of drain water as long as it is useable and blending drainage water with good quality water will reduce the volume of drained water. These ideas of on-farm source control are gaining support. However, these approaches are not the final answer to the problem because they leave unsolved the problem of ultimate disposal.

Several potential sites for disposal of drainage water have been investigated. In the screening process, each site was evaluated on the basis of a number of hydrological, physical, geological, economic and environmental criteria. Finally efforts were focused on in-valley solutions to drainage water disposal problems. Three in-valley disposal options have been evaluated: evaporation ponds; deep well injection; and discharge to the San Joaquin River (San Joaquin Valley Drainage Program, 1987: 27).

Evaporation ponds are presently a common means of drainage water disposal in the Valley. As of 1987, 24 pond systems covering approximately 2700 ha were operating and many more were at a planning stage. The authorities have received applications for ponds covering about 4250 ha and they have estimated that the area requested for ponds could double in the next five to ten years. However, several problems are associated with evaporation ponds, such as: groundwater degradation due to leakage; side effects on the environment; considerable land requirements (1 to 2 ha for 10 ha of drained land); and the ultimate problem of disposal of accumulated salts (San Joaquin Valley Drainage Program, 1987: 28).

Deep well injection is a technology that has been widely used in the Central Valley for many years by the oil and gas industry as a means of disposal of oilfield brines. More limited use of deep well injection has been made for ultimate disposal of toxic or hazardous wastes. An appraisal level study (URS Corporation, 1986) concluded that deep well injection of agricultural drainage water may have merit and warrants additional investigation, such as through pilot studies to determine feasibility.

With respect to disposal to the San Joaquin River, about 31 000 ha of farmland on the west side of the Valley currently have subsurface drainage systems that eventually discharge into the river. The continuation of drainage discharge to the river will depend on water quality objectives. These objectives will serve to prescribe the total load of salts and trace elements (selenium, boron and molybdenum) that can be discharged into surface water bodies.

The desalination option was also investigated. A conceptual level study for a 13.8×10^6 m^3 yr^{-1} reverse osmosis (RO) desalting plant, to treat San Luis drainage water, was undertaken by CH2M Hill, Inc. (1986) as the consultant to the US Bureau of Reclamation. The product water will have a salinity of 550–650 mg L^{-1} TDS and produce a brine with salinity of 55 000 to 67 000 mg L^{-1} TDS. The product water will be virtually free of selenium (greater than 99 per cent removal) and trace elements, but boron will probably not be reduced below 7 to 8 mg L^{-1}. Therefore, reuse of the product water for agricultural uses will require blending with a low boron water supply. The combined wastes from the pretreatment and RO processes will be disposed of in waste ponds. The ponds will be about 113 to 210 ha depending on the type of pretreatment schemes. The desalting plant, including the intake structure and feed reservoir and excluding the waste ponds, will cost US$21.2 to US$28.1 million.

Agroforestry is another option being considered to help the San Joaquin Valley farmers in managing their water and salinity problems. According to the California Department of Food and Agriculture (1988), 21 trial plantations, ranging in size from less than 0.4 ha to more than 11 ha and totalling 73 ha, were planted in 1985–1986 and 1987 in Fresno, Kings and Kern Counties. Major plantings of tree species were *Eucalyptus camaldulensis*, *Casuarina glauca* and *Casuarina cunninghamiana*. Eucalypts were imported from Australia (Lake Albacutya and Alice Springs areas) and casuarina seeds were obtained from both Australia and Egypt. Freshwater was used to irrigate the seedlings during the first year in order to establish the trees. Thereafter, the trees received saline drainage water and/or they directly utilised groundwater from high watertables. Analysis of the collected data indicated that: drainage water can be used to irrigate the trees; once established, trees can use groundwater from high watertables and do not need to be irrigated; trees offered an opportunity for economic use of land; tree production may be a low-cost operation, especially considering the economic aspects of drainage water management; tree plantations attract wildlife; and plantation of trees and salt-tolerant crops may provide an opportunity to diversify production and marketing options.

On-farm options to reduce drainage water volume have been investigated. Numerous irrigation practices, technologies and policies have high potential for drainage reduction (Agricultural Water Management Subcommittee, 1987). These include: modifying existing irrigation practices by developing trickle and sprinkler irrigation systems, irrigation scheduling, land levelling and irrigation with diluted drainage water; appropriate institutional changes and economic incentives such as water marketing, land use change to agroforestry or wildlife habitat, and cost-sharing incentives to growers; implementing educational and informational programs such as developing 'action teams' to educate farmers about drainage reduction practices and programs, and developing materials to support 'action teams'; improving the monitoring of applied drainage water quality and quantity; and researching methods to reuse shallow groundwater.

Further detailed information on drainage, salinity and toxic trace elements in the San Joaquin Valley is available in Boyle Engineering Corporation (1986), Tanji et al. (1987), Lord J.M. Incorporated (1987), Committee of Consultants on Drainage Water Reduction (1988a and b), Committee of Consultants on San Joaquin River Water Quality Objectives (1988a and b), Albasel and Pratt (1989), Jacobs (1989), Albasel et al. (1989), Tracy et al. (1990), Bradford et al. (1990) and Hedlund (1993).

Dryland salinity in the Northern Great Plains

Dryland salinity in the Northern Great Plains, known as saline-seep, was hardly recognised some 50 years ago. The saline-seep problem is caused by the geology of the region, high precipitation periods and farming practices. The major culprit is the change in vegetation from the original grassland to grain-fallow rotation practice. Saline-seeps have developed in the states of Montana, North Dakota, South Dakota and the Canadian Provinces of Alberta, Saskatchewan and Manitoba (Figure 11-10). These are the major grain growing regions in Canada and the United States. Saline-seeps also exist in Colorado, Oklahoma and Texas (Hedlund, 1988b: 29).

Throughout most of the Palaeozoic and Mesozoic eras, thousands of metres of predominantly marine sediments were deposited in this region (Miller et al., 1981). At the end of the Mesozoic and Early Tertiary, the entire area was uplifted, faulted and folded. Continuous erosion during the Tertiary stripped away the uppermost Cretaceous sediments, exposing vast areas of predominantly salt-laden marine shales of Cretaceous age. The erosion products were subsequently redeposited further east as non-marine shale, silt and sand of Tertiary age. As a result, there remains a bedrock landscape essentially composed of four major units: Colorado Shale (oldest); Judith River, Claggett and Eagle Formations; Bearpaw Shale; and Tertiary Fort Union Formation (youngest). The overall soluble salt load contained in each geological unit is reflected in the concentration of total dissolved solids contained in the groundwater at the discharge area. In general, concentrations of salts vary from 20 000 to 55 000 mg L^{-1} TDS in the Colorado and Bearpaw Shale Units; 10 000 to 25 000 mg L^{-1} TDS in the Judith River, Claggett and Eagle Formations; and 3000 to 15 000 mg L^{-1} TDS in the non-marine Fort Union Formation (Miller et al., 1981).

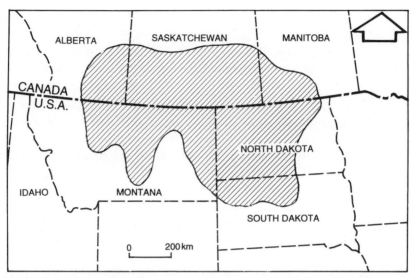

Figure 11-10: Area of potential saline-seep development on the Northern Great Plains (Miller et al., 1981).

During Pleistocene much of the Northern Great Plains region was covered one or more times by glacial ice. The glaciation left a mantle of unconsolidated poorly sorted deposits called glacial till or drift, consisting of clay, silt, scattered sand and gravel lenses and well-rounded pebbles. These glacial deposits filled most of the pre-existing valleys and left a relatively flat to gently rolling glacial plain. In Montana the thickness of glacial till ranges from 1 to 25 m, while till thicknesses exceeding 100 m are not uncommon in Canada. Except for the upper metre, the entire till profile contains abundant calcium and sodium salts.

In the Northern Great Plains, the clay till provides an excellent soil moisture reservoir for dryland farming and the dominant cropping system over the entire region is the alternate grain-fallow farming system. In this region annual precipitation ranges from 250 mm in the western and north-central regions to 500 mm in the eastern and north-eastern regions. About 70 per cent of precipitation is received during the April through August growing season. Although the September through March period accounts for only 30 per cent of the annual precipitation, it frequently accounts for 60 per cent or more of the total soil water stored during the fallow period (Black et al., 1981). Precipitation that penetrates below the root zone during the fallow period is the primary source of the water that mobilises salts and causes most saline-seeps. This problem is aggravated during periods of high precipitation. Other factors contributing water to saline-seeps include: poor surface drainage, snow accumulation, leakage from constructed ponds, artesian water, roadbeds across natural drainage ways, and crop failure (Brown et al., 1983).

Brown et al. (1983) describe seven common types of saline-seeps. These are:

- **Type 1: Geologic Outcrop Seep.** The recharge area is underlain by geologic material of low hydraulic conductivity (HC), such as shale, dense till or clay. The soil material above the low HC layer varies from sand to silty clay.

- **Type 2: Coal Seam Seep.** The recharge area is underlain by coal or lignite which overlies a dense clay. The soil material above the coal seam varies from sandy loam to silty clay loam, and there is no glacial till mantle. Lateral water movement, through the coal related material, is more rapid than with Type 1 seep. The seep occurs where the coal material outcrops at the surface or is truncated by the landscape.

- **Type 3: Glaciated Front Union Seep.** The recharge area is glacial till underlain by sandstone, siltstone, lignite, and dense clay strata of the Front Union Formation. Water from the recharge area passes through the glacial till and enters the more permeable strata to form a watertable above the low HC zone. Water from the watertable moves downslope to a point where the glacial till or low HC truncates the permeable zone. This causes the watertable to the rise to the surface, forming a saline-seep.

- **Type 4: Textural Change Seep.** The recharge area is underlain by geologic material having a low HC. The soil material above the low HC zone is coarse to medium textured. Water moves through the root zone to the low HC zone and laterally downslope where it encounters a soil zone of lower HC, which slows movement and causes the watertable to rise to the soil surface.

- **Type 5: Slope Change Seep.** The recharge area is underlain by geologic material of low HC. The soil material above the low HC zone is variable in texture. Water moves through the root zone to the low HC zone and laterally downslope to a point where the slope gradient decreases. The reduced gradient causes the water movement to slow and the watertable to rise to the soil surface where it produces a saline-seep.

- **Type 6: Hydrostatic Pressure Seep.** The recharge area is underlain by geologic material of low HC. The soil material above the dense layer is variable in texture. Water moves through the root zone to the low HC zone and laterally downslope to a point where it becomes confined by a low HC zone located above the saturated zone. The confined water is under hydrostatic pressure, which often forces the water to the surface to cause a saline-seep.

- **Type 7: Pothole Seep.** The recharge area has potholes or poorly drained areas underlain by low permeable material. The water moves laterally through a zone of higher HC which outcrops at or near the soil surface to form a saline-seep.

In the entire region, some 95 per cent of saline-seeps occur in the glaciated area in Montana. However, several large outbreaks of dryland salinity have appeared in the unglaciated areas characterised by shallow soils derived from underlying Colorado and Bearpaw Shale and Fort Union Formation, on which the alternate grain-fallow farming system is extensively used (Miller et al., 1981).

A 1978 estimate reported by Miller et al. (1981) shows that about 0.8 Mha of dryland farms in the Northern Great Plains were severely affected. This translates to US$120 million of lost annual farm income (Brown et al., 1983). Possibly another 2.7 Mha were also affected to some extent. Estimates of saline-seep affected areas in Montana have risen from 32 000 ha in 1971, to 57 000 ha in 1974, to 69 000 ha in 1976 and 81 000 ha in 1978. However, according to the US Department of Agriculture (1989: 54–55), a 1987 estimate by the Montana Salinity Control Commission set Montana's saline-seep area to 121 000 ha. The US Department of Agriculture (1989: 54–55) also provides an estimate of saline-seep in North and South Dakota of about 58 100 ha in each of the two states for 1981.

In the areas affected by saline-seep, surface and shallow groundwater resources, which in many areas are the primary source of potable water, are affected by increasing salinity and concentration of trace elements. Miller et al. (1981) report a 1978 regional water quality survey encompassing 305 000 km^2 in 42 counties of Montana. Of the 2800 sites evaluated and 452 samples collected, 14 per cent had TDS less than 500 mg L^{-1}; 16 per cent 500–1000 mg L^{-1}; 64 per cent 1000–10 000 mg L^{-1}; and 6 per cent had a salinity of more than 10 000 mg L^{-1} TDS. Significant concentrations of trace elements, particularly selenium, were found in many groundwater samples. Of the 161 samples analysed for selenium, 40 per cent had concentrations greater than 0.01 mg L^{-1} (the limit for potable water). Some values were as high as 2.0 mg L^{-1}. The selenium values were considerably higher in the Colorado Shale region, having a mean value of 0.308 mg L^{-1}, compared to the Judith River region, which had a mean concentration of 0.028 mg L^{-1}. Also numerous fish and livestock kills have been reported throughout the region, particularly in the Colorado Shale area. Most of the fish kills occurred when TDS increased to about 7000 mg L^{-1}.

As mentioned previously, saline-seeps are caused by agricultural practice and climatic and hydrogeological factors. Because the climatic and geological factors are natural and essentially unchangeable, solutions to the problem have to be concentrated primarily around cropping strategies. According to Black et al. (1981), precipitation received in excess of the soil storage capacity, primarily during fallow, is the source of water. One method of controlling saline-seeps involves the establishment of a perennial deep-rooted crop, such as alfalfa (*Medicago sativa* L.), on the recharge area for enough time to use existing and anticipated available water supplies to maximum possible soil depth. Another method is to establish intensive, flexible cropping systems using adaptable crops in combination with proper soil water and crop management practices to improve crop production. Black et al. (1981) discuss the soil, water and crop management strategies needed for the control of saline-seeps. These strategies are based on the specific water requirements and rooting depth of crops, soil water conservation and storage, crop residue management, disease and weed controls, and proper fertilisation.

Brown et al. (1983) also describe a number of measures for the control of saline-seep problems. These include: perennial cropping (alfalfa and grasses); annual cropping and flexible cropping; vegetative snow trapping; and drainage. Hedlund (1988b: 30 and 31) argues that, although a great deal of information on engineering and agronomic options to the saline-seep problem is available, the social and cultural aspects require the greatest attention for long-range solution of the problem. Farmers need to get involved in saline-seep projects, because lack of broad-based concern is the major obstacle to implementing solutions. Also, long-range solutions require good data on the social and economic aspects in order to involve large farmer groups.

In Canada, Henry et al. (1985) investigated the stratigraphy and hydrogeology at 15 sites affected by salinity in Saskatchewan Province. At five sites (Series A) nests of piezometers were installed and at 10 sites (Series B) a single piezometer was installed in or near an aquifer. Piezometric data showed the potential for upward movement in all Series A sites. The electrical conductivity of water from Series A piezometers increased from the deepest to shallowest and there was a general increase in soil salinity towards the soil surface. This study suggests that discharge from deeper aquifers is a significant soil salinisation factor. This leads to an examination of existing management recommendations (avoiding fallow in crop production) for saline soils in Saskatchewan.

Hendry and Buckland (1990) report that in southern Alberta the extent of salinised land is about 1 Mha and examine the causes of soil salinisation affecting 1700 ha of a basin in southern Alberta. Hendry et al. (1991) describe the cause of salinisation in a 21 ha area in east-central Alberta and conclude that the salinisation was derived from evaporation from seasonally high watertables caused by ponded surface water and spring snowmelt in localised depressional areas.

Miller et al. (1993) investigated soil salinisation in a side-hill seep and in a closed basin in southern Alberta, using hydrological, chemical and mineralogical techniques to compare the nature of soil salinisation. Their investigation indicated a strong upward movement of water and soluble salts in the case of the side-hill seep and slight upward movement of water and salt in the closed basin case. They estimated that it would take about 77 years to salinise the soil at the seep via upward groundwater movement, and 6500 years to salinise the soil at the closed basin. Chang and Oosterveld (1981) and Chang et al. (1985) describe the effects of irrigation on soil salinity in southern Alberta.

Salinity problems in other areas

Apart from irrigated salinity problems described in the previous section in the Colorado River Basin and the Central Valley in California and dryland salinity in the Northern Great Plains, human-induced salinity problems exist in other areas of the United States or have been controlled via adequate drainage facilities and appropriate management options. These include the Imperial Valley and Coachella Valley.

Imperial Valley, which is a rich agricultural area in California, is located in the south-east corner of the state on the Mexico and Arizona border (see Part One, Figure 10, p. 96). The valley has a general slope towards the Salton Sea and was initially an uninhabited desert wasteland. It has a hot desert climate characterised by daily temperature extremes. Although the agricultural potential of the valley was recognised as early as 1849, irrigation was begun in the valley in 1901 by importing water from the Colorado River (Kahrl, 1979: 39). Unanticipated flooding in 1905 caused the Colorado River to change its course and pour its waters down the valley, destroying irrigation canals and creating the Salton Sea, which is currently maintained by irrigation drainage water.

Development of the region was slow until the completion of the Hoover Dam in 1935 and the 129 km-long All-American Canal in 1940. The canal provides all the water requirements in the valley for irrigation and domestic purposes. Currently the irrigated area in the valley is about 215 000 ha (Framji et al., 1983: 1502).

As early as 1902, the US Department of Agriculture engineers found that the valley's watertable and salinity needed to be controlled, and by the 1920s it was apparent that a drainage problem existed. Accumulating salts and a rising watertable forced some lands out of production and threatened the productivity of thousands of hectares (Hartshorn, 1986). In 1922 the Imperial Irrigation District began constructing the drainage system. Except for some drains in the north that discharge directly into the Salton Sea, the open drains discharge into the Alamo and New Rivers (see Kaddah and Rhoades, 1975, for the location map). These rivers discharge by gravity flow into the Salton Sea. Annually 1.54×10^9 m^3 of drainage water from the Imperial Valley discharges to the Salton Sea. The valley contains 4800 km of irrigation canals and 2240 km of open drains.

Because of the fine texture of the soil the open drain system was only partially successful and the need for subsurface drains became urgent. A subsurface drainage system was introduced in 1929. It gained popularity and by about 1986 more than 46 000 km of subsurface drains had been installed on more than 80 per cent of the valley's farmland (Hartshorn, 1986). Subsurface drainage systems consist of tile and plastic tubes installed at depths of 1.5–1.8 m and spacings of 15–75 m (Kaddah and Rhoades, 1975). As reported by Holburt (1984), installation of subsurface drains in the Imperial

Valley cost over US$72 million. These drains must also be cleaned every five years or so at a cost of about US$380 per kilometre. The farmers in the Imperial Valley have also been lining their laterals with concrete to decrease the seepage of irrigation water into the soil, thereby reducing the drainage problem. By 1982, 1250 km of district laterals had been concrete lined at a cost in excess of US$23 million and 3800 km of private farm ditches had been lined at a cost of over US$31 million. The Imperial Valley case is the biggest drainage success story in California.

The Coachella Valley in the Colorado Desert of California is located to the north-east of the Salton Sea. Elevations in the valley vary from about 70 m below sea-level to approximately 335 m in the north-east of the valley. The climate of the valley is characterised by long and extremely hot summers. It has had an average annual rainfall of 80 mm since 1877 (Weeks and Levy, 1987). When the irrigation system was completed in 1953, it was the most modern system in the world. It incorporated an underground distribution system to minimise evaporation and seepage and included meters to measure each farm delivery. In the Coachella Valley diverse crops are grown year-round with double and triple cropping.

The Coachella Valley contains 40 500 ha of irrigated land, served by Colorado River water (Framji et al., 1983: 1502). Water for the valley is diverted from Imperial Dam and flows through the All American Canal together with water for Imperial Valley. About 48 km west of Imperial Dam the 197 km long Coachella Canal branches off to the north-west (for the location map see Weeks and Levy, 1987). When constructed in the 1940s, only the last 58 km of the canal was lined.

Farmers have installed an extensive system of subsurface drains which has been successful in controlling the area's drainage problem. According to Hartshorn (1986: 11), more than 3200 km of farm drains serve approximately half of the valley's irrigable area. Annually 148×10^6 m^3 of agricultural drainage water from the Coachella Valley flow into the Salton Sea. This is almost 10 per cent of the drainage discharge from the Imperial Valley to the Salton Sea.

Other areas with human-induced salinity problems in the United States include the Rio Grande Basin of New Mexico and Texas, the Yakima River Basin in Washington, the Snake River Basin in Idaho, and the Arkansas River Basin in Colorado, Kansas and Oklahoma (El-Ashry et al., 1985). In all the affected river basins salinity has progressively increased as water resources have been developed. The water in these rivers also becomes increasingly saline from the headwaters to the mouths. For example, in the Pecos River (New Mexico), salinity increases from 760 to 2020 mg L^{-1} TDS over a distance of about 50 km; in the Arkansas River (Colorado), the increase is from a trace to 2200 mg L^{-1} TDS in 190 km; and in the Rio Grande River (Texas), salinity increases from 870 to 4000 mg L^{-1} in 120 km. Major increases are usually due to seepage and return flows from irrigated land.

Hedlund (1988b) provides a brief description of irrigation with saline water in the Upper Arkansas River and the Triangle Conservation District Saline-Seep Project in Montana. Further information about salinity and its management in the arid and semiarid regions of the western United States is available in French (1984), El-Ashry et al. (1985), and El-Ashry and Gibbons (1986 and 1988a). In particular, El-Ashry and Gibbons (1988b) discuss conservation incentives, water markets, sale of water rights and environmental policies with respect to management of water resources and salinity problems.

References

Agricultural Water Management Subcommittee. 1987. *Farm Water Management Options for Drainage Reduction.* Sacramento: San Joaquin Valley Drainage Program. (Various pagings, about 130 pp.).

Albasel, N. and Pratt, P.F. 1989. Guidelines for molybdenum in irrigation waters. *Journal of Environmental Quality.* 18(3): 259–264.

Albasel, N., Pratt, P.F. and Westcot, D.W. 1989. Guidelines for selenium in irrigation water. *Journal of Environmental Quality.* 18(3): 253–258.

Azimi-Zonooz, A. and Duffy, C.J. 1993. Modeling transport of subsurface salinity from a Mancos Shale hillslope. *Ground Water.* 31(6): 972–981.

Back, W., Rosenshein, J.S. and Seaber, P.R. eds. 1988. *Hydrogeology (The Geology of North America*: v. 0–2). Boulder, Colorado: The Geological Society of America. 524 pp.

Bajwa, R.S., Crosswhite, W.M., Hostetler, J.E. and Wright, O.W. 1992. *Agricultural Irrigation and Water Use.* (Agricultural Information Bulletin no. 638). Washington DC: US Department of Agriculture, Economic Research Service. 116 pp.

Barton, R.V. 1981. Computer simulation of the Colorado River for long-term operation studies. In: Framji, K.K., ed. *State-of-the-Art Irrigation Drainage and Flood Control.* New Delhi: International Commission on Irrigation and Drainage. v.2. 67–92.

Belitz, K. 1988. *Character and Evolution of the Grand-Water Flow System in the Central Part of the Western San Joaquin Valley, California.* Sacramento: US Geological Survey. (Open-File Report 87-573). 34 pp.

Black, A.L., Brown, P.L. and Siddoway, F.H. 1981. Dryland cropping strategies for efficient water-use to control saline seeps in the Northern Great Plains, U.S.A. In. Holmes, J.W. and Talsma, T. eds. *Land and Stream Salinity.* Amsterdam: Elsevier. 295–311.

Bouwer, H., Pyne, R.D.G., and Goodrich, J.A. 1990. Recharging groundwater in the USA. *Civil Engineering.* 60(6): 63–66.

Boyle Engineering Corporation. 1986. *Evaluation of On-Farm Agricultural Management Alternatives.* Fresno, California: Boyle Engineering Corporation. (Various pagings, about 300 pp.).

Bradford, G.R., Bakhtar, D. and Westcot, D. 1990. Uranium, vanadium, and molybdenum in saline waters of California. *Journal of Environmental Quality.* 19(1): 105–108.

Brown, P.L., Halvorson, A.D., Siddoway, F.H., Maylan, H.F. and Miller, M.R. 1983. *Saline-Seep Diagnosis, Control and Reclamation.* (Conservation Research Report no. 30). Washington DC: US Department of Agriculture, Agricultural Research Service. 22 pp.

California Department of Food and Agriculture. 1988. *The Agroforestry Demonstration Program in the San Joaquin Valley: Progress Report.* Sacramento: California Department of Food and Agriculture, Agricultural Resources Branch. (Various pagings, about 200 pp).

Carr, J.E., Chase, E.B., Paulson, R.W. and Moody, D.W. Compilers. 1990. *National Water Summary 1987 – Hydrologic Events and Water Supply and Use.* (US Geological Survey, Water Supply Paper 2350). Washington DC: US Government Printing Office. 553 pp.

CH2M Hill, Inc. 1986. *Reverse Osmosis Desalting of the San Luis Drain: Conceptual Level Study.* (Report prepared for the US Department of Interior, Bureau of Reclamation, Mid-Pacific Region, Sacramento,

California). Emeryville, California: CH2M Hill, Inc. (Various pagings, about 120 pp.).

Chang, C., Kozub, G.C. and Mackay, D.C. 1985. Soil salinity status and its relation to some of the soil and land properties of three irrigation districts in southern Alberta. *Canadian Journal of Soil Science.* 65: 187–193.

Chang, C. and Oosterveld, M. 1981. Effects of long-term irrigation on soil salinity at selected sites in southern Alberta. *Canadian Journal of Soil Science.* 61: 497–505.

Colorado River Basin Salinity Control Forum. 1993. *1993 Review: Water Quality Standards for Salinity, Colorado River System.* Bountiful, Utah: Colorado River Basin Salinity Control Forum. (Various pagings, about 115 pp).

Colorado River Salinity Program Coordinator. 1993. *Salinity Update.* Denver: Bureau of Reclamation. 18 pp.

Committee of Consultants on Drainage Water Reduction. 1988a. *Opportunities for Drainage Water Reduction.* Davis: University of California, Salinity/Drainage Task Force and the Water Resources Center. 28 pp.

Committee of Consultants on Drainage Water Reduction. 1988b. *Associated Costs of Drainage Water Reduction.* Davis: University of California. Salinity/Drainage Task Force and the Water Resources Center. 16 pp. (plus Appendix and 10 Tables).

Committee of Consultants on San Joaquin River Water Quality Objectives. 1988a. *San Joaquin Valley Agriculture and River Water Quality.* Davis: University of California, Salinity/Drainage Task Force and the Water Resources Center. 22 pp.

Committee of Consultants on San Joaquin River Water Quality Objectives. 1988b. *The Evaluation of Water Quality Criteria for Selenium, Boron, and Molybdenum in the San Joaquin River Basin.* Davis: University of California, Salinity/Drainage Task Force and the Water Resources Center. 18 pp. (plus Appendix).

Coppock, R. 1988a. *Resource at Risk: Agricultural Drainage in the San Joaquin Valley.* Davis, California: University of California, Agricultural Issues Center. 13 pp.

Coppock, R. 1988b. *Resource at risk in the San Joaquin Valley: Selenium, Human Health, and Irrigated Agriculture.* Davis, California: University of California, Agricultural Issues Center. 9 pp.

Coppock, R. 1988c. *Resource at Risk in the San Joaquin Valley: Drainage Source Control on the Farm.* Davis, California: University of California, Agricultural Issues Center. 10 pp.

Deverel, S.J. 1989. Geostatistical and principal-component analyses of groundwater chemistry and soil-salinity data, San Joaquin Valley, California. In. Ragone, S. ed. *Regional Characterization of Water Quality.* (IAHS, Pub. no. 182). 11–18.

Deverel, S.J. and Gallanthine, S.K. 1989. Relation of salinity and selenium in shallow groundwater to hydrologic and geochemical processes, Western San Joaquin Valley, California. *Journal of Hydrology.* 109: 125–149.

Deverel, S.J., Gilliom, R.J., Fujii, R., Izbicki, J.A. and Field, J.C. 1984. *Areal Distribution of Selenium and other Inorganic Constituents in Shallow Ground Water of the San Luis Drain Service Area, San Joaquin Valley, California: A Preliminary Study.* (US Geological Survey Water Resources Investigation Report 84–4319). Sacramento: US Geological Survey. 67 pp.

El-Ashry, M.T. 1980. Ground-water salinity problems related to irrigation in the Colorado River Basin. *Ground Water.* 18(1): 37–45.

El-Ashry, M.T. and Gibbons, D.C. 1986. *Troubled Waters: New Policies for Managing Water in the American West.* Washington DC: World Resources Institute. 89 pp.

El-Ashry, M.T. and Gibbons, D.C. eds. 1988a. *Water and Arid Lands of the Western United States.* Cambridge: Cambridge University Press. (A World Resources Institute Book). 415 pp.

El-Ashry, M.T. and Gibbons, D.C. 1988b. New water policies for the West. In. El-Ashry, M.T. and Gibbons, D.C. eds. *Water and Arid Lands of the Western United States.* Cambridge: Cambridge University Press. (A World Resources Institute Book). 377–395.

El-Ashry, M.T. and Gibbons, D.C. 1990. Adverse impacts and alternatives to large dams and interbasin transfers in the Colorado River Basin, U.S.A. In. *The Impact of Large Water Projects on the Environment.* Paris: UNESCO. 137–144.

El-Ashry, M.T., van Schilfgaarde, J. and Schiffman, S. 1985. Salinity pollution from irrigated agriculture. *Journal of Soil and Water Conservation.* 40(1): 48–52.

Food and Agriculture Organization of the United Nations (FAO). 1989. *Production Yearbook.* Rome: FAO. v.42. 350 pp.

Framji, K.K., Garg, D.C. and Luthra, S.D.L. 1983. *Irrigation and Drainage in the World:*

A Global Review. Third edition. New Delhi: International Commission on Irrigation and Drainage. v. III. 1161–1667.

Frederick, K.D. 1988. The future of irrigated agriculture. *Forum for Applied Research and Public Policy.* 3(2): 80–89.

French, R.H. ed. 1984. *Salinity in Watercourses and Reservoirs.* Proceedings of the 1983 International Symposium on State-of-the-Art Control of Salinity. Salt Lake City, Utah, 13-15 July 1983. Boston: Butterworth. 622 pp.

Geraghty, J.J., Miller, D.W., van der Leeden, F. and Troise, F.L. 1973. *Water Atlas for the United States.* Third edition. Port Washington: Water Information Center. 190 pp.

Hartshorn, J.K. 1986. *Layperson's Guide to Agricultural Drainage.* Sacramento: Water Education Foundation. 16 pp.

Heath, R.C. 1984. *Ground-Water Regions of the United States.* (US Geological Survey Water-Supply Paper 2242). Washington DC: US Government Printing Office. 78 pp.

Hedlund, J.D. 1984. USDA planning process for Colorado River Basin salinity control. In. French, R.H. ed. *Salinity in Watercourses and Reservoirs.* Proceedings of the 1983 International Symposium on State-of-the-Art Control of Salinity. Salt Lake City, Utah, 13–15 July 1983. Boston: Butterworth. 63–77.

Hedlund, J.D. 1988a. Wellton-Mohawk farmers deliver water conservation. *Journal of Soil and Water Conservation.* 43 (6): 462–464.

Hedlund, J.D. 1988b. *Project Level Salinity Management Options* (Draft). Portland, Oregon: US Department of Agriculture, Soil Conservation Service. 37 pp.

Hedlund, J.D. 1993. *Conservation Planning for Water Quality Concerns: Toxic Element — Selenium.* (Technical Notes Water Quality Series no. W1). Portland, Oregon: US Department of Agriculture, Soil Conservation Service. 17 pp.

Hendry, M.J. and Buckland, G.D. 1990. Causes of soil salinization: 1. A basin in southern Alberta, Canada. *Ground Water.* 28(3): 385–393.

Hendry, M.J., Chan, G.W. and Harker, D.B. 1991. Causes of soil salinization: 2. A basin in east-central Alberta, Canada. *Ground Water.* 28(4): 544–550.

Henry, J.L., Bullock, P.R., Hogg, T.J. and Luba, L.D. 1985. Groundwater discharge from glacial and bedrock aquifers as a soil salinization factor in Saskatchewan. *Canadian Journal of Soil Science.* 65: 749–768.

Holburt, M.B. 1984. Colorado River salinity – the user's perspective. In. French, R.H. ed. *Salinity in Watercourses and Reservoirs.* Proceedings of the 1983 International Symposium on State-of-the-Art Control of Salinity. Salt Lake City, Utah, 13–15 July 1983. Boston: Butterworth. 13–22.

Howe, C.W. and Ahrens, W.A. 1988. Water resources of the Upper Colorado River Basin: Problems and policy alternatives. In. El-Ashry, M.T. and Gibbons, D.C. eds. *Water and Arid Lands of the Western United States.* Cambridge: Cambridge University Press. 169–232.

Inman, R.R., Olson, D.C. and King, D.L. 1984. Grand Valley salt pick-up calculations. In. French, R.H. ed. *Salinity in Watercourses and Reservoirs.* Proceedings of the 1983 International Symposium on State-of-the-Art Control of Salinity. Salt Lake City, Utah, 13–15 July 1983. Boston: Butterworth. 157–167.

Jacobs, L.W. ed. 1989. *Selenium in Agriculture and the Environment.* Madison: Soil Science Society of America and American Society of Agronomy. 233 pp.

Jensen, E.G. and Leach, R.W. 1984. Salinity control by pumping and deep well injection – The Paradox Valley Unit. In French, R.H. ed. *Salinity in Watercourses and Reservoirs.* Proceedings of the 1983 International Symposium on State-of-the-Art Control of Salinity. Salt Lake City, Utah, 13–15 July, Boston: Butterworth. 349–358.

Kaddah, M.T. and Rhoades, J.D. 1976. Salt and water balance in Imperial Valley, California. *Soil Science Society of American Journal.* 40(1): 93–100.

Kahrl, W.L. ed. 1979. *The California Water Atlas.* Los Altos, California: William Kaufmann. 118 pp.

Liebermann, T.D., Mueller, D.K., Kircher, J.E. and Choquette, A.F. 1989. *Characteristics and Trends of Streamflow and Dissolved Solids in the Upper Colorado River Basin, Arizona, Colorado, New Mexico, Utah, and Wyoming.* (US Geological Survey Water-Supply Paper 2358). Washington DC: US Government Printing Office. 64 pp.

Lord, J.M. Incorporated. 1987. *Study of Innovative Techniques to Reduce Subsurface Drainage Flows.* (Report prepared for the San Joaquin Valley Drainage Program). Fresno, California: J.M. Lord, Incorporated. 91 pp.

McKell, C.M., Goodin, J.R. and Jefferies, R.L. 1986. Saline land of the United States of

America and Canada. *Reclamation and Revegetation Research.* 5: 159–165.

Miller, J.J., Pawluk, S., and Beke, G.J. 1993. Soil salinization at a side-hill seep and closed basin in southern Alberta. *Canadian Journal of Soil Science.* 73: 209–222.

Miller, M.R., Brown, P.L., Donovan, J.J., Bergatino, R.N., Sonderegger, J.L. and Schmidt, F.A. 1981. Saline seep development and control in the North American Great Plains: Hydrogeological aspects. In. Holmes, J.W. and Talsma, T. eds. *Land and Stream Salinity.* Amsterdam: Elsevier. 115–141.

Miller, T.O., Weatherford, G.D. and Thorson, J.E. 1986. *The Salty Colorado.* Washington DC: The Conservation Foundation. 102 pp.

Mueller, D.K. and Moody, C.D. 1984. Historical trends in concentration and load of major ions in the Colorado River System. In French, R.H. ed. *Salinity in Watercourses and Reservoirs.* Proceedings of the 1983 International Symposium on State-of-the-Art Control of Salinity. Salt Lake City, Utah, 13–15 July 1983. Boston: Butterworth. 181–192.

Murphy, R.E. 1978. Face of the land. In. Cayne, B.E. Editor-in-Chief. *The Encyclopedia Americana.* Danbury, Connecticut: American Corporation. v.27. 500–515.

Orlob, G.T. and Ghorbanzadeh, A. 1981. Impact of water resource development on salinization of semiarid lands. In. Holmes, J.W. and Talsma, T. eds. *Land and Stream Salinity.* Amsterdam: Elsevier. 275–293.

Page, R.W. 1986. *Geology of the Fresh Ground-Water Basin of the Central Valley, California, with Texture Maps and Sections.* (US Geological Survey Professional Paper 1401-C). Washington DC: US Government Printing Office. 54 pp.

Pavelis, G.A. 1987a. Economics survey of farm drainage. In. Pavelis, G.A. ed. *Farm Drainage in the United States: History, Status and Prospects.* Washington DC: US Department of Agriculture, Economic Research Service. 110–136.

Pavelis, G.A. ed. 1987b. *Farm Drainage in the United States: History, Status and Prospects.* Washington DC: US Department of Agriculture, Economic Research Service. 170 pp.

Postel, S. 1989. *Water for Agriculture: Facing the Limits.* Washington DC: Worldwatch Institute. (Worldwatch Paper 93). 54 pp.

Sabol, G.V., Bouwer, H. and Wierenga, P.J. 1987. Irrigation effects in Arizona and New Mexico. *Journal of Irrigation and Drainage Engineering.* (American Society of Civil Engineers, Irrigation and Drainage Division). 113(1): 30–68.

San Joaquin Valley Drainage Program. 1987. *Developing Options: An Overview of Efforts to Solve Agricultural Drainage and Drainage-Related Problems in the San Joaquin Valley.* Sacramento: San Joaquin Valley Drainage Program. 28 pp.

Soil Conservation Service. 1977. *On-Farm Program for Salinity Control: Final Report of the Grand Valley Salinity Study.* Denver: US Department of Agriculture, Soil Conservation Service. 69 pp.

Soil Conservation Service. 1981. *Potential for Onfarm Irrigation Improvements: Lower Gunnison Basin Unit Salinity Control Study.* Denver: US Department of Agriculture, Soil Conservation Service. 86 pp.

Soil Conservation Service. 1983. *Onfarm Irrigation Improvements: McElmo Creek Unit Salinity Control Study.* Denver: US Department of Agriculture, Soil Conservation Service. (Various pagings, about 80 pp.).

Solley, W.B., Pierce, R.R. and Perlman, H.A. 1993. *Estimated Use of Water in the United States in 1990.* (US Geological Survey Circular 1081). Washington DC: US Government Printing Office. 76 pp.

Szabolcs, I. 1989. *Salt-Affected Soils.* Boca Raton, Florida: CRC Press. 274 pp.

Tanji, K.K., Knight, A. and Meyer, J. 1987. *1986-87 Technical Progress Report: UC Salinity Drainage Task Force.* Davis: University of California, Division of Agriculture and Natural Resources. 160 pp.

Tanji, K. Läuchli, A. and Meyer, J. 1986. Selenium in the San Joaquin Valley. *Environment.* 28(6): 6–11 and 34–39.

Tracy, J.E., Oster, J.D. and Beaver, R.J. 1990. Selenium in the southern coast range of California: Well waters, mapped geological units and related elements. *Journal of Environmental Quality.* 19(1): 46–50.

Trueman, D.P. and Miller, J.B. 1984. The impact of reservoirs on seasonal and historical salinity of the Colorado River. In. French, R.H. ed. *Salinity in Watercourses and Reservoirs.* Proceedings of the 1983 International Symposium on State-of-the-Art Control of Salinity. Salt Lake City, Utah, 13–15 July 1983. Boston: Butterworth. 171–180.

URS Corporation. 1986. *Deep-well Injection of Agricultural Drain Waters: An Appraisal Level Study with Application to Kesterson Reservoir Problems.* (Summary Report).

Sacramento, California: URS Corporation. (Various pagings, about 140 pp).

US Bureau of Census. 1993. 113th edition. *Statistical Abstract of the United States: 1993*. Washington DC: US Government Printing Office. 1009 pp.

US Bureau of Reclamation 1983. *Colorado River Water Quality Improvement Program*. Denver: Bureau of Reclamation. 126 pp.

US Department of Agriculture. 1989. *The Second RCA Appraisal, Soil Water and Related Resources on Nonfederal Land in the United States, Analysis of Condition and Trends*. Washington DC: US Government Printing Office. 280 pp.

US Department of Agriculture. 1993. *Agricultural Resources: Cropland, Water, and Conservation, Situation and Outlook Report*. Washington DC: US Department of Agriculture, Economic Research Service. 46 pp.

van Schilfgaarde, J. 1987. Summary. In. Pavelis, G.A. ed. *Farm Drainage in the United States: History, Status and Prospects*. Washington DC: US Department of Agriculture, Economic Research Service. iv-x.

Walker, W.R., Skogerboe, G.V. and Evans, R.C. 1979. Developing best management practices for salinity control in Grand Valley, Colorado. *Water Resources Research*. 15(5): 1073–1080.

Weeks, L.O. and Levy, T.E. 1987. Modernization of the Coachella Valley Water District's irrigation system. In. *Transactions, Thirteenth International Congress on Irrigation and Drainage*. Rabat, Morocco, September 1987. New Delhi: International Commission on Irrigation and Drainage. v. I-B. 1441–1454.

World Bank. 1992. *World Development Report 1992: Development and the Environment*. New York: Oxford University Press. 308 pp.

Appendix I Summary data

Country	Area (10⁶ km²)	Population (10⁶)	Average annual rainfall (mm)	Average annual run-off			Annual water use		Cultivated land		Irrigation-induced salt-affected land (10⁶ ha)
				Volume (10⁹ m³)	Depth (mm)	Per cent of rainfall (%)	Total volume (10⁹ m³)	Volume used for irrigation (10⁹ m³)	Total (10⁶ ha)	Irrigated (10⁶ ha)	
Argentina	2.8	32	515	914	330	64	26	19	35.8	1.5	0.58
Australia	7.7	17	465	397	52	11	15	10	47.1	1.8	0.16
China	9.6	1134	630	2600	272	43	444	389	100.0	48.0	6.70
CIS	22.4	291	531	4740	215	40	344	195	232.6	20.5	3.70
Egypt	1.0	52	0	0	0	0	38	34	2.7	2.7	0.88
India	3.3	850	1250	1897	577	46	552	460	169.0	42.1	7.00
Iran	1.6	56	243	71.5	43	18	75	73	14.8	5.7	1.72
Pakistan	0.8	112	348	175	220	63	na	125	20.8	16.1	4.22
South Africa	1.2	36	464	53.5	44	9	12	9	13.2	1.1	0.1
Thailand	0.5	56	1485	190	370	25	33	30	20.0	4.0	0.40
USA	9.4	250	762	1930	206	27	564	189	189.9	18.7	4.16
World	149.0	5292	800	46 768	314	39	4132	2680	1473.70	227.11	45.40

for selected countries

Main affected areas	Major causes	Major management options	Remarks
San Juan, Mendoza, Salta and Rio Negro Provinces.	Extensive irrigation without drainage facilities.	Improving drainage facilities and irrigation systems.	
Murray Basin, south-west of Western Australia, and South Australia	Land clearing, extensive irrigation, construction of hydraulic structures.	In Murray Basin: interception schemes and disposal of saline drainage and groundwater to evaporation basins. In Western Australia: reforestation.	(a) In the irrigated areas 0.65×10^6 ha have watertables shallower than 2 m. (b) Salinisation of water resources in the Murray Basin, Western Australia and South Australia is a major problem.
Huang-Huai-Hai Plain (containing 3.06 Mha salt-affected land out of a total 20 Mha irrigated).	Excessive irrigation without adequate drainage facilities.	Conjunctive use of surface and groundwater resources.	
Central Asia, Ukraine, Caucasus region and Volga Basin.	Excessive irrigation without adequate drainage facilities.	Increasing irrigation efficiency by lining irrigation canals, using efficient irrigation methods, automation of irrigation systems, using subsurface drainage methods (tile drainage and vertical drainage) and changes in agricultural practices.	
Nile Valley and Delta.	Extensive perennial irrigation.	Development of surface and subsurface drainage facilities.	Egypt has practically no surface run-off. The 55.5×10^9 m^3 of Nile water is Egypt's share from the Nile Basin.
Punjab, Haryana, Uttar Pradesh, Bihar, Rajasthan, and Madhya Pradesh.	Excessive irrigation without drainage facilities.	Conjunctive use of surface and groundwater, improved irrigation and drainage.	
Many irrigation projects including: Zarrineh-Rud, Moghan, Khalafabad, Doroudzan and Zayandeh-Rud.	Inadequate irrigation and drainage facilities, irrigation with low quality water.	Improving irrigation and drainage facilities, conjunctive use of surface and groundwater, improved agricultural practices.	
Indus River Basin.	Extensive use of surface water for irrigation and inadequate drainage facilities.	Conjunctive use of surface and groundwater, improving drainage facilities, disposal of saline drainage waters to the sea.	
Breede, Berg, Great Fish and Sundays River Basins.	Irrigation on soils with subsurface salt contents.	Improved irrigation and drainage systems.	Salinisation of surface water resources supplying the main population centres and the industrial, mining and agricultural sectors is a much more important issue compared to land salinisation. Salt-affected rivers include Breede, Berg, Fish and Vaal. Salinisation is mainly due to increased human activity such as the discharge of industrial, mining and municipal effluents to the river system, as well as return flow from irrigation schemes and the phenomenon of dryland salinity.
Khorat and Sakon Nakhon Basins in Khorat Plateau. Lam Pao and Nong Wai irrigated areas.	Land clearing, reservoir construction, salt making and irrigation.	Reforestation, banning salt making, controlling groundwater depth, land levelling, agroforestry and surface mulching.	(a) Dryland salinity affects 2.85×10^6 ha mainly in Khorat Plateau. (b) Irrigated land salinity is mainly limited to Lam Pao and Nong Wai irrigated areas.
Colorado River Basin, San Joaquin Valley, Lower Rio Grande and Northern Great Plains	Excessive irrigation without adequate drainage facilities	On-farm physical improvement and irrigation managemnent, water pricing and irrigation management, blending drainage water with good quality surface water, disposal of drainage water to on-farm evaporation basins.	Salinity of the Colorado River and disposal of irrigation return flow in San Joaquin Valley are among the major issues.
Arid and semiarid regions of the world	Excessive irrigation without adequate drainage facilities and land clearing for agricultural development and grazing.	Improving surface and subsurface drainage facilities, conjunctive use of surface and groundwater resources, and improving farm management.	45.4×10^6 ha of land affected by human-induced salinisation represents only the salt-affected land in the world's irrigated areas. 31.2×10^6 ha are salinised in non-irrigated lands.

Appendix II
Some institutions involved in salinity investigations

Argentina

- Centro Regional de Agua Subterranea (CRAS), Ave Jose Ignocio de la Roza, 125 Este, 5400 San Juan, Provincia de San Juan.
- Centro de Investigaciones de Recursos Naturales, 1712 Castelar-Pcia, Buenos Aires.
- INTA (Instituto Nacional de Tecnologia Agropecuaria), Centro Regional Cuyo, C.C. No 3, 5507 Lujan de Cuyo, Provincia de Mendoza.
- INTA, Centro Regional Tucuman-Santiago del Estero, San Luis 530, 4000 Tucuman, Provincia de Tucuman.
- INTA, Centro Regional Buenos Aires Sur, C.C. No 276, 7620 Balcarce, Provincia de Buenos Aires.
- INTA, Centro Regional Cordoba EGA Manfredi, 5988-Manfredi, Provincia de Cordoba.
- INTA, Centro Regional la Rioja-Catamarca, Buenos Aires 562, 5300 La Rioja, Provincia de la Rioja.
- INTA, Centro Regional la Pampa-San Luis, Avda San Martin 718-1x Piso, Of 108, 6300 Santa Rosa, Provincia de la Pampa.
- INTA, Centro Regional Salta-Jujuy, C.C. No 228, 4400 Salta, Provincia de Salta.
- INTA, Centro Regional Patagonia Norte, C.C. No 277. 8400 San Carlos de Bariloche, Provincia de Rio Negro.
- INTA, Centro Regional Patagonia Sur, C.C. No 88 9100-Trelew, Provincia de Chubut.
- Instituto Nacional de Ciencia y Tecnica Hidricas (INCYTH), Lima 767, 1037-Buenos Aires.
- Instituto Argentino de Investigaciones de las Zonas Aridas (IADIZA), C.C. No 507, 5500 Mendoza, Provincia de Mendoza.
- Sociedad del Estado Agua y Energia Electrica (AEE), L.N. Alem 1134, 1003 Buenos Aires.

Australia

- Australian Geological Survey Organisation, GPO Box 378, Canberra ACT 2601.
- Centre for Groundwater Studies, Private Bag 2, Glen Osmond SA 5064.
- CSIRO (Commonwealth Scientific and Industrial Research Organisation), Division of Forestry and Forest Production, PO Box 4008, Queen Victoria Terrace, Canberra ACT 2600.
- CSIRO Division of Soils (Canberra Laboratory), Black Mountain, Acton, ACT 2601.
- CSIRO Division of Soils (Davies Laboratory), Private Mail Bag, PO, Aitkenvale, QLD 4814.
- CSIRO Division of Soils (Adelaide Laboratory), Private Bag 2, Glen Osmond, SA 5064.
- CSIRO Division of Water Resources (Adelaide Laboratory), Private Bag 2, Glen Osmond SA 5064.
- CSIRO Division of Water Resources (Canberra Laboratory), GPO Box 1666, Canberra ACT 2601.
- CSIRO Division of Water Resources (Griffith Laboratory), Private Bag 3, PO Griffith NSW 2680.
- CSIRO Division of Water Resources (Perth Laboratory), Private Bag, PO Wembley WA 6014.
- Department of Agriculture, South Australia. Soil and Water Conservation Branch, GPO Box 1671, Adelaide, SA 5001.
- Department of Conservation and Environment, PO Box 41 East Melbourne, VIC 3002.
- Department of Conservation and Environment, Centre for Land Protection Research, 22 Osborne Street, PO Box 401, Bendigo, VIC 3550.
- Department of Conservation and Land Management, New South Wales, Soil Conservation Service. PO Box 198, Chatswood, NSW 2057.

APPENDIX II INSTITUTIONS

- Department of Conservation and Land Management, Western Australia, 50 Hayma Road, Como, WA 6152.
- Department of Food and Agriculture, 166 Wellington Parade, East Melbourne, VIC 3002.
- Department of Primary Industries, Salinity and Contaminant Hydrology Section, Land Management Research Branch. Meiers Road, Indooroopilly, QLD 4068.
- Department of Water Resources, New South Wales, PO Box 3720, Paramatta NSW 2150.
- Engineering and Water Supply Department, GPO Box 1751, Adelaide SA 5001.
- Murray-Darling Basin Commission, GPO Box 409, Canberra ACT 2601.
- National Centre for Groundwater Management, University of Technology, Sydney, Broadway, NSW 2007.
- Queensland Forest Service, MS483 Gympie, QLD 4570.
- Rural Water Corporation, 590 Orrong Road, Armadale VIC 3143.
- Salinity Bureau, 7th Floor, 166 Wellington Parade, East Melbourne, VIC 3002.
- South Australian Department of Mines and Energy, PO Box 151, Eastwood SA 5063.
- University of New South Wales Groundwater Centre, Water Research Laboratory, King Street, Manly Vale NSW 2093.
- Water Authority of Western Australia, 629 Newcastle Street, Lederville WA 6007.
- Western Australian Department of Agriculture, Division of Resources Management, South Perth WA 6151.
- Western Australian Water Resources Council, GPO Box S1400, Perth WA 6001.

China

- Beijing Agricultural University, 100094, Beijing.
- Hebei Institute of Hydrotechnics, Shijiazhuang, 050021, Hebei.
- Institute of Geography, Academia Sinica, Building 917, Datun Road Anwai, 100101, Beijing.
- Institute of Hydrogeology and Engineering Geology, 150 Shigang Street, Shijiazhuang City, Hebei.
- Ministry of Geology, Beijing.
- Ministry of Water Resources and Electric Power, Department of Farmland Irrigation and Drainage, Beijing.
- Nanjing Institute of Soil Science, Academia Sinica, 71 East Beijing Road, Nanjing.
- Northwest Institute of Soil and Water Conservation, Xi'an, Shaanxi.
- Shandong Agricultural University, Tai'an, Shandong.
- Shijiazhuang Institute of Agricultural Modernization. Academia Sinica, Shijiazhuang, 050021, Hebei.
- Wuhan College of Hydraulic and Electrical Engineering, Wuhan.
- Xinjiang Institute of Desert, Urumqi, Xinjiang.
- Xinjiang Institute of Geology, Beijing Road, Urumqi, Xinjiang.

Commonwealth of Independent States

- All-Union Institute of Hydrotechnics and Reclamation of Lands, 22 Academic St., Moscow.
- Azerbaijan Institute for Research in Irrigation and Reclamation of Lands, Baku.
- Centre for Ecology and Water Management, Goskompriroda, Prospect A, Kadiry 5a, 700 000 Tashkent.
- Dokuchaev Soil Institute, All-Union Academy of Agricultural Science, 7 Pyzhevshy Lane, 17 Moscow.
- Department of Soil Science, Moscow State University, Moscow.
- GIDROINEGO Research Institute, Uzbekgidrogeologia Corporation, 64 Morozov St, Tashkent.
- Institute of Geography, Academy of Sciences, Staromonetny Per. 29,109 017 Moscow.
- Institute of Soils, Academy of Sciences, Pushchino on Oka River, Moscow.
- Kazakh Institute of Land Mapping, Alma Ata.
- Kazakh Institute of Soil Science, Alma Ata.
- SANIIRI (Central Asian Institute for Irrigation Research), 11 Karasu – 4, 700 187 Tashkent.
- Ukraine Research Institute of Hydrotechnics and Reclamation of Lands, Kiev.
- Water Problems Institute, Academy of Sciences. 13/3 Sadovo-Chernogriazshaya, 103 064 Moscow.

Egypt

- American University of Cairo, Desert Development Centre. 113 Sharia Kasr, El Aini, Cairo.
- Alexandria University, Faculty of Agriculture, Alexandria.
- Desert Research Institute, Al-Matariya, Cairo.
- Drainage Research Institute, PO Box 13621/5, Delta Barrage (El-Kanater), Cairo.
- Minia University, Faculty of Agriculture, Minia.
- Research Institute for Ground Water, Ministry of Public Works and Water Resources, Delta Barrage (El-Kanater).
- Soil and Water Laboratory, Al-Tahrir Street, Dokki, Cairo.
- Soil and Water Research Institute, 9 Gamma Street, Giza.
- Salinity Laboratory, Alexandria University, Alexandria.
- Water Research Centre, 22 El-Galaa Street, Bulaq, Cairo.

India

- All India Coordinated Research Project on Management of Salt Affected Soils and Use of Saline Water for Agriculture, CSSRI, Karnal, Haryana 132001.
- Central Groundwater Board, Ministry of Irrigation, Jamnagar House, Mansingh Road, New Delhi 110001.
- Central Soil Salinity Research Institute, Karnal, Haryana 132001.
- Centre of Advanced Study in Geology, Punjab University, Chandigarh 160014.
- Department of Soils & Water Engineering, Punjab Agricultural University, Ludhiana, Punjab 141004.
- Geological Survey of India, 29 Jawaharlal Nehru Road, Calcutta, West Bengal 70016.
- Gujarat Agricultural University, Arnej, District Ahmedabad 392230.
- Haryana State Minor Irrigation (Tube Wells) Corporation Ltd, The Mall, Karnal, Haryana 132001.
- Irrigation Management and Training Institute, Kota, Rajasthan.
- National Bureau of Soils Survey & Land Use Planning, Indian Council of Agricultural Research (ICAR), Nagpur, Maharashtra.
- National Remote Sensing Agency, Balnagar, Hyderabad 500037.
- Project Directorate of Water Management Research, Indian Council of Agricultural Research (ICAR), M.P.K.V. Rahuri, Dist. Ahmednagar, Maharashtra 413722.
- Soil Salinity Research Centre, Tiruchirapalli, Tamil Nadu.
- Water and Land Management Institute (WALMI), Aurangabad, Maharashtra.
- Water and Land Management Institute, U.P., Canal Colony Okhla, New Delhi 110025.
- Water and Land Management Institute, Gandhinagar, Gujarat.
- Water Technology Centre, Indian Agricultural Research Institute (IARI), New Delhi.

Iran

- Ab-Khan Consulting Engineers, Dr Fatemi Ave, Meydan-e-Golha, Tirgar Ave, 75 Kutche Shahrokh Sharghi, Tehran.
- East-Azerbaijan Regional Water Authority, Tabriz.
- Esfahan Regional Water Authority, Esfahan.
- Fars Regional Water Authority, Baghe-e-Eram Ave, Shiraz.
- Khuzestan Water and Power Authority, Ahwaz.
- Ministry of Energy, 81 North Plestin Ave, Tehran.
- Mahab-Ghodss Consulting Engineers, PO Box 15815-1791, Tehran.
- Pars Consult Consulting Engineers, 15 Aban-e-Jonoubi (Shahid Azodi) Ave, Tehran.
- Research Institute of Forests and Rangelands, PO Box 13185-116, Tehran.
- Sistan and Baluchistan Regional Water Authority, Zahedan.
- Soil and Water Research Institute, Jalal Al-Ahmad Ave, Kargar Shomali, Tehran.
- Yekom Consulting Engineers, 77 Sabay-e-Shomali Ave, 14167 Tehran.

Pakistan

- Ayub Agriculture Research Institute, Faisalabad.
- Centre of Excellence in Water Resources Engineering, University of Engineering and Technology, 31 Lahore.
- Drainage and Irrigation Research Institute, Hyderabad.
- Drainage and Reclamation Institute of Pakistan (DRIP), Tandojam.
- Forest Research Institute, Faisalabad.

APPENDIX II INSTITUTIONS

- Forest Research Institute, Peshawar.
- International Waterlogging and Salinity Research Institute (IWASRI), 20-E-I Gulberg III, Lahore.
- Karachi University, Karachi.
- Land Reclamation Directorate, Irrigation and Power Department, Canal Bank, Moghalpura, Lahore.
- Lower Indus Water Management and Reclamation Research Project, Hyderabad.
- Mehran University of Engineering and Technology, Jamshoro, Hyderabad.
- Ministry of Water and Power, Islamabad.
- Mona Reclamation Experimental Project, Bhalwal.
- Nuclear Institute of Agriculture and Biology (NIAB), Faisalabad.
- Pakistan Agriculture Research Council (PARC), PO Box 1031, Islamabad.
- Pakistan Council of Research in Water Resources (PCRWR), Islamabad.
- Punjab Irrigation Research Institute, Lahore.
- Punjab University, Lahore.
- Quid-e-Azam University, Lahore.
- Soil Salinity Research Institute, Pindi Bhattian.
- Soil Survey of Pakistan, Lahore.
- University of Agriculture, Faisalabad.
- Water and Power Development Authority (WAPDA), 94 Ferozepur Road, Lahore.

South Africa

- Agricultural Research Council, PO Box 8783, Pretoria, 0001.
- Computing Centre for Water Research, PO Box 375, Pietermaritzburg, 3200.
- Department of Agriculture and Water Supply, Private Bag X116, Pretoria 0001.
- Department of Water Affairs, Private Bag X313, Pretoria 0001.
- Directorate Irrigation Engineering, Private Bag X515, Silverton, 0127.
- Division for Water Technology, CSIR, PO Box 395, Pretoria 0001.
- Institute for Soil, Climate and Water, Private Bag X79, Pretoria, 0001
- Ninham Shand Consulting Engineers, PO Box 1347, Cape Town, 8000.
- Soil and Irrigation Research Institute, Private Bag X79, Pretoria, 0001.
- Stewart Scott Inc., PO Box 784506, Sandton, 2146.
- University of Natal, Department of Agricultural Engineering, PO Box 175, Pietermaritzburg 3200.
- University of Natal, Department of Agronomy, PO Box 375, Pietermaritzburg, 3200.
- University of the Orange Free State (OFS), Department of Soil Science, PO Box 339, Bloemfontein, 9300.
- University of the Orange Free State (OFS), Institute for Groundwater Studies, PO Box 339, Bloemfontein, 9300.
- University of Pretoria, Department of Plant and Soil Sciences, Pretoria, 0002.
- University of Stellenbosch, Department of Soil and Agricultural Water Science, PO Box X5018, Stellenbosch, 7600.
- Water Research Commission, PO Box 824, Pretoria 0001.

Thailand

- Asian Institute of Technology, Divisions of Water Resources Engineering and Environmental Engineering, PO Box 2754, Bangkok 10501.
- Department of Agriculture, Bangkhen, Bangkok 10900.
- Geological Survey Division, Department of Mineral Resources, Rama 6 Road, Bangkok 10400.
- Office of the National Environment Board (ONEB), 60/1 Soi Phiboonwattana 7, Rama 6 Road, Bangkok 10400.
- Royal Forest Department, 61 Phahelyothin Road, Bangkok 10900.
- Royal Irrigation Department, Samsen Road, Bangkok 10300.
- Soil Science Department, Faculty of Agriculture, Khon Kaen University, Khon Kaen 40002.
- Soil Salinity Research Section, Land Development Department Pahonyothin Road, Bangkhen Bangkhen, Bangkok 10900.
- Soil Survey and Classification Division, Land Development Department, Pahonyothin Road, Bangkhen, Bangkok 10900.

United States of America

- Arizona Department of Water Resources, 99 East Virginia Street, Room 102, Phoenix, Arizona 85004.

- Bureau of Land Management, Colorado State Office, 1037 20th Street, Denver, Colorado 80202.
- Bureau of Reclamation, Colorado River Salinity Program Coordinator, PO Box 25007, Denver, Colorado 80225.
- Bureau of Reclamation, Lower Colorado Region, Nevada Hwy & Park Street, Boulder City, Nevada 89005.
- Bureau of Reclamation, Lower Colorado Region, Yuma Projects Office, PO Box D, 7301 Calle Agua Salada, Yuma, Arizona 85364.
- Bureau of Reclamation, Mid-Pacific Region, 2800 Cottage Way, Sacramento, California, 95825.
- Colorado River Basin Salinity Control Forum, 106 West 500 South, Suite 101, Bountiful, Utah 84010.
- Colorado River Board of California, 107 South Broadway, Room 8103, Los Angeles, California 90012.
- Colorado River Commission of Nevada, 1515 East Tropicana, Suite 400, Las Vegas, Nevada 89109.
- Department of Natural Resources, State of Colorado, 1313 Sherman, Denver, Colorado 80203.
- Division of Water Development, State of Wyoming, Herscher Building, 3rd Floor East, Cheyenne, Wyoming 82002.
- Environmental Protection Agency, Water Quality Requirements Section, 1860 Lincoln Street, Denver, Colorado 80295.
- Office of the San Joaquin Valley Drainage Program, Room W2143, 2800 Cottage Way, Sacramento, California 95825-1898.
- Salinity Laboratory, 4500 Glenwood Drive, Riverside, California 92501.
- Salt River Project, PO Box 52025, Phoenix, Arizona 85072-2025.
- Soil Conservation Service, Land Treatment Program Division, PO Box 2890, Washington DC 20013.
- Soil Conservation Service, 2121-C 2 Street, Davis, California 95616-5474.
- Soil Conservation Service, BLDGA, 3 Floor DHOC, 2490 West 26 Ave, Denver, Colorado 80211.
- Soil Conservation Service, Midwest National Technical Center, Room 345, Federal Building, 100 Centennial Mall North, Lincoln, Nebraska 68508-3866.
- State Engineer, State of New Mexico, Bataan Memorial Building, Sante Fe, New Mexico 87503.
- University of Arizona, Agricultural Experiment Station, Tucson Arizona 85721.
- University of Arizona, Office of Arid Lands Studies, 845N. Park Ave., Tucson, Arizona 85719-4896.
- University of Arizona, Water Resources Research Center, Geology Building 11, Tucson, Arizona 85721.
- University of California, Agricultural Issues Center, Davis, California 95616.
- University of California, Water Resources Center, Riverside, California 92521.
- University of California, Land, Air and Water Resources Department, Davis, California 95616.
- University of California, Soil and Environmental Science Department, Riverside, California 92501.
- University of California, Cooperative Extension Service, Davis and Riverside, Campuses, California 95616 and 92521, respectively.
- University of California, Dry Lands Research Institute, Riverside California 92521.
- University of Colorado, Armory Building 206, Box B-19, Boulder, Colorado 80309.
- Upper Colorado River Commission, 355 South, 4th East Street, Salt Lake City, Utah 84111.
- Utah Division of Water Resources, 1636 West North Temple, Salt Lake City, Utah 84116.
- Utah State University, Water Research Laboratory, Logan, Utah 84322.
- Water Conservation Laboratory, 4331, East Broadway, Phoenix, Arizona 85040.
- Western States Water Council, 220 South, 2nd East, Suite 200, Salt Lake City, Utah 84111.
- Water Management Research Laboratory, 2021 S. Peach Ave., Fresno, California 93727.

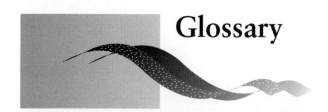

Glossary

Adaptation Change in an organism as a result of exposure to certain environmental conditions which make it react more effectively to these conditions.

Adsorption The condensation of a gas, liquid or dissolved substance onto the surface of a solid. There may or may not be a chemical reaction between the substance adsorbed and the surface of the solid.

Aeolian Caused by or related to wind action.

Agroforestry The integration of commercial tree growing into the operation of a farming enterprise. It involves the commercial utilisation of native trees as well as the planting and management of quicker-growing trees through afforestation. The aim is to ensure a long-term viable enterprise based on timber and timber products, as part of the overall farm operation.

Agronomy The science of soil management and crop production.

Alienation of land The process of opening up Crown land for private ownership.

Alkaline soil See sodic soil.

Alluvium Extensive stream-laid deposits of gravel, sand, silt and/or clay, typically forming a flood plain.

Amendment A substance mixed into the soil to improve its properties.

Anion A negatively charged particle formed when a salt dissolves in water.

Aquifer A formation that contains sufficient permeable material to yield significant quantities of water.

Arable land Land whose ecology and environment are suitable for the production of cultivated crops.

Arboretum A place devoted to the cultivation of a wide selection of trees for scientific or educational purposes.

Available soil moisture The amount of water retained by a soil between field capacity and wilting point.

Bittern Highly concentrated brine beyond halite (NaCl) saturation (350 000 mg L^{-1}).

Brackish water Water with salinity greater than 1500 mg L^{-1} TDS but less than 5000 mg L^{-1} TDS.

Brine Saline water with a salt content higher than sea water (35 000 mg L^{-1} TDS).

Bulge profile A soil profile which has a maximum salinity at an intermediate depth in the profile.

Bulk density The mass of dry soil per unit volume. A unit of measure expressed as gram per cubic cm (g cm^{-3}).

Calibration of a model The procedure of modifying model parameters so that output from the model is some measure of best fit to observations.

Canopy The cover of branches and foliage formed by tree crowns.

Capillary rise Upward movement of water in the soil profile from the saturated zone as a result of surface tension in soil pores.

Capillary zone The zone of soil above a watertable in which capillary movement of water occurs.

Catchment The entire area drained by a river and all its tributaries.

Cation A positively charged particle formed when a salt dissolves in water.

Confined aquifer An aquifer in which the pressure is significantly greater than atmospheric pressure and which is bounded above and below by confining beds. When a confined aquifer is intersected by a well the pressure is sufficient to cause the water to rise significantly above the upper level of the aquifer. When the pressure is sufficient to cause the water to flow at the surface it is known as artesian.

Confining bed A bed of low permeability occurring above and below a confined aquifer and below an unconfined aquifer or perched aquifer.

Critical depth The depth of a watertable, below the ground surface, above which the capillary rise of soil moisture and the salts contained therein occurs at a rate likely to cause salt accumulation in the upper soil.

Cyclic salt The salt derived from oceanic spray, transported inland by winds and deposited by rain.

Dam A wall of earth, rock or concrete that blocks the flow of a river or stream.

Diffusion Movement of dissolved salts from areas of high salt concentration to areas of lower salt concentration.

Discharge zone The area in a catchment where the net movement of water is upwards to the soil surface.

Divertible water resources The average annual volume of water which, using current practice, could be removed from developed or potential surface water or groundwater sources on a sustained basis at rates capable of serving urban, irrigation, industrial or extensive stock uses. It does not include low yielding bores in fractured rock aquifers providing domestic or stock supplies by low yielding pumps such as windmills or surface water sources such as roof run-off or small farm dams.

Dry fallout Salt deposited on the landscape not falling as dissolved salts in rainfall.

Dryland salinity Salinity in areas where irrigation is not present. It may be due to natural or human-induced causes or a combination of both.

Electrical conductivity (EC) The ability of a soil, water sample or solution to conduct electricity. It is proportional to the concentration of salts dissolved in solution and is measured in units of $\mu S\ cm^{-1}$ (microsiemens per centimetre), $mS\ m^{-1}$ (millisiemens per metre) and $dS\ m^{-1}$ (decisiemens per metre) which is equivalent to mmhos cm^{-1} (millimhos per centimetre). The SI unit of the EC is $mS\ m^{-1}$. Converting factors are as follows: $1\ mS\ m^{-1} = 10\ \mu S\ cm^{-1}$ and $1\ dS\ m^{-1} = 1000\ \mu S\ cm^{-1}$. Approximately one mmhos cm^{-1} or $dS\ m^{-1}$ corresponds to about 640 mg L^{-1} total dissolved solids.

EC_e Electrical conductivity of saturated soil past extract.

Ephemeral Lasting for a limited period of time.

Establishment of plants Placing plants in conditions such that they are capable of normal growth and development.

Evaporation Physical process for the change of water from a liquid to a gas (vapour) form and its subsequent loss to the atmosphere.

Evapotranspiration Process of water vapour transfer into the atmosphere from vegetated land surface. It includes water evaporated from the soil surface and water transpired by plants. It is almost equal to the consumptive use (consumptive use also includes water in the plant).

Exchangeable sodium percentage (ESP) The percentage of exchangeable sodium ions of a soil to the total exchangeable cations of all types in the soil sample. It is expressed as:

$$ESP = \frac{\text{Exchangeable sodium ions}}{\text{Soil cation exchange capacity}} \times 100$$

where the ions are expressed in milliequivalent per 100 g of soil.

Fallow Land left unseeded after being cultivated.

Fauna Animals living within a given area or environment or during a stated period. (In Roman mythology, the grandson of the god Saturn, worshipped as god of the fields and of shepherds.)

Field capacity The amount of water remaining in a soil after the removal of free water by gravitational forces.

Flood plain The area covered by water during a major flood; the area of alluvium deposits laid down during past floods.

Flora Aggregate of plants growing in a particular region or period. (In Roman mythology, goddess of flowers and springtime.)

Fodder crops Crops which are suitable as food for livestock.

Fodder trees Trees which are suitable for animal fodder.

Freshwater Water with a salinity of less than 500 mg L^{-1} TDS.

Germination Emergence of shoot and root from a seed.

Grain per gallon (gpg) Imperial measure of solute in solution equivalent to 14.2 mg L^{-1}. One grain per US gallon equals 17.12 mg L^{-1}.

Groundwater Subsurface water contained in a saturated zone of soil or geological strata and capable of moving in response to gravity and hydraulic pressure gradients.

Habitat Place or environment in which specified organisms live.

Halophyte Plant that is adapted to very salty soil.

Human-induced salinisation See secondary salinisation.

Hydraulic conductivity The rate of flow of water through a unit cross-section of soil under a unit hydraulic gradient.

Hydraulic gradient The change in hydraulic head per unit distance.

Hydraulic head The height to which water rises in a well which is not being pumped. It is a measure of the head or pressure in that part of an aquifer open to a well at a specified time.

Hydrological cycle The continuous interchange of water between land, sea or other water surface, and the atmosphere.

Infiltration The downward entry of water into the soil expressed as a one-dimensional vertical flow.

Integrated catchment management (ICM) The integration of water and land management activities and the government agencies involved in these activities within a catchment.

GLOSSARY

Interception scheme A scheme designed to intercept groundwater flow to a river.

Ion An electrically charged atom or group of atoms.

Isohyet A line on a map joining places of equal rainfall amount.

Karst A region underlain by soluble carbonate rocks (limestone or dolomite) and characterised by distinctive surficial and subterranean features (sinkholes, conduits, caverns, etc.) caused by solutional erosion.

Lacustrine Relating to lakes.

Laterite A residual material formed through the prolonged weathering of rocks under warm humid conditions. Generally high in iron, aluminium oxides and silica.

Leaching The process of removal of soluble material by the passage of water through the soil.

Leaching fraction The ratio of the depth of drainage water to the depth of applied water (irrigation plus rainfall).

Leaf area index The ratio of the total surface area of a plant's leaves to the ground area available to the plant.

Lignotuber A woody tuber (mass) which develops at the base of some species of eucalypts and contains dormant buds which may produce shoots if the plant is defoliated or injured (eg. through burning or excessive grazing by herbivores).

Lock An enclosed area of a river adjacent to a weir, with gates at each end which are manipulated to raise or lower boats from one level of the river to another.

Loess Material predominantly of silt size transported and deposited by wind.

Lunette A crescent-shaped ridge of material which accumulates on the leeward rim of lakes, formed by wind erosion of the dry lake bed.

Mallee An Aboriginal term which refers to a growth habit found in several species of eucalypts. Mallee communities are a form of shrubland, scrub or heath, found mainly in low rainfall areas (250–450 mm yr^{-1}). Mallee typically grows to heights of up to 10 m, and is characterised by several slender trunks stemming from a single lignotuber lying just below the soil surface. The trunk structure causes rainfall to be concentrated at the base of trees.

Marginal water Water with salinity greater than 500 mg L^{-1} TDS but less than 1500 mg L^{-1} TDS.

Monotonic profile A soil profile in which salt content increases uniformly with depth.

Mulch Material, usually organic, such as cut grass, straw, foliage, sawdust, bark chips or woodchips used as a covering for the soil to conserve soil-water and check weed growth.

Overgrazing Continued grazing of pasture or rangeland at a level which permanently and adversely affects its plant components. This leads to a reduced capacity to produce forage, deterioration in pasture or range condition and increased erosion hazard.

Pediments Erosional slopes beneath high points in the landscape.

Perched watertable A watertable created by a local impermeable layer which prevents the further downward percolation of water toward the main aquifer.

Percolation The passage of water under hydrostatic pressure through the interstices of a soil or rock, excluding the movement through large openings.

Permeability A measure of the ease with which a porous material can transmit a fluid (water, air) under a potential gradient. It is a property of the material only and is dependent on the size and shape of the pores but is independent of the nature of the fluid.

pH Stands for 'potential of hydrogen'. It is the negative logarithm of the hydrogen-ion activity and is a measure of acidity or alkalinity. A pH of 7 indicates neutrality, less than 7 acidity and more than 7 alkalinity.

Phosphogypsum A byproduct of the phosphate fertiliser industry containing about 90 per cent of gypsum and 3 per cent of phosphate.

Piezometer A tube inserted in and sealed into the ground but with the bottom end open. The pressure of water in the soil at the bottom of the tube causes water to rise to a height in the tube which is a measure of the hydraulic pressure at the bottom of the tube.

Piezometric level The level of water in a piezometer, usually measured with respect to mean sea-level.

Porosity The ratio, usually expressed as a percentage, of the volume of interstices or voids to the total volume of a soil, sediment or rock. In the movement of groundwater only the system of interconnected interstices is significant. This is known as effective porosity, which is expressed as a percentage of the total volume occupied by interconnecting interstices.

Potentiometric surface An imaginary surface representing the total head of groundwater in a confined aquifer that is defined by the level to which water will rise in a well.

Precipitation The deposition of water in a solid or liquid form on the Earth's surface from atmospheric sources.

Perennial plant A plant whose life cycle extends for more than two years and continues to live from year to year.

Primary salinisation A salinisation problem caused by natural conditions.

Provenance The geographical source or place of origin of a given lot of seeds or plants. Within species there are likely to be some genetic differences between provenances.

Pulpwood Wood that is used to make pulp, usually for further processing into paper.

Recharge The addition of water to an aquifer from all sources.

Recharge zone The area in a catchment where the net movement of water is downward to the groundwater.

Reforestation Planting trees as a forest on land previously cleared of native forest overstorey.

Regolith A general term for the entire layer of fragmental and loose, incoherent or unconsolidated rock material of whatever origin (residual or transported) and of very varied character, that nearly everywhere forms the surface of the land and overlies or covers the bedrock.

Rehabilitation The return of a disturbed site to a less disturbed condition, usually being more productive and less degraded.

Salina A saline spring, saline marsh or saline lake commonly dry in summer.

Saline-seep Generally a wet, salty area due to a brackish or saline watertable either at the land surface, or sufficiently near to allow significant capillary flow of water and salt to the land surface.

Saline-sodic soil A soil which has both high soluble salt and high sodium levels.

Salinisation The accumulation of salt in soil or in water, to a level that causes degradation.

Saline water Water with salinity greater than 5000 mg L^{-1} TDS.

Salinity The amount of sodium chloride or dissolved salts in a unit of water. It can be measured in parts per hundred (per cent), parts per thousand, milligrams per litre (mg L^{-1}) or in units of electrical conductivity (EC). In these different scales seawater has a salinity of: 3.5 per cent; 35 parts per thousand; 35 000 mg L^{-1} and about 45 000 μS cm^{-1}.

Saline soil Soil with electrical conductivity of saturated soil extract (EC_e) greater than 4000 μS cm^{-1}, ESP < 15, SAR < 13 and pH of less than 8.5.

Salt flat Land with watertable near the surface, naturally saline in the subsoil or throughout and carrying salt-tolerant plants.

Salt marsh Coastal land with watertable at or near the surface, naturally saline in the subsoil or throughout the soil and carrying salt-tolerant plants.

Salt pan Land with watertable intermittently at the surface and naturally too salty throughout to support plant growth.

Saturated soil extract The solution extracted from a soil sample after being mixed with sufficient water to produce a saturated paste.

Saturated zone Part of the soil in which all voids, large and small, are filled with water.

Secondary salinisation Salinisation due to human activities, such as irrigation, forest clearing and agricultural practices. It is further divided into dryland salinisation and irrigated land salinisation. It is also called human-induced salinisation.

Scald Term used in Australia to define a particular form of dryland salinity caused by the removal of topsoil by wind or water. This usually occurs as a result of denudation of vegetation by overgrazing, drought or fire. When the vegetation is removed wind and water can remove the topsoil, and a crust forms at the topsoil-subsoil boundary.

Seedling Young plant originating from seed.

Seep A wet area due to a watertable at the land surface or sufficiently near to allow significant capillary flow to the land surface.

Semi-confined aquifer An aquifer whose confining beds will conduct significant quantities of water into or out of the aquifer, according to the head distribution.

Silt Soil consisting of particles with a diameter of 0.002–0.05 mm.

Sodic soil Soil with electrical conductivity of saturated soil extract (EC_e) less than 4000 μS cm^{-1}, ESP > 15, SAR > 13 and pH of 8.5 to 10.

Sodium adsorption ratio (SAR) A relation between sodium and divalent cations, for saturated soil extract or irrigation water. It is used to express the relative activity of sodium in exchange reactions with soil. It is expressed as:

$$SAR = \frac{Na^+}{\sqrt{\frac{Ca^{++} + Mg^{++}}{2}}}$$

where ionic concentrations are expressed in milliequivalent per litre.

GLOSSARY

Soluble salts. Inorganic chemicals that are more soluble than gypsum which has a solubility of 2.41 g L^{-1} in water at 0°C.

Solute. A substance dissolved in a solution.

Species. Generally a group of plants or animals with a characteristic shape, size, behaviour and habitat that can breed amongst themselves to produce fertile offspring.

Stem density. Number of tree stems per unit of area.

Stomata. Minute openings in the surface of the leaves through which oxygen, water vapour and other gases leave the leaf and carbon dioxide is taken into the leaf.

Storativity or storage coefficient. The volume of water an aquifer releases or takes into storage per unit surface area of the aquifer per unit change in head.

Surface run-off. That part of run-off which travels over the soil surface to the nearest stream channel, having not passed beneath the surface since precipitation.

Throughflow. Downslope flow of water occurring physically within the soil profile, under saturated or unsaturated conditions.

Total catchment management (TCM). The coordinated use and management of land, water, vegetation and other physical resources and activities within a catchment to ensure minimal degradation and erosion of soils and minimal impact on water yield, quality and other features of the environment.

Total dissolved solids (TDS). A term that expresses the quantity of dissolved material in a sample of water, either the residue on evaporation dried at 180°C, or for many waters that contain more than about 1000 mg L^{-1}, the sum of the chemical constituents.

Transmissivity. The arithmetic product of the hydraulic conductivity and the thickness of an aquifer.

Transpiration. The loss of water that has been absorbed through plant roots and transported through the plants to the atmosphere from the leaves.

Unconfined aquifer. An aquifer in which the water level does not rise above the level that is intersected in a well.

Unsaturated zone. The zone between the land surface and the watertable, including the capillary zone.

Waterlogging. Saturation of soil with water, resulting from overirrigation, seepage or inadequate drainage.

Watertable. The surface of a groundwater body, the pressure of which is equal to atmospheric pressure. The soil below the watertable is fully saturated.

Weir. A construction across a river that dams the water but may be removable during times of flood.

Wetland. Areas of seasonally, intermittently or permanently waterlogged soils or inundated land, whether natural or otherwise.

Wheat belt. The crescent-shaped principal wheat growing region of south-west Western Australia.

Index

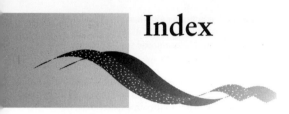

The index is divided into a Subject Index (this page), a Geographical Index (p. 521), and a Plant Index (p.524).

Subject index

Acid 11, 301, 330, 392
Acid sulphate soils 441
Aerial photography 114, 381
Afforestation 75, 142, 235, 236, 312 *see also* Reforestation and Revegetation
Agriculture 6, 7, 14, 18, 35, 78, 83, 84, 86, 133, 135, 157, 162, 167, 168, 171, 174, 201, 203, 214, 233, 272, 279, 281, 288, 293, 327, 346, 359, 371, 384, 385, 416, 439, 440, 441, 471, 486, 488
Agricultural 8, 35, 47, 72, 149, 283, 284, 286, 304, 313, 319, 328, 344, 352, 353, 414, 425, 431, 439, 440, 452, 453, 459, 468, 472, 485, 488, 497, 499
 development 14, 34, 43, 79, 150, 175, 257, 258, 262, 264, 378
 production 15, 133, 139, 263, 279, 283, 307, 345, 353, 359, 365, 366, 377, 389, 431, 435, 441
Agronomic 72, 73, 164, 193, 200, 330, 333
Agroclimate 8, 329
Agroforestry 59, 76, 79, 193, 201, 202, 203, 454, 493, 494
Algae 61, 62, 66
Alienation 168, 175
Alkaline 26, 36, 37, 75, 134, 137, 222, 228, 231, 236, 237, 250, 251, 302, 317, 329–333, 346, 349, 354, 360, 382, 349, 392, 393
Alkalinity 139, 251, 265, 267, 281, 390, 303, 311, 315, 325, 326, 328, 330, 348, 350, 351, 354, 356, 357, 362–364, 366, 382, 383, 392, 393
Alkalisation 31, 32, 41, 222–224, 231–233, 236, 237, 265, 266, 301, 365
Aluminium sulphate 86, 87
Alluvial 226, 227, 292, 305, 308–310, 317, 319, 343, 346, 358, 374
Alluvium 128, 297, 298, 314, 321, 327
Amelioration 166, 257, 332, 333, 392, 441
Amendment 87, 236, 237, 258, 302, 317, 324, 326, 329, 330, 331, 332, 384, 392, 448, 455
Ammonium 333
Anhydrite 443, 472
Anion 69, 301, 302, 316, 334, 363, 364, 393
Aquatic 61, 62, 260, 472
Aquifer 32, 33, 36, 38, 39, 53, 61, 63, 65–68, 91, 132, 135, 146, 148, 167, 183, 190, 197, 220, 221, 226, 229, 230, 233, 234, 246, 248, 266, 275–278, 297, 306, 310, 311, 320, 323, 327, 343–345, 351, 352, 356, 359–361, 364, 380, 402, 403, 409, 413, 437, 438, 443, 448, 455, 464, 465
 confined 31, 33, 66, 89, 184, 226, 248, 319, 487, 488
 unconfined/watertable/ phreatic 31, 34, 141, 184, 195, 226, 305, 319, 374
 semi-confined 33, 487, 488
Arable land 11–13, 15, 40, 156, 232, 248, 264, 279, 299
Arid 8, 11, 32, 33, 57, 131–133, 144, 145, 148, 152, 223, 231, 252, 262, 293, 294, 298, 301, 313, 315, 317, 318, 339, 346, 348, 372, 380, 402, 441, 462, 469, 501
Aridity 11, 272, 407
Arsenic 29, 491
Artesian 52, 148, 196, 246, 276, 438, 446, 495
Atmospheric salt 22, 32, 155, 180, 418

Basalt 297, 327
Basement/Bedrock 36, 111, 196, 220, 276, 305, 327, 413, 414, 446
Bauxite 168
Beryllium 29
Bicarbonate 28, 37, 141, 223, 266, 301, 316, 391
Bioclimate 8
Biodiversity 45
Biophysical 84
Biosphere 22
Bitten 68
Black soil 301, 302, 321, 332
Blending 59, 60, 68, 312, 313, 410, 423, 426, 492, 493 *see also* Mixing
Bore/Borehole 196, 198, 412, 403, 407, 412, 417 *see also* Well and Tube–well
Boron, 28, 29, 62, 135, 136, 490, 491, 493
Brackish 29, 43–45, 153, 154, 158, 162, 167, 230, 306, 356, 391, 409
Brine 65, 66, 425, 451, 482, 492, 493
British Army 299, 377

Cadmium 29, 491
Cainozoic 246
Calcite 66
Calcium 37, 66, 68, 86, 87, 180, 266, 302, 333, 346, 391, 495
Cambrian 220
Canal 26, 27, 80, 151, 231–233, 248, 249, 256, 261, 263, 279, 281, 282, 298, 299, 302, 304–306, 309–312, 316, 322, 323, 325, 345, 347–350, 352, 353, 357, 359, 360, 365, 373, 376–378, 385–387, 401, 411, 412, 426, 440, 468, 499
 lined 64, 365, 500
 unlined 33, 265, 324, 345, 365, 378, 478
Capillary rise 139, 141, 142, 235, 266, 323, 360, 363, 365, 416, 418, 419, 446
Carbonate 28, 29, 66, 68, 87, 180, 220, 223, 301, 316, 346, 391
 rock 220, 344, 437
Catchment 72, 73, 98, 111, 162, 164, 165, 168, 175, 204, 260, 273, 295, 329, 333, 354, 355, 365, 372, 373, 400, 401, 411, 414, 416, 419, 422, 425, 463
 management 85, 175, 205 *see also* Integrated and Total catchment management
Cation 69, 302, 316, 321, 334, 363, 364, 393
Cenomanian 276
Chemical 17, 61, 86, 87, 263, 264, 266, 267, 283, 334, 362, 365, 374, 382, 420
Chernozem 265, 266
Chloride 29, 37, 38, 141, 155, 162, 163, 180, 181, 189, 223, 301, 302, 326, 346, 420, 441, 443
Chromium 29, 62, 490, 491
Clay 25, 31, 32, 51, 63, 66, 106, 132, 137, 138, 156, 225, 277, 302, 305, 310, 321, 324, 326, 327, 329, 351, 354, 355, 392, 417, 437, 438, 441, 487, 495, 496
Clearing 86, 154, 156–159, 161, 167, 168, 171, 174, 175, 182, 204, 445, 446, 448
Climate 20, 26, 30, 57, 78, 106, 130, 144, 168, 179, 182, 214, 216, 220, 224, 242, 243, 246, 263, 272, 293, 301, 313, 317, 318, 325, 371, 397, 399, 403, 432, 460, 461, 485, 500
Climatic change 3, 28, 138, 141, 164, 179
Clogging 66, 67, 68, 286
Coal 496
Coast 32, 180, 397, 398
 coastal area/belt/region/ zone 234, 302, 327, 328, 332, 333, 372, 387, 439, 441
 coastal plain 319, 327, 437, 441
Cobalt 29
Community 86, 98, 100, 103, 104, 105, 106, 201, 202, 203, 204, 205, 206, 307, 402, 404, 427
Computer program 21, 474 *see also* Model
Conflict 93, 94, 97, 205, 275, 374, 471
Conglomerate 132, 298, 442, 443
Conjunctive use 53, 133, 193, 195, 313, 324, 400
Copper 29, 491
Corrosion 427
Cost 23, 24, 47, 52, 64, 71, 77–79, 81, 87, 95, 101, 113, 115, 167, 196–198, 201, 202, 206, 236, 252, 260, 264, 288, 308, 330, 333, 357, 386, 387, 400, 393, 401, 425, 427, 438, 440, 448, 468, 485, 493, 500
 cost-effective 166, 195, 197, 198, 203, 205, 206, 480, 481
 cost-sharing 103, 205, 480, 484, 494
Cretaceous 246, 343, 358, 443, 448, 475, 494
Critical depth 33, 233, 332
Crop 20, 22, 23, 24, 25, 29, 33, 35, 36, 45, 57, 73,

75, 78, 80, 86, 88, 114, 133, 150, 156, 165, 184, 190, 200, 222, 252, 280, 281, 286, 307, 312, 316, 321, 322, 324, 327, 329, 330, 332, 333, 345, 371, 377, 400, 391, 426, 439, 440, 466, 473, 485, 495
rotation 58, 73, 74, 142, 236, 257, 258, 264, 285, 330, 400
water requirement 56, 350, 378
yield 20, 74, 261, 302, 330, 332, 346, 354, 371, 400, 392, 405, 440, 455, 467, 485
Cropland 41, 128, 176, 468–474
Cropping 73, 80, 190, 287, 302, 324, 328, 373, 378, 380, 384, 446, 485, 495, 497, 498, 500
Cultivated area/land 11, 12, 133, 220–222, 224, 228, 231, 232, 233, 248, 270, 272, 283, 299, 390, 316, 319, 322, 362, 403, 439
Cultivation 15, 16, 149, 231, 317, 324, 325, 347, 377, 400, 392, 446

Dam 38, 148, 182, 188, 196, 248, 274, 279, 343, 345, 357, 358, 359, 373, 377, 401, 406, 407, 410, 411, 426, 440, 468, 489
Defoliant 263
Deforestation 18, 445, 446, 447, 452
Degradation 14, 15, 17, 18, 41, 152, 159, 182, 186, 206, 265, 279, 301, 303, 452, 489, 492
Deicing 39
Depletion 7, 14, 319
Deposits see Sediments
Desalination/desalting 68, 198, 234, 237, 266, 423, 425, 479, 493
Desert 130, 214, 223, 251, 252, 270, 272, 279, 292, 313, 318, 339, 344, 346, 369, 461, 471, 499
Desertification 16, 319, 327, 344
Deterioration 17, 47, 98, 158, 319, 327, 344, 365, 384, 405, 411, 420, 421, 422, 487
Dilution 59, 60, 61, 88, 197 flow/water 197, 198, 409, 410, 487 see also Freshening
Discharge 36, 38, 51, 168, 179, 181, 203, 258, 274, 283, 284, 311, 327, 351, 366, 374, 400, 407, 408, 454
area/zone 111, 112, 165, 172, 185, 200, 201, 202, 444
index 111
Disease 168, 489
Disposal 38, 49, 57–69, 71, 93, 135, 167, 178, 182, 186, 193, 195, 197, 198, 286, 307, 308, 313, 325, 326, 387, 389, 425, 426, 489, 492, 493

Divertible 153, 154
Drain 50, 51, 61, 62, 165, 235, 261, 266, 287, 312, 324, 325, 332, 354, 374, 378, 386
mole 51
open 250, 251, 284, 354, 499
tile 51, 62, 64, 151, 186, 285, 286, 288, 386, 389, 492
Drainage 26, 32, 33, 49, 55, 100, 102, 133, 141, 167, 186, 196, 197, 232, 233–237, 250, 251, 257, 258, 261, 280, 281, 284, 288, 302–304, 315–317, 321, 324, 326, 331, 332, 333, 347, 348, 353, 354, 356, 380, 381, 387, 405, 406, 468, 486, 488, 489, 495, 499
horizontal 52, 53, 308, 360
subsurface 26, 62, 151, 186, 192, 198, 234, 235, 250–252, 261, 284–286, 302, 307, 312, 313, 332, 357, 426, 468, 493, 499, 500
surface 48, 50–52, 151, 165, 186, 193, 200, 234, 261, 284–286, 302, 313, 326, 331, 357, 360, 389, 400, 406, 426, 468, 495
tile 193, 198, 285, 288, 347, 415, 488, 489
tube–well 151, 166, 186
vertical 250, 261, 262, 308, 310, 312
water 57, 63, 234, 259, 263, 264, 275, 283, 284, 286, 287, 308, 313, 347, 359, 362, 365, 366, 419, 474, 478, 485, 492, 493, 499, 500
Drought 36, 43, 53, 217, 224, 226, 232, 233, 311, 319, 397, 398, 409, 411, 421, 425, 467
Dryland 11, 33, 35, 36, 49, 76, 84, 110, 112, 190, 195, 204, 416
agriculture/farming 73, 84, 152, 345, 403, 405, 411, 469, 495
salinity/salinisation 35, 36, 47, 77, 85, 152, 154, 156, 165, 167, 182, 186, 187, 193, 200, 203, 204, 206, 257, 417, 421, 494, 497, 499
Dust 263, 354
Dyke 36

EC/EC$_e$ see Electrical conductivity
Ecology 182
Ecological 233, 263, 264, 392, 425
Economic/Economical 14, 49, 66, 68, 84, 154, 165, 166, 205, 248, 263, 264, 286, 326, 333, 384, 400, 391, 403, 423, 439, 492–494
damage/loss 206, 288, 393, 427, 485
Economy 293, 307, 393, 439, 471

Ecosystem 3, 206, 226
Electrical conductivity 37, 44, 106, 108, 110, 113, 141, 190, 192, 283, 307, 308, 310, 311, 312, 316, 319, 326, 235, 346, 347, 349, 350, 353, 356, 359, 360, 362, 364, 365, 391, 392, 418, 444, 449, 450, 498
Electromagnetic method 106–110, 112, 113
Environment 14, 64, 66, 67, 75, 79, 84, 98, 100, 103, 168, 179, 204, 205, 451, 453, 492
Environmental 3, 16, 20, 45, 61–63, 66, 67, 149, 154, 206, 252, 260, 263, 275, 426, 440, 448, 452, 468, 472, 492, 501
Eocene 276, 358, 487
Erosion 16, 17, 36, 50, 78, 98, 166, 182, 200, 219, 474, 494,
ESP see Exchangeable sodium percentage
European settlement 149, 152, 154, 181, 182, 184, 466
Eutrophication 406
Evaporation 23, 32, 33, 45, 146, 200, 231, 232, 257, 278, 306, 316, 339, 342, 344, 345, 348, 353, 355, 358, 361, 363, 364, 371, 374, 401, 410, 411, 434, 442, 463, 472, 474, 476, 479, 500
Evaporation basin/pond 3, 22, 61, 89, 198, 308, 313, 362, 387, 425, 489, 492
Evapotranspiration 11, 22, 23, 27, 33, 35, 36, 56, 72, 73, 133, 165, 252, 255, 272, 275, 286, 310, 341, 346, 403, 405, 488
Evaporite 44, 346, 365, 442
Exchangeable sodium percentage 37, 326, 329

Fallow 72, 257, 258, 286, 317, 330, 355, 400, 416, 441, 494, 495, 497
Farm 50, 63, 86, 203, 345, 364, 406, 426, 468
Off–farm 426
On–farm 64, 65, 81, 203, 345, 389, 400, 406, 422, 426, 451, 480, 484, 492, 493
Farmers 17, 76, 80, 83, 93, 157, 165, 166, 197, 201, 203, 205, 286, 287, 288, 306, 319, 324, 327, 329, 330, 345, 362, 364–366, 380, 385–387, 400, 391, 441, 451, 452, 455, 473, 478, 479, 489, 492–494, 498, 500
Farmland 50, 156, 176
Fauna 3, 4, 45, 84, 182, 204
Feasibility 66, 68, 248, 286, 333, 425, 448, 493
Fertiliser 13, 15, 25, 50, 80, 87, 258, 259, 263, 264, 330, 331, 376, 391, 446, 498
Field capacity 22, 25, 57

Fire 36, 78, 168
Fish 45, 46, 182, 259, 263, 453, 454, 490, 497
Fishery 255, 259, 263, 313
Fishing 259, 263, 275
Flood 98, 217, 224, 232, 260, 274, 279, 296, 318, 352, 398, 399, 411, 425, 436, 455, 471,
plain 45, 230, 231, 265, 275, 308, 343, 353, 375, 412, 415, 437, 438, 443, 447
flooding 55, 137, 138, 139, 182, 183, 224, 231–233, 235, 267, 281, 317, 352, 356, 357, 409, 485, 486, 499
Flora 3, 4, 46, 76, 84, 182, 204
Fluorine 29
Fodder/Forage 20, 75, 76, 78, 142, 165, 166, 254, 331, 377, 391, 392
Food 4, 14, 15, 16, 75, 79, 133, 222, 299, 304, 306, 377, 485
Forest 34, 113, 133, 149, 150, 158, 168, 174, 183, 184, 186, 221, 236, 248, 279, 299, 344, 377, 403, 439, 445, 446, 452, 454, 466,
Forestry 202, 233, 377, 471
Fractured 23, 148, 220, 402, 444
Freshening 413, 414, 415
Freshwater 4, 5, 11, 226, 230, 233, 234, 246, 277, 286, 312, 327, 328, 376, 387, 410, 413, 426, 465, 479, 487, 493
Fuel 75, 452

Glasshouse 78
Gold 488
Granite 128, 220, 475
Grass 20, 36, 73, 75, 142, 165, 236, 317, 331, 332, 455
Grassland 46, 184, 221, 257, 258, 494
Gravel 52, 195, 200, 220, 261, 277, 305, 307, 310, 351, 385, 437, 438, 464, 487, 495
Gravity 51, 284, 313, 342, 499
Grazing 141, 142, 149, 152, 169, 171, 184, 317, 348, 454, 471
Groundwater 5, 7, 14, 23, 26, 32, 36, 38, 43, 47, 50, 53, 63, 88, 111, 132, 135, 148, 161, 164, 167, 168, 171, 196, 197, 200, 219–221, 227, 230, 231, 233, 234, 245–248, 251, 252, 256–258, 260, 262, 264–266, 273, 275–278, 280, 281, 287, 296–298, 301, 302–306, 308–312, 314–316, 319, 320, 324, 326–328, 341, 342, 344, 349, 353, 354, 356, 359, 361, 362, 365, 371, 374–378, 380, 381, 383–387, 400, 392, 393, 402, 407, 409, 421, 437, 438, 443, 447, 451, 453,

464, 465, 467, 468, 476–478, 485, 487, 497
Gypsum 66, 68, 86, 87, 223, 236, 237, 258, 265, 266, 317, 324, 326, 329, 330, 331, 332, 346, 392, 393, 443, 475, 478, 489, 490
  requirement 87, 329, 392

Habitat 45, 47, 63, 489, 494
Halite 346, 474
Halophytes 76, 165, 166
Hazard 3, 29, 33, 48, 63, 167, 168, 315, 330, 384, 492
HC *see* Hydraulic conductivity
Headwater 43, 45, 188, 274, 283, 362, 471, 474, 477, 500
Health 4, 62, 263, 264
Heavy metals 59, 260
Herbicide 169, 258, 259
Holding basin 60, 67
Holocene 374
Horticulture 150, 196
Household 255, 427, 441, 451, 485
Humid 11, 57, 133, 134, 136, 138, 139, 176, 177, 216, 293, 397, 433, 467
Hydraulic conductivity 50, 51, 52, 53, 63, 166, 277, 302, 327, 331, 357, 374, 447, 496
Hydraulic structure 279, 345

Igneous 246, 297
Industrial 7, 14, 43, 61, 68, 83, 132, 148, 359, 422, 438
Industry 6, 7, 38, 65, 84, 214, 263, 358, 408, 427, 439, 464, 492
Infiltration 29, 52, 55, 228, 234, 246, 266, 302, 330, 356, 375, 417, 434, 488
Infrared 57, 114
Injection 66, 198, 482, 492
Integrated catchment management (ICM) 84, 85, 86, 165
Interception 35, 63, 166, 414
  scheme 53, 54, 89, 101, 198
Intrusion 133, 265, 276, 277, 282, 287, 320, 327, 328, 344, 387, 438, 441, 465, 466
Iron 82, 66, 490, 491
  sulphate 66, 86, 87
Irrigation 3, 7, 8, 12, 13, 22, 23, 25, 29, 33, 40, 43, 50, 55–57, 59, 77, 79, 81, 88, 97, 132, 148, 150, 151, 181–184, 193, 221, 228, 232, 233, 236, 237, 240, 244, 245, 247–250, 252, 253, 255–257, 260, 265, 275, 279–282, 287, 292, 293, 296, 299, 303, 305, 306, 308–310, 313, 316, 319–322, 324–326, 333, 334, 345, 346, 355, 356, 360, 362, 373–375, 377, 386, 400, 392, 401, 404–406, 408, 412, 420–422, 435, 436, 439,
440, 445, 453, 465, 466, 471, 473, 477, 486, 488
  efficiency 16, 26, 54, 55, 80, 96, 134, 151, 261, 265, 345, 404, 414, 477, 479
  management 29, 56
  method 20, 23, 24, 55, 261, 405, 468
  method — basin 23, 24, 55, 279, 281
  method — border 23, 24, 55
  method — flood 249, 365, 405, 414,
  method — furrow 23, 24, 26, 55, 82, 353, 478
  method — localised 55, 56
  method — sprinkler 23–26, 55, 56, 60, 312, 400, 405, 468, 482, 493
  method — trickle/drip 23, 25, 56, 59, 255, 312, 389, 414, 468, 482, 493
  scheduling 56, 57, 493
  water 16, 23, 28, 29, 32, 35, 38, 80, 135, 234, 237, 265, 266, 267, 286, 301, 302, 330, 347, 359, 360, 363, 376, 404, 441, 467, 468, 488, 489

Jarosite 66, 246
Jurassic 246, 297, 343

Karst 220, 276
Karstic 220, 246, 343, 344, 402

Lake 5, 61, 178, 179, 243, 256, 258, 273, 279
Land
  clearing 31, 32, 33, 34, 162, 164, 187, 190, 192, 193 *see also* Clearing
  cover 133, 149, 221, 248, 279, 344, 377, 403, 439, 466
  degradation 16, 17, 98, 103, 203
  levelling 52, 55, 236, 261, 379, 389, 400, 448, 455, 479, 482, 494
  resources 43, 85, 436
  retirement/retiring 193, 264, 479
  use 88, 321, 447, 494
Landcare 202, 203
Landholder 78, 79, 84, 93, 103, 167, 186, 203, 206
LANDSAT 115, 308, 312, 321, 325, 444
Lava 129, 475
LBOD *see* Left Bank Outfall Drain
Leaching 8, 23, 26, 29, 32, 56, 60, 79, 156, 180, 234, 235, 257, 283, 287, 302, 317, 324, 326, 329–333, 351, 364, 384, 400–393, 410, 420–422, 424, 455, 491
Lead 29, 62, 490, 491
Leaf area index 73, 114
Leakage 31, 36, 38, 39, 63, 184, 196, 277, 284, 347, 356, 492, 495
Left Bank Outfall Drain 387, 388, 389, 394
Legislative/legislation 168, 328
Lignite 496
Limestone 276, 297, 327, 358, 359, 437, 475
Lining 54, 55, 96, 231, 251, 261, 311, 312, 326, 352, 360, 389, 400, 482, 500
Lithium 29, 490
Littoral 228, 229, 230
Livestock 15, 18, 78, 133, 391, 471, 491, 497
Loam 22, 329, 332, 354, 392, 496
Lock 38, 178
Loess 22, 128, 298
Loss 256, 275, 281, 286, 288, 302, 312, 324, 332, 345, 352, 360, 379, 393, 403, 405, 406, 411, 416, 422
Lunette 179, 411

Magmatic 220
Magnesium 37, 141, 180, 223, 302, 346
Maintenance 7, 50, 60, 98, 197, 198, 287, 326, 345, 386, 448
Mallee 190
Management 16, 35, 43, 49, 59, 61, 62, 66, 67, 76, 79, 84, 85, 103, 104, 106, 114, 151, 157, 164, 167, 201, 233, 236, 390, 317, 328, 332, 352, 357, 365, 385, 392, 414, 426, 441, 444, 445, 447, 454, 498, 501 *see also* Water management
  options 49, 90, 141, 164, 192, 232, 233, 261, 284, 301, 307, 312, 317, 329, 333, 384, 400, 414, 423, 425, 448, 453, 454, 479, 482, 491, 499
  biological/agronomic 49, 72, 165, 167, 498
  engineering 49, 166, 167, 193, 498
  policy 49, 79
Manganese 29, 491
Manure 50, 265
  farmyard 317, 329, 330, 331
  green 236, 324, 329, 330, 392, 455
  organic 236, 324, 326, 330
Marginal 44, 153, 154, 159, 162, 168, 174
Marl 346
Marshland 357
Mercury 62, 491
Mesozoic 220, 246, 297, 442, 487, 494
Metamorphic 220, 246, 297, 437
Microwave sensors 115
Microjet 25, 26
Mining 38, 43, 168, 315, 401, 407, 422, 451–453, 471
Miocene 132, 155, 343, 344, 346
Mitigation 193, 196, 197, 207, 380
Mixing 308, 426
Model 21, 88, 191, 347, 414, 415, 420, 421, 426, 427, 438 *see also* Computer program
Molybdenum 29, 59, 490, 491, 493
Monitoring 62, 66, 67, 103, 110, 114, 161, 172, 185, 188, 203, 248, 312, 328, 405, 421, 425, 494
Monsoon 8, 224, 242, 243, 293, 304, 307, 313, 318, 320, 321, 323, 325–328, 330, 335, 371, 372, 383, 384, 432, 433
Montmorillonite 302, 321, 324
Mosquito fish 489, 490
Mulch 79, 142, 330
  mulching 333, 448
Multispectral scanners 114

Neutron probe 56
Nickel 29, 491
Nitrate 61, 180, 333, 376
Nitrogen 68, 330, 333, 391, 415, 424
Nutrient 17, 38, 61, 302, 330, 415, 423

Oligocene 246, 276
Orchard 25, 155, 254, 327, 350, 361, 412, 416, 420
Ordovician 220
Organic matter 17, 66, 236, 237, 354, 391
Osmotic 37
Overgrazing 18, 36
Overpumping/overdraft 317, 319, 327, 465, 468

Palaeoclimate 179
Palaeozoic 220, 246, 297, 442, 494
Pasture 28, 45, 78, 128, 133, 142, 150, 156, 162, 165, 166, 169, 171, 172, 174, 184, 190, 196, 200–202, 257, 317, 424, 454, 455, 468
  permanent 133, 149, 221, 248, 299, 344, 374, 403, 439, 466
  pastureland 470, 471, 473, 474
Pebble 310, 495
Percolation 33, 52, 80, 281, 286, 306, 477, 478, 488
Permafrost 246
Permeability 50, 64, 66, 106, 138, 316, 317, 321, 354, 391, 402
Permeable 36, 148, 298, 496
Permian 437
Pest 25, 78, 168, 455
Pesticide 15, 25, 61, 62, 258, 259, 263
pH 37, 265, 283, 302, 316, 326, 329, 330, 332, 334, 346, 349, 362, 392, 441
Phosphate 87, 424
Phosphogypsum 236, 258, 266, 329, 330
Piedmont 226, 227, 230, 231, 308, 310
Piezometer 34, 91, 111, 135, 498

Piezometric 276, 277, 278, 319, 498
Pipe 51, 52, 234, 308, 357, 360, 427, 482
Pipeline 52, 54, 151, 197, 198, 401, 425, 426, 468, 482
Plant 8, 22, 25, 28, 37, 45, 49, 56, 57, 75, 264, 302, 313
Plantation 165, 262, 312, 324, 332, 333, 353, 391, 446, 493
Pleistocene 183, 306, 374, 487, 495
Pliocene 132, 183
Plugging 68, 491
Pollution 17, 18, 81, 98, 260, 406, 407, 425
Population 3, 7, 12, 15, 16, 48, 61, 129, 132, 144, 154, 339, 215, 222, 242, 253, 263, 264, 271, 272, 293, 304, 308, 313, 317, 319, 355, 339, 370, 371, 397, 407, 424, 431, 441, 447, 461, 471
Porosity 32, 106, 297
Potable 153, 159, 424, 424, 497
Potash 38, 65
Potassium 180
Potentiometric surface/head 34, 164, 196, 359
Precambrian 246, 276, 297, 346, 402, 417
Precipitation 114, 130, 131, 176, 190, 216, 218, 224, 231, 232, 242, 246, 255, 257, 265, 272, 311, 318, 339, 341, 353, 371, 375, 397, 398, 401, 417, 442, 461, 463, 471, 494, 495
see also Rainfall
Pulpwood 76, 78, 169
Pumping 27, 39, 53, 67, 68, 88, 89, 167, 193, 196, 198, 200, 233, 234, 249, 284, 306, 313, 327, 384, 385, 387, 401, 438, 451, 465, 468, 482, 488
Pyrite 87, 329, 332, 441, 490

Qanat 342, 344, 345, 359, 360
Quartzite 411
Quaternary 132, 135, 136, 178, 179, 183, 220, 226, 246, 297, 298, 308, 438, 443, 448, 487

Rainfall 7, 32, 35, 50, 72–74, 88, 138, 140, 141, 144–146, 156, 161, 162, 164, 168, 169, 172, 177, 184, 185, 216, 224, 228, 234, 242, 272, 273, 293, 298, 301, 304, 307, 308, 313, 317, 319, 320, 325, 327, 339, 345, 346, 351, 352, 355, 356, 358, 361, 362, 365, 371–373, 378, 397–399, 403, 405, 407, 409, 411, 417, 432, 434, 441, 442, 461, 467, 485, 488, 500 *see also* Precipitation
Rainfed 35, 48, 346, 348, 377, 439, 440, 441

Rainforest 179, 433
Recharge 34, 35, 38, 51, 72, 73, 89, 132, 136, 146, 184, 190, 191, 193, 200–203, 228, 234, 266, 277, 281, 298, 307, 308, 311, 314, 315, 319, 327, 328, 341, 344, 351, 352, 356, 359, 374, 403, 417, 446, 448, 455, 465, 468, 488
area/zone 36, 111, 165, 184, 190, 200, 201, 246, 403, 444, 447, 454, 496, 497
artificial 133, 310, 465
Reclamation 12, 86, 87, 167, 201, 236, 279, 390, 301, 317, 329–331, 333, 384, 392
Recycle/Recycling 423, 427
Reforestation 78, 165, 168, 169, 171, 172, 174, 176, 193, 201, 448, 453, 454, 455
Rehabilitation 78, 164, 165, 175, 200, 201, 202, 332, 389, 400
Regolith 113, 447
Remote sensing 113, 313, 320, 321, 325
Reservoir 7, 26, 42, 44, 54, 174, 175, 178, 221, 222, 226, 232, 245, 248, 249, 251, 253, 260, 266, 274, 277, 316, 320, 321, 373, 374, 378, 401, 409, 427, 436–450, 453, 454, 468, 472, 479
Resistivity method 107
Return flow 43, 44, 88, 252, 413–416, 420, 422, 424, 473, 474, 478, 479, 482, 489, 500
Reuse 57–59, 197, 252, 286, 287, 307, 308, 359, 362, 410, 492–494
Revegetation 73, 75, 165, 166, 201–203, 287, 331, 391, 454, 455
Reverse osmosis 68–70, 96, 479, 493
Root zone 8, 23, 33, 36, 58, 88, 218, 219, 252, 258, 264, 265, 307, 317, 332, 364, 365, 400, 414, 420, 448, 495, 496
Run-off 6, 131, 132, 146, 147, 153, 177, 218, 219, 252, 258, 264, 265, 311, 330, 341, 353, 400, 399–401, 411, 422, 434, 436, 463, 471, 486

Salina 180
Saline 26, 36, 38, 39, 43, 46, 49, 53, 61, 76, 134, 158, 167, 228, 233, 250, 333, 346, 353, 354, 376, 381, 382, 385, 387, 392, 393, 441, 474, 476
effluent 54, 62, 63, 89, 167, 199, 410, 423, 425, 426
–seep 36, 42, 73, 152, 168, 169, 417, 470, 494–498
water 7, 26, 59, 60, 69, 227, 282, 306, 312, 313, 316, 317, 328, 344, 376,

409, 449, 451, 454, 465, 492, 501
Salinity credit 101, 193, 195
Salinity criteria 480, 481
Salt
balance 162, 266, 417, 418
cyclic 155, 180
content 23, 47, 48, 60, 141, 156, 174, 223, 234, 236, 250, 266, 275, 283, 301, 314, 332, 334, 342, 374, 405, 408, 417, 419, 426, 427, 478
dome 44, 343, 346, 365, 446, 477
lake 65, 69, 152, 180, 218, 370
load 43, 44, 45, 89, 90, 168, 174, 185, 189, 192, 195, 283, 366, 414, 416, 421, 422, 471, 474, 476–478, 489, 493, 494
making 444, 445, 447, 451, 453, 454
marsh 152, 346
pick–up 186, 419, 477, 478
–tolerant 2, 58, 75, 76, 79, 142, 165, 200, 257, 258, 287, 312, 317, 326, 333, 391, 392, 426, 454, 455, 493
windborne 356, 365
SalinityControl and Reclamation Projects 384–386, 393, 394
Sand 22, 52, 66, 74, 132, 138, 141, 195, 200, 220, 226, 229, 230, 261, 272, 276, 281, 298, 305, 310, 314, 327, 351, 374, 384, 392, 437, 438, 441, 443, 464, 487, 494–496
–dune 138, 141, 304, 308, 327
Sandstone 132, 276, 297, 298, 411, 437, 438, 442–444, 448, 475, 496
SAR *see* Sodium adsorption ratio
Satellite image 140
Savanna 9, 152, 243, 433
Scalds 35, 36
SCARP *see* Salinity Control and Reclamation Projects
Seawater 39, 180, 222, 228, 276, 277, 287, 320, 327, 376, 438, 441
Sedimentary 129, 148, 178, 179, 297, 308, 347, 443, 475, 487
Sediments/deposits 132, 138, 178, 179, 256, 310, 380, 472
aeolian 137, 138, 178–180, 310 310
alluvial 135, 136, 214, 246, 417, 437, 438, 443
consolidated 246, 298, 438
continental 276, 487
deltaic 437
diluvial 220, 226
estuarine 437
fluvial 178, 230, 431
glacial 5, 179, 180, 495
lacustrine 178, 179, 220, 226
marine 220, 230, 276, 487, 494

unconsolidated 132, 148, 220, 226, 246, 464, 495
Seedling 58, 79, 169, 400
Seepage 35, 36, 43, 44, 50, 54, 63, 73, 166, 181, 225, 228, 230, 231, 235, 257, 262, 275, 279, 281, 283,302, 311, 312, 316, 324, 326, 352, 365, 374, 375, 403, 414, 417, 421, 425, 446, 477, 500
Selenium 3, 29, 45, 59, 62, 485, 487, 489–493, 497
Semiarid 11, 32, 33, 131, 132, 133, 134, 148, 150, 152, 180, 216, 223, 224, 231, 232, 293, 301, 308, 339, 346, 372, 380, 402, 404, 408, 411, 460, 462, 469, 471, 486, 501
Semihumid 11, 216, 223, 224, 232
Shale 298, 411, 415, 417, 418, 437, 442–444, 448, 475, 477, 478, 494, 496
Sheep 257, 417
Shrimp 439
Silt 50, 219, 225, 229, 230, 235, 275, 279, 305, 310, 353, 374, 384, 438, 473, 487, 494, 495
Siltation/silting/silted 348, 349, 352, 353, 473
Siltstone 437, 442–444, 496
Silver 491
Snow 39, 339, 372, 373, 462, 495
Snowmelt 365, 372, 470, 498
Social 14, 48, 49, 64, 84, 154, 205, 275, 384, 386, 454, 498
Socioeconomic 26, 28, 84, 302, 454
Sodic 36, 38, 75, 86, 87, 137, 301, 302, 317, 326, 330, 331, 332, 346, 380, 391, 469
Sodicity 28, 29, 469, 470, 474
Sodium 28, 29, 31, 37, 141, 180, 236, 266, 301, 302, 316, 346, 391, 441, 495
adsorption ratio 37, 316, 326, 329, 350, 359, 362, 380, 392
chloride 2, 39, 68, 87, 316
sulphate 87, 316
Soil 17, 18, 24, 25, 45, 60, 64, 78, 110, 137, 182, 223, 233, 234–236, 251, 256, 257, 261, 263, 265, 267, 275, 279, 282, 285, 287, 292, 301, 302, 316, 317, 321, 324–327, 329, 330, 332, 333, 339, 346, 354, 364, 400–393, 412, 424, 441, 446, 455, 487, 489, 490
moisture 22, 56, 57, 110, 203, 307, 442, 455, 495
profile 31, 32, 35, 39, 51, 88, 106, 111, 138, 152, 156, 181, 234, 235, 236, 301, 312, 326, 329, 347, 351, 363, 364, 380, 382, 383, 403, 414, 445, 469
salinity map 106, 107, 453
survey 320, 321, 325, 356, 360, 381

SALINISATION

type 22, 26, 33, 139, 287, 405, 414
water 29, 57, 73, 315, 327, 335, 341, 346
Soluble salts 315, 335, 346, 354, 474–476, 478, 487
Solute 34, 172
Spacing 50, 51, 53, 261, 307, 332, 347, 499
Spring 246, 276, 344, 359, 360, 365, 402, 412, 476, 481
Storage *see* Reservoir
Storage coefficient/storativity 54, 89, 374
Strontium 414
Stream/River flow 6, 44, 174, 181, 218, 244, 245, 274, 296, 341, 358, 366, 372, 373, 374, 400, 403, 436, 463, 471, 488
Subhumid 11, 313, 404
Subsidence 272, 344, 438, 452, 465, 466
Subtropical 176, 293, 325, 462
Sulphate 37, 66, 68, 141, 180, 223, 266, 301, 302, 326, 333, 490
Sulphur 87, 329, 490
Sulphuric acid 86, 87, 329, 330, 392, 441
Surface water 3, 6, 7, 8, 23, 217, 219, 220, 223, 231, 243, 248, 273, 280, 294, 296, 297, 307, 310, 314, 319, 320, 327, 341, 349, 351, 353, 359, 365, 371, 372, 374, 399, 402, 407, 433, 462–464, 467, 468, 485
Swamp 279, 356, 468

Tank 298, 328, 389
TDS *see* Total dissolved solids
Temperature 7, 8, 22, 57, 114, 138, 176, 224, 243, 272, 293, 313, 333, 340, 348, 355, 372, 462, 499
Tensiometer 56
Tertiary 178, 192, 298, 308, 448, 487, 494
Texture 111, 138, 139, 141, 235, 237, 287, 301, 316, 317, 326, 329, 392, 487, 496, 499
Thematic mapper 115, 308
Threshold 20, 21, 60
Tile 261, 307, 332, 354, 499
Timber 76, 78, 455
Titanium 267
Total catchment management (TCM) 84, 85, 206
Total dissolved solids 26, 31, 44, 58, 86, 89, 133, 141, 167, 185, 195, 225, 229, 230, 234, 244, 245, 260, 265, 266, 275, 276, 283, 327, 328, 334, 342, 343, 349, 351, 355, 362, 374, 375, 376, 387, 409, 413, 415, 417, 419, 423–427, 443, 444, 451, 474, 478, 479, 481, 487, 493, 494, 497, 500
Toxic 28, 45, 59, 60, 62, 63, 260, 384, 489, 492, 494
Toxicity 29, 405, 491

Trace elements 29, 45, 49, 63, 406, 487, 489–491, 493, 494, 497
Transferable water entitlement 82, 83, 193, 197
Transmissivity 53, 68, 89, 305, 312
Transpiration 22, 57, 74, 146, 235, 447, 463, 474
Triassic 443
Tropical 293, 399, 432, 433
Tunnel 401

Urea 333
Upconing 312

Vanadium 29, 490
Vegetation 8, 18, 36, 43, 46, 84, 141, 142, 162, 165, 184, 190, 200, 219, 260, 312, 315, 417, 494
native 33, 35, 36, 86, 154, 155, 164, 182, 188, 190, 202, 356
Videography 114
Volcanic 437

Wasteland 76, 133, 166, 180, 455, 499
Water
 balance/budget 35, 73, 278, 311, 341, 418, 478
 management 286, 312, 324, 326, 333, 347, 352, 353, 373, 389, 400, 406, 455, 479, 482, 492, 493
 market 494, 501
 price/pricing 79–81, 193, 197, 261, 264, 345
 quality 7, 28, 29, 85, 88, 93, 98, 100, 154, 168, 206, 226, 237, 245, 275, 278, 319, 328, 334, 355, 359, 365, 407, 411, 413, 415, 420–422, 424, 472, 487, 489, 497
 resources 5, 6, 16, 26, 217, 233, 243, 249, 252, 256, 273, 275, 294, 304, 341, 346, 353, 366, 373, 399, 404, 406, 433, 436, 452–454, 462, 486, 500
 right 82, 83, 501
 supply/use 6, 7, 13, 14, 26, 42, 65, 73, 74, 78, 80, 82, 83, 148, 168, 174, 181, 219, 247, 248, 252, 253, 256, 263, 265, 273, 280, 296, 307, 341, 345, 358, 362, 372, 401, 402, 426, 437, 464, 472, 485, 486
 supply/use — domestic/urban/municipal 61, 219, 248, 255, 260, 265, 277, 297, 298, 314, 341, 345, 401, 402, 464
 supply/use — agricultural/irrigation 219, 248, 255, 260, 261, 265, 277, 296, 297, 298, 314, 341, 345,349, 401, 402, 421, 464
 supply/use — industrial 219, 255, 260, 277, 297, 298, 341, 401, 402, 464

transfer/export 314, 378, 401, 420, 463, 476, 479
Watercourse 31, 34, 38, 49, 155, 167, 181, 374, 378, 380, 386, 389
Waterfowl 45, 182, 259, 489
Waterlogging 14, 40, 41, 45, 47, 49, 55, 63, 75, 78, 79, 86, 100, 112, 138, 139, 141, 166, 168, 176, 185, 193, 195, 197, 206, 224, 232, 234, 235, 261, 266, 281, 284, 286, 288, 298, 302, 303, 306, 307, 310, 311, 315, 320, 321, 324–347, 351, 352, 354, 356, 362, 380–383, 385, 388, 391, 405, 406, 455, 468, 477
Watertable 3, 8, 26, 33, 50–54, 73, 77, 80, 106, 135, 166, 183, 184, 185, 186, 190, 191, 193, 196, 203, 225, 226, 229–236, 257, 282, 286, 302, 306, 307, 312, 315, 316, 319–328, 332, 343, 347, 349, 354, 360, 364, 376, 380, 383–385, 388, 400, 393, 413, 465, 477, 478, 489, 496
 rise 33, 38, 138, 141, 261, 266, 278, 279, 281, 306, 307, 310, 320, 322, 324, 328, 345, 347, 351, 352, 360, 361, 365, 380, 418, 446–448, 485, 489, 491, 496, 499
Weathering 31, 180, 297, 315, 380, 446, 475, 476, 489
Weed 50, 251, 312, 455, 498
Weir 38, 178, 182, 188, 320, 321, 322, 345
Well 52, 67, 89, 97, 132, 186, 196, 221, 233, 234, 248, 261, 276, 315, 316, 322, 327, 328, 344, 359, 374, 388, 389, 400, 393, 404, 438, 443, 451, 464, 466, 478, 490
 dug–well 297, 298
 oil–well 481, 482
 tube–well 52, 53, 286, 297, 298, 306, 308, 310, 312, 327, 374, 377, 378, 381, 384–387, 400, 393
 skiming well 308, 312
 well–point system 53
Wetland 46, 159, 182, 183, 260, 489
Wildlife 59, 61, 259, 490, 491, 493, 494
Wilting point 22, 57
Wind 22, 272, 328, 364, 432
Wood 59, 452, 454
 woodland 46, 133, 149, 150, 152, 165, 190, 221, 248, 279, 299, 305, 344, 377, 403, 439, 447
 woodlot 59, 202
World 3, 6, 7, 11, 12, 15, 41, 142, 148, 169, 203, 274, 293, 372, 401, 440, 446, 459, 485, 500
World War 13, 381, 473

Zinc 29, 490, 491

Geographical Index

Adelaide 43, 154
Afghanistan 362, 369
Africa 3, 4, 6, 11, 12, 14, 18, 19, 31, 270, 273, 276
AHD *see* Aswan High Dam
Ahwaz 347, 353
Alamo River 499
Alaska 459, 462
Alberta 35, 494, 498
Alexandria 272, 288
Alice Springs 493
All–American Canal 499, 500
Alsace 38
Amazon River 131
America
 Central 6, 19, 31
 Latin 4, 14
 North 4, 6, 8, 11, 12, 19, 31, 45, 146, 459
 South 6, 11, 12, 19, 31, 128, 131, 146
Amibara Melka Sadi 47
Amu Darya River/Basin 3, 44, 62, 94, 245, 249, 252, 253, 255, 258, 259, 263
Amur River 245
Andhra Pradesh 294, 299, 301–303, 327, 328
Anhui Province 224
Antarctica 5, 6
Arabian Sea 62, 292, 313, 369, 371–375, 387
Araks River 265, 348 *see also* Aras River
Aral Sea 3, 5, 44, 94, 243, 244, 252, 256, 258, 259, 261–264
Aralsk 263
Aras River 348, 349
Arctic Ocean 218, 240, 243, 244, 246
Argentina 4, 35, 42, 43, 128–134, 137
Arizona 2, 62, 69, 83, 93, 95, 96, 460, 461, 465–469, 470, 474, 475, 477, 479, 485, 499
Arkansas 465–467
Arkansas River 500, 501
Armenia 247, 248, 250
Arvand Rud 341
Ashkhabad 253, 257
Asia 3, 4, 6, 8, 11, 12, 14, 19, 31, 146, 218, 251, 338
Aswan 272–274, 279
 Dam 274
 High Dam 44, 274, 275, 278, 280, 281, 283
Atbara River 274, 275
Atlantic Ocean 133, 399, 416, 459
Atrek River 250
Australia 4, 6, 11, 19, 31, 35, 36, 39, 42, 43, 46, 47, 54, 64, 65, 68, 71, 75, 76, 79, 80, 82–85, 89, 93, 94, 97, 103, 110, 112, 144, 145, 148, 150–152, 187, 201, 206,

258, 288, 346, 391, 455, 493
South Australia 35, 43, 54, 78, 84, 93, 97, 99, 100, 143–146, 150–152, 154, 164, 176, 180, 181, 185, 186, 188, 190, 192, 198, 201, 203
Western Australia 32, 34, 35, 43, 45–47, 51, 73, 76, 78, 85, 86, 111, 113, 143, 144, 150–152, 154–158, 162, 164, 165, 166, 175, 201, 206
Avon River 45, 73
Awash River 47
Axe Creek 72
Azerbaijan 247, 248, 250, 348
Azov Irrigation Systems 266

Babylon 2
Baghdad 44
Bahia Blanca 132, 133
Baltic Sea 243, 244
Baluchistan Province 369, 370, 372, 374, 376, 378, 382, 383, 387, 400
Banashkanta 317, 318
Banas River 295, 313, 318
Band–e–Amir 358, 359, 360
Bangkok 435, 438
Plain 431
Bangladesh 8, 295
Bari Doab 299, 309, 310
Barr Creek 71, 182, 186, 194, 205
Basra 44
Bay of Bengal 292, 371
Beas River 295, 296, 309, 372, 373
Beijing 215, 218, 221, 224, 228
Belgium, 39
Belorussia 247
Berg River 408, 416
Bermejo River 128
Berriquin 151, 206
Bhavnagar 328
Big Sandy River 478, 482
Bihar 301, 303, 330
Bist Doab 309, 310
Black Sea 241, 243, 244, 265
Blackwood River 43, 158, 159, 168
Bloemhof Dam/Reservoir 409, 421, 426
Blue Nile 274, 275, 283
Blue Springs 477
Bo Hai Sea 228, 229
Bolivia 128, 131
Boorowa 186
Borabu District 451, 452
Bourke 181
Brahmaputra River 292, 295, 296, 298
Brandvlei Dam 411, 413, 414
Brazil 8, 128, 131, 218
Breede River 44, 400, 408, 411–415, 426
Bridle Drift Dam 424
Brockman River 158
Budzhak Steppe 265
Buenos Aires 131–134, 136–138, 140, 141
Bufalo River 423–425
Bukhara 253
Bulsar 317, 318

Bundaberg 151
Buronga 194
Burundi 273

Cairo 44, 270, 272, 277, 287
California 3, 21, 45, 47, 58, 59, 61, 62, 64, 65, 81, 83, 93, 96, 106, 459–461, 465, 467–470, 474, 479, 485, 486, 489, 499, 500
Campaspe 205
Canada 35, 39, 213, 218, 258, 469, 494, 495
Capel River 158
Cape Province 397, 398, 408, 411, 414, 416, 420, 462
Caribbean 14, 462
Caspian Sea 5, 241, 242, 243, 244, 245, 250, 257, 264, 339, 341, 342
Castlereagh River 192
Caucasus 249, 251, 267
Cauvery River/Delta 296, 299
Central Asia 8, 216, 243, 244, 248, 249, 251–256, 261, 262
Central Plain 431, 438–441
Central Plateau 338, 339, 341, 344
Central Valley 81, 460, 470, 485, 492, 499
Project (CVP) 486, 487, 489, 490
Chad 8, 276
Chaj Doab 385
Chambal 303, 316
Chambal River 295, 313
Changjiang River 214, 217–220
Chao Phraya River/Basin 431, 433–435, 440
Chardarya Reservoir 246, 253
Chashma Dam 373, 378
Chenab Canal 379, 381
Chenab River 295, 296, 372, 373, 374, 378
Cherrapunji 293
Chiang Mai Basin 438
Chi River 436, 443
Chile 4, 128
China 2, 6, 8, 13, 41, 42, 43, 213–224, 231, 235, 294
Chowilla 65, 98, 195
Dam 98
Chu Basin 246
Coachella Canal 96, 500
Coachella Valley 62, 474, 499, 500
Cohuna 181
Colignan 92
Collie Basin 164
Collie River 34, 158, 162, 168, 175
Colorado River 44, 58, 62, 67, 70, 81, 94, 95, 465, 471–474, 477, 484, 485, 499, 500
Colorado River Basin 8, 46, 47, 66, 67, 93, 94, 469, 470, 473, 474, 476, 477, 478, 481–484, 499
Colorado (State) 83, 93, 465–467, 470, 473, 586, 477–479, 485, 494, 500
Commonwealth of Independent States (CIS)

3, 42, 43, 44, 62, 87, 94, 99, 213, 240–251, 264
Crimea 249, 251, 264
Crookwell 186
Crystal Geyser 480
Cunderdin 72, 73, 74
Curlwaa 151, 194

Damietta Branch 270, 279
Dangs 318, 319
Danube Delta/River 251, 265
Danube River 265
Darling Basin 146, 178, 179, 191, 192
Darling Range 156
Darling River 181, 192
Dartmouth Dam 98, 178, 183
Dasht–e–Kavir 339, 346, 347
Dasht–e–Lut 339, 346, 347
Death Valley 459, 460
Deccan Plateau 292, 293, 295, 302
Delaware 455
Delhi 307, 332
Delta Barrage 44, 47, 60, 279, 283
Delta–Mendota Canal 486, 490, 492
Delta–Suisun Bay 61
Deniliquin 185
Denmark River/Catchment 86, 158, 165, 168, 174
Denver 472
Dez River 44, 342, 353
Dirty Devil 481
Dnepr River/Basin 245, 249, 251, 265
Dniester River 265
Dolores River 67, 477, 480, 482
Don River/Basin 249, 265, 266
Dons Catchment 162, 163, 164
Dorudzan 347, 358, 359
Duchesne River 478
Dulce River 128
Durban 397
Dwarka 328

Egypt 39, 40, 42, 44, 47, 60, 62, 270–281, 284, 286–288, 391, 493
Ekasatsuntorn Reservoir 449
Emba River 250
Entebbe 273
Ernies Catchment 162, 163
Ertix River 219, 223
Estonia 240, 247
Ethiopia 47, 273, 274, 275
Euphrates River/Valley 2, 40
Euphrates Valley 40
Europe 4, 6, 11, 12, 18, 19, 31, 38, 39, 250
Euston 92, 190

Faisalabad 392
Farafra 276
Far East 242, 243, 248
Faridkot 310, 311
Fars Province 345, 358
Federally Administered Tribal Areas 369, 370
Federally Service Area 489, 490

Fergana Basin/Valley 246, 258
Firozpur 311
Fletcher Lake 183
Florida 459, 461, 462, 465–468
France 38, 39
Frankland River 158

Gandak Project 303
Gang Canal 319
Ganga Canal 299
Ganga River 292, 295, 296, 298
Georgia (Republic) 240, 247, 250
Georgia (State) 461
Germany 39, 65
Ghaggar River 296, 304, 309
Gila River 95, 470, 478
Giza 283
Glen Canyon Dam 95, 472
Glenelg River 43
Glenwood Dotsero 477
Glenwood Springs 477
Gobi Desert 214, 223
Godavari River/Delta 296, 299, 302, 335
Golodnaya Steppe 253, 262
Goulburn 200, 205
Goulburn Valley 151, 203
Grand Canal 218
Grand Valley 474, 477, 478, 480, 482, 483
Great Artesian Basin 148
Great Fish River 401, 408, 419–422, 426, 427
Great Indian Desert 292, 313, 370
Great Plains 35, 459, 460, 462
Green River 470, 471, 473
Gujarat 75, 294, 297, 298, 390–303, 317, 319–321, 326–328,
Gulf of Mexico 459, 460, 470
Gulf of Oman 339, 341
Gulf of Thailand 431, 433

Haft Tappeh 353
Haihe River 214, 219, 224, 225, 232
Hamilton Hume 181
Haryana 75, 294, 298, 390–304, 306–309, 328, 330, 332, 333
Hawaii 459, 462
Hebei Province 221, 224, 228, 230
Helena River 158
Henan Province 218, 221, 224, 228, 230, 234
Heilongjian Province 222
Hirmand (Helmund) River 362
Hisar District 307
Hoover Dam 472, 474, 475, 480, 499
Hotham River 158
Huaihe River 214, 219, 221, 224, 232
Huanghe River 214, 217, 218, 219, 223–226, 230, 232, 235
Huang–Huai–Hai Plain 214, 219–221, 223–229, 232–234, 236, 237

SALINISATION

Hub Dam 373
Hume Reservoir 178
Hunter Valley 186
Hyderabad 371

Idaho 465–467, 469, 500
Ili River 219, 259, 260
Illinois 468
Imperial Dam 45, 67, 95, 473, 474, 475, 480, 500
Imperial Valley 58, 474, 499, 500
India 3, 8, 13, 35, 40–43, 75, 291–306, 308, 319, 328, 329, 333–335, 370–372, 378, 391
Indiana 468
Indian Ocean 204, 398, 400
Indira Gandi Canal 292, 314, 315
Indo–Gangetic Plains 292, 301, 305, 331, 371
Indonesia 13, 39, 218
Indus Plain 47, 369–372, 374–376, 380–382, 387, 393
Indus River 2, 8, 62, 292, 295, 296, 298, 369, 372–374, 376–378, 380, 384, 388
Inner Mongolia 217, 218, 222, 232
Iowa 468
Iran 13, 39, 40, 42, 43, 338–349, 365, 366
Iraq 43, 44
Irtysh River 251
Ismailia Canal 277
Israel 39, 75

Jachal Valley 135, 136
Jacobabad 371
Jaj–Rud River 343, 365
Jamu and Kashmir 294
Jarrahi River 354–357
Jhelum River 295, 296, 372–374, 378
Jiangsu Province 224, 230

Kabul River 372
Kackchh 296, 318, 319, 328
Kakrapar 320, 321, 322, 323, 324
Kalahari Basin 397
Kalmykia 267
Kansas 465, 467, 469, 500
Kapchagay Reservoir 246, 259
Karachi 371
Karakalpak 257, 263
Kara Kum Canal 253, 257
Kara Kum Desert 241, 243
Karkheh River 347, 353
Karnal 75
Karnataka 293, 294, 301, 302, 303, 328, 332
Karshi Steepe 253
Karun River 44, 341, 343, 353
Kazakhstan 94, 240, 243, 244, 247, 248, 251, 255–257, 262
Kent River 168, 174, 158
Kenya 8, 273
Kerala 293, 301, 302, 327
Kerang 71, 151, 181, 182, 185, 186

Kesterson Reservoir 45, 61, 62, 65, 489–492
Khairpur 386, 387
Khalafabad 354–357
Khartoum 283
Khon Kaen 444, 453, 454
Khorat 443, 447
Khorat Basin 443, 447
Khorat Plateau 8, 431, 441–444
Khorezm Oblast 251, 257
Khuzestan Plain 339, 353
Khuzestan Province 340, 345, 347, 353, 354, 366
King William's Town 424, 425
Kolyma River 245
Kondut 72, 73, 74
Korbal Plain 358, 360
Kor River 44, 342, 358, 359, 360
Krishna River 296, 302
Kuibyshev Reservoir 246
Kura River 267
Kyrgyzstan 94, 247, 248, 251, 256
Kyzyl Kum Desert 241, 243
Kzyl–Orda 44, 257, 258

Lahore 303
Laing Dam 424, 425
Lake Albacutya 493
Lake Albert 283
Lake Aydarkul 62, 256, 258
Lake Baikal 244, 245, 246
Lake Balkhash 244, 245, 259
Lake Bungunnia 179
Lake Burullus 270, 287
Lake Eyre 143, 145
Lake Hamoun 341
Lake Hawthorn 194
Lake Issyk-Kul 245
Lake Kyoga 283
Lake Ladoga 245
Lake Margaret 145
Lake Mentz 44, 409, 420, 421
Lake Nasser 274, 275
Lake Onega 245
Lake Orumieh 341, 342, 351
Lake Ranfurly 89
Lake Sarykamysh 62, 256, 258, 259
Lake St Lucia 398
Lake Victoria 283
Lakhpat 328
Las Vegas Wash 480, 482
Latvia 240, 247
LaVerking Springs 477
Lebanon 39
Lees Ferry 474, 475
Leeton Region 186
Lemon Catchment 162, 163, 164
Lena River 240, 241, 243, 245
Libya 39, 276
Limpopo River 399, 400
Lithuania 240, 247
Loess Plateau 225
Los Angeles 472
Los Banos 489
Louisiana 466
Lower Gunnison 474, 482, 483

MacIntyre River 192
Madhavpur 327, 328

Madhya Pradesh 301, 302, 303, 313, 328
Mae Klong River Basin 436, 437
Mahanadi River 295, 296
Maharashtra 294, 390, 301, 302, 328
Maha Sarakham Province 451–453
Mahi–Kadana 319, 325
Mahi River 296, 313, 318
Malia 328
Mallee Cliffs 54, 68, 195
Mallee Region 179, 186, 190, 199
Mangla Dam 373, 378
Manitoba 35, 494
Mar del Plata 133
Marun River 353, 354
Marvdasht Plain 358, 360
McElmo Creek 474, 482, 483
MDB see Murray–Darling Basin
Mediterranean Climate 8, 33
Mediterranean Coast 272, 277
Mediterranean Sea 44, 62, 270, 273, 283, 284, 286, 287
Meghalaya 293
Mehsana 318, 319
Mekong River 431, 433, 436, 440, 443
Mendota 487
Mendoza 132, 133, 134, 135
Menindee Lakes 192
Merbein 81, 82
Mesopotamia 2, 128, 131, 281, 339
Mexicali Valley 95
Mexico 4, 13, 69, 71, 94, 95, 97, 470, 471, 473, 479, 480, 485, 499
Middle East 4, 6
Mildura–Merbein 54, 81, 82, 88, 89, 92, 181, 193, 195
Minjiang River 219
Mississippi 461
Mississippi River 459, 462
Moghan 347, 348, 349
Moldavia 247, 251
Montana 35, 467, 469, 494, 495, 497, 501
Morelos Dam 95, 479
Morgan 43, 100, 101, 181, 188, 189, 190, 191, 193, 195
Mosul 44
Mt Lofty Ranges 164
Muktsar 310, 311
Mundaring Catchment 174
Mun River 436, 443, 453
Murgab 251
Murray Basin 62, 65, 100, 101, 146, 150, 168, 178, 179, 181, 191, 195, 197, 198
Murray Bridge 188
Murray–Darling Basin 46, 47, 79, 85, 93, 94, 97–99, 101–103, 146, 150, 154, 176–178, 180, 186–188, 192, 193, 203, 204, 206
Murray River 158
Murrumbidgee Irigation Area 151, 177, 185, 198
Muynak 263

Nakhon Ratchasima Province 444, 451, 452, 454
Namaqualand 398
Namoi River/Valley 151, 192
Nangiloc–Colignan 205
Nan River 435
Nanyang Basin 220
Narmada River 296, 318, 335
Naryn Basin 246
Natal Province 397, 399, 407
Nebraska 465, 466, 467
Netherlands 39, 52
Nevada 83, 93, 461, 467, 469, 470, 485
New Mexico 83, 93, 94, 465–467, 469, 470, 479, 485, 500
New River 499
New South Wales 64, 68, 78, 84, 85, 93, 97, 99, 101, 112, 143, 144–146, 150–152, 176, 178, 185, 186, 192, 193, 197, 198, 201, 203, 206
Nile Delta 8, 44, 60, 61, 270, 272, 277–279, 282, 284–286
Nile River 44, 273–275, 279, 281, 283, 284
Nile Valley 47, 270–272, 275, 277–279, 281, 283
Ningxia Province 222
Nira Irrigation Project 302
Nong Bo Reservoir 449, 451, 453
Noora Evaporation Basin 65, 194
North Carolina 459, 461
North Dakota 35, 469, 494, 497
Northern Great Plains 73, 469, 470, 494, 495, 497, 499
Northern Territory 143, 150, 152
North Stirling District 166
North West Frontier Province 46, 47, 369, 370, 374, 376, 378, 382, 383, 386, 387, 400, 393
Nowrouzlu 351
Nukus 44
Nurek Reservoir 246, 253

Ob River 243, 245, 251
Oceania 4, 6, 12
Odessa 251
Ogallala Aquifer 465
Ohio 468
Oklahoma 465, 494, 500
Olifants River 400, 422
Orange–Fish Tunnel 401
Orange Free State 397
Orange River 399–401, 404, 420, 421, 427
Oregon 467
Orissa 390, 301, 328

Pacific Ocean 61, 218, 240, 241, 243, 244, 460
Painted Rock Dam 95
Pakistan 3, 8, 41, 42, 46, 47, 53, 62, 75, 295, 299, 313, 369–373, 367–380, 382–384, 388–392
Palo Verde 481
Panjnad Abbasia 386

Paradox Valley 66, 67, 477, 480, 482
Parana River 128, 131, 132, 133
Parker Dam 475, 480
Pa Sak River 435
Patagonia 129, 130
Pechora River 245
Pecos River 500
Persian Gulf 339–341, 346, 354, 365,
Perth 113, 148
Perup River 158
Phoenix 472
Pilcomayo River 128, 131
Ping River 435
Poesjesnels River 415, 416
Port Nolloth 397
Potwar Plateau 369
Preston River 158
Pretoria 47, 407, 426, 427
Price River 477
Price–San Rafael 481, 482
Punjab Province 369–371, 374, 376, 378, 382–384, 386, 387, 389, 400, 393
Punjab State 46, 75, 294, 298–302, 308, 311, 312, 314, 328, 330, 333, 369

Qaidam Basin 213, 218
Qattara Depression 276
Qinghai Province 222
Queensland 84–86, 93, 99, 100, 143–146, 150–153, 176, 178, 201
Qurna 44

Rajasthan 293, 294, 297, 298, 390–304, 309, 313, 315, 317, 328
Rajasthan Canal 315 *see also* Indira Gandi Canal
Ramshir 354, 355
Rann of Kachchh 318
Ravi River 295, 296, 309, 372, 373, 378
Rechna Doab 381, 382, 384, 386
Red Cliffs 81, 82, 92, 190
Red Flag Irrigation Canal 218
Renmark 181, 194
Rhine River 38
Rio Colorado 129
Rio de la Plata 131
Rio Grande River 479, 500
Rio Negro 133, 134
Riverine Plain 81, 179, 185, 186, 195, 196, 198, 200
River Murray 43, 54, 89, 91, 92, 97–100, 102, 154, 181, 182, 185, 186, 188–193, 195, 196
Robinvale 81, 82
Roiet Province 447, 452–454
Rosseta Branch 270, 279
Rufus River 194
Russia 247, 248, 251

Sabarmati River 296, 313, 318
Sacramento River 485
Sakon Nakhon Basin/Province 443, 452, 453

Salado del Norte 128, 131
Salmon catchment 34, 162, 163
Salton Sea 62, 474, 499, 500
Salt Range 369
Salween River 433
San Diego 472
Sandspruit River 416, 417
San Francisco Bay 61
San Joaquin River 45, 59, 485, 487, 489, 492, 493
San Joaquin Valley 3, 46, 47, 61, 64, 65, 67, 68, 106, 135, 469, 485–488, 490, 491, 493
San Juan Province 132, 133, 134, 135
San Juan River 470, 471
San Luis Drain 62, 489, 490, 492
San Luis Service Area 489, 490, 491
San Rafael River 477
Santa Clara Slough 70
Santa Fe 131, 133, 137, 138
Santiago del Estero 133, 134, 135
Sarysu Basin 246
Saskatchewan 35, 110, 494, 498
Sasolburg 427
Sasyk Lake 265
Saurashtra Peninsula 296, 317, 318, 327, 328,
Senegal 455
Shadgan 354
Shandong Province 221, 224, 228, 230, 236
Shanghai 215
Shapur–Dalaki River Basin 365
Shepparton 71, 185, 186, 192, 195, 203, 205
Siberia 242, 243, 248, 251, 264
Siberian Lowland 240
Siberian Plateau 240
Siberian Uplands 241
Sierra de Cordoba 128
Sierra Nevada 460, 488
Siew River 453
Simineh–Rud 351, 352
Sind Province 369, 370, 371, 374–376, 378, 382–384, 386, 387, 400, 393
Sirhind Canal 309, 310
Sisaket Province 452, 453
Sistan Province 347, 362
Sistan River 362
Siwa 276
Snake River Basin 500
Song Khram River 443
South Africa 35, 38, 42–47, 54, 396–410, 421–423, 425, 428
South China Sea 436,
South Dakota 35, 469, 494, 497
Soviet Union 218, 240 *see also* USSR *and* CIS
Sriramsagar Project 303
Sudan 8, 270, 273, 274, 275, 276
Sudd 283
Suisun Bay 61, 489
Sundays River 45, 401, 408, 409, 420, 421, 427
Sunraysia 54, 81, 151, 205
Sutlej River 295, 296, 309, 312, 313, 372, 373

Syr Darya River 3, 44, 62, 94, 245, 249, 252, 257, 258, 263
Syria 40

Tadjikistan 94, 247, 248, 251, 255, 256
Tailem Bend 191, 301, 327, 328
Tamil Nadu 301, 327, 328
Tanzania 273
Tapi River 296, 321
Tarbela Dam 372, 373, 378
Tashkent 255
Tasmania 84, 143, 144, 145, 150–152
Tennessee 461
Texas 465–469, 494, 500
Thailand 8, 13, 35, 39, 42, 43, 44, 431–441, 443–446, 449–455
Peninsular Thailand 431, 433
Thal Doab 386
Thar Desert 313, 370
Thomson River 158
Tianjin 215, 221, 224
Tibet Plateau 218, 223, 298, 373, 436
Tigris 2, 44
Toktogul Reservoir 246, 253
Tragowel Plain 205
Transvaal Province 397, 398, 402, 407
Tucuman Province 133, 134
Tullakool 193, 194, 198.
Tully 145
Tulum Valley 135
Tungabhadra Project 303
Tung Kula Ronghai (TKR) 446–448, 454
Turkey 35, 339
Turkmenistan 247, 251, 255, 256, 257
Tutchewop Evaporation Basin 65

Uganda 273
Uinta Basin 474, 482
Ukai 318, 321
Ukai–Kakrapar 303
Ukraine 247, 248, 251, 264, 265
Una 327, 328
Uncompahgre River 477
United Arab Emirates 69
United Kingdom 39, 75
United States 4, 6, 35, 39, 41–44, 46, 47, 49, 62, 64, 65, 66, 69, 71, 73, 75, 80, 82–84, 87, 94–96, 218, 219, 455, 459–470, 475, 480, 494, 499, 501
Ural River 249, 250
Uruguay River 131
USSR 4, 11, 12, 13, 41, 349
Utah 83, 93, 461, 467, 469, 470, 473, 475, 477, 478, 479, 485
Uttar Pradesh 75, 297, 299–302, 328, 330, 332, 333
Uzbekistan 8, 94, 243, 247, 248, 251, 255–257, 261

Vaal Barrage 44, 47, 422, 423, 426, 427

Vaal River 399, 400, 408, 421, 426
Vereeniging 44, 426, 427
Vernalis 487
Victoria 43, 46, 54, 71, 73, 76, 78, 80, 81, 93, 97, 99, 101, 103, 105, 143–146, 150–153, 176, 181–183, 185, 187, 192, 195–198, 200–206
Virgin River 477
Volga River/Basin 245, 249, 250, 251, 264, 267

Wagin River
Waikerie 54, 195
Wakool 64, 65, 68, 151, 185, 186, 193, 194, 198, 206
Wang River 435
Wargan Evaporation Basin 65
Warren River/Catchment 158, 174
Washington 467, 500
Wellington Catchment 174, 175
Wellington Dam 175, 176
Wellton–Mohawk 69, 95, 474, 478, 479
West Bengal 390–302, 327, 328, 332, 333
Wheat Belt 165, 166, 167
White Nile 274, 283
Wights Catchment 34, 162, 163, 164
Wilgarup River 158
Williams River 158
Witwatersrand 47, 407, 426, 427
Woolpunda 54, 65, 68, 195
Woorollo Catchment 158
Wyoming 83, 93, 467, 469, 470, 473, 475, 485

Xinjiang Province 218, 222, 224, 231, 232
Xizang Plateau *see* Tibetan Plateau

Yakima River Basin 500
Yamuna Canal 299, 302
Yamuna River 304, 306, 307
Yass 186
Yenisei River 240, 243, 245
Yornaning 113
Yuma 62, 95, 478
Yuma Desalting Plant 69, 71, 95, 96

Zabol 362–364
Zarrineh–Rud River 44, 343, 347, 351, 352
Zayandeh–Rud River 44, 343, 347, 361

Plant Index

Acacia acuminata 77
 A. ampliceps 77
 A. auriculiformis 454
 A. catechu 76
 A. cyanophylla 76
 A. decurrens–dealbata 76
 A. decurrens–mollis 76
 A. mearnsii 77
 A. melanoxylon 77
 A. nilotica 75, 76, 331, 332
 A. salicina 77
 A. saligna 77, 288
 A. stenophylla 77
Agropyron elongatum 142
 A. scabrifolium 142
 A. spp. 392
Ailanthus excelsa 76
Albizzia lebbek 76
Alfalfa 21, 23, 58, 72, 73, 236, 412, 497
Amorpha fruticosa 236
Apricot 21
Aristida spp. 317
Atriplex 76
Atriplex halimus 288
 A. nummularia 288
 A. spp. 392, 455
Azadirachta indica 76, 454

Banana 322
Barley 2, 8, 21, 72–74, 237, 257, 331, 333, 345, 346, 348, 350, 355, 359, 361, 362, 377, 391, 466
Bean 21, 23, 333
Bermuda grass (Cynodon dactylon) 317, 331, 332
Berseem (Trifolium alexandrinum) 330, 331
Black gram (Phaseolus mungo) 331
Broad bean 21
Butea monosperma 76

Cabbage 333
Cantaloupe 58
Capparis aphylla 76
Carrot 21
Cassava (Manihot utilissima) 439–441, 446, 447
Cassia siamea 454
Casuarina 5, 9, 493
Casuarina cunninghamiana 59, 76, 77, 493
 C. equisetifolia 76
 C. glauca 59, 76, 77, 288, 454, 493
 C. obesa 77, 168
Citrus 26, 420
Clover 72–74, 169, 280
Coconut 439, 440
Corn 21, 23, 133, 280, 350, 455, 466
Cotton 8, 21, 23, 28, 58, 59, 222, 224, 237, 251–255, 262–264, 279, 280, 285, 286, 325, 331, 333, 345, 359, 371, 377, 388, 440, 466
Cowpea (Vigna unguiculata) 331
Crotalaria juncea 330
Cucumber 21
Cyamopsis tetragonoloba 330

Dactyloctenium sindicum 317
Dalbergia sissoo 76
Date palm 333
Dhaincha (Sesbania aculeata) 330, 331
Dichanthium annulatum 317
Diplachne fusca see Kallar grass
Dipterocarp 454
Distichlis spicata 455

Echinochloa spp. 317
Eleusine compressa 317
Eragrestis tremula 317
Eucalypt 59, 152, 165, 167, 169, 170, 190, 493
Eucalyptus astringens 77
 E. botryoides 169
 E. camaldulensis 59, 76, 77, 78, 167, 168, 170, 182, 183, 288, 392, 454, 455, 493
 E. camphora 77
 E. citriodora 76
 E. cladocalyx 77, 167, 169
 E. globulus 77, 78, 167, 169, 170, 174
 E. grandis 77
 E. hybrid 331
 E. kondininensis 77
 E. largiflorens 77
 E. leucoxylon 77, 169
 E. maculata 169 E. microtheca 77
 E. occidentalis 76, 77, 168
 E. ovata 77
 E. resinifera 174
 E. rudis 170
 E. saligna 169
 E. sargentii 77, 168, 170
 E. sideroxylon 77
 E. spathulata 77
 E. tereticornis 75–77, 332
 E. viminalis 169
 E. wandoo 169, 170

Fraxinus sp. 236

Gram (Cicer arietinum) 331
Grape 252, 254
Grapefruit 21
Green gram (Vigna radiata) 13, 331
Groundnut 324, 331 see also Peanut

Halophyte 288, 417, 455
Hay 466

Kallar grass (Leptochloa fusca, synonym: Diplachne fusca) 75, 331, 332, 391, 392
Karnal grass see Kallar grass
Kenaf 440, 441, 446, 447

Lentil 331
Leptochloa fusca see Kallar grass
Lettuce 21, 58
Leucaena leucocephala 76, 332
Lucerne 72, 74, 200, 257, 350
Lupins 72, 73, 74

Maize 8, 257, 285, 286, 331, 377, 439, 440, 446
Mangrove 439
Marous sp. 236
Melaleuca halmaturorum 76, 77
Melilotus albus 142
 M. indica 331
 M. officinalis 142
 M. spp. 258
Millet 280

Oat 8, 331, 466
Onion 21
Orange 21
Para grass (Brachiaria mutica) 331, 332
Parkinsonia aculeata 76
Pea 331, 333
Peach 21
Peanut 23, 440, 441
Pearl millet (Pennisetum typhoides) 331
Pinus halepensis 76, 169
 P. pinaster 170
 P. radiata 169, 170, 174
Pongamia pinnata 332
Populus diversifolia 236
 P. euphratica 76
 P. nigra 236
 P. spp. 77
Potato 21, 23, 345, 359, 455
Prosopis juliflora 76, 317, 331, 332, 392

Raya (Brassica juncea) 331
Rhodes grass (Chloris gayana) 331, 332
Rice 21, 23, 28, 222, 224, 233, 235, 236, 253, 254, 257, 262, 264, 280, 281, 285, 286, 317, 324, 325, 329, 330, 331, 333, 345, 346, 359, 361, 371, 377, 392, 439–441, 447, 448, 451, 453–455, 466
Robinia pseudoacacia 236
Rubber 440

Salicornia spp. 392
Salix spp. 77, 236
Saltbush 76, 166
Salvadora persica 317
Senji (Melilotus indica) 331
Sesbania 455
Sesbania aculeata 317, 330, 392
 S. sesban 332
Shrub 21, 75, 76, 77, 79, 164, 165, 236, 287, 332, 455
Sorghum 21, 23, 133, 466
Sorghum bicolor 392
 S. sudanense 392
Soybean 21, 23, 133, 440, 441, 466
Sporobolus sp. 317
 S. virginicus 455
Stipa sp. 257
Strawberry 21
Sugar beet 21, 23, 58, 59, 331, 345, 350, 359, 361, 377
Sugarcane 21, 23, 28, 133, 279, 281, 288, 322, 324, 331, 333, 353, 354, 371, 377, 388, 440
Sunflower 133, 333

Tall wheat grass see Agropyron elongatum
Tamarix aphylla 392
 T. articulata 76, 317
 T. dioica 332
 T. gallica 76
 T. spp. 77
Terminalia arjuna 76, 332
Tobaco 24, 325, 377, 441
Tomato 21, 333, 441, 455
Tree 20, 21, 25, 59, 75, 77, 78, 79, 164, 165, 167, 168, 169, 171, 172, 180, 190, 200, 201, 202, 203, 236, 262, 287, 288, 312, 317, 331, 332, 391, 446, 448, 454, 455, 493
Trifolium alexandrinum see Berseem

Vegetable 20, 25, 28, 133, 252, 253, 254, 280, 322, 325, 327, 345, 355, 359, 361, 377, 439
Vine 24, 25, 412, 416

Watermelon 441
Wheat 2, 8, 21, 23, 58, 72, 73, 74, 133, 154, 222, 224, 257, 258, 280, 285, 317, 325, 329, 330, 331, 333, 345, 346, 348, 350, 355, 359, 361, 362, 371, 377, 388, 392, 416, 466

Zizyphus jujuba 76
 Z. spina–vulgaris 76